Food Insecurity & Hydroclimate in Greater Horn of Africa

Joseph Awange

Food Insecurity & Hydroclimate in Greater Horn of Africa

Potential for Agriculture Amidst Extremes

Springer

Joseph Awange
School of Earth and Planetary Sciences
Curtin University
Bentley, WA, Australia

ISBN 978-3-030-91001-3 ISBN 978-3-030-91002-0 (eBook)
https://doi.org/10.1007/978-3-030-91002-0

This Springer imprint is published by the registered company Springer Nature Switzerland AG
The registered company address is: Gewerbestrasse 11, 6330 Cham, Switzerland

To the memories of my late grandmothers Amelea Gwada (Ere tekone dani nyar Mur) and Rosalina Omulo Odera who tirelessly toiled the land in scorching sun to provide food for us. It is because of you two that I am whom I am today. You are greatly missed.

Joseph L. Awange

Perth Australia, September 2021

Foreword

The book represents a significant effort by Dr. Awange in trying to offer a comprehensive overview of the hydroclimate in the Greater Horn of Africa (GHA). He has stitched together several papers into four areas: 1. A hydroclimate overview of the GHA, 2. Water resources, 3. Extreme climate drought, and 4. Potential of irrigated agriculture in GHA; and within these parts are sub-sections based on his and other researchers. The GHA consists of Djibouti, Eritrea, Ethiopia, Kenya, Somalia, Sudan (now North and South Sudan), and parts of Uganda. While the book has substantial information from a scientific perspective, it fails in laying out the reasons why the people of the GHA are suffering from drought. I will try and fill in some of the policy and military conflicts that have impacted the GHA people. First, the GHA suffers from the combination of significant population growth (currently over 3%) and persistent military strife among various groups that resulted in great suffering among the people. Unfortunately, the climate variability, especially in the southern portion of the GHA, can't be addressed unless the poverty and migration from military strife are addressed. This point is crucial because the globalization of conflicts in the Horn led directly to militarization and its attendant consequences. As *Agyeman-Duah* (1966) states: "Arguably, governance and ethnic relations, as bad as they have been in the region, were seriously jaundiced by the alacrity with which the global contestants were prepared to be dragged into the conflicts."

As listed in Wikipedia, post-1960, there have been 17 conflicts including the current 2020 Tigray military intervention starting November 2020. Some examples from the post-colonial era (late 1950s to the present) show almost continuing and devastating inter-state wars that include the Ethiopian-Somali wars (1964, 1977–78, 2006–9), the Kenyan-Somali war (1963), the Ugandan-Tanzanian war (1978–79), the Ethiopian-Eritrean border war (1998–2000), the South Sudan-North Sudan conflict, and now the Tigray intervention. The *Daily Maverick* reports that this conflict is escalating out of control, threatening regional stability. The people have fled the area to neighboring states, stuck in camps and suffering terribly. How can they build up agricultural infrastructure to survive hydroclimate variability? They can't. Much can be said for the division of Sudan into northern and southern Sudan—lots of fighting, innocent people fleeing and stuck in camps that the UN (UN Office for the Coordination of Humanitarian Affairs), Doctors Without Borders, and other aid groups are providing help to. I could discuss this about every conflict in GHA. The problem in the GHA is that the United States and Russia (and Sweden, France, China, Czech Republic, Austria, and other countries) are providing arms to each side. The African Union (AU) has tried without much success for the cessation of hostilities.

My grandmother was 13 when she and her younger sister went with their mother (my Great Grandmother) from what is now Poland to Alberta, Canada, where she had an uncle. Her mother was given a ¼ section of land (about 250 acres), had to build a house and start farming to keep the land. How could they do this? Fundamentally, there was political and military stability. This allowed them to secure the land and grow crops within a region that has significant climate variability. So political and military stability is necessary.

The GHA countries, like other African countries, have short and/or fragmented hydroclimate (water, temperature, rainfall, and soil moisture) records, often as a result of armed conflicts at various times over the past 50 years and the sheer size of the region ($> 6{,}000{,}000$ km^2). In some cases, the available records are inaccessible due to governmental red-tapes, and where accessible, some are too short, compounded by missing data lacking consistency or sparsely and unevenly distributed to be useful for adequate hydroclimate analysis. The book uses remote sensing to estimate the GHA hydroclimate. It's my sense that the remotely sensed data is also too short, ending in 1999 or early 2000s, to make statistically significant conclusions. Also, the tools being used have a variety of spatial and temporal resolutions that are used to estimate indices such as Total Water Storage (TWS) using GRACE and MODIS-2. Section 13.3.2 provides a list of the remote sensing products that are described in Sects. 13.3.3–13.3.5. What I find missing are careful validation studies that would indicate the best data for drought studies, both in a historical setting and in future projections. As Tierney et al (2015) stated

> *In contrast to 20th century drying, climate models predict that the Horn of Africa will become wetter as global temperatures rise. The projected increase in rainfall mainly occurs during the September-November "short rains" season, in response to large-scale weakening of the Walker circulation.*

It seems that such projections should be central to any GHA study on drought and its impact on food production.

December 2020

Eric F. Wood
NAE (USA); FRSC (Canada); Foreign member, ATSE (Australia)
Professor of Civil and Environmental Engineering
Princeton University
Princeton, USA

References

Agyeman-Duah, B. (1996). The Horn of Africa: Conflict, Demilitarization and Reconstruction, Journal of Conflict Studies, 16(2). https://journals.lib.unb.ca/index.php/JCS/article/view/11813

Tierney, J.E., Ummenhofer, C.C., Demenocal, P.B. (2015), Past and future rainfall in the Horn of Africa, Science Advances, 1(9): e1500682-e1500682, DOI: 10.1126/sciadv.1500682

Preface

This book, which will benefit various users in the fields of environment, agriculture, and water provides the first comprehensive analysis of the Greater Horn of Africa (GHA)'s food insecurity and hydroclimate (temperature, precipitation, drought extremes, and total water storage changes) using the state-of-the-art *Gravity Recovery and Climate Experiment (GRACE) and its Follow-on (GRACE-FO)'s water storage changes, centennial precipitation, hydrological models' as well as reanalysis'* products. This is informed by under utilization of remote sensing data to support the formulation of food security measures in the GHA region. For instance, FAO states[1]:

> Knowledge and information systems underlie a broad range of fields, including social safety net policies, agricultural knowledge, the environment, health and education, administration, marketing, and even political information. Their poor state of development in the region handicaps both households and communities in their efforts to survive and prosper under difficult conditions. It also limits the capacity of governments to formulate appropriate policies and programmes that address the problem of food insecurity. Knowledge enhancement services, early warning systems and management information systems underpin all other efforts to address food security. Information systems have been geared almost exclusively to the collection of performance data that are relevant to crop production areas, using a combination of remote sensing and field data-gathering networks to provide early warning of emerging food insecurity situations. In some countries, there is a multiplicity of early warning and vulnerability systems, operated by governments, donors and NGOs. Systems for providing a similar warning of impending disaster in pastoral systems have emerged only recently and are being tested on a pilot scale. Over time, there has been increased capacity to provide accurate early warning information. However, the ability or willingness to respond adequately to the warnings that are produced has not improved. The recent crisis has demonstrated that there are weak links in the chain between early warning, pledges of food aid, ultimate delivery and properly directed distribution. It has also highlighted the one-way nature of current information systems in the vulnerable areas, where the capacity to disseminate knowledge and information in order to improve the coping abilities of the population remains poorly developed.

GHA, a region bedevilled by poor governance and the donor syndrome, is of late (2020–2021) faced by the so-called "triple threat" of *desert locust infestation,*

[1] http://www.fao.org/3/x8406e/X8406e02.htm.

the *impact of climate change*, and the *COVID-19 pandemic*. Its climate extremes (floods and droughts) are becoming the new normal given its heavy reliance on rain-fed agriculture and as such, one of the most food insecure regions in the world whenever these extremes strike. GHA, a region prone to climate extremes (droughts and floods) and conflicts, has seen its meagre water resources increasingly coming under threat specifically from drought leading to perennial food insecurity. In fact, for the GHA region, rather than responding successfully to the frequent recurrent droughts that afflict the region, the communities are invariably devastated by famine crisis, instabilities in national economies, and political tensions. For example, the Ethiopian *biblical* famines of 1973–74 and 1984–85 left about 200,000 and 400,000, people dead, respectively. The 1973–74 famine resulted in the overthrow of Emperor Haile Selassie who fed his dogs as starvation raged and the 1984–85 famine marked the end of Mengistu Haile Mariam who was celebrating his decadal hold of power while people died.

In fact, for the GHA region, rather than responding successfully to the frequent recurrent drought, which is a fact of life in many parts of the GHA having been recorded from as far back as 253 B.C. and afflicts the region, the communities are invariably devastated by famine crisis, instabilities in national economies, and political tensions. GHA's climate extremes such as droughts due to low rainfall and the aridity nature of much of the region influence its meagre water resources leading to perennial food insecurity. This, coupled with frequent regional and local conflicts, high population growth rate, low crop yield due to poor water control, climate change and/or variability, invasion of migratory pests, contagious human and livestock diseases such as HIV/AIDs and currently (2020–2021) COVID-19, and poverty in the region, simply makes life for more than 310 million inhabitants unbearable. Alarming is the fact that drought-like humanitarian crises in the GHA are increasing despite recent progress in drought monitoring and prediction efforts. Notwithstanding these efforts, there remain challenges stemming from uncertainty in drought prediction, and the inflexibility and limited buffering capacity of the recurrent impacted systems.

Food security of the region is hugely connected to the agricultural sector, a major economic endeavor that on the one hand provides employment while on the other hand provides the nutrition needs of the people within the region. However, this vital ingredient, "agriculture", is increasingly coming under threat from climate extremes given the fact that Greater Horn of Africa (GHA) is heavily reliant on rain-fed agriculture and as such, one of the most food insecure regions in the world whenever these climate extremes (droughts and floods) strike. This is perplexing given GHA's freshwater (surface and groundwater) potential! It is the home to the world's second largest freshwater lake (Victoria) and endowed with towers in the continent (Ethiopian Highlands) all of which could be tapped in a sustainable way to support irrigated agriculture. First, however, the obsolete Nile treaties that hamper the use of Lake Victoria (White Nile) and Ethiopian Highland (Blue Nile) have to be done away with.

Furthermore, food insecurity in GHA is linked to low productivity, which is attributed to the fact that only less than 1% of the cultivatable land is irrigated due to lack of proper utilization of groundwater. With climate extremes and high population growth projected to increase in future, food insecurity situation in the GHA region is also expected to worsen. To achieve greater food security, therefore, in addition to boosting GHA's agricultural output, the UN Office for the Coordination of Humanitarian Affairs opines that *inhabitants must create more diverse and stable means of livelihood to insulate themselves and their households from external shocks*, a task that they acknowledge will not be easy as the path ahead is "strewn with obstacles—two of the most important being *natural hazards* and *armed conflict*. Understanding GHA's hydroclimate, therefore, is a good starting point towards tackling the natural hazard on the one hand while understanding the impacts associated with extreme climate (drought) on the GHA's water resources and assessing the potential of its groundwater to support irrigated agriculture in the region would be the first step towards *coping with drought* on the other hand. This book looks at the hydroclimate of GHA with the view to assess extreme climate (drought), considers drought characteristics associated with coupled ocean-atmosphere phenomena in the region, and discusses food insecurity issues. This book will not be complete without thanking Associate Professor Freddie Mpelasoka (Curtin University, Australia), Professor Richard Anyah (Connecticut University, USA), the three anonymous reviewers and my wife Naomi Akoth Awange for their valuable comments that enriched the book. Thanks too to various publishers who granted permission to re-use the publications.

Perth, Australia Joseph L. Awange

Contents

Part I
Food Insecurity in GHA: Potentials and Challenges

Chapter 1
Food Insecurity: Causes and Eradication

The main categories of chronically food-insecure people that emerged from discussions at the country level are: a) pastoralists and agro-pastoralists in arid and semi-arid areas; b) small-scale, resource-poor farmers; and c), the urban poor. The report identifies the underlying causes of long-term food insecurity as a dangerous conjunction of different factors. There is a high risk of natural hazards, especially drought, because of the aridity of much of the region and the fact that rainfall is low, unreliable and unevenly distributed. There is also evidence that the climate is becoming more unstable. Widespread regional and local conflict also triggers food insecurity. It drives people from their homes and disrupts marketing and distribution systems. Governments are using scarce resources on arms and, in 1997, the countries of the region devoted US$2 billion to the military. This discourages donors, who are prepared to support people in need but want to avoid indirectly financing warfare. All this is compounded by high rates of population growth. The population of the Horn of Africa has more than doubled since the first of the modern droughts hit the region in 1974, and it is projected to increase by a further 40 percent by 2015. This puts intense pressure on natural resources. Many of the causes of food insecurity are in rural areas, where 80 percent of the population and most of the food-insecure are to be found. The natural resource base is fragile and degraded. The agriculture practised by almost all farmers is characterized by perhaps the lowest productivity in the world. A mere 1 percent of the cultivable area is irrigated, compared with 37 percent in Asia, denying farmers protection from the vagaries of the climate. The pastoral systems, which are well adapted to the vast arid lands, are nonetheless fragile and susceptible to climatic cycles and population pressure. For almost all rural people, household economies are narrowly based, and they have limited access to technology, knowledge and markets. Being only weakly connected to the market, few of the farmers have benefited from liberalization of the economy or from globalization. Indeed, they may well have suffered adverse consequences, having to pay more for inputs such as fertilizer, and receiving lower prices for their crops. All these factors serve to undermine the capacity of the people of the area to feed themselves or to be able to buy the food they need. – Food and Agriculture Organization of the United Nations (FAO) [37].

J. Awange, *Food Insecurity & Hydroclimate in Greater Horn of Africa*,
https://doi.org/10.1007/978-3-030-91002-0_1

1.1 Greater Horn of Africa: Background

Greater Horn of Africa (GHA, Fig. 1.1a), one of the most food insecure regions in the world comprises of 11 countries; Burundi, Djibouti, Ethiopia, Eritrea, Kenya, Rwanda, Somalia, Sudan, Tanzania, Uganda and South Sudan, majority of which are classified as least developed where most of the societies survive on less than one dollar per day [2, 16]. Abshir [1] puts the area of Horn of Africa (which excludes Tanzania, Rwanda and Burundi) to 5.2 million square kilometers with a population of 230 millions. If Tanzania, Rwanda and Burundi are included to form the Greater Horn of Africa, it covers an area of over 6.2 million square kilometers and a population of over 310 million people with a growth rate that averages 3% a year. It is identified by the European Union [36] as a region of geo-strategic importance whose climate change exacerbates existing pressures in the region, including poverty, food insecurity and population growth, despite the fact that the countries of the Horn have little to no control over global carbon emissions [1].

Its climate may be classified as arid and semi-arid (Fig. 1.1b) with frequent recurrences of droughts (Fig. 1.1c and 1.3a) and floods (Fig. 1.1b and 1.3b), e.g., [4, 11, 20, 24, 27, 34, 35, 41, 42, 48, 58, 60, 81, 87]. The recurrences of floods and droughts have been associated with many socio-economic miseries. Furthermore, the region is often faced with serious food insecurity and resource-based conflicts. For example, 2010–2011 was shown to be the driest period in 60 years with more than 12 million people in need of emergency relief [9]. Intergovernmental Panel on Climate Change (IPCC) assessments [10] showed that climate change is real and the poor are the most vulnerable due to the already high level of vulnerability and low coping capacity. The vulnerability is amplified by the fact that many of GHA's livelihoods are dependent on farming and livestock; two sectors that are especially sensitive to perturbations in the climate system. For instance, drought, whose major episode has occurred at least once in each decade, is a fact of life in many parts of the GHA having been recorded from as far back as 253 B.C. [37]. Its impacts on agriculture include, e.g., crop losses, lower yields in both crops and livestock productions, increased livestock deaths, increase in insect infestation and plant and animal diseases [22]. Its impact on human health ranges from increased risk of food and water shortages to increased risk of malnutrition and a higher risk of water and food borne diseases [22, 38, 59, 73]. Climate change is, therefore, likely to set back development and food production in many of the predominantly agro-based economies of most communities.

The importance of agriculture in GHA is underpinned by the fact that it engages about 65% of the labour force (i.e., contributing to 42% of GHA's gross domestic products), where smallholder farmers who depend on rain-fed agriculture for subsistence farming contributes about 90% [3, 22, 73]. According to [22], the per capita food availability over Sub-Sahara Africa, to which GHA belongs, is below the world's average due to high population growth rate, slow and sometimes negative growth rate in agricultural production, climate variability, and poverty [51]. GHA's food insecurity, therefore, is a result of [22] natural hazards (frequent droughts and occasional floods), conflicts, population growth, fragile ecosystems, low agricultural

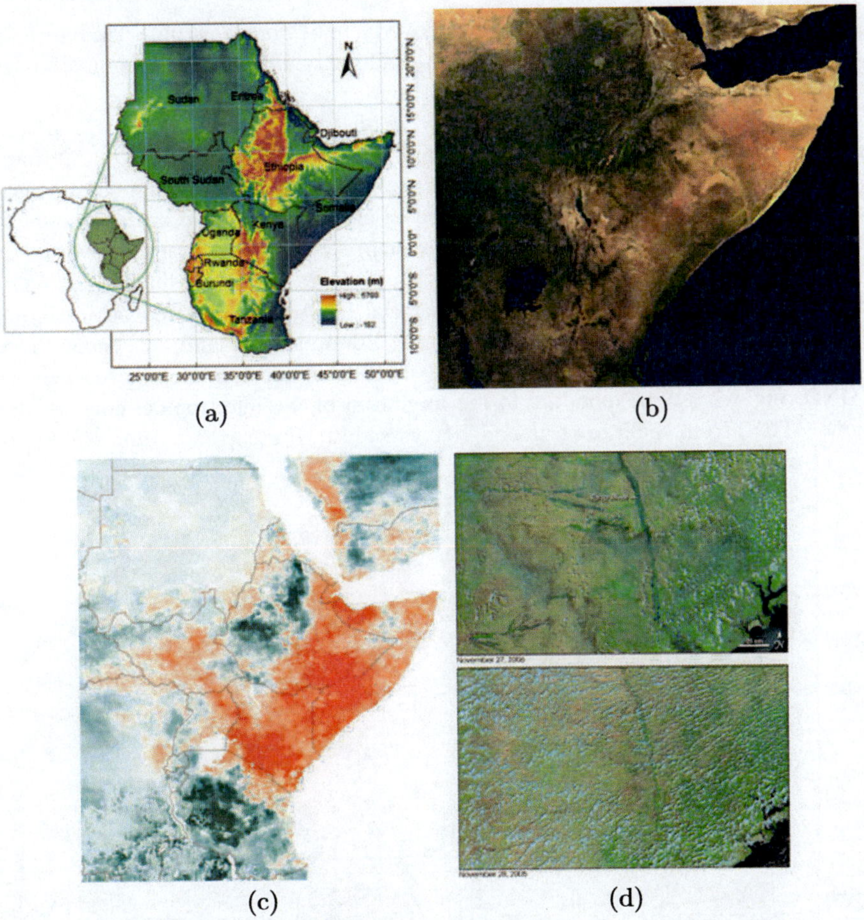

Fig. 1.1 **a** The African continent (left), indicating the location of the Greater Horn of Africa (GHA). The zoomed out image indicates the countries within GHA (*Source* [74]-https://www.mdpi.com/2073-4433/9/3/112), **b** the location of the major surface water bodies within GHA, specifically the dominant Lake Victoria Basin (LVB) (*Source* NASA, https://upload.wikimedia.org/wikipedia/commons/2/23/Nasa_Horn_of_Africa.JPG), **c** soil moisture anomalies in April 2019 in GHA depicting drought and its effect on growing crops (*source* https://earthobservatory.nasa.gov/images/145116/food-crisis-grows-from-dry-soils), and (**d**), flooding of River Tana in GHA on November 27, 2006 (top), compared to its normal state on November 28, 2005 (bottom). *Source* https://www.earthobservatory.nasa.gov/images/7164/flooding-in-the-horn-of-africa

productivity, climate variability/change, poverty, and "*poor governance-donor help syndrome*" see e.g., [2, 30, 38, 51, 59, 78]. Indirectly, it has led to economic losses in growth of countries in the region [73].

Most regions of GHA (with the exception of the Ethiopian highlands, southern parts of Kenya and Uganda, and equatorial Sudan) experience low (c.a. 500 mm) and highly unreliable inter-annual and intra-annual rainfall that cannot sustain agricultural activities. These areas, designated as arid and semi-arid areas, amount to about 67% of the total area [37]. Spatial variability of seasonal rainfall over GHA from 1979 to 2010 is shown in Fig. 1.2, with Fig. 1.2a indicating the dry season of January-March. The long rains over equatorial GHA region mostly occur during March-May (MAM, Fig. 1.2b) while over the Ethiopian highland, this occurs from June to September (JJAS, Fig. 1.2c). The short rains occur in October–December (OND, Fig. 1.2d) corresponding to the migration of the intertropical convergence zone (ITCZ) from south to north and vice versa [12]. Furthermore, Ethiopia, South

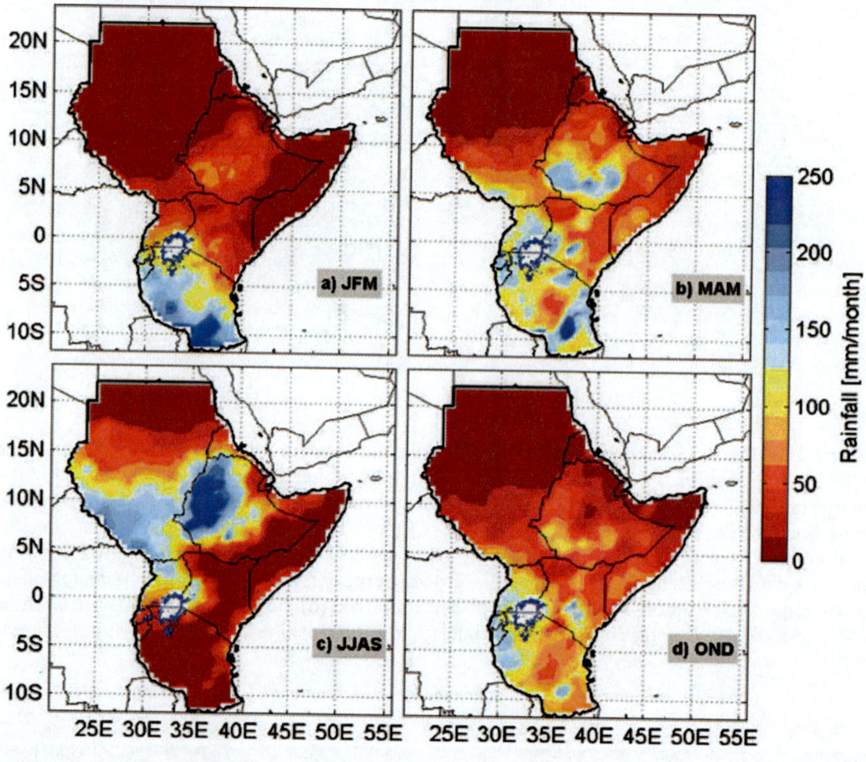

Fig. 1.2 Spatial variability of seasonal rainfall over GHA from 1979–2010 shown for different period of change based on GPCC v6 data: **a** January–March (JFM), **b** March–May (MAM), **c** June–September (JJAS), and (**d**), October–December (OND). Note that the signals over Lake Victoria has been masked out in the analysis to avoid spurious trends due to varying lake levels. *Source* [4]

(a)

(b)

Fig. 1.3 **a** Drought in the Greater Horn of Africa is widespread, triggering a regional humanitarian crisis with food insecurity skyrocketing, particularly among livestock-owning communities, and devastating livelihoods, **b** floods in GHA, *Source* [37]

Sudan, Sudan, and parts of Uganda experience a single rainy season from June–September (JJAS; Fig. 1.2c).

Apart from its seasonal differences, rainfall variability over the GHA is closely associated with the large-scale regional and global circulations such as El Niño

Southern Oscillation ENSO [50], fluctuation of the Indian Ocean and Atlantic Ocean Sea Surface Temperature (SST), and moisture fluxes over the Congo region [11, 13, 15, 86]. Mean temperature patterns over GHA follow the annual rainfall pattern. The major water bodies found within GHA are shown in Fig. 1.1b.

1.2 Causes of Food Insecurity

1.2.1 Poor Governance and the Donor Syndrome

Reading most scholarly articles and organizational bodies' reports on the causes of famine/food insecurity in the GHA region, the usual culprits (climate, conflicts, and population growth) are mentioned. Seldom is the elephant in the room "poor governance and the donor syndrome" mentioned. For example, Food and Agricultural Organization (FAO) discusses it not under causes for food insecurity but rather under the supportive environment where it states [37]:

> Although most governments of the region have food security policies or poverty reduction strategies and programmes that encompass food security issues, the allocation of national resources to achieving food security does not reflect the level of commitment that is needed. For example, throughout the region, budget allocations to supporting the agricultural sector are small and declining while, in many cases, expenditures on arms have soared. There has been a tendency to increase dependence on external assistance for meeting food security goals, especially when humanitarian considerations play a part.

Now, before we blame the usual culprits (conflicts, climate and population growth rate), it could be argued that the greatest cause of food insecurity in the GHA region could be the issue of poor governance and the perennial reliance on donor support. As a start, let us get some historic perspective in order to understand the argument that poor governance and reliance on donor support causes food insecurity and exacerbates famine in the region. Most countries within GHA are now almost 60 years since they attained independence. Since then, most lessons on the traditional causes of famine should have been learned and strategies for copying up formulated. Take the case of the perennial floods in Kenya as an example. Year in, year out, the rivers Nyando and Nzoia feeding into Lake Victoria bursts causing floods that destroys crops and displaces the riparian peasants living within the banks of these rivers. Sixty years along the line, has no successive government seen the need to address the problem? Another problem is the sustained killing of the local/rural agricultural activities. In Kenya for example, cotton used to be grown but the encouragement by the successive governments of second-hand imported clothes known locally as "mitumba" ultimately killed the cotton industry. On the chop board appears to be the sugarcane growing, where greed has crippled in and the "who is who" in the pecking order allowed to import foreign refined sugar at the expense of the local industry. The list goes on and on.

Then comes the donor. During one of the floods episode of river Nyando in Kenya mentioned above, a woman was captured begging the Kibaki government "Serikali saidia sisi", a Swahili saying, which means government please help us. Her statement/voice is now popularly used by the Kenyan mobile giant Safaricom. But this seems to echo what the respective governments within the GHA are perennially telling donors. "Donors, please help us!" And the donors respond. Due to the so called "triple threat" of locust invasion, climate change impacts and the COVID-19 pandemic from 2020, the European Union donated about EUR 149 Million in humanitarian aid to the Greater Horn of Africa.[1] Will this money ever be reported to have achieved its aim? And yet this is not something new to the donors. Times without number, they know very well that the help they have been giving most often than not does not achieve its aims, yet they choose to turn a blind eye. Reports abound in newspapers where food meant to help the famine stricken communities have been found in the markets being sold or the money given for aid finding its way to fund armed conflicts. Could it be a symbiotic relationship between the donors and the respective governments? That 60 years along the line, the nations still have not learned any lessons to ameliorate the perennial food insecurity is baffling.

1.2.2 Natural Hazards

GHA, a region that is either arid or semi-arid with unreliable low and unevenly distributed rainfall, consist of many subsistence farmers who rely heavily on rain-fed agriculture, which in turn relies on the good will of the weather and climate. Climatic extremes (droughts and floods) are, therefore, one of the major factors contributing to vulnerability to food insecurity in GHA, where there is no year or season in which the whole region receives normal rainfall and is free from climatic anomalies such as flood or drought [37], (see also Fig. 1.3). For instance, just two years after the 2016/2017 drought and one year after flooding in 2018, back-to-back droughts and floods in 2019 led to rising needs and compounded the humanitarian consequences of conflict and violence in multiple locations. In addition to loss of lives, livestock and crops (Fig. 1.3a), as well as population displacement, the above normal rains and cyclonic activity in late-2019 and early-2020 contributed to a desert locust upsurge that affected Djibouti, Eritrea, Ethiopia, Kenya, Somalia, Sudan and spread to Uganda, South Sudan and Tanzania.[2] With sufficient rainfall, crops will perform well and food production will increase where farmers see bumper harvests. This good will, is however, not forthcoming with rain-fed agriculture increasingly becoming vulnerable to drought events, see e.g., [41, 45, 49, 63, 67, 76], leading to food insecurity [46, 49, 61, 62]. Between the two climate extremes (floods and

---

[1] https://ec.europa.eu/echo/news/eu-allocates-149-million-humanitarian-aid-greater-horn-africa-region_en.

[2] https://reliefweb.int/report/ethiopia/greater-horn-africa-region-humanitarian-snapshot-february-2020.

droughts), [56] opine that drought, which has a unique impact on agricultural systems due to its duration that often extends over several seasons, is the main cause of food insecurity over GHA leading to malnutrition and famine, and affecting all the four dimensions of food security; availability, accessibility, utilization, and stability, see also [38, 78]. This is captured by FAO who state [37]:

> Drought has a perhaps unique impact on agricultural systems because of its duration, which often extends over several seasons. The people of the region have, over centuries, evolved mechanisms for coping with the risks of the environment in which they live. Farmers have, up to a point, learned to cope with late rains or with the mid-season cessation of rains, spreading risk by planting different crops and at different times, through on-farm storage and by resorting to hunting and gathering at times of stress. For the pastoralists, travelling with their herds and flocks to follow the rains and the growth in pasture is a natural part of their system, while setting areas aside for grazing reserves and splitting herds to minimize risk are elements of their coping mechanism. Increases in population have, however, disturbed the equilibrium between people and natural resources.

For the GHA region that experienced servere droughts in 1973/74, 1984/85, 1987, 1992 to 1994 and 1999/2000, [56] notes that rather than responding successfully to the frequent recurrent droughts that afflict the region, the communities are invariably devastated by famine crisis, instabilities in national economies and political tensions. For example, [56] point out that the Ethiopian "biblical" famines of 1973–74 and 1984–85 left about 200,000 and 400,000, people dead, respectively, with the former disaster resulting in the overthrow of Emperor Haile Selassie. The latter contributed to the end of the Marxist regime of Mengistu Haile Mariam. These numbers differ from those provided by Food and Agricultural Organization (FAO) that states [37]:

> In Ethiopia alone, the 1984 drought affected 8.7 million people, about 1 million died and 1.5 million livestock perished. In the Sudan 8.5 million people were affected by the same drought, and about 1 million people and 7 million livestock died. In 1987, about 2 million people in the Sudan, more than 5.2 million in Ethiopia, 1 million in Eritrea and 200 000 in Somalia were severely affected. The current drought, which started in 1998, is affecting about 16 million people in the Horn of Africa. Drought is, therefore, a recurring phenomenon in the region and there will always be certain locations experiencing localized drought conditions.

To underscore this point, climate variability/change is projected to increase irrigation water demand on the one hand [73], and on the other hand, accelerate drought frequency, severity, spatial extents, and duration and its impacts [11, 35, 38, 41, 45, 56, 59, 60, 63, 71, 73, 84, 85]. Floods, when they occur (see Fig. 1.3b), as witnessed during the 1997/98 El Niño Southern Oscillation (ENSO; [5, 50]), lead to vulnerability in food security as they cause animal death (see Fig. 1.3a), siltation of reservoirs and destruction of crops, all which impact on the livelihood of both pastoralist and peasants of GHA. In this regard, therefore, increase in frequency and severity of climate extremes (floods and droughts), and increased irrigation water demand are likely to further decrease crop water availability and threaten the productivity of rain-fed agriculture over the region, e.g., [73], thus increased food insecurity [22].

1.2.3 Conflicts: Regional and Local

Although the region is largely stable politically, Greater Horn of Africa (GHA) is not new to conflicts, see e.g., [43, 44]. Be it the ship piracy in the Indian ocean, frequent militia attacks in Somalia, or the disputed Grand Ethiopian Renaissance Dam (GERD) being built by Ethiopia, GHA will always be on news. The dire conflict situation in GHA is captured by FAO who states [37]:

> The Horn of Africa has been plagued by conflict since time immemorial. Although the war between Ethiopia and Eritrea has attracted the most media attention, the region has suffered from almost continuous civil conflicts over the last 30 years in Ethiopia (as formerly defined), the Sudan, Somalia and Uganda, and these have spilled over into Djibouti. The countries of the region devote between 8 and 50 percent of central government expenditure, or between 2 and 8 percent of gross national product (GNP), to the military, totalling US$2 billion in 1997. These figures rise substantially, of course, whenever conflict flares up. Conflicts in the region undoubtedly exacerbate the famine and food insecurity triggered by drought. Even before the recent hostilities between Ethiopia and Eritrea, more than 1 million people from the region were refugees. Large populations of internally displaced persons (IDPs) were to be found in the Sudan, Somalia and Uganda. Conflict removes able-bodied men from agricultural production and, incidentally, places an extra work burden on women. It also diverts resources, directly and indirectly, from more productive and socially beneficial uses, and tests the willingness of the international community to provide assistance.

Somalia and South Sudan have experienced internal armed conflicts, e.g., within Somalia, there exists the Alshebab threat while within South Sudan fighting have occurred between supporters of the current President Silver Kirr and his Vice President Riek Machar. Armed conflicts between countries have also occurred in the region, e.g., within/between Ethiopia and Eritrea. Abshir [1] points out that sometimes countries often intervene in their neighbours' conflicts, either directly by sending troops or indirectly by sponsoring proxies or supporting rebel groups. Besides cross-border conflicts, another common feature in the region is the regularity of ethnic overlaps, affinities and loyalties that transcend national borders (e.g., Somalis in Ethiopia-Kenya-Djibouti; Karamajong in Uganda, Kenya, and South Sudan; Afars in Djibouti-Eritrea-Ethiopia; Borans in Kenya-Ethiopia) [1]. Abshir [1] states: "*On one hand, these overlaps facilitate informal trade and commerce, but on the other they are seen by state authorities as a liability and a potential source of insecurity*". At the time of writing this book (2021), three agencies; The Food and Agriculture Organization of the United Nations (FAO), the United Nations World Food Programme (WFP), and United Nations International Children's Emergency Fund (UNICEF) have sounded an alarm over acute food insecurity in northern Ethiopia's conflict-ravaged Tigray region, where more than 350,000 people have been afflicted by famine. In fact, the Integrated Food Security Phase Classification (IPC), a system used by humanitarian aid agencies and governments to determine the scale of a hunger crisis placed Tigray in Phase 5 (i.e., IPC5, i.e., catastrophe/famine),[3] and blamed the conflict for triggering massive population displacement, widespread destruction of livelihoods

[3] https://www.nytimes.com/2021/06/10/world/africa/ethiopia-famine-tigray.html.

and critical infrastructure, loss of employment, and limiting access to markets.[4] The IPC Acute Food Insecurity (IPC AFI) classification provides strategically relevant information to decision makers that focuses on short-term objectives to prevent, mitigate or decrease severe food insecurity that threatens lives or livelihoods, and provides differentiation between different levels of severity of acute food insecurity in five distinct phases; IPC1 (minimal/none), IPC2 (stressed), IPC3 (crisis), IPC4 (emergency), and IPC5 (catastrophe/famine). Each of these phases has important and distinct implications for where and how best to intervene, and therefore influences priority response objectives.[5] IPC data showed that of 5.5 million people facing food insecurity in Tigray and neighboring zones during May and June, 350,000 were now in Phase 5 as a result of cascading effects of conflict, including population displacements, movement restrictions, limited humanitarian access, loss of harvest and livelihood assets, and dysfunctional or nonexistent markets.[6] These armed conflicts impact on agricultural productivity, see e.g., [66, 69, 78].

On the one hand, some of these conflicts are lethal and impact on establishing hydroclimate monitoring stations and/or collection of the recorded data. On the other hand, the link of conflicts to food insecurity means that each can trigger the other as scramble over resources is associated with population increase. As food security is linked to smallholder rain-fed subsistence agriculture, any factor that prevents the population from day to day farming e.g., population displacements witnessed in South Sudan and Somalia and/or deviation of agricultural allocated funds to military use is likely to put additional pressure on food resources of those countries thus making them vulnerable to food insecurity [22]. Pastoral areas are not spared either nor are the internally displaced people (IDPs). FAO states [37]:

> Pastoral areas, which are under pressure from the expansion of cropping into marginal areas and increasingly degraded rangelands, are especially susceptible to local conflict and cattle raids, which break out when people have ready access to modern weapons. Northern Kenya and northern Uganda have been particularly prone to prolonged outbreaks of such violence. Such tendencies are exacerbated when drought hits and the scramble for limited grazing and water intensifies. Conflict, whether transboundary or internal, exacerbates the vulnerability of poor people, displacing them from their homes and depleting their assets. It makes emergency relief operations directed towards IDPs difficult and dangerous for those involved. Conflict also has a much more insidious impact on long-term development efforts, diverting scarce resources, both national and external, away from development activities and into war. The fungibility of funds means that donors face the risk of funding conflict when their intention is to alleviate poverty through development programmes.

---

[4] https://www.unicef.org/press-releases/un-agencies-concerned-looming-famine-northern-ethiopia-call-urgent-life-saving.

[5] http://www.ipcinfo.org/ipcinfo-website/ipc-overview-and-classification-system/ipc-acute-food-insecurity-classification/en/.

[6] https://www.nytimes.com/2021/06/10/world/africa/ethiopia-famine-tigray.html.

1.2.4 Population Growth: Rural Urban Migration

The population of the 11 countries that comprise GHA, which has more than doubled since 1974 is currently more than 310 million [1, 37], with an average population growth rate for the region being 3% [6], and where the population is expected to double every 23 years [17], see Fig. 1.5. This rapid increase in population growth where young people form the majority [17], will certainly exert pressure on the natural resources (water and land) on the one hand, and increase in rural urban migration on the other hand leading to impact on food security [3, 40, 51, 53, 59, 73, 78, 82]. Traditionally, land in GHA is passed largely to the sons through subdivision and as such, a rapid increase in population, therefore, would see increase in land fragmentation that is unhealthy for subsistence agricultural productivity and subsequently food security, see e.g., [55].

Another problem, which is closely associated with population growth is that of environmental degradation, i.e., land and vegetation. Population increase puts pressure on energy demand [64, 65], which is supplied mainly by wood and animal manure. This in turn leads to land degradation as the forests that have been harvested to provide fuel leads to exposed land that favour soil erosion during rainy seasons. Eventually, this overall degradation leads to increase in rainwater losses through runoff that triggers soil erosion, which in turn exacerbates drought impact resulting in further land productivity decline, loss of biodiversity and continuing desertification [37]. Furthermore, soil erosion leads to siltation of water bodies, e.g., Lake Victoria, which are essential for livelihood. Besides deforestation for fuel purposes, there is also the problem of urban encroachment on arable land that could be used for crop production. The encroachment takes the form of the land being converted to other uses, e.g., biofuel production [2] and the expansion of cities. Morgan et al. [54] have shown (see Fig. 1.4) that for the Ugandan cities of Jinja, Kampala, Masaka and Mbarara; Kigali in Rwanda, and Kisii in Kenya, the outward urban expansion was the primary contributor for the long-term vegetation decline. To this end, FAO states [37]:

> To the extent that there has been any increase in the area of land being farmed, this has taken place largely in marginal areas, using systems that may not be sustainable. Shrinking land resources have not been compensated for by increases in land productivity. Average cereal yields are a mere 860 kg/ha and, where comparative data are available, statistics confirm the general impression that yields are declining. For example, in the Sudan and Uganda, average yields have dropped by 12 and 18 percent, respectively, over the last decade.

1.2.5 Poverty

Poverty is connected to food insecurity [2]. This connection is more important in GHA where most food is produced in rural areas where a majority of the population is poor. For these poor, subsistence farming and pastoralism form their main source

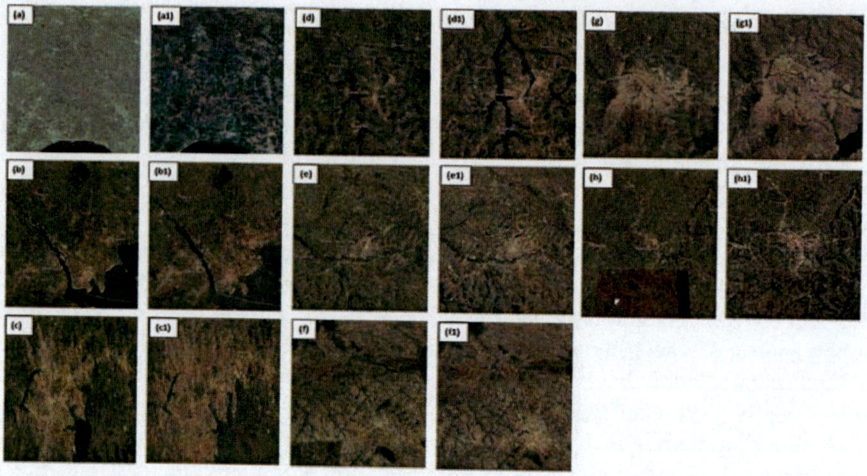

Fig. 1.4 Google Earth Pro imagery with Landsat images as base map **a** Butundu, Uganda 2004 and (a1) Butundu, Uganda 2017; **b** Jinja, Uganda 2003 and (b1) Jinja, Uganda 2016; **c** Kampala, Uganda 2003 and (c1) Kampala, Uganda 2016; **d** Masaka, Uganda 2003 and (d1) Masaka, Uganda 2016; **e** Mbarara, Uganda 2003 and (e1) Mbarara, Uganda 2016; **f** Katoro, Tanzania 2003 and (f1) Katoro, Tanzania 2016; **g** Kigali, Rwanda 2003 and (g1) Kigali, Rwanda 2016; **h** Kisii, Kenya 2003 and (h1) Kisii, Kenya 2018. *Source* Morgan et al. [54]

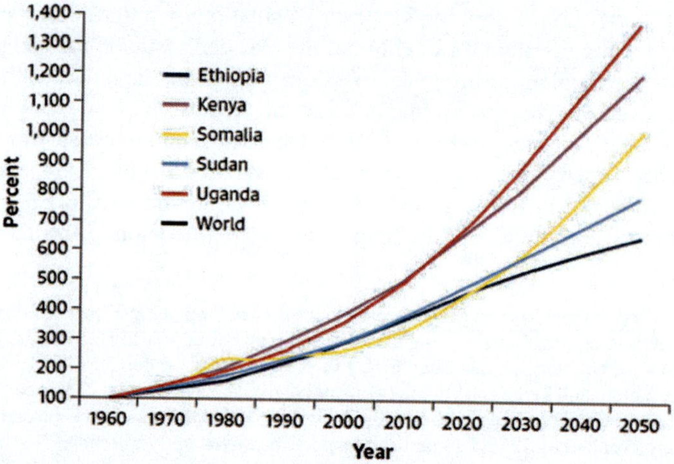

Fig. 1.5 Greater Horn of Africa's projected population growth since 1960. *Source* [17]-https://reliefweb.int/sites/reliefweb.int/files/resources/The-Borderlands-of-the-Horn-of-Africa.pdf

of livelihood. In this regard, poverty alleviation strategies, e.g., [2] plays a significant role in reducing food insecurity. A long-term solution to food insecurity therefore goes beyond its increased production to include measures that address rural poverty. FAO captures the difficulties faced in the rural areas, which are partly attributed to poverty, i.e. [37]:

> Agriculture in the region is, for the most part, characterized as being low-input/low-output. The level of technology is generally basic, and productivity per hectare and per person employed are perhaps the lowest in the world. In the parts of the region with higher potential (i.e. those areas with high and reliable rainfall), in which crop-based systems predominate and population densities are highest, productivity is constrained by lack of knowledge, lack of financing, and poorly articulated markets. In these areas a substantial proportion of farmers live at the edge of subsistence, and are food-insecure simply because they have limited access to land. For example, in Ethiopia, almost 40 percent of farm households have less than 0.5 ha of land, and more than 60% have no more than 1 ha from which to support a family of between six and eight people.

1.3 Famine Eradication: Proposed Strategies

Eradicating famine in GHA will not be easy especially where governance is poor, with a weak policy and institutional framework in many of the countries. Most important of all, there is inadequate commitment to addressing the problems of food insecurity by the governments of the region [37]. FAO suggest that eliminating famine would include [37];

> Actions for disaster preparedness, restructuring and strengthening early warning systems and basing them on active two-way communications between local communities and national and international decision makers. The complex issue of strategic grain reserves and protected funds set aside for imports would be addressed, as well as means of moving quickly from emergency relief to rehabilitation and development. The need for protecting the most needy would be addressed through cash- or food-for-work schemes for the able-bodied. For the elderly, the handicapped and orphans, safety net mechanisms would be needed, but these would have to be community-based in order to be sustainable.

In this section, two of the possible strategies to eliminate poverty are discussed. For detailed exposition, the reader is directed to [37].

1.3.1 Good Governance and Donor Awakening

Section 1.2.1 dealt with the issues of poor governance at length. If meaningful outcome is to be realized in terms of eradicating food security, then the respective governments have to bite the bullet and realize that 60 plus years of independence is enough to stop the joke of relying on donor subsidies. In equal measures, the donors too have to realize that the extended joke has to come to an end. Without these two (governments and donors) stopping the poker game, then the cycle of food insecurity and famine will continue. There is absolutely no harm in donors funding the projects in GHA as long as the intended outcome are achievable. The truth of the matter, however, and the donors themselves know this too well, is that several non-governmental organizations formed in GHA to address food insecurity thrive in the very issue they are supposed to address. This funding quagmire is captured by [1] who states[7]:

> In 2016, the EU funded a project (within a broader €31 million initiative to improve Kenya's ecosystem services) called the WaTER programme focused on protecting the "Water Towers" or high-elevation forests that serve as natural reservoirs. Much of the country draws its water supply from these towers; their conservation is thus vital to the country's resilience to drought and climate change. However, the EU was forced to suspend the programme because of the Kenya Forest Service's forcible evictions of communities of indigenous peoples from one of the Water Towers areas, resulting in one fatality. Human rights organisations subsequently warned that the WaTER program could encourage abuses against indigenous communities. In January 2018, the programme was suspended.

If no lesson has been learned over the past 60 years, will any be learned now without good will? Is it not time for alternative measures to be tested?

It all starts with good governance. A government that thrives in its citizens reduced to baggers and does not address health, education and food security issues is surely in a web that is hard to attain freedom. Much of the donor received money plus local income from revenues should go towards technologically tested food enhancing measures, infrastructure development and service provision rather than funding armed conflicts, which in themselves further fuel famine and food insecurity. Yet these issues being addressed in this section are not new nor unknown. It appears to be chronic in that the more no action is taken by the respective governments to address food insecurity, the more the famine, the more the cash income into few pockets coming from the donors.

1.3.2 Broadening and Maximizing Opportunities

Most of GHA inhabitants live in the rural areas, and as such, depend on agriculture. To overcome food insecurity in the region, a combination of home-based technology to farming, diversification of income generating activities and improved services have to be invoked. To this end, FAO proposes that [37]:

> Programmes to address long-term, chronic food insecurity would focus on broadening the opportunities for sustainable livelihoods. The immediate focus would be on enhancing the livelihoods of small-scale resource-poor farmers, through a combination of agricultural technologies and support services, access to markets and credit, along with rural enterprises and agroprocessing. For many farmers, this would mean making better use of water, through, for example, small-scale irrigation, building on the experiences of FAO's Special Programme

[7] Source: https://eeas.europa.eu/delegations/kenya/14465/protecting-and-increasing-forest-coverkenya_en.

> for Food Security. In the drier areas, the focus would be more on the promotion of drought-resistant crops, as well as the conservation of both soil and water - more "crop per drop". Pastoralists could achieve greater security if they had better marketing and information systems for their stock, as well as broader opportunities for investment instead of simply buying more livestock. They could also boost their incomes by processing milk and meat, as well as hides and skins, into products for sale. All farmers should be looking to diversify their sources of income, for example by rearing more short-cycle livestock, taking advantage of non-timber forest products and, in some places, developing ecotourism. Underpinning these actions is the need to create an enabling environment for the economy and to enhance food security. Policy and institutional measures to resolve problems of governance would be part of the programme, as well as proposals for conflict prevention and resolution, in collaboration with the regional intergovernmental bodies. Infrastructure must be developed, especially rural roads and livestock markets, to provide better access to trading opportunities, basic services, especially health, water and sanitation, as well as both formal education and skills training. Civil society must be allowed to play a greater role in achieving food security, using the skills and experience at community level of NGOs and rural producers' organizations. There is clearly a need for concerted action at the regional level to address problems such as conflict, trade, transboundary human and animal health issues and early warning systems. This has led the report to call for the formulation of a Regional Food Security Programme (RFSP), which would complement and strengthen the programmes prepared by each country.

Improved technology would lead to sustainable utilization of the available water both for the rainfall prone areas as well as arid regions. As stated in Sect. 1.2.1, Nyando and Budalangi regions in Kenya experience perennial floods during heavy rains. Excess water, which displaces people and destroys crops could be tapped into some dams and used during dry period for small scale irrigation to produce vegetables, which could be sold in cities such as Kisumu (Kenya's third largest city). It should be pointed out that during dry seasons, Kisumu obtains its vegetable supply from more than 200 km away, e.g., from the Rift Valley province. Where peasant farmers go out of their way to produce vegetables from irrigable Nyando waters for example, lack of accessible markets renders them pray to middle men who come all the way from Nairobi (Kenya's capital city >500 km) and rent their farms for less than USD 150 for the entire year. Improved services, therefore, includes not only infrastructure but also accessibility to both markets and credit lending facilities.

Yet, farming (crop and animal production) is just part of the empowering mechanism for the more than 80% of the population living in the rural areas but not exclusive. Diversifying of local economic activities would play a great deal in eliminating food insecurity as people will have alternative means to fall back to during hard times, caused e.g., by crop failures due to extreme climate. According to FAO [37], in some areas, this can be achieved by diversifying the farming system, particularly by expanding the use of short-cycle livestock such as poultry, sheep, goats, pigs and, where water resources allow, fish. In pastoral areas, the processing of milk and meat products, hides and skins may provide opportunities for supplementing incomes. Options for raising additional earnings from non-wood forest products have also been noted. In the long term, it is essential that conditions be created whereby people have increasing access to employment opportunities outside agriculture. The ingredients for this include a combination of improved education, better transport and communications, easier access to markets and financial services and, in some cases, a reduction in the legal and bureaucratic barriers to entry into business [37].

Take Lake Victoria Basin (LVB) for example, which supports a population of more than 40 million [8]. The Lake itself is a sleeping resource giant that has not been fully tapped [7]. The governments within the LVB should provide; services (e.g., health services, fish cooling plants and roads), educate the locals on the need to achieve standards set, e.g., by foreign importers of fish such as the European Union, facilitate proper fishing technologies that prevents depletion of sub juvenile fish, and facilitate access to both local and international markets in order to foster capacity for growth and the efficiency of LVB's economy. These measures would enable the more than 40 million to diversify their income by tapping into the lake's resources and go a long measures in cushioning them when there are shocks that threaten food security. The example given above is just but one on LVB. Other areas within the broader GHA context include; small stock-keeping, artisanal fisheries and domestic processing and crafts, all which will contribute to increased incomes that will ensure wider access to the available food [37].

1.4 Potentials and Challenges

1.4.1 Freshwater Potential

Of the developing world regions and the African continent at large, no place is endowed with freshwater resources as the Greater Horn of Africa (Fig. 1.1b). Yet, it is the region that perennially suffers from famine! It is the mother to Lake Victoria, the world's second largest freshwater lake, second only to Lake Superior in the US [5, 7]. Other major lakes dotting the region include lakes Tanganyika, Edward, Kivu, Albert, Kyoga, Naivasha, Turkana, and Tana among others (Fig. 1.1b). These lakes together with numerous rivers in the region provide source of freshwater that could ensure sustainable exploitation in addition to groundwater to support irrigated agriculture in order to alleviate food insecurity. Indeed this potential is recognised by FAO who state [37]:

> As rainfall diminishes and becomes less reliable, the options open to farmers tend to become fewer, although farm area may be less of a constraint. The priorities are likely to be to maximize the returns on available moisture, to reduce interannual variations in output and to improve labour productivity. Irrigation is intuitively the most appealing solution to the problems of crop production in marginal areas, but the practical difficulties of irrigation development and operation are considerable. One of the unfortunate aspects of the region's geography is that there is, in general, a poor coincidence between available water resources and suitable land for large-scale irrigation. Where both land and water are available, constraints on irrigation development may be imposed by transboundary water-sharing agreements. There are, however, important opportunities in most of the concerned countries for improving the performance of existing major irrigation schemes by raising management standards and rehabilitating infrastructure. Such measures, however, while contributing importantly to

> national food supplies, will tend to benefit those rural populations that are least at risk. For the more drought-prone populations, the main opportunities for improving water use include small-scale irrigation, water harvesting and, above all, better use of available moisture in the rainfed farming systems, on which the bulk of farmers will continue to depend.

This calls for the need for continuous monitoring of these stored water potential in order to understand long-term impacts of climate variability/change as well as unsustainable exploitation resulting in depletion e.g., [23, 33, 39, 47].

To unlock the giant water body of the region, however, a two-pronged approach has to be adopted, i.e.,

(i) Exploitation of the state-of-the-art monitoring tools that can measure changes in the total water storage (surface water, groundwater, soil moisture and vegetation water) GHA-wide. Owing to the sheer size of GHA, i.e., >6,000,000 km^2 [22], "boots on the ground" in-situ based approach of collecting the data has to give room to Gravity Recovery and Climate Experiment (GRACE), which is discussed in Chap. 3 and is hugely exploited in this book.

(ii) For a complete and comprehensive utilization of the Nile Rivers (both Blue and White), it is important to address the question of the Nile Treaties that were signed in 1929 and 1959 between Britain, Egypt and Sudan, which prohibit the use of the Nile by 8 upstream countries in order to safeguard the interest of Egypt. Any meaningful eradication of food insecurity in GHA must take into consideration the need for equitable use of the Nile River Basin (NRB)'s waters by all 11 countries within its basin, and doing away with obsolete Nile treaties given that the other 8 riparian countries were not party to the treaties yet are the source of the Nile's waters. It is in this light that controversies have arisen in recent years, with other riparian states demanding their abrogation. Awange [5, 6] extensively detail the Nile Treaties and their implications for the economic development of the riparian states upstream. Specifically, these treaties prohibits the use of giant Lake Victoria's surface water and the Ethiopian Highland's water tower for irrigated agriculture without express permission from Egypt thus definitely impacting on the potential of irrigated agriculture, which is one of the means to alleviate food insecurity in GHA. This realization is acknowledged by FAO who state [37]:

> Many of the available freshwater resources are in river basins and lakes that extend beyond the boundaries of individual nations. Shared water resources include lakes Victoria, Albert, Edward, Kivu and Turkana and major rivers such as the Blue Nile, White Nile, Atbara, Awash and Shebele. The potential for developing irrigation from these sources is constrained by the problem of achieving agreement on sharing the resources and avoiding conflict.

To this extent, Ethiopia has unilaterally taken upon itself to build the Grand Ethiopian Renaissance Dam (GERD), the largest in the continent, which they hope to utilize to among others, to boost its agricultural potential through the produced energy. The solution could lie within the Nile Basin Initiative, which was launched in 1999 following consultations among the 10 riparian countries (Sudan was still one). Although Egypt is today backing a new initiative among the Nile Basin riparian states, which might redistribute more equitably the river's water usage rights, it is not clear how far this will go. Furthermore, Egypt and Ethiopia could still come up with a way to tap on to the potentials of both GERD and Lake Nasser for a mutual benefit of both countries. A comprehensive assessment of the law governing the use and management of the Nile and the development of a legal and institutional framework for cooperation over shared freshwater resources is presented in [83] while [70] introduces an analytic framework constructed upon the iterated Prisoners' Dilemma game to model and analyze transboundary water interactions along the Nile River.

1.4.2 Potential for Agriculture

If Australia and Israel, which are largely arid areas can produce food, certainly Greater Horn of Africa, a food-insecure region with over 70 million people living in areas prone to food shortages [78] but endowed with fertile land within Eastern Africa (particularly Uganda, South Sudan, Rwanda, Burundi, Kenya and Ethiopia) can! The region certainly has agricultural potential, which can be both a source of food and income, thereby alleviating food insecurity in the region. Agriculture, therefore, directly determines the food security vulnerability of the majority of GHA population.

As discussed in Sect. 1.4.1, the first step that GHA could take to address food insecurity is to unlock its water potential for irrigated agriculture. This will shield it from, e.g., impacts of climate variability/change on rainfall essential for rain-fed agriculture, and in so doing, minimise low productivity as evidenced, e.g., in [39, 80].

Agutu [22] sees the potential growth of irrigated agriculture in GHA driven primarily by small scale farmers (i.e., having less than 2 ha), who have been attracted to groundwater irrigated agriculture due to the following benefits; groundwater respond to farmers demand for reliable and flexible irrigation water supply, better and more appropriate and cost effective drilling and pump technology, and services and markets making groundwater irrigated agriculture feasible. Agutu [22] further opines that the response of crops to fertilizer is higher where the supply of irrigation water (groundwater) is assured compared to rain-fed agriculture on the one hand while on the other hand, the reliability and flexibility of groundwater allows farmers to take risk of investing in fertilizer, which in return substantially increases their crop productivity [52].

1.4.3 Hydroclimate Monitoring Network: Simply Insufficient

The GHA countries, like other African countries, have short and/or fragmented hydroclimate (water, temperature, rainfall and soil moisture) records, often as a result of armed conflicts at various times over the past 50 years and the sheer size of the region (>6,000,000 km^2). In some cases, the available records are inaccessible due to governmental red-tapes, and where accessible, some are too short, compounded by missing data or lack consistency where available [58, 60, 67], or sparsely and unevenly distributed to be useful for adequate hydroclimate analysis. Nonetheless, they are still valuable in providing a baseline for future analyses, as well as monitoring inter-annual climate variability.

To enhance and unlock the agricultural prowess of the region, adequate hydrological (soil moisture and groundwater) monitoring units are essential. For instance, monitoring of agricultural drought would require soil moisture data, e.g., [21, 24, 57, 67, 75], which are directly related to rainfall that are hampered by insufficient and/or lack of in-situ rainfall measurements in the region to begin with [14, 19, 25, 86]. These sparseness and/or unevenness distribution of rain gauges in GHA leads to inadequate capture of the spatial variability of precipitation and subsequently soil moisture. Further, although soil moisture is a good indicator of agricultural drought, its utilization in the region has rather been low due largely unavailable in-situ data. This has led to the consideration of satellite, models and reanalysis data [29] discussed in Chap. 3. Satellite and reanalysis products are functions of the satellites and the forcing model products, which in themselves come with uncertainties that further diminish their utilization, see e.g., [18]. Besides these uncertainties, satellite, model and reanalysis products are influenced by topography and rainfall gauge density especially over Upper GHA (Sudan, South Sudan, Ethiopia, and Somalia; e.g., [31, 32, 68]. For a region that faces food insecurity each time a drought event occurs, and due to the magnitude and severity of the impacts of drought-induced food insecurity, it is of utmost importance to evaluate the utility of available drought indicators in order to provide comprehensive, clear, reliable, accurate, and consistent spatial extent of drought monitoring in order to aid policies, planning and mitigation of agricultural drought impacts [21]. Agutu et al. [21] evaluated the utility of satellite-gauge merged precipitation, reanalysis and model soil moisture, vegetation condition index, and terrestrial water storage in characterizing agricultural drought over GHA.

For surface and groundwater, keys to unlocking irrigated agriculture in the GHA, inadequate monitoring infrastructure remains a bottleneck. This hampers monitoring of total water storage (TWS; surface water, groundwater, soil moisture and vegetation) changes and the associated impacts of climate variability/change in the region, see e.g., [23, 26, 28], resulting in an inadequate understanding of the long-term sustainability of groundwater (aquifers) as sources of long-term water supply [22]. Fortunately, the Gravity Recovery and Climate Experiment (GRACE) satellite mission launched in 2002 and its Follow on GRACE-FO launched in 2017 provides the means of monitoring GHA-wide total water storage changes, see e.g., [5, 6].

1.5 Objectives and Aims of the Book

Extreme weather events due to climate crisis are becoming the new normal in the Greater Horn of Africa (GHA), a region that is heavily reliant on rain-fed agriculture and as such, one of the most food insecure regions in the world whenever climate extremes (droughts and floods) strike. This acute food insecurity often occur with catastrophic consequences that claim millions of lives. GHA's climate extremes, coupled with frequent conflicts, high population growth, low crop yield due to poor water control, climate change and/or variability, and poverty in the region simply makes life for more than 310 million inhabitants unbearable. With climate extremes and high population growth projected to increase in future, food insecurity situation in the GHA region is also expected to worsen. To achieve greater food security, therefore, in addition to boosting GHA's agricultural output, UN Office for the Coordination of Humanitarian Affairs opines that inhabitants must create more diverse and stable means of livelihood to insulate themselves and their households from external shocks, a task that they acknowledge will not be easy as the path ahead is strewn with obstacles—two of the most important being natural hazards and armed conflict [37]. Understanding its hydroclimate regime, therefore, is a starting point towards tackling the natural hazard component. *This book provides the first comprehensive analysis of Greater Horn of Africa (GHA)'s hydroclimate (temperature, precipitation, drought extremes as well as water storage changes) using the state-of-the-art Gravity Recovery and Climate Experiment (GRACE) and GRACE-Follow on (GRACE-FO) water storage* [88], *centennial precipitation, hydrological models' as well as reanalysis' products. Specifically, it looks at the GHA's drought (meteorological, hydrological and agricultural) and the potential for irrigated agriculture through groundwater exploitation.*

1.6 Concluding Remarks

GHA's livelihood is closely intertwined with climate variability through the dependence of large portion of the population on subsistence rain-fed agriculture, understanding of its hydroclimate, e.g., through drought monitoring and the provision of its timely information is essential for drought risk reduction [72] on the one hand. On the other hand, the enhancement of drought monitoring systems through accurate and consistent drought indicators together with forecast information as will be demonstrated in this book will support droughts to be predicted earlier, their impact areas to be delineated more accurately, and impacts on crops diagnosed before harvest [22, 79]. Indeed, [22] has demonstrated that making drought monitoring tools (i.e., drought indices) a part of comprehensive early warning systems (EWS) can provide decision makers with improved and timely information that allows decision makers to assess food security indicators to detect major changes in food availability and advice on the likely occurrence of food crises due to drought in advance of a severe event [77].

Alternatively, adopting a risk management approach to drought management could drastically reduce its impacts. The risk management approach stresses the development of monitoring and EWS, the assessment of drought risks and implementation of drought mitigation and preparedness activities before the occurrence of the drought [77]. All these measures eventually contribute to the enhancement of food security through drought monitoring. These impacts of drought on food security over the region are exacerbated by the effects of high population growth rate [22]. For detailed discussion of the GHA's food security, the reader is referred to FAO [37].

References

1. Abshir S (2020) Climate Change and Security in the Horn of Africa: can Europe help to reduce the risks? https://www.eip.org/wp-content/uploads/2020/10/csen_policy_paper_climate_change_and_security_in_the_horn_of_africa.pdf. Accessed 10 Sep. 2021
2. Agola NO, Awange JL (2014) Globalized poverty and environment. 21st century challenges and innovative solutions. Springer, Berlin. ISBN 978-3-642-39733-2, https://doi.org/10.1007/978-3-642-39733-2
3. Adhikari U, Nejadhashemi AP, Woznicki SA (2015) Climate change and Eastern Africa: a review of impact on major crops. Food Energy Secur 4(2):110–132. https://doi.org/10.1002/fes3.61
4. Awange JL, Khandu Forootan E, Schumacher M, Heck B (2016) Exploring hydro-meteorological drought patterns over the Greater Horn of Africa (1979–2014) using remote sensing and reanalysis products. Adv Water Res 94:45–59. https://doi.org/10.1016/j.advwatres.2016.04.005
5. Awange JL (2021) Lake Victoria monitored from space. Springer Nature International
6. Awange JL (2021) Nile Waters. Weighed from space. Springer Nature International
7. Awange JL, Ong'ang'a O (2006) Lake Victoria-Ecology. Resource of the Lake Basin and Environment. Springer, Berlin
8. Awange JL, Saleem A, Sukhadiya RO, Ouma YO, Hu K (2019) Physical dynamics of Lake Victoria over the past 34 years (1984–2018): is the lake dying? Sci Total Environ 658:199–218. https://doi.org/10.1016/j.scitotenv.2018.12.051
9. CRS Report (2012) Congressional research service report on Horn of Africa: the humanitarian crisis and international response. http://www.fas.org/sgp/crs/row/R42046.pdf
10. IPCC (2007) Climate change 2007: the physical science basis. In: Solomon S, Qin D, Manning M, Chen Z, Marquis M, Averyt KB, Tignor M, Miller HL (eds) Contribution of working Group I to the fourth assessment report of the Intergovernmental Panel on Climate Change. Cambridge University Press, Cambridge, NY
11. Lyon B (2014) Seasonal drought in the Greater Horn of Africa and its recent increase during the March-May long rains. J Clim 27(21):7953–7975. https://doi.org/10.1175/JCLI-D-13-00459.1
12. Marthews TR, Otto FEL, Mitchell D, Dadson SJ, Jones RG (2015) The 2014 drought in the horn of Africa: attribution of meteorological drivers. Bull Am Meteor Soc 96:83–88. https://doi.org/10.1175/BAMS-EEE_2014_ch17.1
13. Nicholson SE (1997) An analysis of the enso signal in the tropical atlantic and western INDIAN oceans. Int J Clim 17(4):345–375. https://doi.org/10.1002/(SICI) 1097-0088(19970330)17:4 (345::AID-JOC127) 3.0.CO;2-3
14. Omondi P, Awange JL et al (2014) Changes in temperature and precipitation extremes over the Greater Horn of Africa region from 1961 to 2010. Int J Climatol 34:1262–1277. https://doi.org/10.1002/joc.3763
15. Tierney JE, Smerdon JE, Anchukaitis KJ, Seager R (2013) Multidecadal variability in East African hydroclimate controlled by the Indian Ocean. Nature 493:389–392. https://doi.org/10.1038/nature11785

16. UNDP (2001) Disaster profiles of the least developed countries. Third United Nations Conference on Least Developed Countries Brussels, 14–20 May 2001. https://reliefweb.int/sites/reliefweb.int/files/resources/2B808FD25701476285256A69006CE8D0-undp_ldc_12jun.pdf. [Accessed 5/01/2022]
17. World Bank Group (2020) From isolation to integration. The borderlands of the Horn of Africa. International Bank for Reconstruction and Development/The World Bank. https://reliefweb.int/sites/reliefweb.int/files/resources/The-Borderlands-of-the-Horn-of-Africa.pdf. Accessed 10 Nov 2020
18. AghaKouchak A (2015) A multivariate approach for persistence-based drought prediction: application to the 2010–2011 East African Drought. J Hydrol 526:127–135. https://doi.org/10.1016/j.jhydrol.2014.09.063
19. AghaKouchak A, Nasrollahi N, Habib E (2009) Accounting for uncertainties of the TRMM satellite estimates. Remote Sens 1:606–619. https://doi.org/10.3390/rs1030606
20. AghaKouchak A, Farahmand A, Melton FS, Teixeira J, Anderson MC, Wardlow BD, Hain CR (2015) Remote sensing of drought: progress, challenges and opportunities. Rev Geophys 53(2):452–480. https://doi.org/10.1002/2014RG000456, 2014RG000456
21. Agutu N, Awange J, Zerihun A, Ndehedehe C, Kuhn M, Fukuda Y (2017) Assessing multi-satellite remote sensing, reanalysis, and land surface models' products in characterizing agricultural drought in East Africa. Remote Sens Environ 194:287–302. https://doi.org/10.1016/j.rse.2017.03.041
22. Agutu NO (2017) A remote sensing based approach to enhance food security in the greater Horn of Africa. Doctor of Philosophy of Curtin University
23. Alley WM (2007) Sustainable management of groundwater in Mexico: proceedings of a workshop (Series: Strengthening Science-Based Decision Making in Developing Countries), Chap. The importance of monitoring to groundwater management. The National Academies Press, Washington, DC, pp 76–85. https://doi.org/10.17226/11875
24. Anderson WB, Zaitchik BF, Hain CR, Anderson MC, Yilmaz MT, Mecikalski J, Schultz L (2012) Towards an integrated soil moisture drought monitor for East Africa. Hydrol Earth Syst Sci 16:2893–2913. https://doi.org/10.5194/hess-16-2893-2012
25. Awange J, Ferreira V, Forootan E, Khandu S, Andam-Akorful N. Agutu, He X (2016) Uncertainties in remotely sensed precipitation data over Africa. Int J Climatol 36(1):303–323. https://doi.org/10.1002/joc.4346
26. Becker M, Lovel WL, Cazenave A, Güntner A, Crétaux J (2010) Recent hydrological behaviour of the East African great lakes region inferred from GRACE, satellite altimetry and rainfall observations. C R Geosci 342:223–233. https://doi.org/10.1016/j.crte.2009.12.010
27. Clark CO, Webster PJ, Cole JE (2003) Interdecadal variability of the relationship between the Indian Ocean Zonal Mode and East African Coastal Rainfall Anomalies. J Clim 16:548–554. https://doi.org/10.1175/1520-0442(2003)016 < 0548:IV OTRB > 2:0.CO;2
28. Comte JC, Cassidy R, Obando J, Robins N, Ibrahim K, Melchioly S, Mjemah I, Shauri H, Bourhane A, Mohamed I, Noe C, Mwega B, Makokha M, Join JL, Banton O, Davies J (2016) Challenges in groundwater resource management in coastal aquifers of East Africa: investigations and lessons learnt in the Comoros Islands, Kenya and Tanzania. J Hydrol: Region Stud 5:179–199. https://doi.org/10.1016/j.ejrh.2015.12.065
29. Damberg L, AghaKouchak A (2014) Global trends and patterns of drought from space. Theore App Climatol 117:441–448. https://doi.org/10.1007/s00704-013-1019-5
30. Demombynes G, J Kiringai (2011) The drought and food crisis in the Horn of Africa: impacts and proposed policy responses for Kenya. Poverty Reduct Econ Manag (PREM) Net (71):1–4. http://siteresources.worldbank.org/INTPREMNET/Resources/EP71.pdf. Accessed 27 Apr 2017
31. Dinku T, Ceccato P, Grover-Kopec E, Lemma M, Connor SJ, Ropelewski CF (2007) Validation of satellite rainfall products over East Africa's complex topography. Int J Remote Sens 28(7):1503–1526. https://doi.org/10.1080/01431160600954688
32. Dinku T, Chidzambwa S, Ceccato P, Connor SJ, Ropelewski CF (2008) Validation of high resolution satellite rainfall products over complex terrain. Int J Remote Sens 29(14):4097–4110. https://doi.org/10.1080/01431160701772526

33. Döll P, Müller Schmied H, Schuh C, Portmann FT, Eicker A (2014) Global-scale assessment of groundwater depletion and related groundwater abstractions: combining hydrological modeling with information from well observations and GRACE satellites. Water Res Res 50(7):5698–5720. https://doi.org/10.1002/2014WR015595
34. Dutra E, Magnusson L, Wetterhall F, Cloke HL, Balsamo G, Boussetta S, Pappenberger F (2013) The 2010–2011 drought in the Horn of Africa in ECMWF reanalysis and seasonal forecast products. Int J Climatol 33:1720–1729. https://doi.org/10.1002/joc.3545
35. Edossa DC, Babel MS, Gupta AS (2010) Drought analysis in the Awash River Basin Ethiopia. Water Res Manag 24:1441–1460. https://doi.org/10.1007/s11269-009-9508-0
36. European Union Council (2011) Conclusions on the Horn of Africa, 3124th foreign Affairs council meeting, Brussels, 14 Nov 2011; Annex: a strategic framework for the Horn of Africa. https://www.consilium.europa.eu/uedocs/cms_data/docs/pressdata/EN/foraff/126052.pdf. Accessed 10 Sep 2021
37. Food and Agriculture Organization of the United Nations (2000) The elimination of food insecurity in the Horn of Africa. A strategy for concerted government and UN agency action FINAL REPORT. http://www.fao.org/3/x8406e/X8406e00.htm. Reproduced with permission. Accessed 3 Jul 2021
38. FAO (2011) Drought related food insecurity: a focuss on the Horn of Africa, Emergency Ministerial Level Meeting, FAO, Rome, pp 1–7. http://www.fao.org/crisis/28402-0f9dad42f33c6ad6ebda108ddc1009adf.pdf. Accessed 27 Apr 2017
39. Foster S, Shah T (2012) Groundwater resources and irrigated agriculture-making a beneficial relation more sustainable. Int Assoc Hydrogeol-Strat Overv Ser (4). http://www.gwp.org/en/ToolBox/PUBLICATIONS/Perspectives-Papers/. Accessed 01 Mar 2017
40. Funk CC, Brown ME (2009) Declining global per capita agricultural production and warming oceans threaten food security. Food Secur 1(3):271–289. https://doi.org/10.1007/s12571-009-0026-y
41. Gebrehiwot T, van der Veen A, Maathuis B (2011) Spatial and temporal assessment of drought in the Northern highlands of Ethiopia. Int J Appl Earth Obser Geoinform 13(3):309–321. https://doi.org/10.1016/j.jag.2010.12.002
42. Gedif B, Hadish L, Addisu S, Suryabhagavan KV (2014) Drought risk assessment using remote sensing and GIS: the case of southern zone, Tigray Region, Ethiopia. J Nat Sci Res 4(23):87–94
43. Healy S (2009) Peacemaking in the midst of war: an assessment of IGAD's contribution to regional security, working paper no. 59, Crisis States Research Centre, London School of Economics and Political Science, London. http://eprints.lse.ac.uk/28482/1/WP59.2.pdf. [Accessed 05/01/2022]
44. Healy S (2011) Hostage to conflict: prospects for building regional economic cooperation in the Horn of Africa, A Chatham House Report. https://www.chathamhouse.org/sites/default/files/public/Research/Russia%20and%20Eurasia/251211summary.pdf. [Accessed 05/01/2022]
45. Ibrahim F (1988) Causes of the famine among the rural population of the Sahelian zone of the Sudan. Geo J 17(1):133–141. https://doi.org/10.1007/BF00209083
46. IFRC (2011) Drought in the Horn of Africa: preventing the next disaster. Int Federat Red Cross Red Crescent Soc Geneva. http://www.ifrc.org. Accessed 15 Mar 2015
47. Konikow LF, Kendy E (2005) Groundwater depletion: a global problem. Hydrogeol J 13(1):317–320. https://doi.org/10.1007/s10040-004-0411-8
48. Kurnik B, Barbosa P, Vogt J (2011) Testing two different precipitation datasets to compute the standardized precipitation index over the Horn of Africa. Int J Remote Sens 32(21):5947–5964. https://doi.org/10.1080/01431161.2010.499380
49. Loewenberg S (2011) Humanitarian response inadequate in Horn of Africa crisis. Lancet 378(9791):555–558. https://doi.org/10.1016/S0140-6736(11)61276-2
50. Lyon B (2004) The strength of El Niño and the spatial extent of tropical drought. Geophys Res Lett 31(21). https://doi.org/10.1029/2004GL020901
51. McCalla AF (1999) Prospects for food security in the 21st Century: 1Developed from a speech given at the World Food Prize Symposium, Des Moines, Iowa, 16 October 1997.1 with special emphasis on Africa. Agric Econ 20(2):95–103. https://doi.org/10.1016/S0169-5150(98)00080-2

52. Moench M, Burke J, Moench Y (2003) Rethinking the approach to groundwater and food security. Rome: food and agriculture organization of the United Nations. ftp://ftp.fao.org/agl/aglw/docs/wr24e.pdf. Accessed 01 Mar 2017
53. Moore P, Williams SDP (2014) Integration of altimetric lake levels and GRACE gravimetry over Africa: inferences for terrestrial water storage change 2003–2011. Water Res Res 50:9696–9720. https://doi.org/10.1002/2014WR015506
54. Morgan B, Awange JL, Saleem A, Hu K (2020) Understanding vegetation variability and their "hotspots" within Lake Victoria Basin (LVB: 2003–2018). Appl Geograp 122. https://doi.org/10.1016/j.apgeog.2020.102238
55. Morton JF (2007) The impact of climate change on smallholder and subsistence agriculture. Proc Nat Acad Sci 104(50):19680–19685. https://doi.org/10.1073/pnas.0701855104
56. Mpelasoka F, Awange JL, Zerihun A (2018) Influence of coupled ocean-atmosphere phenomena on the Greater Horn of Africa droughts and their implications. Sci Total Environ 610–611:691–702. https://doi.org/10.1016/j.scitotenv.2017.08.109
57. Mwangi E, Wetterhall F, Dutra E, Di Giuseppe F, Pappenberger F (2014) Forecasting droughts in East Africa. Hydrol Earth Syst Sci 18(2):611–620. https://doi.org/10.5194/hess-18-611-2014
58. Naumann G, Dutra E, Pappenberger F, Wetterhall F, Vogt JV (2014) Comparison of drought indicators derived from multiple data sets over Africa. Hydrol Earth Syst Sci 18:1625–1640. https://doi.org/10.5194/hess-18-1625-2014.225
59. Niang I, Ruppel OC, Abdrabo MA, Essel A, Lennard C, Padgham J, Urquhart P (2014) Climate change 2014: impacts, adaptation, and vulnerability. Part B: regional aspects. Contribution of working Group II to the fifth assessment report of the Intergovernmental Panel on Climate Change, chap Africa. Cambridge University Press, Cambridge, UK, New York, NY, USA, pp 1199–1265
60. Nicholson SE (2014) A detailed look at the recent drought situation in the Greater Horn of Africa. J Arid Environ 103:71–79. https://doi.org/10.1016/j.jaridenv.2013.12.003
61. OEA (2011a) Eastern Africa Drought humanitarian report, OCHA Eastern Africa, number 3, 01–31 May 2011. http://reliefweb.int/report/burundi/eastern-africa-drought-humanitarian-report-no-3. Accessed 17 Mar 2015
62. OEA (2011b) Eastern Africa Drought humanitarian report, OCHA Eastern Africa, number 4, 01 June 15–July 2011. http://reliefweb.int/report/burundi/eastern-africa-drought-humanitarian-report-no-4. Accessed 17 Mar 2015
63. Olsson L (1993) On the causes of famine: drought, desertification and market failure in the Sudan, Ambio 22(6):395–403. http://www.jstor.org/stable/4314110
64. Othieno H, Awange JL (2016) Energy resources in Africa. Distribution, opportunities and challenges. Springer International Publishing AG
65. Otieno H, Awange JL (2006) Energy resource in East Africa. Springer, Berlin, Heidelberg, New York
66. Peace Bulletin (2003) Conflicts impacting negatively on food security in Greater Horn of Africa. ITDG Practical Action-EA Peace Bulletin. https://practicalaction.org/peace3-food-security
67. Rojas O, Vrieling A, Rembold F (2011) Assessing the drought probability for agricultural areas in Africa with coarse resolution remote sensing imagery. Remote Sens Environ 115:343–352. https://doi.org/10.1016/j.rse.2010.09.006
68. Romilly TG, Gebremichael M (2011) Evaluation of satellite rainfall estimates over Ethiopian river basins. Hydrol Earth Syst Sci 15(5):1505–1514. https://doi.org/10.5194/hess-15-1505-2011
69. Rowhani P, Degomme O, Guha-Sapir D, Lambin EF (2011) Malnutrition and conflict in East Africa: the impacts of resource variability on human security. Clim Change 105(1):207–222. https://doi.org/10.1007/s10584-010-9884-8
70. Samaan MM (2019) The Nile development game. Tug-of-War or Benefits for All? Springer International Publishing
71. Schmidhuber J, Tubiello FN (2007) Global food security under climate change. Proc Nat Acad Sci 104(50):19703–19708. https://doi.org/10.1073/pnas.0701976104

72. Sheffield J, Wood EF, Chaney N, Guan K, Sadri S, Yuan X, Olang L, Amani A, Ali A, Demuth S, Ogallo L (2014) A drought monitoring and forecasting system for Sub-Sahara African water resources and food security. Bull Am Meteorol Soc 95(6):861–882. https://doi.org/10.1175/BAMS-D-12-00124.1. 4385
73. Shiferaw B, Tesfaye K, Kassie M, Abate T, Prasanna B, and Menkir A (2014) Managing vulnerability to drought and enhancing livelihood resilience in Sub-Saharan Africa: technological, institutional and policy options. Weather Clim Extrem 3:67–79. https://doi.org/10.1016/j.wace.2014.04.004. High Level Meeting on National Drought Policy
74. Shiferaw A, Tadesse T, Rowe C, Oglesby R (2018) Precipitation extremes in dynamically downscaled climate scenarios over the Greater Horn of Africa. Atmosphere 9(3):112. https://doi.org/10.3390/atmos9030112
75. Shukla S, McNally A, Husak G, Funk C (2014) A seasonal agricultural drought forecast system for food-insecure regions of East Africa. Hydrol Earth Syst Sci 18(10):3907–3921. https://doi.org/10.5194/hess-18-3907-2014
76. Stampoulis D, Andreadis KM, Granger SL, Fisher JB, Turk FJ, Behrangi A, Ines AV, Das NN (2016) Assessing hydro-ecological vulnerability using microwave radiometric measurements from WindSat. Remote Sens Environ 184:58–72. https://doi.org/10.1016/j.rse.2016.06.007
77. Tadesse T, Haile M, Senay G, Wardlow BD, Knutson CL (2008) The need for integration of drought monitoring tools for proactive food security management in sub-Saharan Africa. Nat Res Forum 32:265–279
78. Technical Cooperation Department (2000) The elimination of food insecurity in the Horn of Africa: a strategy for concerted government and UN agency action—FINAL REPORT, FAO Corporate Document Repository, pp 1–21. http://www.fao.org/docrep/003/x8406e/X8406e05.htm. Accessed 27 Apr 2017
79. Thenkabail P, Gamage M, Institute I (2004) The use of remote sensing data for drought assessment and monitoring in Southwest Asia. IWMI Research Report, International Water Management Institute
80. Villholth KG (2013) Groundwater irrigation for smallholders in Sub-Saharan Africa—a synthesis of current knowledge to guide sustainable outcomes. Water Int 38(4):369–391. https://doi.org/10.1080/02508060.2013.821644
81. Viste E, Korecha D, Sorteberg A (2013) Recent drought and precipitation tendencies in Ethiopia. Theor Appl Climatol 112:535–551. https://doi.org/10.1007/s00704-012-0746-3
82. Voll J, Voll S (2016) The Sudan: unity and diversity in a multicultural state. Sudan, Taylor & Francis, Routledge Library Editions
83. Wehling P (2020) Nile water rights. An international law perspective. Springer, Berlin, Heidelberg
84. Wilhite DA, Pulwarty RS (2005) Drought and water crises: science, technology, and management issues, chap Drought and water crises: lessons learned and the road ahead. CRC Press, Boca Raton,Florida, pp 389–398
85. Williams AP, Funk C (2011) A westward extension of the warm pool leads to a westward extension of the Walker circulation, drying Eastern Africa. Clim Dyn 37(11):2417–2435. https://doi.org/10.1007/s00382-010-0984-y
86. Williams AP, Funk C, Michaelsen J, Rauscher SA, Robertson I, Wils THG, Koprowski M, Eshetu Z, Loader NJ (2012) Recent summer precipitation trends in the Greater Horn of Africa and the emerging role of Indian Ocean Sea surface temperature. J Clim Dyn 39:2307–2328. https://doi.org/10.1007/s00382-011-1222-y
87. Yang W, Seager R, Cane MA (2014) The East African long rains in observations and models. J Clim 27:7185–7202. https://doi.org/10.1175/JCLI-D-13-00447.1
88. Yang Y, Long D, Guan H, Scanlon BR, Simmons CT, Jiang L, Xu X (2014) GRACE satellite observed hydrological controls on interannual and seasonal variability in surface greenness over mainland Australia. J Geophys Res: Biogeosci 119:2245–2260. https://doi.org/10.1002/2014JG002670.239

Chapter 2
Food Security in Blue Nile: Ethiopian GERD

Transboundary cooperation is essential amongst the riparian countries to reduce trade-offs and ensure that the Grand Ethiopian Renaissance Dam (GERD) project reduces hunger, boosting food production with a reduced impact on smallholder farmers' livelihoods, ecosystem, and water access to vulnerable people in the downstream countries.–Olufemi Samson Adesina[1]

2.1 Summary

In recent years, many researchers, government agencies, and world leaders have critiqued the possibility of achieving sustainable food systems in the face of climate change and increasing population without significant trade-offs affecting other industrial activities and potential economic gain [1, 2]. Again, there has been much argument around the definition of the term "sustainability" and how it can be achieved in this era. Countries in the global south are fast becoming industrious and keenly interested in economic growth to enable them to improve their standard of living and compete well with the global north for political power and global relevance. This is perceived as reasons why the controversial Grand Ethiopian Renaissance Dam (GERD) situated along the Nile River has gathered more attention from foreign bodies and international organizations [3]. Besides, the construction of GERD has raised much concern with regards to the possibility of a future war amongst the countries along the river Nile basin. This is due to the predicted impacts of the Dam on available water for agriculture and food production for the downstream countries, particularly Sudan and Egypt [4]. Therefore, achieving a sustainable balance of water, energy, and food security nexus becomes more challenging as the world

[1] This is an invited chapter from Olufemi Samson Adesina, Sustainability Research Institute, School of Earth and Environment, University of Leeds, UK.

J. Awange, *Food Insecurity & Hydroclimate in Greater Horn of Africa*,
https://doi.org/10.1007/978-3-030-91002-0_2

battles with several shocks such as climate change and extreme weather conditions, COVID19 pandemic, and population explosion. As it stands, assessing the possible future impacts of GERD on food production becomes highly imperative. Exploring the future impacts of the Dam on food production will help give birth to sustainable policy intervention, which can help ensure future food security in the Nile region. Besides, it will help reduce tension and promote peaceful coexistence in the region.

2.2 The Grand Ethiopian Renaissance Dam: Background

The Grand Ethiopian Renaissance Dam (GERD), initially referred to as the Millennium Dam, is the largest African hydroelectric dam project in Ethiopia, East Africa. The nearly completed GERD is constructed on the Blue Nile (Abay) river, which contributes about 60% to the entire Nile flow [5]. Furthermore, the GERD project on the Blue Nile is documented to be around 20km from the Ethiopian-Sudan border (Fig. 2.1). The Dam is a mix of about 1.8-kilometer-high gravity dam, a 5-km rockfill saddle dam, and a 300-m separate spillway between the main and saddle dams (Fig. 2.2) [5]. Once completed, the GERD project will be home to the 8th largest hydroelectric infrastructure in the World. It is projected that the Dam will produce about 16000 GWh/y for an estimated 109 million population. In addition, it intends to produce a reservoir with over 63 million cubic meters, which accounts for an estimated 1.3 times the annual flow of the Blue Nile [6].

After a series of fruitless negotiations among the Nile riparian countries, the Ethiopian government decided to initiate the construction of GERD in 2011. The GERD project is expected to triple current Ethiopia's national power supply and capacity. As a result, it is projected that Ethiopia will become a primary source of hydropower in the region, providing about 4000 MW of power to the riparian countries in the region [1, 4]. Besides, the Dam was designed to symbolize economic success and growing political advantage, intended to promote national security and unity through nationalism. Furthermore, the Dam is pictured as a multiplier effect for Ethiopia's development, providing jobs for the builders and dam operators. Moreover, the distribution of energy to the rural areas is envisaged to make agriculture and farming more productive. According to Rocky Mountain Institute, there exist about 4 billion dollars potential economic opportunity from electrifying small rural farms and making their activities more efficient and productive [5, 6].

However, the construction of GERD has faced severe criticism from political opposition and has faced severe challenges in securing additional funds from international development agencies. This issue is the reason why the people of Ethiopia decided to self-fund the project. Egypt believes the location and size of the GERD project will reduce the amount of water supply from the Nile, which is depended on for agriculture and other uses in the downstream countries [7]. Over time, the control and reshaping of the flow of Nile such as the opening of Aswan low dam and several other dams and the creation of the single largest irrigation system in Sudan will impact the Nile River. This event brought about treaties giving Egypt 66% significant

Fig. 2.1 Map showing the GERD project on the Blue Nile basin of Ethiopia and Sudan [5]

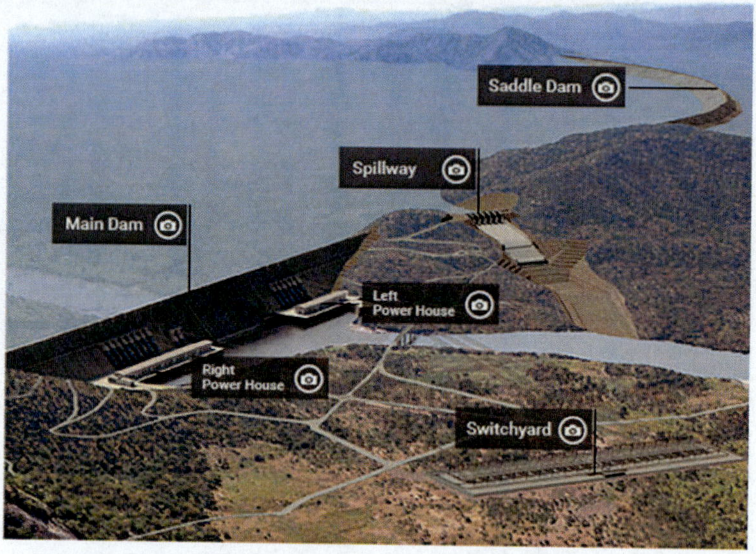

Fig. 2.2 Image showing an overview of the GERD main and saddle dam [7]

power over the water and Sudan 22% over upstream activities in the region without Ethiopia actively participating or ratifying the agreements of 1929 and 1959 [5]. On the other hand, the Ethiopian government has repeatedly assured that the GERD will have minimal impact on downstream flow and will significantly benefit downstream countries concerning flood control and increase in the irrigated area [3]. Nonetheless, Egypt often refers to the British colonial treaties signed between 1882 and 1959 as a binding document that gives Egyptians the 'historical right' and control over the Nile, preventing the upstream countries from initiating infrastructural development on the water resources [6].

With several impact assessments done, the results have been mixed. While some research supports the benefits highlighted by the Ethiopians, others suggest that the impact of the construction and location of the Grand Ethiopian Renaissance Dam (GERD), to a large extent has the potentials to reduce water availability to Lake Nasser, thus negatively affecting energy generation in Egypt [6, 7]. Moreover, the project host Ethiopia remains the most populated amongst the riparian states, with a growth rate of about 3.18% per annum (see Fig. 2.3) [3, 8]. Hence, population growth and uncertainty in climate and weather conditions pose severe threats to the Nile basin water resources, which the GERD project could exacerbate if global acceptable and sustainable standards are compromised, thus affecting the source of water that contributes significantly (about 40–60%) to the gross domestic product (GDP) and food security of the countries along the Nile basin [3, 9].

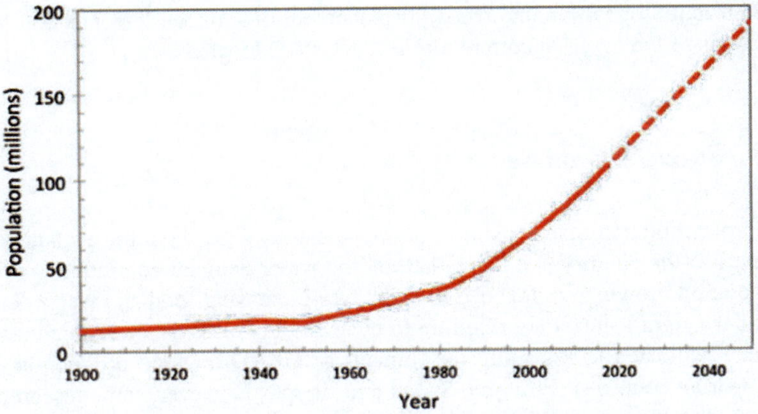

Fig. 2.3 Graph showing Ethiopia's population growth and projected trend [12]

2.3 Impacts on Food Security

2.3.1 Ethiopia's Food Security

Over the past decades, Ethiopia has been observed to have depended less on the Nile river despite housing the highest volumetric share contribution to the Nile flow compared to Sudan and Egypt [6]. Nonetheless, Ethiopia still faces the challenge of poverty and food insecurity, just like many other African countries. Ethiopia's deteriorating state of food security can be attributed to a growing population, which is predicted to increase (Fig. 2.3), recurrent drought, and semi-regular famines, including crop failure, plant and disease outbreak, and poor infrastructural facility coupled with post harvest loss [10]. Moreover, the Ethiopian highlands around the GERD watershed are remarkable for watershed degradation. Therefore, scholars have raised concerns about keeping sediments out of the reservoir to ensure the sustainability of the GERD and maintain soil health for food production in the area with an accelerated rate of erosion in Ethiopia [11, 12].

However, Ethiopia believes that overturning Egypt's hegemony in the Nile basin region with the birth of the GERD project will improve its hydropower generation and as such, have managed to make Sudan see reasons and accept the benefits of the project. The attempt to restructure Ethiopia's productivity and rural development strategies can be noticed in the term "renaissance", which is synonymous with renewing interest and independence. The Dam is expected to increase Ethiopia's available electric power supply by about 200% [6]. Furthermore, this dam project will benefit agriculture by powering agricultural innovations, storage facilities and cold rooms, irrigation pumps, agro-processing industries, and manufacturing industries, thus improving food availability and accessibility. Besides, it is envisioned that the GERD could help control and manage extreme floods by reducing erosion and destruction of farmlands, crops, displacement of farm settlements, and loss of livelihoods of the agrarian community in the GERD located area [1, 4].

2.3.2 Sudan's Food Security

Food production in Sudan consumes the verse majority (97%) of the available water resources in the country [1]. Nevertheless, the water demand continues to rise due to population growth and the need to meet ever-increasing food and water demand. Despite the significance of agriculture to the economy of Sudan, not much research has been dedicated to assessing the impacts of GERD on food production in the three riparian countries (Ethiopia, Sudan and Egypt). However, some research suggests that the construction of GERD will promote the regulation of water supply all year round, which could aid the expansion of agricultural activities in Sudan [13]. However, water losses to evaporation and uncertainties due to climate changes cannot be overlooked as they affect dam projects, influence decision-making, and seriously cause significant damage to food production. The GERD operation is expected to cover the water supply shortage, which occurs once the annual filling of Roseires and Sennar reservoirs occur (Fig. 2.1). Besides, it is also foreseen to reduce about 86% of the sediment and silt transportation along the Nile basin leading to downstream countries such as Sudan suffering from reduced sediments often relied upon to replenish and maintain the fertility of farmlands needed to enhance food production [14, 15]. Furthermore, some researchers have predicted a high possibility that the GERD could fail at some point in the future due to the meteorological nature and geographical location of the Dam [1, 13, 16]. Its failure could be a partial or total collapse of the structure that holds the water, significantly affecting food production in Sudan. In addition, extreme weather conditions such as long periods of rainfall and flooding could damage dam structures. Thus, with climate change, there are many uncertainties as heavy downpours in Ethiopia could flood farmlands, particularly in the eastern parts of Sudan.

Moreover, the GERD is reportedly located around a plate tectonic, responsible for about 15,000 earthquakes in Ethiopia [16]. Mohamed and Elmahdy [17] reckons that the possibility of earthquakes in the area could break the GERD with a devastating effect on major dams (e.g., Sennar dam; Fig. 2.1) in Sudan. As such, [16] reports that the failure of GERD could submerge an estimated 25,400 km^2 of agricultural lands and roads (particularly in Khartoum city), which are essential to achieving food security in Sudan. Thus, the GERD is expected to have significant environmental impacts if not well managed. Furthermore, several models regarding the filling process of the GERD have indicated a possible reduction in the velocity of the Blue Nile, with some models also indicating changes in water temperatures that could become colder (0.5–1.5 °C) in downstream countries such as Sudan (including changes in water quality and chemical composition and the amount of dissolved oxygen available in downstream waters), hence affecting aquatic ecosystems, which is essential for the survival of aquatic foods [15].

GERD is envisaged to positively foster the expansion of food production and agricultural activities with the possibilities of power supply from Ethiopia, which will power water pumps for irrigation. Nonetheless, the GERD is foreseen to reduce the hydropower generation capacity of Sudan, reducing electricity generation by 28% during the impounding phase of the initial draft of the GERD operation [18]. This significant trade-off could impact local power generation for crop irrigation and storage facilities [19]. Besides, farming activities that depend on the Nile River for irrigation could face reduced water levels whenever water is impounded in Ethiopia, which will increase the cost of growing crops in Sudan. Subsequently, fish farming sector in Sudan might suffer more from a reduction in water level and increased pollutants in the water. In addition, farming communities along the river might need to resettle, thus affecting the socio-cultural aspects of the Sudanese farmers in the river basin [20].

2.3.3 Egypt's Food Security

Egypt is one of the countries located in one of the driest and hottest areas of the Sahara Desert in Northern Africa. Food production in Egypt depends mainly on water resources from the Nile river, which is used to irrigate crops. This event explains why the country took a dominant role in controlling and managing the River Nile using ancient treaties [21]. Egypt is one of the downstream countries that would be significantly impacted by the construction of the GRED. The GERD is foreseen to bring about severe changes to the Nile river, which has remained the primary source of water for growing crops, domestic and industrial purposes in Egypt since time immemorial. Currently, Egypt is experiencing shrinking access to strategic resources, arable land, and population growth. These challenges are further complex as countries' industrialization, and urbanization in the river Nile basin have affected the annual share of water resources previously available to the Egyptians. In addition, Egypt has recently experienced a surge in food importation as wheat tops the chart of imported foodstuff. Besides, the impact of climate change and extreme weather conditions on the Nile river cannot be overemphasized. The study conducted by Farrag [22] resolved that the construction of GERD will reduce surface water and impact groundwater along the Nile basin. Recently, Sherien Abdel Aziz et al. [23] reported that new policies in Egypt influence crop patterns by reducing the planting of crops that require a large amount of water and reduction in water use for cultivation during the irrigation process. Furthermore, research findings agree that the GERD project could reduce Ground Water Level (GWL) significantly, rated as Egypt's secondary water source as there is a significant relationship between the GWL and Surface Water Level (SWL), see e.g., [23]. This relationship means that a reduction in GWL by 3 m also reduces SWL by almost 50 in some vulnerable areas in Egypt, reducing soil quality due to salt accumulation. Thus, the operation of GERD could potentially reduce the planting of economic crops, which rely on a large amount of water (e.g., rice) and could later affect the economy and increase food insecurity across Egypt.

2.4 Recommendations on Sustainable Utilization

The GERD has awakened a landmark dimension to the Nile Basin's social, political, economic, and environmental controversies and discourse. It is expected to resuscitate Ethiopia's lost glory and strengthen its economy and political stand in the Nile region. However, the construction and operation of GERD could impact on food production and security in the downstream countries (Sudan and Egypt). The filling process and practices adopted by Ethiopia will significantly determine the Dam's impact on food security. This event is foreseen to be exacerbated by increasing population, climate variability/change, soil degradation, and evaporative losses. The World is currently faced with many uncertainties and shocks, and prolonged hostility in the Nile basin will only worsen the situation. Therefore, the downstream countries may have to accept that the GERD has come to stay, but then the following recommendations should be made to achieve sustainable development and ensure that poverty and hunger are reduced to the bare minimum in the Nile region.

- Transboundary cooperation is essential amongst the riparian countries to reduce trade-offs and ensure that the GERD project contributes to ending hunger, boosting food production with a reduced impact on farmers' livelihoods, ecosystem, and water access to vulnerable people in the downstream countries.
- A well-structured water resource and quality management tailored to benefit the riparian countries regarding water safety, equitable access to water for agriculture, and improved knowledge of water use efficiency to boost food production.
- It is essential that the jointly agreed filling process and operation scenarios be upheld and followed without any compromise. For instance, GERD should be filled when dams and reservoirs in downstream countries are at total capacity.
- Out of the three possible scenarios (unilateral, coordination, and collaboration) for GERD operation, the collaboration scenario will best support food production across the countries if adequately adopted and implemented. This means an adequate and timely share of information about the Blue Nile and river flow state. Besides, downstream water demand would be a top factor for releasing water from the GERD.

References

1. Elnour M (2019) The impact of the Grand Ethiopian Renaissances Dam on the Water-Energy-Food security nexus in Sudan
2. Kansara P, Li W, El-Askary H, Lakshmi V, Piechota T, Struppa D, Sayed MA (2021) An assessment of the filling process of the grand Ethiopian Renaissance Dam and its impact on the downstream countries. Remote Sens 13(4):711
3. Pemunta NV, Ngo NV, Fani Djomo CR, Mutola S, Seember JA, Mbong GA, Forkim EA (2021) The Grand Ethiopian Renaissance Dam, Egyptian National Security, and human and food security in the Nile River Basin. Cogent Soc Sci 7(1):1875598

4. Elsayed H, Djordjević S, Savić DA, Tsoukalas I, Makropoulos C (2020) The nile water-food-energy nexus under uncertainty: impacts of the grand Ethiopian Renaissance Dam. J Water Res Plan Manag 146(11):04020085
5. Abtew W, Dessu SB (2019) The Grand Ethiopian Renaissance Dam on the Blue Nile. Springer Geograp. https://doi.org/10.1007/978-3-319-97094-3_1
6. Nasr H, Neef A (2016) Ethiopia's challenge to Egyptian hegemony in the Nile River basin: the case of the Grand Ethiopian Renaissance Dam. Geopolitics 21(4):969–989
7. Handiso BW (2018) The challenges and Opportunities of the Grand Renaissance Dam for sustainable Energy - Water - Food - Ecosystem services Nexus in Ethiopia. (Dissertation). Retrieved from https://urn.kb.se/resolve?urn=urn:nbn:se:uu:diva-360827. Accessed 27 May 2021
8. World Bank (2017) World development report 2017: Governance and the law. The World Bank
9. Nile Basin Initiative (2012) State of the Nile Basin 2012 report
10. Birara E, Mequanent M, Samuel T (2015) Assessment of food security situation in Ethiopia. World J Dairy Food Sci 10(1):37–43
11. Zeleke G (2015) Exit strategy and performance assessment for watershed management: a guideline for sustainability. Water and Land Resource Center of Addis Ababa University, Addis Ababa, Ethiopia
12. International Hydropower Association (2021) Sediment management. https://www.hydropower.org/sediment-management-case-studies/ethiopia-grand-ethiopian-renaissance-dam-gerd. Accessed 29 May 2021
13. Basheer M, Wheeler KG, Ribbe L, Majdalawi M, Abdo G, Zagona EA (2018) Quantifying and evaluating the impacts of cooperation in transboundary river basins on the Water-Energy-Food nexus: the Blue Nile Basin. Sci Total Environ 630:1309–1323
14. Tayie MS (2018) The Grand Ethiopian Renaissance Dam and the Ethiopian challenge of hydropolitical hegemony on the Nile Basin. In Grand Ethiopian Renaissance Dam Versus Aswan High Dam. Springer, Cham, pp 485–517
15. Elsanabary MH, Ahmed AT (2018) Impacts of constructing the grand Ethiopian Renaissance Dam on the Nile River. In: Grand Ethiopian Renaissance Dam Versus Aswan High Dam. Springer, Cham, pp 75–93
16. Soliman AH, El Zawahry A, Bekhit H (2017) GERD failure analysis and the impacts on downstream countries. In: Grand Ethiopian Renaissance Dam Versus Aswan High Dam. Springer, Cham, pp 149–171
17. Mohamed MM, Elmahdy SI (2017) Remote sensing of the Grand Ethiopian Renaissance Dam: a hazard and environmental impacts assessment. Geomat Nat Hazards Risk 8(2):1225–1240
18. Wheeler KG, Basheer M, Mekonnen ZT, Eltoum SO, Mersha A, Abdo GM, Zagona EA, Hall JW, Dadson SJ (2016) Cooperative filling approaches for the grand Ethiopian renaissance dam. Water Int 41(4):611–634
19. Thengius S, Preston O (2018) Identifying Synergies and Trade-offs between the Grand Ethiopian Renaissance Dam and the Sustainable Development Goals (Dissertation). Retrieved from https://urn.kb.se/resolve?urn=urn:nbn:se:kth:diva-229702. Accessed 28 May 2021
20. Batisha AF (2015) Sustainability assessment in transboundary context: Grand Ethiopian Renaissance Dam. Model Earth Syst Environ 1(4):1–16
21. Abdalla IH (1971) The 1959 Nile Waters Agreement in Sudanese-Egyptian relations. Middle Eastern Stud 7(3):329–341
22. Farrag AA (2005) The hydraulic and hydrochemical impacts of the Nile system on the groundwater in upper Egypt. Assute Univ Bull Environ Res 8(1):87–102
23. Aziz SA, Zeleňáková M, Mésároš P, Purcz P, Abd-Elhamid H (2019) Assessing the potential impacts of the Grand Ethiopian Renaissance dam on water resources and soil salinity in the Nile Delta Egypt. Sustainability 11(24):7050

Chapter 3
Earth Observation Remote Sensing

More than 150 Earth-observation satellites are currently in orbit, carrying sensors that measure different sections of the visible, infrared and microwave regions of the electromagnetic spectrum. The majority of Earth-observation satellites carry "passive" sensors, measuring either reflected solar radiation or emitted thermal energy from the Earth's surface or atmosphere. Newer satellites also employ "active" sensors that emit energy and record the reflected or backscattered response, from which information about the Earth can be inferred'—American Scientist.[1] In GHA, satellites and new technology have proven to be game changers in FAO's emergency efforts over the past years. Remote-sensing tools continue to help FAO to lead the way and break new ground in terms of reaching remote rural or previously inaccessible areas, especially in times of COVID-19. These tools not only inform rescue and response operations, they promote flood preparedness and contingency planning.[2] In addition, information systems have been geared almost exclusively to the collection of performance data that are relevant to crop production areas, using a combination of remote sensing and field data-gathering networks to provide early warning of emerging food insecurity situations [13].

3.1 GHA's Hydroclimate: Monitoring Products

Greater Horn of Africa (GHA) is a large area comprising of 11 countries (Fig. 1.1a). It's overall water movement expressed through the hydrological cycle is characterised by a simple water balance equation: $\Delta S = P - E - Q$, where ΔS is the total water storage (TWS representing the sum of groundwater, soil moisture, vegetation, and surface water), P the precipitation, E its evapotranspiration, and Q its runoff. Due to its sheer size of area of 6,000,000 km^2 [10], monitoring changes in ΔS, P, E, Q

[1] https://www.americanscientist.org/article/fifty-years-of-earth-observation-satellites

[2] http://www.fao.org/emergencies/fao-in-action/stories/stories-detail/en/c/1402112/

J. Awange, *Food Insecurity & Hydroclimate in Greater Horn of Africa*,
https://doi.org/10.1007/978-3-030-91002-0_3

through "boots on the ground" in-situ observations is practically impossible and a daunting task indeed. On the one hand, the in-situ (ground-based, see e.g., [101]) data may be inaccessible while on the other hand, the meagre accessible data where available, might be inconsistent or suffer from missing data.

To supplement in-situ observations and water resource management over large and poorly gauged GHA, ***satellite remote sensing (SRS)*** offers the only possibility of rigorously monitoring changes in these components (ΔS, P, E, Q) of the hydrological cycle as will be shown in the Chapters ahead. To this end, GHA's total water storage changes $\Delta S/dt$, where dt is the temporal differences, e.g., changes between consecutive months can now consistently be measured using the Gravity Recovery And Climate Experiment (GRACE, discussed in Sect. 3.3.3) mission launched in 2002 and its follow on GRACE-FO launched in 2017, see e.g., [2]. It enables the closure of GHA's terrestrial water budget by providing large-scale quantitative estimates of changes in TWS fields ΔS on an approximately monthly basis.

GHA's precipitation P can be remotely sensed using both infra-red (IR) and microwave (MW) sensors on board geostationary satellites and low-Earth orbiting satellites, e.g., Tropical Rainfall Measuring Mission, TRMM, Kummerow et al., [27]. TRMM satellite, jointly managed by National Aeronautics and Space Administration (NASA) of the United States and Japan Aerospace Exploration Agency (JAXA) was designed to monitor and study tropical rainfall. It has the advantage of providing four dimensional (X, Y, Z, t) rainfall and latent heat data over inaccessible areas such as the ocean, un-sampled terrains, etc. The primary rainfall parameters on TRMM are the TRMM Microwave Imager (TMI), the Precipitation Radar (PR) and the Visible and Infrared Radiometer System (VIRS) [27]. Infrared-based rainfall estimates have a longer historical record and have higher spatio-temporal resolution than the more accurate (and shorter temporal record) Microwave-based estimates but suffer from higher uncertainty due to various problems [115, 116]. Awange et al. [4] discusses uncertainties in satellite based derived precipitation over Africa while Awange et al. [5] compares the performance of merged products in Africa and Australia. In this book, various satellite based precipitation will be employed and as such, they are discussed at their point of use. Evapotranspiration E on its part can be monitored using MODerate Resolution Imaging Spectroradiometer (MODIS) onboard NASA's Terra and Aqua satellites [34], which is also useful for monitoring vegetation [6, 7, 78]. Details of MODIS will be covered in those chapters where they are used. For surfaces runoff Q, useful for measuring changes in levels of the lake and rivers, satellite altimetry missions [12, 16] such as Topex/Poseidon, Jason I and II, and the European Space Agency (ESA) discussed in Sect. 3.5.1 are employed.

Besides satellite data, ***reanalyses*** have made significant contributions to many global/regional hydrological and climatic studies. With the release of many new high-resolution reanalyses in the past decade, see e.g., [47, 48, 50, 51], their application into regional- and basin-scale studies have become increasingly valuable [115]. Yet certain elements of the analyzed fields (e.g., precipitation) remain highly uncertain at global and regional scale both in terms of trends and interannual variabilities. The reliability of reanalysis fields can considerably vary in space and time due to lack of adequate observational data, instrumental changes, changing mix of observations, biases in observations, etc., which can introduce spurious variability and trends into reanalysis fields [115]. Khandu [115] opines that since reanalysis products are increasingly used as regional climate forcing data and hydrological model inputs, it is vital to estimate their skills in a region. A reanalysis system consists of (i) a "data assimilation system" that combines available observations from various data sources (e.g., SRS-based observations), and (ii), a "forecast model" consisting of an atmospheric model at its core, which is often coupled to a land surface model and/or ocean model [47, 48, 51]. The performance of reanalyses have significantly improved with the assimilation of additional data from SRS-based observations, e.g., [50, 51, 53–55]. Reanalysis are employed in various chapters of this book to complement satellite-based remotely sensed data. In this regard, the individual reanalysis product used will be discussed in the respective chapters. Although this book employed ERA-Interim reanalysis, it should be pointed out that its newest version (ERA-5, see e.g., [26, 31]) has just been released and will be considered in future works.

Climate models and hydrological models add knowledge to the key climate and hydrologic process and has been rigorously applied to understand the contributions of climate change (both natural and man-made) and other drivers (e.g., land use change, water use) on water resources [55, 115]. SRS-based observations are also assimilated into hydrological models that are widely used to quantify climate change impacts on various hydrological components as will be demonstrated for GHA. Vishwakarma et al. [117] categorise hydrological models into *land surface models* (LSMs), e.g., Global Land Data Assimilation System (GLDAS) and *global hydrological and water resource models* (WGHMs) such as water gap hydrological model (WGHM). The difference between LSMs and WGHMs is that the former does not incorporate groundwater and are poorly constrained in data deficient regions resulting in large uncertainties that vary in space and time [118] while the latter incorporates both groundwater and human abstraction estimates [25]. Both climate models and hydrological models are employed in subsequent chapters to complement satellite remote sensing and help in understanding of the GHA hydrological process. Since various chapters employ different products of different spatio-temporal resolutions and different time spans, they are discussed in those chapters, what somewhat is repetitive but nonetheless essential in understanding the contents of the respective chapters.

3.2 Optical and Microwave Remote Sensing

The multi-spectral Landsat and Sentinel-2 data have a wide range of spectrum bands that includes, visible to Near-Infrared (VNIR) and shortwave Infrared (SWIR) wavelengths. Landsat satellite system collect multi-spectral images that are useful for monitoring and understanding human activities, e.g., detecting, measuring and highlighting changes in landscape patterns over time including those of coastal studies [11, 14]. Since changes in the landscape are dynamic, these imagery can be used to study and capture extent of changes at large scale within short time frames and at no cost since the images are free. Without remotely sensed data, the task of capturing and mapping changes in landscape within GHA would be much more difficult and time consuming. The imagery from Landsat platform, which can be accessed freely from USGS website (https://earthexplorer.usgs.gov/) are powerful data source providing accurate and timely spatial information on changes in the Earth's landscape. Because of its long-term digital archive since 1972, it has been used in many land use and land cover change studies over different parts of the world, see e.g., [15].

Over the past 46 years, there have been numerous studies that have used Landsat imagery for agriculture forecasting and management (production and conservation), see e.g., [17–19]. Its other applications are reported, e.g., in [20–24, 29, 30, 32, 33]. Landsat data, therefore, are important when there are no other accurate way of mapping and detecting changes due to cost, time, and access constraints. Landsat datasets obtained from the USGS are Level 1 products, which have been corrected for terrain effects, calibrated radiometrically, and geographically georeferenced [11].

To validate the used Landsat images whose spatial resolution is 30 m, the Sentinel-2 (MSI) with 10 m spatial resolution, officially launched by ESA on June 23, 2015, with a 5-day temporal resolution or repeat cycle [35] are employed. The corporation between ESA and the USGS provides Sentinel-2 (MSI) data in Level-1C product to end users and can be accessed from the USGS website and the official website of the Copernicus Open Access Hub for the European Space Agency (ESA) (https://scihub.copernicus.eu/). Sentinel-2 data has been processed by ESA and USGS to include radiometric and geometric corrections along with ortho-rectification to generate highly accurate geo-located products for researchers [36]. Landsat and Sentinel-2 have been validated for GHA in [14] and are employed in the chapters ahead are discussed therein.

3.3 Remote Sensing of Gravity Variations

In the subsections that follow, it is explained how satellites (particularly the low earth observing-LEO satellites) are used to monitor variations in gravity field, which are in turn used to remotely sense the changes in stored waters of GHA. The most significant success of a LEO satellite is evidenced in the Gravity Recovery and Climate Experiment (GRACE) satellites discussed in Sect. 3.3.3. A possible use of the global navigation satellite systems (GNSS, see [37]) to measure variations in water mass is illustrated by Tregoning et al. [38] whose predictions derived from GRACE measured fields show a correlation with GNSS measured deformations, suggesting the possible use of such deformations to infer changes in stored water potential on much shorter temporal and spatial scales than GRACE provides (and with low-latency), while averaging over much larger spatial scales than afforded by multipath amplitude measurements [39]. In this book, GRACE satellites are employed to monitor changes in GHA's stored water (see Sect. 3.1). To understand how GRACE satellites work, let us first discuss the relationship between mass variation and gravity next.

3.3.1 Mass Variation and Gravity

Two types of gravity field variation exists. The *first* is the long-term, also known as mean gravity field or static gravity field, which is due to the static part of the gravity field. The variation is constant over a very long time interval. To simplify the understanding of the mean gravity field, consider a case where you frequently go to a given destination and along the way, you pass a given mountain. Every time you pass by, the mountain is stationary at the same place. Come months and years, the mountain, which represents concentration of mass is stationary and does not change its position. In simple terms, one can visualize this mountain as representing the long-term changes in gravity field as its mass concentration does not seem to change. Study of long-term (mean gravity field) is useful in understanding the solid structure of the Earth, ocean circulation, and in achieving a universal height measuring system. In this respect, the GOCE (Gravity field and the steady state-of-the ocean circulation explorer, Fig. 3.1) satellite[3] that completed its mission in mid-October 2011 used the gradiometer with improved accuracy to provide data that are useful in mapping changes in gravity, see e.g., Hirt [40]. GOCE data is expected to benefit other studies such as those concerned with earthquakes, changes in sea level, and volcanoes.[4]

The *second* type of variation of the Earth's gravity field is associated with those processes that occur over shorter time scales, such as atmospheric circulation or the hydrological cycle. In the illustration of the mean gravity field above, we used the mountain as an example. To visualize time variable gravity field, instead of a mountain, consider now a small pond along your frequently used route to a given destination. As you frequently pass this pond, you will notice that its water level changes depending on climatic or human induced impacts. If it rains heavily and over a long period of time, the level rises while if there is a prolonged drought or massive use by humans, the level of the pond falls. This change in pond level over time symbolizes changes in mass with time. This is known as the *time-varying gravity field* and is the component, which enables the monitoring of, for example, changes in water levels of GHA as will be demonstrated in Chaps. 12 and 13.

By removing the effects of the other processes that cause changes in the time variable gravity field, changes in *terrestrial water storage* can thus be inferred from the observed temporal changes in the terrestrial gravity field. By assuming the density of water as 1.00g/cm^3, and following the relation of [84], Ellet et al. [41] present the relationship between changes in stored water and gravity as (see [49] for the derivations)

$$\Delta S = 0.419 \Delta g, \tag{3.1}$$

where water storage change ΔS is given in units of cm of water and gravity change Δg is in units of microGal (10^{-6} cm/s^2). From Eq. (3.1), it is seen therefore, that monitoring variations in the gravity field can enable hydrological changes to be monitored.

[3] http://www.esa.int/Our_Activities/Observing_the_Earth/GOCE.

[4] see, e.g., http://www.esa.int/esaCP/SEMV3FO4KKF_Germany_0.html.

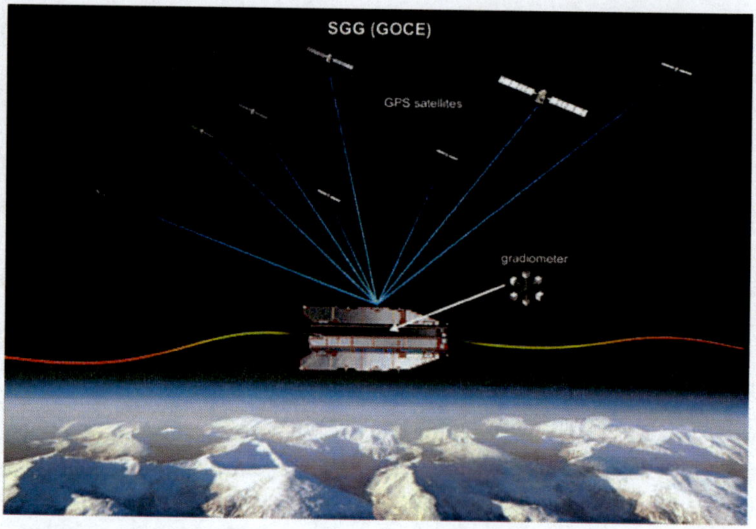

Fig. 3.1 Global Navigation Satellite System (GNSS) satellites track the GOCE satellite in space, thus contributing to the determination of its position (©ESA). The GOCE satellite's accurate determination of the static gravity field is expected to contribute towards studies of changes in sea level, earthquakes, and volcanoes. Figure modified by Rieser [45]

3.3.2 High and Low Earth Orbiting Satellites

At the broadest conceptual level, Low Earth Observing (LEO) satellites' gravity field missions observe (either directly or indirectly) gradients in the Earth's external gravitational field. This is essentially done through differential measurements between two or more points, thus largely eliminating spatially correlated errors. When done from space, two approaches can be used, e.g., [42, 43]:

1. Satellite-to-satellite tracking (SSTr), or
2. A dedicated gravity gradiometer on board a satellite, coupled with SSTr.

The SSTr methods can use either low-low inter-satellite tracking (ll-SSTr, see Fig. 3.2, right), where two LEO satellites track one another and additional observations in terms of high precision ranges and range rates between the two satellites are taken, or high-low inter-satellite tracking (hl-SSTr, see Fig. 3.2, left), where high-Earth orbiting satellites (notably GPS) track a LEO satellite. The low-low mode, compared to the high-low mode, has the advantage of signal amplification leading to a higher resolution of the obtained gravity variations, up to the medium wavelength spectrum of a few hundred km in spatial extent [42]. Taking this further, a combination of ll-SSTr and hl-SSTr is conceptually better still, as demonstrated by the GRACE mission (Fig. 3.2, right) with a baseline length between the two satellites of about 220 km. This is treated in detail in the next section.

GNSS tracking a low satellite, e.g., CHAMP.	GNSS tracking 2-low satellites, which are tracking each other, e.g., GRACE.
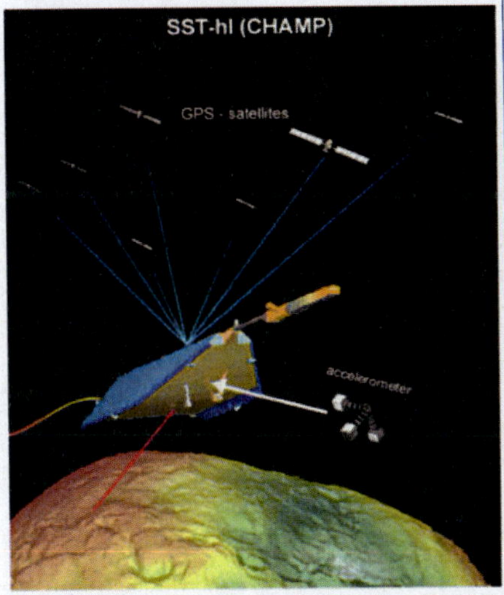	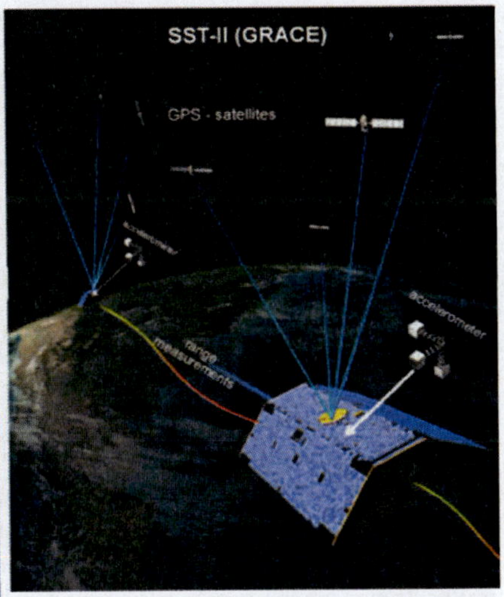

Fig. 3.2 Left: SSTr-hl realized with CHAMP (©GFZ Potsdam ([2.2]). Right: A combination of ll-SSTr and hl-SSTr realized with GRACE and GNSS satellites(©GRACE—CSR Texas ([2.2]). Figures modified by D. Rieser [45]). GNSS satellites are used in determining the positions of these satellites in space. For the GRACE satellites (right) inter-satellite distances can be computed from these positions and compared to the measured K-band distances, thus providing additional independent information

In order to detect temporal gravity field variations at smaller spatial scales, the satellite(s) being tracked must be in as-low-as-possible orbits (close to the mass source), with the satellites being as free as possible from the perturbing effects of atmospheric drag [42]. In addition, so-called de-aliasing models (for correcting short-term - 6 h - variations due to atmosphere and ocean mass variations) have to be used to mitigate the propagation of unwanted signals (e.g., leakage from the oceans) into the derived gravity solutions, e.g., [44].

3.3.3 Gravity Recovery and Climate Experiment

The GRACE satellite mission, launched on 17th of March 2002, whose lifespan ended in mid-October 2017 and was followed by GRACE-Follow-On (GRACE-FO) mission launched on 22nd May 2018, consisted of two near-identical satellites following one another in nearly the same orbital plane (about 400 km altitude) separated by a distance of 220 km; the so-called tandem formation (see Fig. 3.2, right). The ll-SSTr inter-satellite distances was measured using K-band ranging, coupled with hl-SSTr tracking of both satellites by Global Navigation Satellite System (GNSS) (Fig. 3.2, right). GNSS receivers were placed on GRACE satellites to measure occulted signals, see [37], and also to determined the orbital parameters of GRACE satellites required in order to determine gravity changes. On-board accelerometers monitored orbital perturbations of non-gravitational origin, see, e.g., [37].

GRACE mission processes GNSS data to contribute to the recovery of long-wavelength gravity field, remove errors due to long-term onboard oscillator drift, and aligns measurements between the two spacecraft [46, p. 200]. The timing function of GNSS for precision orbit determination, in terms of position and velocity as a function of time, enabled orbits to be determined within an accuracy better than 2 cm in each coordinate [46, p. 200]. The precise locations of the two satellites in orbit allowed for the creation of gravity maps approximately once a month.[5] These gravity maps, when converted to total water storage maps, are useful for monitoring changes in stored water potential, e.g., those of GHA as demonstrated in Chap. 13.

The Earth's gravity field is mapped by making accurate measurements of changes in the distance between the satellites, using GNSS and a microwave ranging system. GRACE-FO employs laser ranging to measure the inter-satellite distances. These changes in the distances between the two satellites occur due to the effect of the gravity (mass concentration) of the Earth. As the lead satellite passes through a region of mass concentration, it is pulled away from the trailing satellite (Fig. 3.2, right). As the trailing satellite passes over the same point, it is pulled towards the lead satellite thus changing the distance between the satellites.

Time-variable gravity field solutions are obtained by the exploitation of GRACE observation data over certain time intervals, i.e., every month [52, 56], or less, e.g., [57, 59]. There are a number of institutions delivering GRACE products, each applying their own processing methodologies and, often, different background models. GRACE mission, which lasted for 15 years and replaced by GRACE-FO provide scientists with an efficient and cost-effective way to monitor time-varying component of the gravity field with unprecedented accuracy and in the process yield crucial information about the distribution and flow of mass within the Earth's system. The process causing gravity variations that are being studied by GRACE include [60];

- changes due to surface and deep currents in the ocean leading to more information about ocean circulation, e.g., [61, 62],
- changes in groundwater storage on land masses such as that of GHA, relevant to water resource managers, e.g., [60, 63–65, 100, 106],
- exchanges between ice sheets or glaciers and the oceans, needed for constraining the mass balance of the global ice regime and sea level change, e.g., [66, 67],
- air and water vapour mass change within the atmosphere, vital for atmospheric studies, e.g., [68, 69], and
- variations of mass distribution within the Earth arising from, e.g., on-going glacial-isostatic adjustments and earthquakes, e.g., [38, 70].

---

[5] http://www.csr.utexas.edu/grace/publications/brochure/page11.html.

Currently, rivers and lakes' basins of the order of 200,000 km^2 and above in area can be successfully studied using the GRACE products [71]. Furthermore, MASCON (Mass Concentration) solution [3] and downscaling techniques [28, 91] offer GRACE solutions with improved spatio-temporal resolution that will benefit hydroclimatic monitoring in GHA. In general, to understand how the GRACE satellites monitor changes in freshwater (all groundwater, soil moisture, snow, ice, and surface waters), first, the larger effect of the mass of the Earth, i.e., the static gravity field discussed in Sect. 3.3.1, which is always a constant G_0 corresponding to nearly 99% of the total field, is computed from a static model (e.g., *GGSM01S* [58]) and removed by subtracting it from the monthly gravity field ($G(t)$) measured by GRACE at a time t [72], i.e.,

$$\Delta G(t) = G(t) - G_0, \tag{3.2}$$

to give the monthly time-variable gravity field $\Delta G(t)$. Alternatively, instead of computing the static gravity field G_0, mean gravity field over a given number of years can also be used. Changes mostly related to the atmosphere and ocean, which occur over time scales shorter than one month, are then removed using models, see e.g., Wahr et al. [73]. Remnant atmospheric and oceanographic effects that last for more than one month can be removed using atmospheric and ocean circulation models before water storage change can be analyzed. The resulting difference in Eq. (3.2), which is called the gravity field anomaly is usually due to changes in stored water. If we consider $\overline{\Delta C}_{lm}(t)$ and $\overline{\Delta S}_{lm}(t)$ to be the normalized Stoke's coefficients expressed in terms of millimeters of geoid height, with l and m being degree and order respectively, and t the time interval of a month. The time-variable geoid in (3.2) is then expanded in-terms of spherical harmonic coefficients (see [74]) as

$$\Delta G(t) = \sum_{n=1}^{N}\sum_{m=0}^{l}(\overline{\Delta C}_{lm}(t)cos(m\lambda) + \overline{\Delta S}_{lm}(t)sin(m\lambda))\overline{P}_{lm}\left(cos(\theta)\right), \tag{3.3}$$

where N is the maximum degree of expansion, θ is the co-latitude, λ the longitude and $\overline{P}_{lm}$ the fully normalized Legendre polynomial [74]. From the gravitational spherical harmonic coefficients (3.3), the equivalent water thickness is computed using the following steps:

1. The gravitational residual coefficients are converted into the surface density coefficient differences by [73]

$$\begin{pmatrix}\Delta\check{C}_{lm}(\mathrm{M}_j)\\ \Delta\check{S}_{lm}(\mathrm{M}_j)\end{pmatrix} = \frac{\rho_{\mathrm{avg}}}{3\rho_{\mathrm{w}}}\frac{2l+1}{1+k_l'}\begin{pmatrix}\overline{\Delta C}_{lm}(\mathrm{M}_j)\\ \overline{\Delta S}_{lm}(\mathrm{M}_j)\end{pmatrix}, \tag{3.4}$$

   where k_l' is the *load Love* number of degree l, $\rho_{\mathrm{avg}} = 5517\,kg/m^3$ the average density of the Earth, M_J the month J, and $\rho_{\mathrm{w}} = 1000\,kg/m^3$ the density of water.
2. The spatial variation of the surface density is then computed through

$$\Delta\sigma(\theta, \lambda, \mathrm{M}_j) = R\rho_{\mathrm{w}} \sum_{l=1}^{l_{\max}} \sum_{m=0}^{l} [\Delta\check{C}_{lm}(\mathrm{M}_j) \cos m\lambda + \Delta\check{S}_{lm}(\mathrm{M}_j) \sin m\lambda] \bar{P}_{lm}(\cos\theta), \quad (3.5)$$

where $R = 6378137\,m$ is the radius of the Earth and $\Delta\sigma$ is in kg/m^2.

3. Finally, the changes in total water storage (TWS) are calculated by

$$\mathrm{TWS}(\phi, \lambda, \mathrm{M}_j) = \frac{\Delta\sigma(\theta, \lambda, \mathrm{M}_j)}{\rho_{\mathrm{w}}} = \frac{\Delta\sigma(\theta, \lambda, \mathrm{M}_j)}{1000} \quad \text{[meters]}. \quad (3.6)$$

The first steps in the analysis of GRACE data would provide an estimate of the changes in total water storage. In the second step, the changes can then be separated into their various components as discussed, e.g., in [60, 72] to obtain changes in the respective components (e.g., groundwater, surface water, soil moisture, and ice).

As stated previously, the GRACE satellites exceeded their planned 5 year life-span and ended their mission in mid-October 2017. However, given the excellent results that have been delivered so far, see e.g., [75] and also Fig. 4.2 in Section 4.3, GRACE follow-on mission (GRACE-FO) was launched on the 22nd of May 2018. Although GRACE-FO satellites, like its predecessor, will use the same kind of microwave ranging system giving a similar level of precision, they will also test an experimental instrument using lasers instead of microwaves, which promises to make the measurement of their separation distance at least 20 times more precise.[6] An example of GRACE's application is presented in Yang [76], where it is used to constrain recent freshwater flux from Greenland where the data show that Arctic freshwater flux started to increase rapidly in the mid-late 1990s, coincident with a decrease in the formation of dense Labrador Sea Water, a key component of the deep southward return flow of the Atlantic Meridional Overturning Circulation (AMOC). Recent freshening of the polar oceans may be reducing formation of Labrador Sea Water and hence may be weakening the AMOC [76].

3.4 Gravity Field and Changes in Stored Water

In the discussion that follows, the concept of *gravity field variations* discussed in Sect. 3.3 is related to hydrological processes. Measurements of the time-varying gravity field by LEO satellites, e.g., GRACE discussed in Sect. 3.3.3 are the key to monitoring of changes in water levels at basin scales. Such techniques now enable the monitoring of groundwater recharge, which is the most important element in groundwater resources management and could also be applicable to monitoring changes in GHA's water levels. For example, in 2009, GRACE satellites showed that north-west of India's aquifers had fallen at a rate of 0.3048 m yr^{-1} (a loss of about 109 km^3 per year) between 2002 and 2008 (see Fig. 4.2 in Sect. 4.3).[7]

[6] http://gracefo.jpl.nasa.gov/mission/.

[7] The Economist, September 12th 2009, pp. 27-29: Briefing India's water crisis.

3.4.1 Gravity Field Changes and the Hydrological Processes

The hydrological cycle (Fig. 3.3) refers to the pathway of water in nature, as it moves in its different phases through the atmosphere, down over and through land, to the ocean and back to the atmosphere [77]. The associated variations in gravity field are therefore caused, e.g., by

- the redistribution of water in the oceans, including e.g., El Niño and Southern Oscillation (ENSO) events,
- movement of water vapour and other components in the atmosphere,
- seasonal rainfall; snow and subsequent drying and melting,
- groundwater extraction, or
- drying and filling of lakes, rivers, and reservoirs.

3.4.2 Monitoring Variation in Stored Water Using Temporal Gravity Field

The potential of using the relationship between temporal gravity changes and hydrology (Fig. 3.3) was first recognized by Montgomery [79] who estimated specific yield through a correlation between gravity and water-level changes [80]. In 1977, Lambert and Beaumont [81] used a gravity meter to correlate groundwater fluctuations and

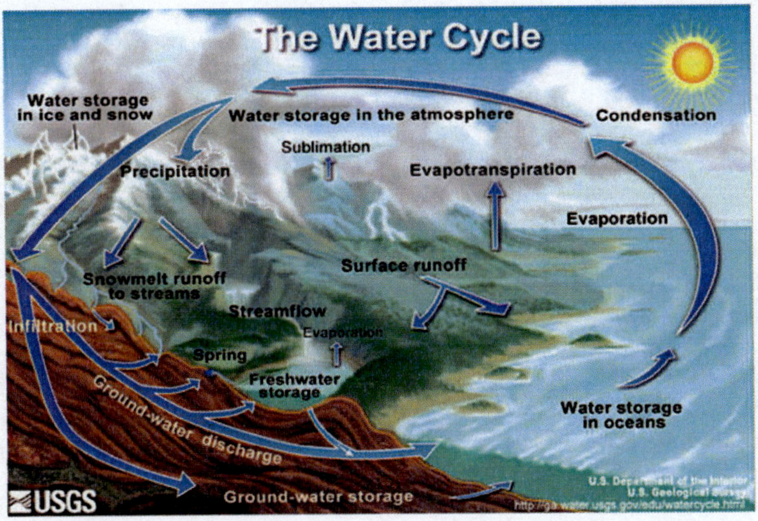

Fig. 3.3 Components of hydrological cycle that lead to temporal variations in the gravity field. *Source* US Geological Survey (USGS)

temporal changes in the Earth's gravity field. Goodkind [82] recorded observations from seven super conducting gravimetric stations to examine non-tidal variations in gravity and noted that at one of the stations (Geysers geothermal station), much of the variation could be correlated with rainfall and seismic activity. Such measurements had not been possible before the advent of super conducting gravimeters, thus providing evidence of the existence of temporal variation in gravity.

In 1995, while estimating the atmospheric effects on gravity observations around Kyoto, Mukai et al. [83] noted that changes in gravity around the station could have been caused by changes in underground water. In the same year, Pool and Eychaner [84] assessed the utility of temporal gravity-field surveys to directly measure *aquifer-storage* changes and reported gravity changes of around 100–134 μGal, equivalent to 2.4–3.2 m of water column, considering infinitely extended sheet approximation. Their results from the analysis of changes in stored water in the aquifer indicated an increase in the gravity field of 158 ± 6 μGal when the water table rose by about 17.7 m, providing further evidence of the possibility of using temporal gravity-field surveys to monitor changes in stored aquifer water. In fact, according to Bower and Courtier [85] who analyzed the effect of precipitation on gravity and well-levels at a Canadian absolute gravity site, 90% of the gravity variation was found to be due to the effects of precipitation, evapotranspiration and snow-melt.

The last decade has also recorded increased use of temporal gravity field studies in monitoring changes in stored water, see e.g., [41]. It saw the beginning of satellite missions dedicated to monitoring temporal variations in the gravity field. Smith et al. [86] investigated the ability of ground-based gravity meters to monitor changes in soil moisture storage.

Moving from local tests to regional, a different application of gravity surveys was investigated by Damiata and Lee [87], who simulated the gravitational response to aquifer hydraulic testing. The synthetic system was composed of an unconfined shallow aquifer and the purpose of the investigation was to assess the potential of the gravity measurements for detecting groundwater extraction. Drawdown due to pumping causes a decrease in mass and consequently in gravity measured at the surface. The results showed that the gravitational response to aquifer testing could be used to monitor the spatial development of the drawdown cone. For the configuration considered in the investigation, the signal was of the order of tens of μGals and could be detected up to several hundred meters away from the pumping well.

Water storage changes, such as changes in soil moisture, snow and ice cover, surface and groundwater, including deep aquifers, can be monitored either by in-situ observations or indirectly through changes in gravity [56]. While in-situ observations provide valuable localized information, they suffer from limited spatial coverage for regional to continent-wide studies [88]. Any change in water storage also manifests itself in a change in the gravity field. This property can be used to infer water-storage changes from time-variable gravity observations as demonstrated by Rodell and Famiglietti [63] for 20 globally distributed drainage basins of sizes varying from 130,000 km^2 to 5,782,000 km^2 to assess the detectability of hydrological signals with respect to temporal and spatial variations. Space borne techniques provides time-variable gravity observations on a regional and global scale, thus allowing for large-scale water storage monitoring and the ability to close the 'gaps' between locally limited in-situ observations [89].

Since the launch of the GRACE satellite mission in 2002 (see Sect. 3.3.3) until the end of its mission in mid-October 2017, it provided a new and powerful tool for studying temporal gravity field changes as evidenced from numerous articles assessing the potential of GRACE recovering hydrological signals, see e.g., Awange et al. [42, and the references therein]. Tapley [56] provided early results of the application of the GRACE products for detecting hydrological signals in the Amazon-Orinoco basin. Following these results, numerous other authors have subsequently applied GRACE to detect hydrological signals in various situations and locations, see references in [42].

For instance, Ramillien et al. [60, 72] and Andersen et al. [92] investigated the potential of inferring inter-annual gravity field changes caused by continental water storage changes from GRACE observations between 2002 and 2003, and compared these changes to the output from four global hydrological models. It was possible to correlate large scale hydrologic events with the estimated change in the gravity field for certain areas of the world to an accuracy of 0.4 μGal, corresponding to 9 mm of water, see also [92–94].

Syed et al. [95] examined total basin discharge for the Amazon-Orinoco and Mississippi river basins from GRACE, while Rodell et al. [90] estimated groundwater storage changes in the Mississippi basin. Crowley et al. [96] estimated hydrological signals in the Congo basin, while Schmidt et al. [97] and Swenson et al. [71, 98] used GRACE to observe changes in continental water storage. Winsemius et al. [99] compared hydrological model outputs for the Zambezi River Basin with estimates derived from GRACE. Monthly storage depths produced by the hydrological model displayed larger amplitudes and were partly out of phase compared to the estimates based on GRACE data. Likely reasons included leakage produced by the spatial filtering used in the GRACE data, and the difficulty to identify the time of satellite overpass as opposed to simply averaging over the whole period. Awange et al. [2, 6, 100] used GRACE to study the fall of Lake Victoria's water level in Africa. This last example will be elaborated upon in more detail in Chap 5. The Nile Basin and water changes in West Africa have been studied, e.g., by [103, 104] while its application to study agricultural droughts and changes in aquifer storage over the Greater Horn of Africa was pioneered in the studies of Agutu et al. [105, 106].

As already discussed in Sect. 3.3.3, GRACE satellites detect changes in the Earth's gravity field by measuring changes in the distance between the two satellites at a 0.1 Hz sampling frequency. The variation in the distance between the two twin satellites caused by gravitational variations above, upon, and within the Earth all have an effect on the satellites. This variation in gravity could be due to *rapid* or *slow* changes caused, for example by the redistribution of water in the oceans, the movement of water vapor and other components in the atmosphere, the tidal effect of the Sun and Moon, and the displacement of the material by earthquakes and glacial isostatic adjustment. The data therefore must be processed to isolate these effects so as to retain only those which correspond to the process of interest, in this case, terrestrial water storage changes, see e.g., [107]. Equation (3.6) is used to compute changes in stored water.

3.5 Satellite Altimetry

3.5.1 Remote Sensing with Satellite Altimetry

Satellites altimetry (Fig. 3.4) operates in two steps:

- *First*, the precise orbit of the satellite, i.e., its position, is determined. Through this, its *height* above the Earth is obtained.
- *Second*, range measurements are made by obtaining the time an emitted signal (radar or laser) travels to the Earth's surface and reflected back to the satellite.

GNSS satellites, see [37, 102], contribute to the *first step* where height is determined through GNSS receivers onboard the space satellites that enables monitoring of ranges and timing signals from GNSS satellites. The observed GNSS ranges provide precise and continuous tracking of the spacecraft, thereby delivering its position $\{\phi, \lambda, h\}$ at any time. The height component h is useful in determining the measured

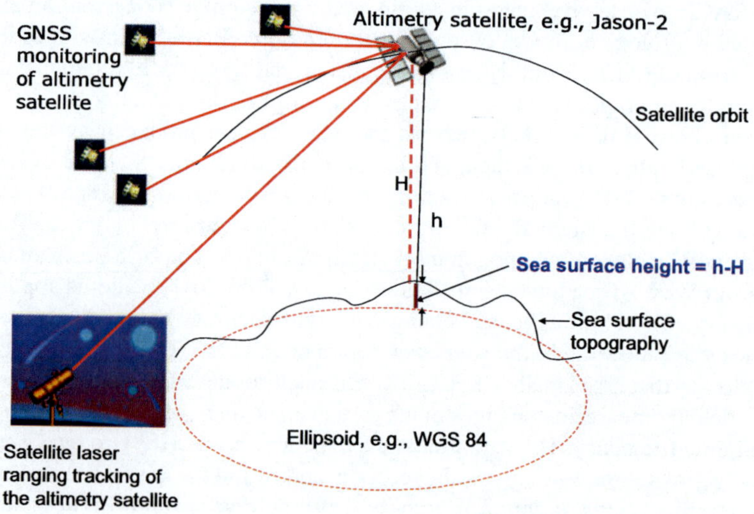

Fig. 3.4 Global Navigation Satellite System (GNSS) in support of monitoring changes in sea/lake level through the determination of the altimetry satellites' precise orbit. From the precise orbital parameters, the height component h is useful in determining changes in sea/lake level through the difference $\{h - H\}$, where H is measured by multiplying the speed of light with the time taken by the signals to travel from and to the satellite divided by 2, since the same distance is covered twice

height (see Fig. 3.4). Besides GNSS tracking, other approaches such as satellite laser ranging (SLR) and DORIS (Doppler Orbitography and Radio positioning Integrated by Satellite) are also used to ensure that precise orbit determination is achieved.

In the *second step*, the Earth's surface heights (e.g., lake surface, ocean surface, glaciers, and ice sheets) are measured using ranges from the space altimetry satellite to the surface of interest. Radar altimeters send microwave signals to the Earth's surface and measures the time taken by the reflected signals to travel back. Using the velocity of light, the distance from the satellite to the Earth's surface is derived. Since the signals pass through the atmosphere from and to the satellites, they are affected by the atmosphere and as such, atmospheric corrections have to be made. The sea surface height is then obtained by subtracting the measured ranges in step 2 from the GNSS-derived satellite heights in step 1 (Fig. 3.4).

Example 3.1 (Satellite altimetry monitoring of Lake Victoria water level).

As an example, Lake Victoria water levels for the period 1993 to 2006 are plotted for both TOPEX/Poseidon altimetry-derived heights and tide gauge (in-situ) data in Fig. 3.5. The figure indicates a close relationship between the two data sets. Crétaux et al. [108] compares water levels of Lake Victoria from the Jinja tide gauge and those from Jason-1 altimetry satellite and obtain correlation value of 0.99 with a standard deviation of 2.7 cm for the period 2004 to 2007. This supports the fact that satellite altimetry provides useful information on changes in the lake level. Satellite altimetry are largely used for observing changes in sea level. By averaging the few-hundred thousand measurements collected by the satellite in the time it takes to cover the global oceans (i.e., 10 days for TOPEX/Poseidon), global mean sea

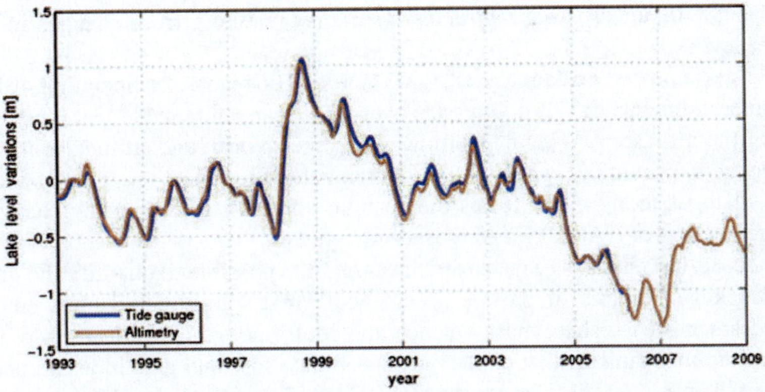

Fig. 3.5 A comparison of water gauge readings at Jinja station in Uganda (near Lake Victoria's outlet, see Fig. 5.2b) and water Levels from Topex/Poseidon and Jason-1 altimetry satellites. The figure shows a close match between the tide gauge and satellite altimetry data (cf. Fig. 5.7 on Sect. 5.5 obtained from GRACE)

level can be determined with a precision of several millimeters [12, 16, 112]. Such information is vital for mitigation of disasters related to sea level changes. Detailed satellite altimetry study on East African lakes (e.g., Fig. 3.5) and the Nile are given e.g., in [1, 26, 108–111, 114].

End of Example 3.1.

3.5.2 *Satellite Altimetry Missions*

The first altimetry mission was TOPEX/Poseidon, developed by NASA and the Centre National d'Etudes Spatiales (CNES) and launched on 10 August 1992. Its mission ended in 2006 after 13 years of operation, providing 11 years of data. It was followed by Jason-series (Jason-1 was launched on 07/12/2001 and Jason-2 on 20/06/2008). Both TOPEX/Poseidon and Jason-1 were dedicated to measuring global mean sea level from space. TOPEX/Poseidon orbited at 1336 km above the Earth and covered the global oceans every 10 days, measuring the heights of the ocean surface directly underneath the satellite with an accuracy of 2-4 cm or better when averaging over several measurements [112]. Jason-2 is expected to be replaced by Jason-3 launched on 17th of January 2016, and subsequently Sentinel-6, which will continue high precision ocean altimetry measurements in the 2020-2030 time-frame using two successive, identical satellites (Jason-Continuity of Service); Jason-CS-A and Jason-CS-B, and as a secondary objective, collect high resolution vertical profiles of temperature using the GNSS Radio-Occultation sounding technique discussed in Awange [37, 102].[8] Combined, all these satellites will provide long-term series of data capable of undertaking *hydroclimate studies* resulting from changes in lakes and sea level.

ICESat (launched on January 12, 2003) uses a 1064 nm-laser operating 40 Hz to make measurements at 172-m intervals over ice, ocean, and land [113]. It combines state-of-the-art laser ranging capabilities with precise orbit and attitude control and knowledge to provide very accurate measurements of ice sheet topography and elevation changes along track. It has the specific objective of measuring changes in polar ice as part of NASA's Earth Observing System.

By observing changes in ice sheet elevation, it is possible to quantify the growth and shrinkage of parts of the ice sheets with great spatial detail, thus enabling an assessment of ice sheet mass balance and contributions to sea level. Moreover, because the mechanisms that control ice sheet mass loss and gain in accumulation, surface ablation, and discharge presumably have distinct topographic expressions, ice sheet elevation changes also provide important insights into the processes causing the observed changes [113]. ICESat-2 launched on 15th September 2018 is expected to be a follow-on mission to ICESat (Fig. 3.6) with improved laser capability com-

[8] https://eospso.nasa.gov/missions/sentinel-6.

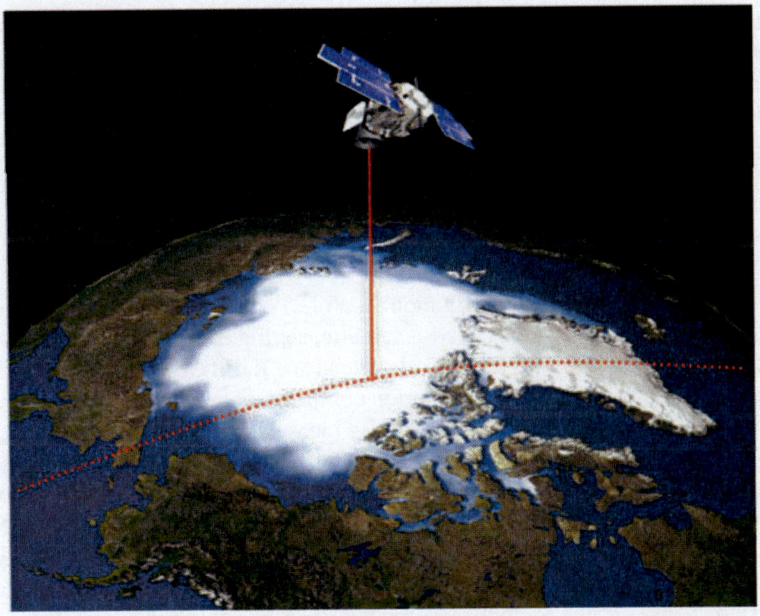

Fig. 3.6 Schematic diagram of ICESat on a transect over the Arctic. ICESat uses a 1064 nm-laser operating 40 Hz to make measurements at 172-m intervals over ice, oceans, and land. *Source* Abdalati et al. [113]

pared to ICESat and will have the objectives of measuring ice sheet changes, sea ice thickness, and vegetation biomass. Achieving these objectives will contribute to the following [113]:

- Contribute to the development of predictive models that capture both dynamic and surface processes.
- Since the thickness distribution of sea ice controls energy and mass exchanges between the ocean and atmosphere at the surface, and the freshwater fluxes associated with melting ice serve as stabilizing elements in the circulation of the North Atlantic waters, basin-scale fields of ice thickness are therefore essential to improve our estimates of the seasonal and interannual variability in regional mass balance, the freshwater budget of the polar oceans, and the representation of these processes in regional and climate models.
- The capability of producing a vegetation height surface with 3-m accuracy at 1-km spatial resolution, assuming that off-nadir pointing can be used to increase the spatial distribution of observations over terrestrial surfaces. This sampling, combined with a smaller footprint of 50 m or less, would allow characterization of vegetation at a higher spatial resolution than ICESat, and is expected to provide a new set of global ecosystem applications.

- In addition, the atmospheric measurement capability of ICESat-2, even at near-IR wavelengths, will enable global measurements of cloud and aerosol structure to extend the record of these observations beyond those provided by the current lasers onboard ICESat.

3.6 CHAMP Radio Occultation Satellite

Radio occultation with Global Navigation Satellite System (GNSS) takes place when radio signals from a transmitting GNSS satellite, setting or rising behind the Earth's limb, are received by a GNSS receiver aboard a Low Earth Orbiting (LEO) satellite. The meteorological products of occultation comprise the air pressure P, air temperature T and the water vapour pressure P_w. By measuring the occulting GNSS signal due to the effect of the atmosphere, LEO satellites such as Challenging Minisatellite Payload (CHAMP), Gravity Recovery and Climate Experiment (GRACE) and Constellation Observing System for Meteorology, Ionosphere, and Climate (COSMIC) are capable of providing accurate tropospheric measurements to sub-kelvin accuracy, see e.g., [8, 9, 37, 102]. In particular, they have been found to be suitable in measuring changes of the tropopause, which in-turn is vital for understanding global warming and its impact on water resources essential for GHA. Currently, these techniques are used to study tropical expansion, impacts of climate variability/change etc. This technique is applied in Chap. 5.

3.7 Concluding Remarks

Earth observation and its application to water resource monitoring, particularly GRACE, is a new and active area of research. The data that has been collected from GRACE so far has provided information that was hitherto difficult to fathom. The new technique clearly promises to contribute significantly to water resource studies of GHA and other freshwater lakes the world over. When the life span of the various missions (e.g., GRACE) is reached, thousands of data sets will have been collected that will help to unravel some of the complex nature of the hydrological cycle.

References

1. Awange JL, Saleem A, Sukhadiya RM, Ouma YO, Kexiang H (2019) Physical dynamics of Lake Victoria over the past 34 years (1984–2018): is the lake dying? Sci Total Environ 658:199–218. https://doi.org/10.1016/j.scitotenv.2018.12.051
2. Awange JL, Sharifi MA, Ogonda G et al (2008) The Falling Lake Victoria water level: GRACE, TRIMM and CHAMP satellite analysis of the lake basin. Water Res Manage 22:775. https://doi.org/10.1007/s11269-007-9191-y

3. Awange JL, Fleming KM, Kuhn M, Featherstone WE, Anjasmara I, Heck B (2011) On the suitability of the 40 × 40 GRACE mascon solutions for remote sensing Australian hydrology. Remote Sens Environ 115:864–875. https://doi.org/10.1016/j.rse.2010.11.014
4. Awange JL, Ferreira VG, Forootan E, Khandu Andam-Akorful SA, Agutu NO, He XF (2016) Uncertainties in remotely sensed precipitation data over Africa. Int J Climatol 36(1):303–323. https://doi.org/10.1002/joc.4346
5. Awange JL, Hu K, Khaki M (2019) The newly merged satellite remotely sensed, gauge and reanalysis-based Multi-Source Weighted-Ensemble Precipitation: evaluation over Australia and Africa (1981–2016). Sci Total Environ 670:448–465
6. Awange JL (2021) Lake Victoria monitored from space. Springer Nature International
7. Awange JL (2021) Nile Waters. Weighed from space, Springer Nature International
8. Awange JL, Kiema JBK (2013) Environmental geoinformatics. Monitoring and Management, Springer, Berlin, New York
9. Awange JL, Kiema JBK (2018) Environmental geoinformatics. Extreme hydro-climatic and food security challenges
10. Agutu NO (2017) A remote sensing based approach to enhance food security in the Greater Horn of Africa. Doctor of Philosophy of Curtin University
11. USGS (2015) Landsat data. USGS, USA
12. Berry PAM, Garlick JD, Freeman JA, Mathers EL (2005) Global inland water monitoring from multi-mission altimetry. Geophys Res Lett 32(16). https://doi.org/10.1029/2005GL022814
13. Food and Agriculture Organization of the United Nations (2000) The elimination of food insecurity in the Horn of Africa. A strategy for concerted government and UN agency action
14. Saleem A, Awange JL (2019) Coastline shift analysis in data deficient regions: exploiting the high spatio-temporal resolution Sentinel-2 products. Catena 179:6–19. https://doi.org/10.1016/j.catena.2019.03.023
15. Saleem A, Awange JL, Corner R (2021) Exploiting a texture framework and high spatial resolution properties of panchromatic images to generate enhanced multi-layer products: examples of Pleiades and historical CORONA space photographs. Int J Remote Sens 42(3):929–963. https://doi.org/10.1080/01431161.2020.1820617
16. Shum CK, Ries JC, Tapley BD (1995) The accuracy and applications of satellite altimetry. Geophys J Int 121:321–336. https://doi.org/10.1111/j.1365-246X.1995.tb05714.x
17. Kauth R, Lambeck P, Richardson W, Thomas G, Pentland A (1979) Feature extraction applied to agricultural crops as seen by Landsat
18. Stow D, Tinney L, Estes J (1980) Deriving land use/land cover change statistics from Landsat-A study of prime agricultural land
19. Peterson U, Aunap R (1998) Changes in agricultural land use in Estonia in the 1990s detected with multitemporal Landsat MSS imagery. Landsc Urban Plan 41:193–201
20. Cihlar J (2000) Land cover mapping of large areas from satellites: status and research priorities. Int J Remote Sens 21:1093–1114
21. Jusoff K, Senthavy S (2003) Land use change detection using remote sensing and geographical information system (GIS) in Gua Musang district, Kelantan, Malaysia. J Trop Forest Sci 15:303–312
22. Porter-Bolland L, Ellis EA, Gholz HL (2007) Land use dynamics and landscape history in La Montaña, Campeche, Mexico. Landsc Urban Plan 82:198–207
23. Dewan AM, Yamaguchi Y (2009) Land use and land cover change in Greater Dhaka, Bangladesh: using remote sensing to promote sustainable urbanization. Appl Geogr 29:390–401. https://doi.org/10.1016/j.apgeog.2008.12.005
24. Kesgin B, Nurlu E (2009) Land cover changes on the coastal zone of Candarli Bay, Turkey using remotely sensed data. Environ Monit Assess 157:89–96
25. Forootan Khandu EE, Schumacher M, Awange J, Müller Schmied H (2016) Exploring the influence of precipitation extremes and human water use on total water storage (TWS) changes in the Ganges-Brahmaputra-Meghna River Basin. Water Res Res 52(3):2240–2258. https://doi.org/10.1002/2015WR018113/full

26. Khaki M, Awange J (2021) The 2019–2020 rise in Lake Victoria monitored from space: exploiting the state-of-the-art grace-fo and the newly released ERA-5 reanalysis products. Sensors 21(13):4304. https://doi.org/10.3390/s21134304
27. Kummerow C, Barnes W (1998) The tropical rainfall measuring mission (TRMM) sensor package. J Atmos Ocean Technol 15:809–817
28. Miro ME, Famiglietti JS (2018) Downscaling GRACE remote sensing datasets to high-resolution groundwater storage change maps of California's Central Valley. Remote Sens 10:143. do1: 10.3390/rs10010143
29. Vittek M, Brink A, Donnay F, Simonetti D, Desclée B (2014) Land cover change monitoring using landsat MSS/TM satellite image data over West Africa between 1975 and 1990. Remote Sens 6:658–676
30. Yadav P, Kapoor M, Sarma K (2012) Land use land cover mapping, change detection and conflict analysis of Nagzira-Navegaon corridor, Central India using geospatial technology. Int J Remote Sens GIS 1:90–98
31. Yuan P, Hunegnaw A, Alshawaf F, Awange J, Klos A, Teferle FN, Kutterer H (2021) Feasibility of ERA5 integrated water vapor trends for climate change analysis in continental Europe: an evaluation with GPS (1994–2019) by considering statistical significance. Remote Sens Environ 260. https://doi.org/10.1016/j.rse.2021.112416
32. Otukei JR, Blaschke T (2010) Land cover change assessment using decision trees, support vector machines and maximum likelihood classification algorithms. Int J Appl Earth Obs Geoinf 12:S27–S31. https://doi.org/10.1016/j.jag.2009.11.002
33. Moran EF (2010) Land cover classification in a complex urban-rural landscape with QuickBird imagery. Photogr Eng Remote Sens 76:1159
34. Justice CO, Townshend JRG, Vermote EF, Masuoka E, Wolfe RE, Saleous N, Roy DP, Morisette JT (2002) An overview of MODIS Land data processing and product status. Remote Sens Environ 83:3–15. https://doi.org/10.1016/S0034-4257(02)00084-6
35. ESA (2017) European space agency, copernicus open access Hub. In: SERCO
36. Drusch M, Del Bello U, Carlier S, Colin O, Fernandez V, Gascon F, Hoersch B, Isola C, Laberinti P, Martimort P (2012) Sentinel-2: ESA's optical high-resolution mission for GMES operational services. Remote Sens Environ 120:25–36
37. Awange JL (2018) GNSS environmental sensing. Revolutionizing environmental monitoring. Springer, Berlin, New York
38. Tregoning P, Watson C, Ramillien G, McQueen H, Zhang J (2009) Detecting hydrologic deformation using GRACE and GPS. Geophys Res Lett 36:L15401. https://doi.org/10.1029/2009GL038718
39. Hammond WC, Brooks BA, Bürgmann R, Heaton T, Jackson M, Lowry AR, Anandakrishnan S (2010) The scientific value of high-rate, low-latency GPS data. A white paper. https://www.unavco.org/highlights/2010/RealTimeGPSWhitePaper2010.pdf. Accessed 27 Aug 2021
40. Hirt C, Gruber T, Featherstone WE (2011) Evaluation of the first GOCE static gravity field models using terrestrial gravity, vertical deflections and EGM2008. Quasigeoid heights. J Geodesy 85:723–740. https://doi.org/10.1007/s00190-011-0482-y
41. Ellett KM, Walker JP, Western AW, Rodell M (2006) A framework for assessing the potential of remote sensed gravity to provide new insight on the hydrology of the Murray-Darling Basin. Aust J Water Res 10(2):125–138. https://doi.org/10.1080/13241583.2006.11465286
42. Awange JL, Sharifi MA, Baur O, Keller W, Featherstone WE, Kuhn M (2009) GRACE hydrological monitoring of Australia. Current limitations and future prospects. J Spatial Sci 54(1):23–36. https://doi.org/10.1080/14498596.2009.9635164
43. Rummel R, Balmino G, Johannessen J, Visser P, Woodworth P (2002) Dedicated gravity field missions—principles and aims. J Geodyn 33(1):3–20. https://doi.org/10.1016/S0264-3707(01)00050-3
44. Schrama EJO, Visser PNAM (2007) Accuracy assessment of the monthly GRACE geoids based upon a simulation. J Geodesy 81(1):67–80. https://doi.org/10.1007/s00190-006-0085-1
45. Rieser D (2008) Comparison of GRACE-derived monthly Surface Mass Variations with Rainfall Data in Australia. MSc Thesis. Graz University of Technology

46. Prasad R, Ruggieri M (2005) Applied satellite navigation using GPS. GALILEO and Augmentation Systems, Artech House, Boston, London
47. Kalnay E, Kanamitsu M, Kistler R, Collins W, Deaven D, Gandin L, Iredell M, Saha S, White G, Woollen J, Zhu Y, Chelliah M, Ebisuzaki W, Higgins W, Janowiak J, Mo KC, Ropelewski C, Wang J, Leetmaa A, Reynolds R, Jenne R, Joseph D (1996) The NCEP/NCAR 40-year reanalysis project. Bull Am Meteorol Soc 77(3):437–470. https://doi.org/10.1175/1520-0477(1996)077<3c0437:TNYRP>3e2.0.CO;2
48. Onogi K, Tsutsui J, Koide H, Sakamoto M, Kobayashi S, Hatsushika H, Matsumoto T, Yamazaki N, Kamahori H, Takahashi K, Kadokura S, Wada K, Kato K, Oyama R, Ose T, Mannoji N, Taira R (2007) The JRA-25 reanalysis. Quart J R Meteorol Soc 85(3):369–432. https://doi.org/10.2151/jmsj.85.369
49. Pool DR, Eychaner JH (1995) Measurements of aquifer-storage change and specific yield using gravity surveys. Ground Water 33(3):425–432
50. Saha S, Moorthi S, Pan H-L, Wu X, Wang J, Nadiga S, Tripp P, Kistler R, Woollen J, Behringer D, Liu H, Stokes D, Grumbine R, Gayno G, Wang J, Hou YT, Chuang HY, Juang H-MH, Sela J, Iredell M, Treadon R, Kleist D, Delst PV, Keyser D, Derber J, Ek M, Meng J, Wei H, Yang R, Lord S, Dool HVD, Kumar A, Wang W, Long C, Chelliah M, Xue Y, Huang B, Schemm J-K, Ebisuzaki W, Lin R, Xie P, Chen M, Zhou S, Higgins W, Zou C-Z, Liu Q, Chen Y, Han Y, Cucurull L, Reynolds RW, Rutledge G, Goldberg M (2010) The NCEP climate forecast system reanalysis. Bull Am Meteorol Soc 91(8):1015–1057. https://doi.org/10.1175/2010BAMS3001.1
51. Dee DP, Uppala SM, Simmons AJ, Berrisford P, Poli P, Kobayashi S, Andrae U, Balmaseda MA, Balsamo G, Bauer P, Bechtold P, Beljaars ACM, van de Berg L, Bidlot J, Bormann N, Delsol C, Dragani R, Fuentes M, Geer AJ, Haimbergere L, Healy SB, Hersbach H, Holm EV, Isaksen L, Kållberg P, Köhler M, Matricardi M, McNally AP, Monge-Sanz BM, Morcrette J-J, Park B-K, Peubey C, de Rosnay P, Tavolato C, Thépaut J-N, Vitarta F (2011) The ERA-Interim reanalysis: configuration and performance of the data assimilation system. Q J R Meteorolog Soc 137:553–597. https://doi.org/10.1002/qj.828
52. Luthcke S, Rowlands D, Lemoine F, Klosko S, Chinn D, McCarthy J (2006) Monthly spherical harmonic gravity field solutions determined from GRACE inter-satellite range-rate data alone. Geophys Res Lett 33:L02402. https://doi.org/10.1029/2005GL024846
53. Rodell M, Houserand PR, Jambor U, Gottschalck J, Mitchell K, Meng C-J, Arsenault K, Cosgrove B, Radakovich J, Bosilovich M, Entin JK, Walker JP, Lohmann D, Toll D (2004) The global land data assimilation system. Bull Am Meteorol Soc 85:381–394. https://doi.org/10.1175/BAMS-85-3-381
54. Rienecker MM, Suarez MJ, Gelaro R, Todling R, Bacmeister J, Liu E, Bosilovich MG, Schubert SD, Takacs L, Kim GK, Bloom S, Chen J, Collins D, Conaty A, da Silva A, Gu W, Joiner J, Koster RD, Lucchesi R, Molod A, Owens T, Pawson S, Pegion P, Redder CR, Reichle R, Robertson FR, Ruddick AG, Sienkiewicz M, Woollen J (2011) MERRA: NASA's Modern-Era retrospective analysis for research and applications. J Clim 24(14):3624–3648. https://doi.org/10.1175/JCLI-D-11-00015.1
55. Döll P, Schmied MH, Schuh C, Portmann FT, Eicker A (2014) Global-scale assessment of groundwater depletion and related groundwater abstractions: combining hydrological modeling with information from well observations and GRACE satellites. Water Res Res 50(7):5698–5720. https://doi.org/10.1002/2014WR015595
56. Tapley BD, Bettadpur S, Ries JC, Thompson PF, Watkins MM (2004) GRACE measurements of mass variability in the Earth system. Science 305:503–505. https://doi.org/10.1126/science.1099192
57. Bruinsma S, Lemoine J, Biancale R, Valès N (2010) CNES/GRGS 10-day gravity field models (release 2) and their evaluation. Adv Space Res 45(4):587–601. https://doi.org/10.1016/j.asr.2009.10.012
58. Tapley BD, Bettadpur S, Watkins MM, Reigner C (2004) The gravity recovery and climate experiment: mission overview and early results. Geophys Res Lett 31:L09607. https://doi.org/10.1029/2004GL019920

59. Lemoine F, Luthcke S, Rowlands D, Chinn D, Klosko S, Cox C (2007) The use of mascons to resolve time-variable gravity from GRACE. In: Tregoning P, Rizos C (eds) Dynamic planet. Springer, Berlin, pp 231–236
60. Ramillien G, Cazenave A, Brunau O (2004) Global time variations of hydrological signals from GRACE satellite gravimetry. Geophys J Int 158(3):813–826. https://doi.org/10.1111/j.1365-246X.2004.02328.x
61. Chambers D, Wahr J, Nerem R (2004) Preliminary observations of global ocean mass variations with GRACE. Geophys Res Lett 31(L13310). https://doi.org/10.1029/2004GL020461
62. Wahr J, Jayne S, Bryan F (2002) A method of inferring changes in deep ocean currents from satellite measurements of time-variable gravity. J Geophys Res 107(C12):3218. https://doi.org/10.1029/2002JC001274
63. Rodell M, Famiglietti JS (1999) Detectability of variations in continental water storage from satellite observations of the time dependent gravity field. Water Res Res 35(9):2705–2724. https://doi.org/10.1029/1999WR900141
64. Tiwari V, Wahr J, Swenson S (2009) Dwindling groundwater resources in northern India, from satellite gravity observations. Geophys Res Lett 36:L18401. https://doi.org/10.1029/2009GL039401
65. Werth S, Güntner A, Petrovic S, Schmidt R (2009) Integration of GRACE mass variations into a global hydrological model. Earth Plan Sci Lett 27(1–2):166–173. https://doi.org/10.1016/j.epsl.2008.10.021
66. Baur O, Kuhn M, Featherstone W (2009) GRACE-derived ice-mass variations over Greenland by accounting for leakage effects. J Geophys Res 114(B06407). https://doi.org/10.1029/2008JB006239
67. Velicogna I (2009) Increasing rates of ICE mass loss from the Greenland and Antarctic ice sheets revealed by GRACE. Geophys Res Lett 36:L19503. https://doi.org/10.1029/2009GL040222
68. Boy J-P, Chao B (2005) Precise evaluation of atmospheric loading effects on Earth's time-variable gravity field. J Geophys Res—Solid Earth 110(B8):4–12. https://doi.org/10.1029/2002JB002333
69. Swenson S, Wahr J (2002) Estimated effects of the vertical structure of atmospheric mass on the time-variable geoid. J Geophys Res 107(B9):2194. https://doi.org/10.1029/2000JB000024
70. Barletta V, Sabadini R, Bordoni A (2008) Isolating the PGR signal in the GRACE data: impact on mass balance estimates in Antarctica and Greenland. Geophys J Int 172(1):18–30. https://doi.org/10.1111/j.1365-246X.2007.03630.x
71. Swenson S, Wahr J, Milly PCD (2003) Estimated accuracies of regional water storage variations inferred from the Gravity Recovery and Climate Experiment (GRACE). Water Res Res 39(8):1223. https://doi.org/10.1029/2002WR001736
72. Ramillien G, Frappart F, Cazenave A, Güntner A (2005) Time variations of land water storage from an inversion of two years of GRACE geoids [rapid communication]. Earth Plan Sci Lett 235(1–2):283–301. https://doi.org/10.1016/j.epsl.2005.04.005
73. Wahr J, Molenaar M, Bryan F (1998) Time variability of the Earth's gravity field: hydrological and oceanic effects and their possible detection using GRACE. J Geophys Res (Solid Earth) 103(B12):30205–30230. https://doi.org/10.1029/98JB02844
74. Heiskanen WA, Moritz H (1967) Physical geodesy. WH Freeman and Company San Francisco
75. Arras C, Jacobi C, Wickert J, Heise S, Schmidt T (2010) Sporadic E signatures revealed from multi-satellite radio occultation measurements. Adv Radio Sci 8:225–230. https://doi.org/10.5194/ars-8-225-2010
76. Yang Q (2016) Applications of satellite geodesy in environmental and climate change. Graduate Theses and Dissertations. http://scholarcommons.usf.edu/etd/6440 Accessed 26 Jan 2017
77. Brutsaert W (2005) Hydrology. An introduction, 4th edn. Cambridge University Press, New York
78. Morgan B, Awange JL, Saleem A, Hu K (2020) Understanding vegetation variability and their "hotspots" within Lake Victoria Basin (LVB: 2003–2018), 122. https://doi.org/10.1016/j.apgeog.2020.102238

79. Montgomery EL (1971) Determination of coefficient of storage by use of gravity measurements. PhD thesis, University of Arizona, Tucson
80. Leirião S, He X, Christiansen L, Andersen OB, Bauer-Gottwein P (2009) Calculation of the temporal gravity variation from spatially variable water storage change in soils and aquifers. J Hydrol 365(3–4):302–309. https://doi.org/10.1016/j.jhydrol.2008.11.040
81. Lambert A, Beaumont C (1977) Nano variation in gravity due to seasonal groundwater movements: implications for the gravitational detection of tectonic movements. J Geophys Res 82(2):297–306. https://doi.org/10.1029/JB082i002p00297
82. Goodkind JM (1986) Continuous measurement of nontidal variations of gravity. J Geophys Res 91(B9):9125–9134
83. Mukai A, Higashi T, Takemoto S, Nakagawa I, Naito I (1995) Accurate estimation of atmospheric effects on gravity observations made with a superconducting gravity meter at Kyoto. Phys Earth Plan Inter 91(1–3):149–159
84. Pool DR, Eychaner JH (1995) Measurements of aquifer-storage change and specific yield using gravity surveys. Groundwater 33(3):425–432. https://doi.org/10.1111/j.1745-6584.1995.tb00299.x
85. Bower DR, Courtier N (1998) Precipitation effects on gravity measurements at the Canadian Absolute Gravity Site. Phys Earth Plan Inter 106:353–369. https://doi.org/10.1016/S0031-9201(97)00101-5
86. Smith AB, Walker JP, Western AW, Ellett KM (2005) Using ground based measurements to monitor changes in terrestrial water storage. In: 29th hydrology and water resources symposium (CD Rom). Institute of Engineers Australia
87. Damiata BN, Lee TC (2002) Gravitational attraction of solids of revolution—part 1: vertical circular cylinder with radial variation of density. J Appl Geophys 50(3):333–349. https://doi.org/10.1016/S0926-9851(02)00151-9
88. Rodell M, Famiglietti JS, Chen J, Seneviratne SI, Viterbo P, Holl S, Wilson CR (2004) Basin scale estimates of evapotranspiration using GRACE and other observations. Geophys Res Lett 31:L20504. https://doi.org/10.1029/2004GL020873
89. Ellett KM, Walker JP, Rodell M, Chen JL, Western AW (2005) GRACE gravity fields as a new measure for assessing large-scale hydrological models. In: Zerger A, Argent RM (eds) MODSIM 2005 international congress on modelling and simulation. The modelling and simulation society of Australia and New Zealand, Dec 2005, pp 2911–2917. ISBN: 0-9758400-2-9
90. Rodell M, Chen J, Kato H, Famiglietti JS, Nigro J, Wilson CR (2006) Estimating groundwater storage changes in the Mississippi River basin (USA) using GRACE. Hydrogeol J 15(1):159–166. https://doi.org/10.1007/s10040-006-0103-7
91. Vishwakarma BD, Zhang J, Sneeuw N (2021) Downscaling GRACE total water storage change using partial least squares regression. Sci Data 8:95. https://doi.org/10.1038/s41597-021-00862-6
92. Andersen OB, Seneviratne SI, Hinderer J, Viterbo P (2005) GRACE-derived terrestrial water storage depletion associated with the 2003 European heat wave. Geophys Res Lett 32(L18405):2–5. https://doi.org/10.1029/2005GL023574
93. Neumeyer J, Barthelmes F, Dierks O, Flechtner F, Harnisch M, Harnisch G, Hinderer J, Imanishi Y, Kroner C, Meurers B, Petrovic S, Reigber C, Schmidt R, Schwintzer P, Sun HP, Virtanen H (2006) Combination of temporal gravity variations resulting from superconducting gravimeter (SG) recordings, GRACE satellite observations and global hydrology models. J Geodesy 79(10–11):573–585. https://doi.org/10.1007/s00190-005-0014-8
94. Yan JP, Hinderer M, Einsele G (2002) Geochemical evolution of closed-basin lakes: general model and application to Lakes Qinghai and Turkana. Sediment Geol 148(1–2):105–122. https://doi.org/10.1016/S0037-0738(01)00212-3
95. Syed T, Famiglietti J, Rodell M, Chen J, Wilson C (2008) Total basin discharge for the Amazon and Mississippi River basins from GRACE and a land-atmosphere water balance. Water Res Res 44:W02433. https://doi.org/10.1029/2006WR005779
96. Crowley JW, Mitrovica JX, Bailey RC, Tamisiea ME, Davis JL (2006) Land water storage within the Congo Basin inferred from GRACE satellite gravity data. Geophys Res Lett 33:L19402. https://doi.org/10.1029/2006GL027070

97. Schmidt R, Schwintzer P, Flechtner F, Reigber C, Güntner A, Döll P, Ramillien G, Cazenave A, Petrovic S, Jochmann H, Wünsch J (2006) GRACE observations of changes in continental water storage. Global Plan Change 50(1–2):112–126. https://doi.org/10.1016/j.gloplacha.2004.11.018
98. Swenson S, Yeh PJ-F, Wahr J, Famiglietti J (2006) A comparison of terrestrial water storage variations from GRACE with in-situ measurements from Illinois. Geophys Res Lett 33:L16401. https://doi.org/10.1029/2006GL026962
99. Winsemius HC, Savenije HHG, van de Giesen NC, van den Hurk B, Zapreeva EA, Klees R (2006) Assessment of gravity recovery and climate experiment (GRACE) temporal signature over the upper Zambezi. Water Res Res 42:W12201. https://doi.org/10.1029/2006WR005192
100. Awange JL, Anyah R, Agola N, Forootan E, Omondi P (2013) Potential impacts of climate and environmental change on the stored water of Lake Victoria Basin and economic implications. Water Res Res 49:8160–8173
101. Awange JL, Ogallo L, Kwang-Ho B, Were P, Omondi P, Omute P, Omulo M (2008) Falling Lake Victoria water levels: is climate a contribution factor? J Clim Change 89:287–297. https://doi.org/10.1007/s10584-008-9409-x
102. Awange JL (2012) Environmental monitoring using GNSS Global Navigations Satellite Systems. Springer, Heidelberg, Berlin
103. Awange JL, Forootan E, Kuhn M, Kusche J, Heck B (2014) Water storage changes and climate variability within the Nile Basin between 2002 and 2011. Adv Water Res 73:1–25. https://doi.org/10.1016/j.advwatres.2014.06.010
104. Ndehendehe C, Awange J, Agutu N, Kuhn M, Heck B (2016) Understanding changes in terrestrial water storage over West Africa between 2002 and 2014. Adv Water Res 88:211–230. https://doi.org/10.1016/j.advwatres.2015.12.009
105. Agutu N, Awange JL, Zerihun A, Ndehedehe C, Kuhn M, Fukuda Y (2017) Assessing multi-satellite remote sensing, reanalysis, and land surface models' products in characterizing agricultural drought in East Africa. Remote Sens Environ 194:287–302. https://doi.org/10.1016/j.rse.2017.03.041
106. Agutu NO, Awange JL, Ndehedehe C, Kirimi F, Kuhn M (2019) GRACE-derived groundwater changes over greater Horn of Africa: temporal variability and the potential for irrigated agriculture. Sci Total Environ. https://doi.org/10.1016/j.scitotenv.2019.07.273
107. Bettadpur S (2007) UTCSR Level-2 Processing standards document for level-2 product release 0004. Gravity recovery and climate experiment (GRACE) Rev 3.1, GRACE 327–742 (CSR-GR-03-03). Center for Space Research, The University of Texas at Austin
108. Crétaux J-F, Jelinski W, Calmant S, Kouraev A, Vuglinski V, Bergé-Nguyen M, Gennero M-C, Nino F, Abarca Del Rio R, Cazenave A, Maisongrande P (2011) SOLS: a lake database to monitor in the Near real-time water level and storage variations from remote sensing data. Adv Space Res 47:1497–1507. https://doi.org/10.1016/j.asr.2011.01.004
109. Khaki M, Awange J (2020) Altimetry-derived surface water data assimilation over the Nile Basin. Sci Total Environ. https://doi.org/10.1016/j.scitotenv.2020.139008
110. Khaki M, Awange JL (2019) Improved satellite remotely sensed products for studying climate-induced changes over Lake Victoria. Sci Total Environ 652:915–926. https://doi.org/10.1016/j.scitotenv.2018.10.279
111. Khaki M, Awange J, Forootan E, Kuhn M (2018) Understanding the association between climate variability and the Nile's water level fluctuations and water storage changes during 1992–2016. Sci Total Environ 645:1509–1521. https://doi.org/10.1016/j.scitotenv.2018.07.212
112. Pugh D (2004) Changing sea levels. Effect of tides, weather and climate. Cambridge Univeristy Press
113. Abdalati W, Zwally HJ, Bindschadler B, Csatho B, Farrell SL, Fricker HA, Harding D, Kwok R, Lefsky M, Markus T, Marshak A, Neumann T, Palm S, Schutz B, Smith B, Spinhirne J, Webb C (2010) The ICESat-2 laser altimetry mission. Proc IEEE 98(5):735–751. https://doi.org/10.1109/JPROC.2009.2034765
114. Becker M, Llowel W, Cazenave A, Güntner A, Crétaux J-F (2010) Recent hydrological behaviour of the East African Great Lakes region inferred from GRACE, satellite altimetry

and rainfall observations. Comptes Rendus Geosci 342(3):223–233. https://doi.org/10.1016/j.crte.2009.12.010

115. Khandu (2016) Assessing climate change impacts on water resources in the Ganges-Brahmaputra-Meghna River Basin. PhD thesis, Curtin University

116. Ebert EE, Janowiak JE, Kidd C (2007) Comparison of near-real-time precipitation estimates from satellite observations and numerical models. Bull Am Meteorol Soc 88(1):47–64. https://doi.org/10.1175/BAMS-88-1-47

117. Vishwakarma BD, Sneeuw N, Westaway RM, Bamber JL (2021) Re-assessing global water storage trends from GRACE time series. Environ Res Lett 16:034005. https://doi.org/10.1088/1748-9326/abd4a9

118. Scanlon BR, Zhang Z, Save H, Sun AY, Schmied HM, Beek LPH, Wiese D, Wada Y, Long D, Reedy R, Longuevergne L, Döll P, Bierkens MFP (2018) Global models underestimate large decadal declining and rising water storage trends relative to GRACE satellite data Proc. Natl Acad Sci 115:E1080–E9

Part II
Water Resources

Chapter 4
Global Freshwater Resources

Several authors, politicians, leaders of international organizations and journalists have cautioned the world community that the increasing scarcity of freshwater resources might lead to national and international conflicts. When relating this to climate change forecasts – most of which indicate that climate change will have a significant impact on the availability of freshwater resources, on water quality, and on the demand for water – this is alarming news for humankind as it threatens human security. – Molen and Hildering [3]

Accounting for water for environmental requirements shows that abstraction of water for domestic, food and industrial uses already have a major impact on ecosystems in many parts of the world, even those not considered "water scarce" [4]. In the absence of coordinated planning and international cooperation at an unprecedented scale, therefore, the next half century will be plagued by a host of severe water-related problems, threatening the well being of many terrestrial ecosystems and drastically impairing human health, particularly in the **poorest regions** of the world. [5].

4.1 Diminishing Freshwater Resources

4.1.1 Status

Freshwater, influenced globally by climate variability/change [6–11], human use [12, 14, 15, 98] and knowledge deficiency resulting from inadequate hydrometeorological observation stations [16–20], is one of the basic necessities without which human beings cannot survive since it is key to the sustainability of all kinds of life forms. Water has multiple uses namely; nutritional, domestic, recreational, navigational, waste disposal and ecological as it is a habitat for living and non-living organisms (biodiversity), etc. And, because it is indispensable to different sectors including manufacturing, agriculture, fisheries, wildlife survival, tourism and hydroelectric power generation, just to list but a few, it is a vital factor of economic production. For many

J. Awange, *Food Insecurity & Hydroclimate in Greater Horn of Africa*,
https://doi.org/10.1007/978-3-030-91002-0_4

countries, most freshwater endowments encompass surface waters, groundwater, wetlands and glaciers. Surface water bodies include lakes, rivers, swamps, springs, dams and water pans dispersed within different basins. In general, people living in the vast arid and semi-arid parts of the world rely heavily on groundwater resources. Gleeson et al. [21] puts the number of people living in regions where groundwater is threatened to be 1.7 billion. Furthermore, groundwater is also an important supplementary source of water for many urban households in most developing countries and plays important role in irrigated agriculture.

At a global scale, although much of the Earth is covered by water, most of it is unsuitable for human consumption, since 96% of it is found in the saline oceans. According to the U.N., only 2.5% of the roughly 1.4 billion cubic kilometers of water on Earth is freshwater, and approximately 68.9% of the freshwater is trapped in glacial ice or permanent snow in mountainous regions—the Arctic and Antarctica. Roughly 30.8% is groundwater, much of which is inaccessible to humans, and the remainder 0.3% comprise surface waters in lakes and rivers [22]. Of the 0.3% available for human and animal consumption, much is inaccessible due to unreachable underground locations and depths [23]. Jury and Vaux [5] caution that focusing on the global freshwater storage resource alone is misleading because much of the water is inaccessible. They suggest that humanity's freshwater resource consisting of rainfall used to grow crops, accessible groundwater, and surface water be considered.

4.1.2 Water Scarcity

Although no common definition of water scarcity exists, Rijsberman [4] defines a water insecure person as an individual who does not have access to safe and affordable water to satisfy his or her needs for drinking, washing, or livelihood. A water scarce area is then said to be an area where a large number of people are water insecure for a significant period of time, e.g., Rijsberman [4]. Rijsberman [4] points out that an area qualifies as a water-scarce area depending on, e.g., (i) the definition of people's needs and whether the definition takes into account the needs of the environment, the water for nature, (ii) the fraction of the resource that is made available (or could be made available) to satisfy these needs, and (iii), the temporal and spatial scales used to define scarcity.

Even though Rijsberman [4] argues that at a global scale, and from a supply and demand perspective, it is still debatable whether water scarcity is fact or fiction, it is incontestable that freshwater is increasingly becoming a scarce resource and shortages could drive conflict as well as negatively hit food and energy production [5, 24, 25] (see Fig. 4.1). That water shortage is emerging as one of the leading challenges of the 21st century has been documented, e.g., in [4, 26–28]. To underscore the seriousness of water shortage, the World Bank in its 1992 World Development Report pointed out that 22 countries faced severe water shortage while further 18 were in danger of facing shortages if fluctuation in rainfall persisted [29]. Recent studies, e.g., [30–33, 83, 84] point to fluctuation in rainfall in parts of Greater Horn of Africa

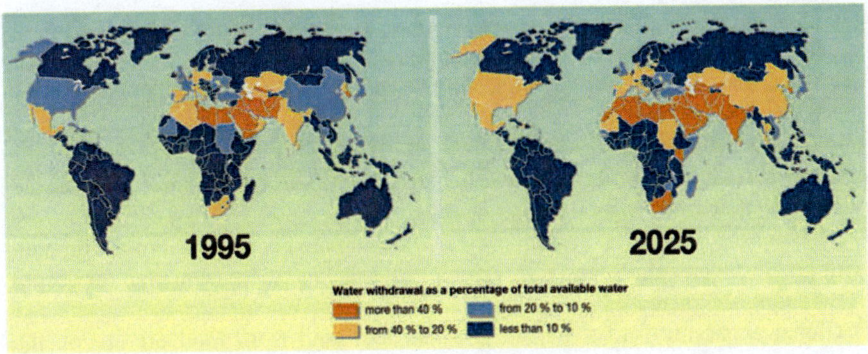

Fig. 4.1 Increased global water stress. The Greater Horn of Africa can be categorized as a low to medium water withdrawal area. It can be seen from the figure that by 2025, Sudan and Kenya are likely to move from low (1995) to medium-high water withdrawal areas. *Source* [40]

where most low income earners depend on rain-fed agriculture for food production, thus suggesting that the water shortage problem is not fading away any time soon. A more gloomy picture is the estimate that by 2050, about two billion people will be short of water, a potential cause of conflict [34]. That a large population of the world will face water scarcity is supported by several studies, e.g., [7, 35–38], with the most likely to be affected dwelling in Africa, Asia and the Middle East, see, e.g., [26, 28, 39, 85]. Already, model projections, see e.g., [9] suggest a 40% of the population languishing under water scarcity, with 57% likely to be water stressed by 2015 and 69% by 2075 [27, 35].

By reviewing several publications on water in relation to poverty and environmental degradation nexus, Duraiappah [29] presents activities that lead to water shortages. ***First***, there is the issue of the commercial interest of the rich and wealthy driven primarily by power, greed and wealth that benefits from market and institutional failures (e.g., absence or misuse of water property rights). ***Second***, Duraiappah [29] attribute over usage of water supply by the small holders (the poor) to water subsidies that provide incentives. Of the two groups, the poor are more likely to be affected by water shortages as compared to the rich, a situation that could contribute to environmental degradation on the one hand, which would lead to poverty on the other hand, i.e., endogenous poverty causing environmental degradation, see, e.g., Duraiappah [29].

This picture leads Jury and Vaux [5] to warn that "without immediate action and global cooperation, water supply and water pollution crisis of unimaginable dimensions will confront humanity, limiting food production, drinking water access, and the survival of innumerable species on the planet". They list the following four factors to support their hypothesis that the world is headed towards a future where billions of people will be forced to live in places where their food and water requirements will not be met [5]:

1. Unlike estimates of the global supply of scarce minerals or underground fuels, which are surrounded by uncertainty, planetary supplies of water are relatively well characterized. No large deposits of groundwater await human detection in readily accessible locations, so that any new resources discovered will be very expensive to develop.
2. Many vital human activities have become dependent on utilizing groundwater supplies that are being exhausted or contaminated.
3. Much of the population growth projected for the next century will occur in areas of greatest water shortages, e.g., Africa and Asia, and there is no plan for accommodating the increases.
4. Global economic forces are luring water and land from food production into more lucrative activities such as biofuel, see, e.g., [41, 42, 82], while at the same time encouraging pollution that impairs drinking water quality for a large and ever-growing segment of the population.

Evidently, the management of water resources' conflicts, focusing on negotiation, mediation and decision-making processes, in order to prevent, manage and resolve water conflicts is emerging as a contemporary and topical research issue.

Physical water scarcity is evident in densely populated arid areas in many parts of Central and West Asia, and North Africa with projected availabilities of less than 1000 m^3/capita/year [4]. This has a wide range of negative impacts and ramifications. For instance, it results in higher incidences of waterborne, water-related or sanitation-related diseases such as malaria, diarrhoea and skin infections. In addition, as has been pointed out before, increasing cases of water conflicts, especially between pastoralist and farming communities along lower and upper river basins, as witnessed in water stressed countries like Kenya [44] (with a renewable freshwater per capita endowment estimated at about 548 m^3/capita/year [43]), are also likely to heighten food insecurity. Alcamo [35] point to the fact that 12% of the global basins are already vulnerable to water withdrawals. On the gender scale, since women are the primary collectors, users and managers of water for domestic use in most developing countries, water scarcity disproportionately affects them because it is they who have to trek long distances, often all day, in search of water. Against this background, is the growing realization today that availability of sufficient, accessible and quality potable water is not only a matter of great socio-economic and political importance, but also one of fundamental human rights, see e.g., [45–48].

4.1.3 Impacts of Climate Variability/Change on Freshwater

Intergovernmental panel of climate change [86] projects a temperature increase of 1.5 °C by 2030, an increase which will definitely have an enormous impact on the world's freshwater. That temperature has been on the rise due to human's carbon footprint had been noted in the previous IPCC (2007 and 2013) reports [87]. Infact, [88] points to the fact that the global and ocean temperatures rose by 0.85 °C [0.65 °C–

1.06 °C] over the period 1880–2012, 0.89 °C [0.69 °C–1.08 °C] from 1901–2012, and 0.72 °C [0.49 °–0.89 °C] between 1951 and 2012. Change in climate resulting in the temperature rise above as well as climate variability exemplified through indicators such as global teleconnections, i.e., ENSO [89, 90] impact on the availability of freshwater. For instance, climate variability influence extreme whether/climate conditions resulting, e.g., in increased frequency and severity of droughts [84, 91, 92]. Droughts majorly affect surface water although a long spell could also impact on groundwater [93, 94]. Moreover, the influence of climate variability/change on the hydrological cycle impacts on precipitation, which is the main recharge of freshwater.

4.1.4 Water-Poverty-Environment Nexus

Water plays a significant role both in poverty alleviation and environmental related issues. With regard to poverty, food policy contributes toward the overall goal of eradicating extreme poverty and hunger, and is dependent on water as one of the drivers for realizing such a goal, see e.g., Hanjra and Qureshi [49]. It provides a key component of agricultural requirement for food production as well as for domestic use. If people are able to provide food, then absolute poverty measure of a society could arguably be low. Indeed, that water and food security issues are closely linked has been pointed out, e.g., by [12, 49], see also Chap. 1. Rijsberman [4] is even more direct by stating that water will be the major constraint for agriculture in the coming decades, more so in the continents that experience high percentage of poverty such as Asia and Africa.

The problem, however, is that more focus is placed on the water crisis, which is likely to be fuelled by increased population leading to about 1.2 billion lacking water. To the contrary, less documentation exist for the large part of the population that live in rural areas below the poverty line. Duraiappah [29], through literature review, identified

- water shortage, and,
- water pollution,

as the two major issues within the water sector that plays an important role in the poverty-environmental degradation nexus. Understanding the link between poverty and environment in relation to water, therefore, calls for a more closer look at the whole issue of water scarcity in relation to the poor rather than a holistic approach where such scarcity is viewed in terms of the total population. This in essence calls for poverty eradication measures to incorporate issues that will address water scarcity and insecurity. This is partly because if the poor are unable to access clean and safe water, food productivity will be hampered leading to hunger and malnutrition hence less productivity (see Chap. 1). Also, lack of access to safe drinking water by the poor will only aggravate the risk of water borne diseases such as cholera. When the low income group are faced by such water-borne diseases, their productivity deteriorates and they risk losing their jobs, and hence sources of income. The expected outcome of loss in income is that the low income earners will experience economic and social hardship, which over time results in poverty, thus exemplifying how environmental degradation causes poverty, e.g., [26, 29].

4.2 Water Resource Monitoring

4.2.1 *Need for Monitoring*

The importance of water as a resource, therefore, calls for sound environmental conservation measures that enhance its protection and management. It is in relation to this that the World Bank, as an emerging priority of its lending framework, decided to broaden the development focus in its 1993 "Water resource management policy paper" to include the *protection* and *management* of water resources in an environmentally sustainable, socially acceptable, and economically efficient manner [43]. The protection and management of water resources calls for an elaborate and well established management and monitoring program, e.g., [50, 51].

Information about water resources and the environment is inherently geographic. Maps, whether on paper or in digital Geographical Information System (GIS, see e.g., [50, 51]) formats, continue to be the medium for the expression of engineering plans and designs. This is because we are basically concerned about the spatial distribution and character of the land and its waters. Johnson [52] argues that weather patterns, rainfall and other precipitation, and resultant water runoff are primary driving forces for land development, water supplies, and environmental impacts and pollution. Our water resources systems comprise dams and reservoirs, irrigated lands and canals, water supply collection and distribution systems, sewers and storm water systems, and flood plains. These systems are designed in response to a complex mix of topography and drainage patterns, population and land use, sources of water, and related environmental factors [52, 95].

In general, the planning and engineering design processes used in the development and management of water resources involve different levels of data abstraction. Data are collected and used to characterize the environment at some level of detail, or scale. In seeking to make decisions about plans and designs, data must be collected to describe the resource, and procedures or models must be developed to predict the resultant changes. These data and models help us understand the real world, and this understanding guides our decision making [52].

According to Taylor and Alley [53], essential components of a water level monitoring program include; *selection of observation wells*, *determination of the frequency of water level measurements*, *implementation of quality assurance*, and *establishment of effective practices for data reporting*. In selecting the observation wells, the authors state that the decisions made about the number and locations of observation wells are crucial to any water-level data collection program [53], see also [94]. In regard to locations, Global Navigation Satellite Systems (GNSS) satellites [54, 55] could contribute in generating a fast and accurate survey of well location-based data. These data could then be integrated with other information such as water levels in a GIS system to enhance the accessibility of water level data, where the GIS plays the role of depicting the locations of the observed wells relative to pertinent geographic, geologic, or hydrologic features, e.g., [53].

Taylor and Alley [53] present areas where the monitored ground water levels could be used. Some of these include: determination of the hydraulic properties of aquifers (aquifer tests); mapping of the altitude of the water table or potentiometric surface; monitoring of the changes in groundwater recharge and storage, e.g., [94]; monitoring of the effects of climatic variability; monitoring of the regional effects of groundwater development; statistical analysis of the water level trends; monitoring of the changes in groundwater flow directions; monitoring of the groundwater and surface water interaction; and numerical (computer) modeling of groundwater flow or contaminant transport.

Information on the spatial and temporal behaviour of terrestrial water storage, therefore, is crucial for the management of local, regional and global water resources. This information will [56]:

- Enhance sustainable utilization of water resources by, e.g., farmers, urban consumers, miners, etc.
- Guide water resource managers and policy makers in the formulation of policies governing its sustainable use, conservation and management. In particular, State water managers are more informed in regulating the utilization of water, e.g., for industrial and irrigation purposes.
- Benefit local environmental monitoring, management policies and practices that ensures a balance between sustainable utilization and environmental conservation and protection. Changes in water availability impacts upon the environment in several ways, e.g., any significant imbalance in its level affects the ecological system by influencing salinity, land subsidence, and the vulnerability of wetlands ecosystem among others.
- Benefit various government agencies at various levels (national, provincial, and local) by providing data that enhance and compliment their works. Such agencies include departments of *water*, *agriculture*, *weather forecasting* and *climate* studies, and so forth.

The conservation and management of water is of paramount importance in areas with arid or semi-arid climates, which include many parts of Australia, especially in times of severe drought, as experienced in Murray Darling Basin [56]. In 2006, Australia faced its worst drought in a century as was seen from daily reports that were emerging in both the local and international media. A more grim picture of the future of the water situation for Australia was to follow from the IPCC [57] report, which stated that Australia's water crisis will worsen in the coming years due to drought! There clearly exists an urgent need to have efficient monitoring technique(s) that will enhance the analysis of water scarcity at river basin or more localized scales. Indeed, this argument is supported by Rijsberman [4] who states that the global analysis of water scarcity is of very limited use in assessing whether individual or communities are water secure. To this effect, Rijsberman [4] states:

> The river basin is more and more adopted as the appropriate scale to understand the key processes with increasing water scarcity as human use goes up to the point where basins "close".

One such technique that has emerged supreme in monitoring changes in stored water at river basin scales, is the Gravity Recovery and Climate Experiment (GRACE) satellites [58] discussed in details in Sect. 3.3.3; see details in [50, 51, 54, 55, 93].

Timely and precise information on the changes in stored water at smaller (localized) scales of economical values, e.g., urban consumption, agriculture, industries, and mining to within 10 to 14 days (so far achievable by GRACE satellite) will enhance sustainable conservation and management of this precious *dwindling* resource.

The availability of techniques that delivers information on the changes in stored water at a more local scale, is the first step towards realizing an efficient water society.

Water resource managers are able to make decisions based on timely and accurate knowledge; thereby saving considerable resources that are often spent as a penalty of inefficient decisions based on lack of information. In the south-western wheat belt of Australia, for example, accurate knowledge of changes in stored water will be beneficial to the sustainable utilization of water, while at the same time realizing the economic contribution of wheat farming to the overall Gross Domestic Product (GDP). A blind focus on the GDP's growth without paying attention to the state of salient contributors such as water stored in aquifers is detrimental, since a fall in the amount of the available water in such areas would definitely mean reduced yields.

Since the entire system of stored water is coupled within the hydrological cycle (Fig. 3.3), hydrologists will be in a position to better understand their local hydrological cycle, thanks to information at localized levels. Hydrologists will also be able to use such information to refine and calibrate local-scale models, e.g., rainfall runoff models [59], for further improvement in their hydrological cycles. This will also contribute to our understanding of the impacts of climate change on regional and global hydrological cycles.

Environmental studies have a chance of greatly benefiting from information about changes in stored water. It is widely acknowledged that stored water (surface and groundwater) plays a key role in sustaining natural biodiversity and the functioning of the environment as a whole. Knowledge of the changes in water level is therefore essential for the very survival of the entire ecosystem, which could be adversely affected by extreme changes in stored water. In wetlands, for example, some vegetation and ecosystems have been known to respond to water level fluctuations [60].

Accurate monitoring of changes in stored water at smaller wetland scales will thus help in the preservation and conservation of such wetland ecosystems. Changes in water level also brings with it environmental phenomena such as salinity, compacting of aquifers due to the removal of water causing land subsidence, and changes in

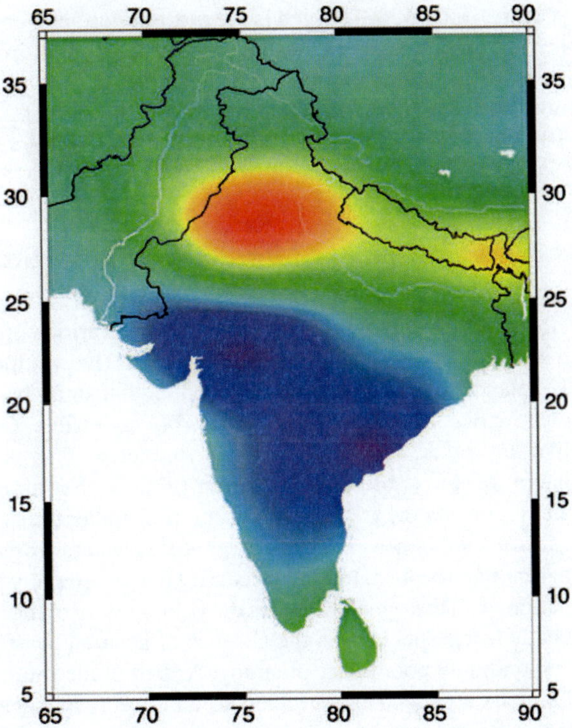

Fig. 4.2 Groundwater changes in India (2002–2008) with losses in red and gains in blue, based on GRACE satellite observations. The estimated rate of depletion of groundwater in northwestern India was 4.0 centimeters of water per year, equivalent to a water table decline of 33 centimeters per year. Increases in groundwater in southern India was due to above-average rainfall, whereas rain in northwestern India was close to normal during that study period. *Source* NASA (I. Velicogna/UC Irvine)

the properties of the top 5 cm of soil. Information on changes in stored water thus contributes enormously to the environmental conservation and protection. Remote sensing and GIS [50, 51] can be used to monitor the water quality. This is possible by employing multispectral, multi-temporal image data and analyzing parameters such as the distribution of suspended sediment, turbidity and chlorophyll. These indicators can be determined through regression analysis.

4.2.2 Monitoring of Stored Water at Basin Scales

In Awange and Kiema [50, 51, 54, 55], the *geoid* is introduced as a fundamental physical surface to which all observations are referred to if they depend on *gravity*, and whose shape is influenced by inhomogeneous mass distribution within the interior of the Earth [61, p. 29]. In Chap. 3, the concept of *gravity field variations* is related to hydrological processes. Measurements of the time-varying gravity field by LEO (low earth orbiting) satellites, e.g., GRACE discussed in Sect. 3.3.3 are the key to the contribution of space monitoring of changes in water levels at basin scales, see e.g., [62]. Such techniques now enable the monitoring of groundwater recharge, see e.g., [63, 64, 93], which is the most important element in groundwater resources management and could also be applicable to monitoring salinity management measures at the catchment level. For example, in 2009, GRACE satellites showed that north-west of India's aquifers had fallen at a rate of 0.3048 m yr^{-1} (a loss of about 109 km^3 per year) between 2002 and 2008, see Fig. 4.2.[1]

4.3 Importance of Monitoring GHA's Stored Water

Greater Horn of Africa (GHA) is home to various great lakes (e.g., Victoria, Tanganyika, Kyoga, Tana; Fig. 1.1b) as well as the continent's water towers (Ethiopian Highlands). It has groundwater potential that so far has not been fully tapped. However, unlike surface water that is readily accessible, GHA's groundwater in storage like any other, cannot all be fully extracted. This is further compounded by the nature of the aquifers in the region [93]. Furthermore, these stored water (surface and groundwater) are increasingly coming under threat from extreme climate as well as human influence [1, 2]. Yet these stored waters are essential in propelling irrigated agriculture to alleviate food insecurity that perennially bedevils the region. A combination of all these factors shows the vulnerability of the region in facing water scarcity challenges, especially in the absence of accurate water monitoring system and correspondingly poor future planning. As part of the solution to water scarcity problem, quick fix approach has revolved around supply management approach where infrastructures, e.g., dams have been constructed or expanded to increase the available water supply. Rijsberman [4] argue, however, that although the quick fix approach has largely succeeded in producing cheap food, water supply, and sanitation to a large number of people, many people still do not have access to safe and affordable drinking water despite huge investments. Rijsberman [4] states:

> ... close to half the world population lacks access to sanitation, many rural poor do not have access to water for productive purposes, groundwater levels in key aquifers are falling rapidly, many rivers are no longer reaching the sea, etc.

These quick fix measures are only temporary and do not provide long term solution to the locals. In addition, damming of the rivers contribute to geomorphological changes of the same rivers. As a shift from quick fix school of thought, Rijsberman [4] points to the emergence of integrated water resource movement that has brought about organizations such as the World Water Council and the Global Water Partnership that is pushing for a demand management approach seeking;

1. to involve users more in the management of water, often through the establishment of forms of water user associations;
2. to price water and/or make it a trade-able commodity. This arguably would help with its management in that people will use water efficiently when they know that they will eventually pay for it. In Australia for example, water use for irrigating

[1] The Economist, September 12th 2009, pp. 27–29: Briefing India's water crisis.

gardens and lawns are regulated and the use costed in order to efficiently use water, a measure that seems to be working in guarding against excessive water use. In countries where individuals are unable to pay for water, governmental subsidies should be in place so that no one misses out on this precious resource; and

3. to establish river basin authorities that integrate the usually fragmented government responsibilities for water into a single authority responsible for a hydrographically defined area, the river basin.

In support of the third item above, an elaborate monitoring is affordable only through the use of satellites. It is in support to item (3) above that satellites have been recognized as having the potential to provide space-based estimate of changes in terrestrial water storage. In essence, they are tools that assist water managers in conserving and controlling the utilization of dwindling water resources in a sustainable way. Water is arguably one of the most precious resource in the world, therefore, it is logical to try to monitor its distribution as efficiently as possible, and space techniques offer such opportunity [12, 73–76]. This is because one of the environmentally important signals detected by satellites such as GRACE is the temporal gravity field, e.g., gravity field variation induced by changes in the distribution of water on and below the Earth's surface, i.e., hydrology, e.g., [76]. Satellite altimetry discussed in Sect. 3.5 provides the possibility of monitoring sea or lake surface heights as was demonstrated for Lakes Naivasha and Victoria within GHA [77]. Other studies undertaken with respect to use of GRACE to monitor hydrology include, e.g., [56, 63, 78–81, 93].

Moreover, ***Lake Victoria*** (Fig. 5.1), the region's giant and the world's second largest freshwater lake, and the largest in the developing world, is the source of the White Nile (also known as the Victoria Nile) and a resource shared by the three East African countries: Kenya, Uganda and Tanzania. It is a source of water for irrigation, which if sustainably utilized, can enormously contribute to enhancing irrigated agriculture and help towards alleviating perennial food insecurity in the region. It is also vital for transportation, domestic and livestock uses, and supports the livelihood of more than 42 million people who live around it [1, 2, 65, 96, 97]. Its fish products, (i.e., Tilapia and Nile Perch) are exported the world over [65]. Its role as an indicator of environmental and climate change on long-term scales together with its global significance are documented, e.g., in Nicholson et al., [68], Awange and Ong'ang'a [65], and [1, 2]. Since the 60s, the lake level has exhibited fluctuations as pointed out by Nicholson [66, 67] . The sharpest rise in the lake water level occurred during the El'Niño rains of early 60s and 1997/1998. Some reports, e.g., [69] suggest that the lake level rose by 2.5 m following the 1960s floods. Kite [70] attributed this rise to over-lake precipitation.

Although the lake has continued to attract worldwide attention due to its significance and other environmental phenomenon such as water hyacinth (see Fig. 5.5 on Sect. 5.4), in the last decade, and perhaps most threatening, Lake Victoria water level receded at an alarming rate causing concerns as to whether the lake was actually drying up as it happened in the pre-historic time, see e.g., [12]. According to Kull [71], the lake levels dropped to more than 1.1 m below the 10-year average. Water levels have remained above average for more than 40-years, but the 2002–2006 water levels were below normal and the lowest level since September 1961. The socio-economic impacts of this drastic fall in Lake Victoria water level have been reported in Awange et al. [12, 72]. At the time of writing this book (2020–2021), the lake level has risen to a point where people are wondering whether it is retracing its shorelines of the 60s.[2] Khaki and Awange [13] attribute this to the effect of the Indian Ocean Dipole (IOD) climate variability index. Continuous monitoring of the lake, therefore, is paramount to its sustainable use. A special coverage of monitoring Lake Victoria from space is presented in Awange [1, 2].

The aim of this book as already pointed out in Sect. 1.5, therefore, among others, is to provide a space (satellite) monitoring of entire GHA's total water storage (surface, groundwater, soil moisture and vegetation water) changes.

References

1. Awange JL (2021) Lake Victoria monitored from space. Springer Nature International
2. Awange JL (2021) Nile waters. Weighed from space. Springer Nature International
3. Molen I, Hildering A (2005) Water: cause for conflict or co-operation? ISYP J Sci World Affairs 1(2):133–143
4. Rijsberman FR (2006) Water scarcity: fact or fiction. Agricultural water management, vol 80. Elsevier, pp 5–22
5. Jury WA, Vaux HJ Jr (2007) The emerging global water crisis: managing scarcity and conflict between water users. Advances in agronomy, vol 95. Elsevier Inc, 77p. https://doi.org/10.1016/50065-2113(07)95001-4
6. Anyah RO, Forootan E, Awange JL, Khaki M (2018) Understanding linkages between global climate indices and terrestrial water storage changes over Africa using GRACE products. Sci Total Environ 635:1405–1416. https://doi.org/10.1016/j.scitotenv.2018.04.159
7. Vörösmarty CJ, Green P, Salisbury J, Lammers RB (2000) Global water resources: vulnerability from climate change and population growth. Science 289(5477):284–288. https://doi.org/10.1126/science.289.5477.284
8. Phillips T, Nerem RS, Fox-Kemper B, Famiglietti JS, Rajagopalan B (2012) The influence of ENSO on global terrestrial water storage using GRACE. Geophys Res Lett 39:L16705. https://doi.org/10.1029/2012GL052495
9. Schewe J, Heinke J, Gerten D, Haddeland I, Arnell NW, Clark DB, Dankers R, Eisner S, Fekete BM, Colón-Gonzlez FJ, Gosling SN, Kim H, Liu X, Masaki Y, Portmann FT, Satoh Y, Stacke T, Tang Q, Wada Y, Wisser D, Albrecht T, Frieler K, Piontek F, Warszawski L, Kabat P (2013) Multimodel assessment of water scarcity under climate change. PNAS 111(9):3245–3250. https://doi.org/10.1073/pnas.1222460110
10. Hurkmans R, Troch PA, Uijlenhoet R, Torfs P, Durcik M (2009) Effects of climate variability on water storage in the Colorado River Basin. J Hydrometeorol 10:1257–1270. https://doi.org/10.1175/2009JHM1133.1
11. Ndehedehe CE, Awange J, Kuhn M, Agutu N, Fukuda Y (2017) Climate teleconnections in Uence on West Africa's terrestrial water storage. Hydrol Process 31(18):3206–3224. https://doi.org/10.1002/hyp.11237

[2] http://news.trust.org/item/20200819141141-rb3c8/.

12. Awange JL, Sharifi M, Ogonda G, Wickert J, Grafarend EW, Omulo M (2008) The falling Lake Victoria water level: GRACE. Water resource management. TRIMM and CHAMP Satellite Analysis. https://doi.org/10.1007/s11269-007-9191-y
13. Khaki M, Awange J (2021) The 2019–2020 rise in Lake Victoria monitored from space: exploiting the state-of-the-art grace-fo and the newly released ERA-5 reanalysis products. Sensors 21(13):4304. https://doi.org/10.3390/s21134304
14. Van Loon AF, Stahl K, Di Baldassarre G, Clark J, Rangecroft S, Wanders N, Gleeson T, Van Dijk AIJM, Tallaksen LM, Hannaford J, Uijlenhoet R, Teuling AJ, Hannah DM, Sheffield J, Svoboda M, Verbeiren B, Wagener T, Van Lanen HAJ (2016) Drought in a human-modiffied world: reframing drought deffinitions, understanding, and analysis approaches. Hydrol Earth Syst Sci 20(9):3 631–3650. https://doi.org/10.5194/hess-20-3631-2016.
15. Konar M, Evans TP, Levy M, Scott CA, Troy TJ, Vörösmarty CJ, Sivapalan M (2016) Water resources sustainability in a globalizing world: who uses the water? Hydrol Process 30(18):3330–3336. https://doi.org/10.1002/hyp.10843
16. Hall JW, Grey D, Garrick D, Fung F, Brown C, Dadson SJ, Sadoff CW (2014) Coping with the curse of freshwater variability. Science 346(6208):429–430. https://doi.org/10.1126/science.1257890
17. Alsdorf DE, Lettenmaier DP (2003) Tracking freshwater from space. Science 301(5639):1491–1494. https://doi.org/10.1126/science.1089802
18. Alsdorf D, Lettenmaier D, Vörösmarty C (2003) The need for global, satellite based observations of terrestrial surface waters. EOS Trans Am Geophys Union 84(29):269–276. https://doi.org/10.1029/2003EO290001
19. Alsdorf DE, Rodrguez E, Lettenmaier DP (2007) Measuring surface water from space. Rev Geophys 45(2):RG2002. https://doi.org/10.1029/2006RG000197
20. Vörösmarty C, Askew A, Grabs W, Barry RG, Birkett C, Döll P, Goodison B, Hall A, Jenne R, Kitaev L, Landwehr J, Keeler M, Leavesley G, Schaake J, Strzepek K, Sundarvel SS, Takeuchi K, Webster F (2001) Global water data: a newly endangered species. EOS Trans Am Geophys Union 82(5):54–58. https://doi.org/10.1029/01EO00031
21. Gleeson T, Wada Y, Bierkens MFP, van Beek LPH (2012) Water balance of global aquifers revealed by groundwater footprint. Nature 488:197–200. https://doi.org/10.1038/nature11295
22. UNEP (2002) A world of salt: total global saltwater and freshwater estimates. http://www.unep.org/dewa/assessments/ecosystems/water/vitalwater/freshwater.htm. Accessed 25 Aug 2010
23. Hofman AR (2004) The connection: water and energy security. http://www.iags.org/n0813043.htm. Accessed 25 Aug 2010
24. Gleick PH (1993) Water and conflict: freshwater resources and international security. Int Secur 18(1):79–112. https://doi.org/10.2307/2539033
25. Gleick PH (2000) The world's water: Biennial report on freshwater resources 2000–2001. Island Press, Washington, DC
26. Agola NO, Awange JL (2014) Globalized poverty and environment 21st century challenges and innovative solutions. Springer, Berlin, Heidelberg
27. Kuylenstierna JL, Bjrklund G, Najlis P (1997) Sustainable water future with global implications: everyone's responsibility. Nat Res Forum 21(3):181–190. https://doi.org/10.1111/j.1477-8947.1997.tb00691.x
28. Vörösmarty CJ, Douglas EM, Green PA, Revenga C (2005) Geospatial indicators of emerging water stress: an application to Africa. Ambio
29. Duraiappah A (1998) Poverty and environmental degradation: a review and analysis of the nexus. J Hydrol 113(1–4):297–306
30. Awange JL, Khandu Forootan E, Schumacher M, Heck B (2016) Exploring hydro-meteorological drought patterns over the Greater Horn of Africa (1979–2014) using remote sensing and reanalysis products. Adv Water Res. https://doi.org/10.1016/j.advwatres.2016.04.005
31. Awange JL, Ferreira VG, Forootan E, Khandu Andam-Akorful SA, Agutu NO, He XF (2016) Uncertainties in remotely sensed precipitation data over Africa. Int J Climatol 36(1):303–323. https://doi.org/10.1002/joc.4346

32. Omondi P, Awange J, Ogallo L, Ininda J, Forootan E (2013) The influence of low frequency sea surface temperature modes on delineated decadal rainfall zones in Eastern Africa region. Adv Water Res https://doi.org/10.1016/j.advwatres.2013.01.001
33. Omondi P, Awange J, Ogallo LA, Okoola RA, Forootan E (2012) Decadal rainfall variability modes in observed rainfall records over East Africa and their relations to historical sea surface temperature changes. J Hydrol 464–465, 140–156. https://doi.org/10.1016/j.jhydrol.2012.07.003
34. IRIN humanitarian news and analysis, UN office for coordination of humanitarian affairs (2006) Global: the global water crisis: managing a dwindling resource. http://www.irinews.org. Accessed 25 Sept 2011
35. Alcamo J, Döll P, Kaspar F, Siebert S (1997) Global change and global scenarios of water use and availability: an application of WaterGAP 1.0. University of Kassel, CESR, Kassel, Germany
36. Alcamo J, Henrichs T, Rosch T (2000) World water in 2025: global modeling and scenario analysis. In: Rijsberman FR (ed) World Water Scenar Anal. World Water Council, Marseille, France
37. Wallace JS (2000) Increasing agricultural water efficiency to meet future food production. Agric Ecosyst Environ 82:105–119
38. Wallace JS, Gregory PJ (2002) Water resources and their use in food production. Aquat Sci 64:363–375
39. Yang H, Reichert P, Abbaspour K, Zehnder AJB (2003) A water resources threshold and its implications for food security. Environ Sci Technol 37:3048–3054
40. Rekacewicz P (2006) Increased global water stress. Vital water graphics 2. Le monde diplomatique. http://www.grida.no/publications/vg/water2/page/3289.aspx. Accessed 15 Apr 2012
41. Neff RA, Parker CL, Kirschenmann FL, Tinch J, Lawrence RS (2011) Peak oil, food systems, and public health. Am J Public Health 101(9):1587–1597
42. Payne WA (2010) Farming systems and food Security in sub-saharan Africa. In: Lal R, Stewart BA (eds) Food security and soil quality. Taylor and Francis, New York
43. World Bank (2003) Water resource and environment. In: Davis R, Hirji R (eds), Technical note G.2, Lake Management
44. Carolina BF (2002) Competition over water resources: analysis and mapping of water-related conflicts in the catchment of Lake Naivasha (Kenya). MSc thesis, ITC
45. Birnie P, Boyle A (1993) International law & the environment. Cambridge Law J 52(3):540p. Cambridge University Press
46. Gleick PH (1999) The human right to water. Water Policy 1:487–503
47. Schelton D (1991) Human rights, environmental rights, and the right to environment. Stanford J Int Law 28:103–138
48. Zehnder AJB, Yang H, Schertenleib R (2003) Water issues: the need for action at different levels. Acquat Sci: Res Across Bound 65(1):1–20. https://doi.org/10.1007/s000270300000
49. Hanjra MA, Qureshi ME (2010) Global water crisis and future food security in an era of climate change. Food Policy 35(5):365–377
50. Awange JL, Kiema JBK (2013) Environmental geoinformatics. Monitoring and management. Springer, Berlin, New York
51. Awange JL, Kiema JBK (2018) Environmental geoinformatics. Extreme hydro-climatic and food security challenges: exploiting the big data, 2nd edn. Springer International Publishers
52. Johnson LE (2009) Geographic information systems in water resources engineering. CRC Press Taylor & Francis Group. ISBN 978-1-4200-6913-6
53. Taylor CJ, Alley WM (2001) Ground-water-level monitoring and the importance of long-term water-level data. US Geological Survey Circular 1217, Denver, Colorado
54. Awange JL (2012) Environmental monitoring using GNSS. Global navigation satellite system. Springer, Berlin, New York
55. Awange JL (2018) GNSS environmental sensing. Revolutionizing environmental monitoring. Springer International

56. Rieser D, Kuhn M, Pail1 R, Anjasmara IM, Awange J (2010) Relation between GRACE-derived surface mass variations and precipitation over Australia. Australian J Earth Sci 57(7):887–900. https://doi.org/10.1080/08120099.2010.512645
57. IPCC (Intergovernmental Panel on Climate Change) (2007) Contribution of working Group I to the fourth assessment report
58. Tapley BD, Bettadpur S, Ries JC, Thompson PF, Watkins MM (2004) GRACE measurements of mass variability in the Earth system. Science 305:503–505. https://doi.org/10.1126/science.1099192
59. Ellett KM, Walker JP, Rodell M, Chen JL, Western AW (2005), GRACE gravity fields as a new measure for assessing large-scale hydrological models, in MODSIM 2005 international congress on modelling and simulation. In: Zerger A, Argent RM (eds) The modelling and simulation society of Australia and New Zealand, Dec 2005, pp 2911–2917. ISBN: 0-9758400-2-9
60. Casanova MT (1994) Vegetative and reproductive responses of charophytes to water-level fluctuations in permanent and temporary wetlands in Australia. Australian J Marine Freshw Res 45:1409–1419
61. Leick A (2004) GPS satellite surveying, 3rd edn. Wiley, New York
62. ILEC (International Lake Environment Committee) (2005) Managing lakes and their basins for sustainable use. A report for the lake basin managers and stakeholders. International Lakes Environmental Committee Foundation: Kusatsu, Japan
63. Awange LJ, Gebremichael M, Forootan E, Wakbulcho G, Anyah R, Ferreira CG, Alemayehu T (2014) Characterization of Ethiopian mega hydrogeological regimes using GRACE, TRMM and GLDAS datasets. Adv Water Res 74:64–68. https://doi.org/10.1016/j.advwatres.2014.07.012
64. Hu K, Awange JL, Khandu Forootan E, Goncalves RM, Fleming K (2017) Hydrogeological characterisation of groundwater over Brazil using remotely sensed and model products. Sci Total Environ 599–600:372–386. https://doi.org/10.1016/j.scitotenv.2017.04.188
65. Awange JL, Ong'ang'a O (2006) Lake Victoria-Ecology. Resource of the Lake Basin and environment. Springer, Berlin
66. Nicholson SE (1998) Historical fluctuations of Lake Victoria and other lakes in the Northern Rift Valley of East Africa. In: Lehman JT (ed) Environmental change and response in East African lakes. Kluwer, Dordrecht, pp 7–35
67. Nicholson SE (1999) Historical and modern fluctuations of lakes Tanganyika and Rukwa and their relationship to rainfall variability. Clim Change 41:53–71. https://doi.org/10.1023/A:1005424619718
68. Nicholson SE, Yin X, Mamoudou BA (2000) On the feasibility of using a lake water balance model to infer rainfall: an example from Lake Victoria. Hydrol Sci J 45(1):75–95. https://doi.org/10.1080/02626660009492307
69. Aseto O, Ong'ang'a O, Awange JL (2003) Poverty. A challenge for the Lake Victoria basin. OSIENALA Series 5, Printed by Africa Herald Publishing House, Kendu Bay, Kenya
70. Kite GW (1982) Analysis of Lake Victoria levels. Hydrol Sci J 27(2):99–110. https://doi.org/10.1080/02626668209491093
71. Kull D (2006) Connections between recent water level drops in Lake Victoria, dam operations and drought. http://www.irn.org/programs/nile/pdf/060208vic.pdf. Accessed 25 Sept 2011
72. Awange JL, Aluoch J, Ogallo L, Omulo M, Omondi P (2007) Frequency and severity of drought in the Lake Victoria region (Kenya) and its effects on food security. Clim Res 33:135–142. https://doi.org/10.3354/cr033135
73. Awange JL, Forootan E, Kuhn M, Kusche J, Heck B (2014) Water storage changes and climate variability within the Nile Basin between 2002 and 2011. Adv Water Res 73:1–25. https://doi.org/10.1016/j.advwatres.2014.06.010
74. Awange JL, Anyah R, Agola N, Forootan E, Omondi P (2013) Potential impacts of climate and environmental change on the stored water of Lake Victoria Basin and economic implications. Water Res Res 49:8160–8173

75. Awange JL, Fleming KM, Kuhn M, Featherstone WE, Heck B, Anjasmara I (2011) On the suitability of the 4° × 4° GRACE mascon solutions for remote sensing Australian hydrology. Remote Sens Environ 115:864–875. https://doi.org/10.1016/j.rse.2010.11.014
76. Awange JL, Sharifi MA, Baur O, Keller W, Featherstone WE, Kuhn M (2009) GRACE hydrological monitoring of Australia. Current limitations and future prospects. J Spat Sci 54(1):23–36. https://doi.org/10.1080/14498596.2009.9635164
77. Awange J, Forootan E, Kusche J, Kiema JKB, Omondi P, Heck B, Fleming K, Ohanya S, Gonçalves RM (2013) Understanding the decline of water storage across the Ramser-lake Naivasha using satellite-based methods. Adv Water Res 60:7–23. https://doi.org/10.1016/j.advwatres.2013.07.002
78. Andersen OB, Seneviratne SI, Hinderer J, Viterbo P (2005) GRACE-derived terrestrial water storage depletion associated with the 2003 European heat wave. Geophys Res Lett 32(L18405):2–5. https://doi.org/10.1029/2005GL023574
79. Forootan E, Awange J, Kusche J, Heck B, Eicker A (2012) Independent patterns of water mass anomalies over Australia from satellite data and models. Remote Sens Environ 124:427–443. https://doi.org/10.1016/j.rse.2012.05.023
80. Forootan E, Khaki M, Schumacher M, Wulfmeyer V, Mehrnegar N, van Dijk AIJM, Brocca L, Farzaneh S, Akinluyi F, Ramillien G, Shum CK, Awange J, Mostafaie A (2019) Understanding the global hydrological droughts of 2003–2016 and their relationships with teleconnections. Sci Total Environ 650:2587–2604. https://doi.org/10.1016/j.scitotenv.2018.09.231
81. Khandu E, Forootan M, Schumacher J. Awange, Miler Schmied H (2016) Exploring the influence of precipitation extremes and human water use on total water storage (TWS) changes in the Ganges-Brahmaputra-Meghna River Basin. Water Res Res 52(3):2240–2258
82. Othieno H, Awange JL (2016) Energy Resources in Africa. Distribution, opportunities and challenges. Springer International Publishing AG. https://doi.org/10.1007/978-3-319-25187-5
83. Agutu NO, Awange JL, Zerihun A, Ndehedehe CE, Kuhn M, Fukuda Y (2017) Assessing multi-satellite remote sensing, reanalysis, and land surface models' products in characterizing agricultural drought in East Africa. Remote Sens Environ 194:287–302. https://doi.org/10.1016/j.rse.2017.03.041
84. Mpelasoka F, Awange JL, Zerihun A (2018) Influence of coupled ocean-atmosphere phenomena on the Greater Horn of Africa droughts and their implications. Sci Total Environ 610–611:691–702. https://doi.org/10.1016/j.scitotenv.2017.08.109
85. Awange JL, Hu K, Khaki M (2019) The newly merged satellite remotely sensed, gauge and reanalysis-based Multi-source weighted-ensemble precipitation: evaluation over Australia and Africa (1981–2016). Sci Total Environ 670:448–465. https://doi.org/10.1016/j.scitotenv.2019.03.148
86. IPCC (2018) Global Warming of 1.5°C. An IPCC Special Report on the impacts of global warming of 1.5°C above pre-industrial levels and related global greenhouse gas emission pathways, in the context of strengthening the global response to the threat of climate change, sustainable development, and efforts to eradicate poverty [Masson-Delmotte V, Zhai P, Pörtner H-O, Roberts D, Skea J, Shukla PR, Pirani A, Moufouma-Okia W, Péan C, Pidcock R, Connors S, Matthews JBR, Chen Y, Zhou X, Gomis MI, Lonnoy E, Maycock T, Tignor M, Waterfield T (eds)]
87. IPCC (2014) Climate change 2014: synthesis report. Contribution of working Groups I, II and III to the fifth assessment report of the intergovernmental panel on climate change [Core Writing Team, Pachauri RK and Meyer LA (eds)]. IPCC, Geneva, Switzerland, 151 pp
88. Hartmann DL, Klein Tank AMG, Rusticucci M, Alexander LV, Brönnimann S, Charabi Y, Dentener FJ, Dlugokencky EJ, Easterling DR, Kaplan A, Soden BJ, Thorne PW, Wild M, Zhai PM (2013) Observations: atmosphere and surface. In: Stocker TF, Qin D, Plattner G-K, Tignor M, Allen SK, Boschung J, Nauels A, Xia Y, Bex V, Midgley PM (Hrsg) Climate change 2013: the physical science basis. Contribution of working Group I to the fifth assessment report of the intergovernmental panel on climate change. Cambridge University Press, Cambridge, UK, New York, NY, USA

89. Saji NH, Goswami BN, Vinayachandran PN, Yamagata T (1999) A dipole mode in the tropical Indian Ocean. Nature 401:360–363
90. Trenberth KE (1997) The definition of El Niño. Bull Am Meteorol Soc 78:2771–2777
91. Awange JL, Mpelasoka F, Goncalves R (2016) When every drop counts: analysis of Droughts in Brazil for the 1901–2013 period. Sci Total Environ 566–567:1472–1488. https://doi.org/10.1016/j.scitotenv.2016.06.031
92. Mpelasoka F, Awange JL, Goncalves RM (2018) Accounting for dynamics of mean precipitation in drought projections: a case study of Brazil for the 2050 and 2070 periods. Sci Total Environ 622–623:1519–1531. https://doi.org/10.1016/j.scitotenv.2017.10.032
93. Agutu NO, Awange JL, Ndehedehe C, Kirimi F, Kuhn M (2019) GRACE-derived groundwater changes over greater Horn of Africa: temporal variability and the potential for irrigated agriculture. Sci Total Environ. https://doi.org/10.1016/j.scitotenv.2019.07.273
94. Hu K, Awange JL, Kuhn M, Saleem A (2019) Spatio-temporal groundwater variations associated with climatic and anthropogenic impacts in South-West Western Australia. Sci Total Environ. https://doi.org/10.1016/j.scitotenv.2019.133599
95. Agutu NO, Awange JL, Ndehedehe C, Mwaniki MW (2020) Consistency of agricultural drought characterization over Upper Greater Horn of Africa (1982–2013): topographical, gauge density, and model forcing influence. Sci Total Environ 709. https://doi.org/10.1016/j.scitotenv.2019.135149
96. Awange JL, Saleem A, Sukhadiya RM, Ouma YO, Kexiang H (2019) Physical dynamics of Lake Victoria over the past 34 years (1984–2018): is the lake dying? Sci Total Environ 658:199–218. https://doi.org/10.1016/j.scitotenv.2018.12.051
97. Morgan B, Awange JL, Saleem A, Kexiang H (2020) Understanding vegetation variability and their "hotspots" within Lake Victoria basin for the 2003–2018 period. Appl Geograp 122. https://doi.org/10.1016/j.apgeog.2020.102238
98. Vishwakarma BD, Sneeuw N, Westaway RM, Bamber JL (2021) Re-assessing global water storage trends from GRACE time series. Environ Res Lett 16:034005. https://doi.org/10.1088/1748-9326/abd4a9

Chapter 5
GHA's Greatest Freshwater Source: Victoria

> Lake Victoria is a living Lake. From Kisumu (Kenya) to Mediterranean, millions of people are being fed by it! —Pastor Martin Mbandu, former pastor, Kisumu Pentecostal Church.

> It is here opined that its proper use could benefit agricultural efforts of GHA in multiple ways. First, however, the obsolete Nile treaties that were signed by Britain, Egypt and Sudan in the 1920s and 1950s, have to be done away with.—Joseph Awange.

5.1 Summary

Lake Victoria (Fig. 5.1a), Greater Horn of Africa (GHA)'s giant is the world's second largest freshwater lake after Lake Superior in the USA and the largest in the developing world. It is here opined that if properly utilized, it could provide more than adequate water essential to support both irrigated agriculture in terms of surface water irrigation and rain-fed agriculture in terms of modulating the regions hydroclimate as discussed in Sect. 5.2.3. It could therefore shield the region against food insecurity. It is the source of the White Nile, which flows from Jinja in Uganda (Figs. 5.1b and 5.2) to join the Blue Nile in Khartoum (Sudan) to form one Nile and provides water for irrigation, transport, domestic and livestock uses, thereby supporting the livelihood of more than 76 million people who live around it [6, 13, 19]. The presentation of this book, therefore, would not be complete without a special coverage of the Lake, which is so vital to the livelihood of the GHA's inhabitants. Nicholson [15, 16] documents its significance as an indicator of environmental and climate change over long-term scales. Since the 1960s, the lake level has experienced significant fluctuation, see e.g., [15, 16], which impacts on the overall water budget of the Nile. From 2001 to 2006, for example, its water level showed a dramatic fall that alarmed water resource managers as to whether the lake was actually drying up [23]. Kull [14] reported that the lake's levels fell by more than 1.1 m below the 10 year average. In 2019–2020,

J. Awange, *Food Insecurity & Hydroclimate in Greater Horn of Africa*,
https://doi.org/10.1007/978-3-030-91002-0_5

the lake level rose again to a point where it was thought to be tracing its roots to the 60s. Khaki and Awange [17] established that the cause of the 2019–2020 rise was due to the Indian Ocean Dipole climate variability influence. This Chapter presents the Lake and its challenges, and argues that its proper use could benefit agricultural efforts of GHA in multiple ways. First, however, the obsolete Nile treaties that were signed by Britain, Egypt and Sudan in the 1920s and 1950s, see e.g., [7] have to be done away with and an all inclusive utilization of the lake's resource agreed upon. A more detailed elaborate coverage of this precious lake, its potentials, and challenges are presented in the books of Awange [6, 7, 13].

5.2 Features of the Lake and Its Environs

5.2.1 The Origin

It is important to know, for historical and scientific purposes, the possible origins and age of Lake Victoria (Fig. 5.1). How old is Lake Victoria? How did it come about? Has it changed much during its existence? If so, how? Why should we be interested in knowing the age of Lake Victoria? These are some of the pertinent questions, which have occupied the attention of scientists for centuries; and rightly so. The origins and age of the Lake are still shrouded in deep mystery. Although scientists have grappled with these questions, no definitive answers have been forthcoming. As it is, the origins of Lake Victoria are still the subject of scientific dispute [3]. It seems likely that it is much more recent in its origins than the other great Lakes of Eastern Africa. But there is an interesting aspect of the direction of flow of waters into the Lake. There is a scientific contention that many of the rivers now flowing east into Lake Victoria (including Kagera) once flowed west.

Hickman et al. [2, p. 36] suggests that Lake Victoria was formed due to warping and tilting of the land. They note that lower reaches of some river valleys have been raised, so that they flooded, and the directions of the rivers changed at least in the Miocene, Pliocene, and part of the Pleistocene eras; within the past 2 million years, possibly eventually into the Nile system. A more recent upthrust of the western side of the basin is thought to have however reversed these rivers, and caused Lake Victoria to form by flowing eastwards [3]. Although Fuggle [4] suggest that the Lake formed sometimes during the last 400,000 years, Aseto and Ong'ang'a [3] and Ong'ang'a [9] suggest the possibility of the Lake having formed as recently as 25,000 to 35,000 years ago and dried up completely between 10,000 and 14,000 years. Samples of Earth taken from the bottom of the Lake corroborate this suggestion by indicating that the Lake may have temporarily but completely dried up 12,000 to 14,000 years ago. But this suggestion raises a host of other questions. For instance, *what happened to make the Lake dry up? What happened to make the dry Lake become a freshwater Lake? Could it dry up again in the future?* In attempting to address the issue of the drying of the Lake, Johnson et al. [8] suggest that the Lake probably dried up

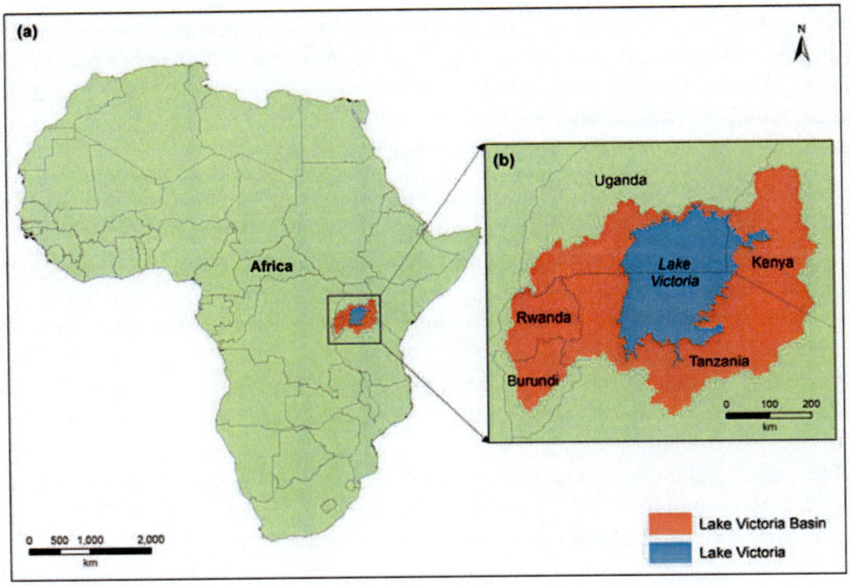

(a)

(b)

Fig. 5.1 **a** Lake Victoria Basin (LVB). *Source* Morgan et al. [1], and (**b**), The White Nile flowing from lake Victoria through Nalubale dam in Uganda (Fig. 5.2b) entirely about 15,000 years ago due to climatic phase of extreme dryness. Be that as it may, some scientists maintain that tectonic movement that swept across the Eastern African plateau about 14,600 years ago caused the formation of shallow basins, which filled with waters to become what is today ***Lake Victoria***. However, Some unresolved issues requiring further research include:

- Resolving the age of Lake Victoria since the conflicting explanations as to the age of the Lake still persists.
- Finding out whether the Lake can die under stress, as was supposed to have happened thousands of years ago?
- Continuous monitoring of fluctuation of its water level to establish the trend as to whether the lake is dying.
- Establishing whether the Lake is getting deeper or shallower with the passage of time and also whether it is expanding or decreasing, thus increasing or reducing its area of coverage.
- The Lake's parameters, e.g., length, area, etc., vary according to various authors as will be seen in Sect. 5.2.3. Use of satellite techniques discussed in Chap. 3 can resolve this once and for all as demonstrated by Awange et al. [5].

5.2.2 *The Name "Lake Victoria"*

We elaborate in this section how the name Lake Victoria was coined. Is the name Lake Victoria a proper name for the Lake? What is in a name? Does a name matter? If so, how and why? From time immemorial, these questions have been asked over and over again, and in different contexts. The general answer is that giving a name to an entity is an extremely important decision that people can make, in that, it is a means of providing an identity. A good name is given to a good thing; and a bad name is usually given to a bad thing. In the Luo language in Kenya, it is taken axiomatically that "Nying' ema ichiemogo". which is translated to mean "one eats by the fame of one's name". It is for this reason that human beings name their children after important persons, or give their children the names of their parents and fore parents to maintain their genealogy, or after important events, or in honour of important events or entities. It is also the reason why human beings give names to animals, hills, valleys, and stars. And even at an advanced age, individuals can acquire new names. Some are given names such as "sibuor" (Luo name for a lion) in recognition of one's bravery, etc.

The inhabitants around the Lake have followed suit; they always had different names for the Lake. The Lake is also known as Victoria Nyanza. This appears to have been derived from the Basuba (a tribe in Kenya) name for the Lake: "Nyancha" or "Nyanja" by certain communities in Uganda. The Luos on the Kenya side call it "Nam (Namb) Lolwe" (endless waters). The Baganda in Uganda name it "Nalubaale", the Sukuma in Tanzania call it "Sukuma Lake". This being the case, why then is the Lake called Lake Victoria?

Fig. 5.2 **a** Pillar signifying the source of the White Nile erected by John H. Speke in 1858 in Lake Victoria, **b** exit of the White Nile through Nalubaale dam, (**c**), flow of the White Nile downstream, and (**d**), off goes the White Nile from Jinja towards Lake Kyoga all the way to Sudan, Egypt and into the Mediterranean Sea

The name *Victoria* is "foreign" in that it was named in honour of an individual in a foreign land. "Foreigners" had different names for the Lake. The Arabs called it *"Ukerewe"*. But an Englishman, explorer John H. Speke, stumbled onto its southern shore and proclaimed he had discovered the fabled source of River Nile in 1858. He built a pillar at the site (Fig. 5.2a) and being the first European to sight it, he decided to honour Queen Victoria of England by naming it after her. One recalls that during 1890s, Great Britain colonized Kenya and Uganda while the Germans had a stronghold of Tanzania's mainland in 1899. In 1875, an American journalist, Henry Morton Stanley, seeking to confirm Speke's claim and looking for the British missionary Dr. David Livingston, circumnavigated the Lake. He spent two weeks spinning tales of God and England to curry favour with Mtesa, king of Buganda and ruler of the northern Lake region. Then Stanley sent word back to England, calling for missionaries. They came with soldiers and traders. He popularized the name Lake Victoria in the Western world. And the name Lake Victoria has continued to-date. Nevertheless, African countries around the Lake have not been content with the present name of Lake Victoria. Thus, soon after the independence of the three surrounding East African states of Uganda, Kenya and Tanzania, there were attempts to rename the Lake so as to give it a proper African name; but such attempts have not yet succeeded, and the name Lake Victoria has been retained. It may be just a matter of time before the Lake is given its rightful appropriate African name that would give its true African identity and character. After all, changing the name of the Lake from a 'foreign' tag to an African name should not be a problem. Several examples of name change abound in East Africa and elsewhere in the continent and in the world. In the case of Kenya, African names have replaced foreign names. For instance, upon the country's independence on December 12, 1963, all landmarks, monuments and streets in Kenya that bore foreign names were given local, or to put it bluntly, African names. Lord Delamere Avenue in Nairobi was renamed Kenyatta Avenue. Queen's Way was renamed Mama Ngina Street. Government Road became Moi Avenue. And Duke Street became Ronald Ngala Street. Similar examples abound in the country. Thus changing the name of the Lake from a foreign one to an African name is in order and is consistent with the tradition of name changing initiated since independence.

Could the name "***Nyanza***" be a good candidate for the changed name, given that originally, the natives in the Suba District called it by that name? Already, the communities around the Lake expressed their desire to rename the Lake. At a cultural conference held in Mwanza, Tanzania, in 2001, the cultural leaders selected from major ethnic communities around Lake Victoria questioned the rationale for retaining the name Victoria. Following extensive discussions on the matter, the leaders recommended that the Lake be renamed. Their argument was that the current name makes their Lake sound foreign and detached from their cultures.

5.2.3 Lake Victoria Basin: Physical Description

Here, only a brief background of lake Victoria is presented. For more details, we refer the reader to the books and article of [3, 5, 7, 13]. Lake Victoria is the largest freshwater Lake in Africa (also largest tropical Lake) and the second largest in the world after Lake Superior in the United States. The Lake is also the largest in the developing world. It is located at Longitude 31° 39'E–34° 53' E and Latitude 0° 20'N–3° S, and has a surface area of c.a. 69,295 km^2 [5]. Its greatest length is about 400 km and its breadth is about 320 km.[1] Using a high spatio-temporal resolution Sentinel-2 remotely sensed images (see Sect. 3.2), Awange et al. [5] established the 2018 values to be 386 km (length) and 362 km (width).[2] It contains about 2,760 cubic kilometers of water and is situated at an altitude of 1135 m above mean sea level. The shoreline is very irregular and totals some 3,300 km in length. Hughes and Hughes [10] have put the shoreline close to 3,500 km in length, which is less than

[1] Other sources, e.g., Britannica Concise put the length at 337 km and width at at 240 km.

[2] The mean values of Awange et al. [5] from the 1984, 2002, 2017 and 2018 images were 388 km for length and 364 km for width respectively.

the 2018 value of 4.572 km established by Awange et al. [5]. The shoreline on the Kenyan side is 760 km. Much of the Lake is relatively shallow, reaching a maximum depth of about 80 m (Fuggle [4] puts the figure to 85 m), and an average depth of about 40 m; the deepest zone (60–90 m)[3] lies toward the shore. There is little annual variation in water temperatures, the mean surface being about 24 °C, and that of deeper waters to be 23 °C.

Its basin (Fig. 5.1) spreads from Longitude 30° E–35 ° 50' E to Latitude 0 ° N–3 ° 20'S. The ***shoreline*** is comprised as follows: 17% in Kenya, 33% in Tanzania and 50% in Uganda. Of the total catchments area of 193,000 km^2, Burundi accounts for 7.2%, Kenya 21.5%, Rwanda 11.4%, Tanzania 44.0% and Uganda 15.9%. The entire drainage basin covers an area of 258,000 km^2. Awange et al. [5] show the lake's mean surface area to be 69,295 km^2 (i.e., 812 km^2 or 1.2% more than that of the 37 publications that they reviewed) and its 2018 value to be 69,216 km^2 (i.e., ~733 km^2 (1.1%) more than the mean of the 37 publications). Details of this study are presented in [5, 6]. The Lake is also the source of the river Nile (i.e., the source of the White Nile while the Blue Nile originates from Ethiopia), arguably the world's longest river. The Lake's main physical parameters have been summarized by [5, 6, 11] in Table 5.1.

Because the Lake is shallow, its volume is substantially less than that of other Eastern African lakes with much smaller surface areas. The volume of Lake Victoria's water is only 15% of the volume of Lake Tanganyika, even though the latter has less than half the surface area. The Lake straddles the Equator and touches the Equator in its northern reaches. It is shared by Kenya (6%), Uganda (45%) and Tanzania (49%). The Lake's shoreline is long and convoluted, enclosing innumerable small, shallow bays and inlets, many of which include swamps and wetlands, which differ a great deal from one another and from the Lake itself.

Lake Victoria contributes much to the regional rainfall and subsequently the hydrological cycle. Of particular importance to Lake Victoria are the South-East trade winds, which picks moisture from the lake while passing. They move to the northern and western shores of lake Victoria where they eventually condense to give heavy rainfall. Uganda benefits more due to the fact that winds carry more moisture to the *northwest* shores of Lake Victoria, with major rainfall season normally between March, April and May (MAM), with low rainfall season occurring between September, October and November (SON). There exist no remarkable dry months.

The climate of Lake Victoria catchment area is mild with small variations in monthly average air temperature between 19 °C and 25 °C throughout the year. Daily temperature fluctuates more widely, ranging from 15 °C to 30 °C. Rainfall in this catchment area averages about 1,300 mm annually, varying from 2,000 mm in highlands to 1,000 mm in the north, southwest and lowlands along the Lakeshore.

---

[3] 82 m—Britannica Concise.

Table 5.1 Morphoedaphic characteristics of Lake Victoria [5, 6, 11]. The mean width*, length* and area* are with respect to the Landsat (1984, 2002, 2017) and Sentinel-2 (2018) images

Characteristic	Measure
Position: Latitude	00° 20' N to 03° 00' S
Longitude	31° 39' E to 34° 53' E
Altitude [m above mean sea level]	1134
Catchment Area [km^2]	184,000
Lake Surface Area in 2018 [km^2]	69,216
Mean Area* [km^2]	69,295
Lake Area as Percentage of Catchment	37
Shoreline in 2018 [km]	4,572
Maximum Length in 2018, North-South [km]	386
Maximum Width in 2018, East-West [km]	362
Mean Width* [km]	364
Mean Length* [km]	388
Maximum Depth [m]	84
Mean Depth [m]	40
Volume [km^3]	2,760
Inflow [km^3/yr]	20
Outflow [km^3/yr]	20
Precipitation [km^3/yr]	114
Annual Fluctuations in level [m]	0.4–1.5
Flushing time [yrs]	138
Residence Time [yrs]	21

5.3 Population and Demographic Features

Lake Victoria Basin (LVB) had a population of approximately 21 million in 1997, 26 million in 1999 (above 30 million if Rwanda and Burundi are considered), with a growth rate of around 3% per annum with considerably higher rates in urban areas. Interesting results are presented by [20] who puts the 2010 population to over 42 million and projects the population to increase to over 60 million in 2020, over 76 million in 2030 and tripple to over 113 million in 2050. Indeed the population estimate of Lake Victoria basin inhabitants by 2017 is placed at above 76 million, with a population growth above 4%, e.g., [7]. On the commonly used figure of 30 million population of LVB, [20] writes:

> Interestingly, our population estimates derived for the LVB reveal that the total population that most stakeholders for the basin, including the Lake Victoria Basin Commission, were commonly citing - about 30 million inhabitants likely underestimated by more than 10 million people. While the number of 30 million people was being used in reports and other communications about the threat of population growth, we could identify no original source for this estimate.

The region is increasingly characterized by growing variability of ethnic composition. However the larger ethnic groups dwelling around Lake Victoria are the Luo, Samia, and the Suba people in Kenya; the Buganda and Busoga in Uganda; In Tanzania the distribution is as follow; in Mwanza region are the Sukuma, Kerewe, Jita, Kara, and Zinza. Sukuma constitute the overwhelming majority in this region. In Mara region are the Jita, Luo, Kuria, Zanaki, and Ruli. In Kagera there are Haya, Hangaza, Nyambo, Subi, and Sukuma. This means that there is a very great cultural diversity in the basin requiring a complex analytical framework as regards harmonization towards better and sustainable use of the natural resources of the basin for improving people's livelihoods.

The five countries in the Lake Region (i.e., Kenya, Uganda, Tanzania, Rwanda and Burundi) cover a total land area of 170, 270 km^2, supporting one of the densest and poorest rural populations in the world. Women constitute just over 50% of the population in the region. In the Lake Victoria region as a whole, compared to the 2010 population of over 42 million, Bremner et al. [20] estimate the population increase by 43% in 2020, 80% in 2030 and 167% in 2050, see also [7] for the projected growth rates). About 50% of the population in the region comprise youth who need more gainful employment opportunities to avoid the risks of social insecurity. The region already experiences some insecurity because of internally displaced persons as well as refugees from neighbouring countries.

In practically all the East African countries, increasing attention is being directed towards the status of women in society. Programmes aimed at mainstreaming gender issues in development activities are at various stages of development. To date, however, the socio-economic status of women is lower than that of men in a number of ways. For instance, women have limited access to productive resources such as land and capital. The literacy levels are lower among women than among men and the number of girls in secondary schools is considerably lower than that of boys. At the primary school level, the numbers are about the same. Women bear a heavier burden of the widespread poverty. Commercial fishing and processing factories employs more men than women and has impacted negatively on women who were previously engaged in fish processing and marketing. Moreover, when men migrate from rural to urban areas in search of employment, they leave behind women and children to bear the brunt of agricultural activities. The heavy work load women bear on farms, in their reproductive and community roles is a major source of poor health. Due to lack of money, women have limited access to health facilities [21, p. 8].

There is a strong cultural dimension to environmental degradation in the region. The dictates of cultural practices of sons inheriting their fathers' lands and wives owning lands to cultivate are reinforcing the need to subdivide land into small units, which are uneconomic for meaningful farming needed to cushion people from food insecurity. Such practices continue to generate a population of landless youth who must migrate elsewhere to earn a living and the cycle of poverty and food insecurity created continues to cause further environmental degradation that does not support food production. Population pressure on limited land leads to rapid land degradation. It is, therefore, imperative that conservation measures are adopted on a massive scale, if not, then it may not be possible to control the rate of environmental degradation as well as food security in the region.

Livelihood standards of the area have deteriorated as already noted by [22], not just because of the consequences of the population increases, but through a host of limiting factors and livestock production that lead to declining land productivity. There are also the people of Rwanda and Burundi in the wider Lake catchment region. Population density in the Lake basin is above the national averages found in all three riparian countries and the populations within these riparian communities grow at rates that are among the highest in the world. Population density in the region ranks among the highest in the world for rural areas where the majority of the people live [7].

5.3.1 *Historical Perspective of Early Settlements*

East Africa is largely considered the cradle of mankind, since the first paleontologic traces of humans have been found there. Most of the present Bantu-speaking inhabitants of the LVB started entering the basin around the beginning of the second century. The communities presently settled around Lake Victoria were immigrants from other parts of Africa. As early as the 13th and 14th centuries, Lake Victoria was well surrounded by organized and settled communities, which had and still enjoy a lot of interactions. The Luos arrived via the River Nile from the North of Africa and settled in the Lake Victoria basin in Uganda before moving to Kenya. When the Luos arrived, they forced out other tribes like Elgon Masai and Bantus that had earlier settled in the Lake basin. They interacted with ethnic groups that lived in Uganda. Some of the Luos later moved to North Mara in Tanzania in search of grazing ground when drought destroyed their crops. The Bantus who now occupy the Lake Victoria basin in Uganda and some parts of Tanzania trace their origins in Central Africa. Their settlement developed into indigenous kingdoms in the 14th century. Among them were the Baganda, Banyoro, Toro, Ankole and Basoga. The Baganda people created a dominant kingdom that could not be penetrated until the 19th century (see Fig. 5.3 for the Buganda Kingdom's palace[4]). The Kabaka king Mutesa II reigned until his overthrowal in 1966 by Milton Obote who himself was overthrown in 1971 by the butcher of Uganda (Idi Amin, see Fig. 5.4).

The other most important group was the Basoga who later established their kingdom around Jinja. Before colonization, the larger tribes that occupied the southern shores of Lake Victoria in Tanzania were the Wasukuma and Wanyamwezi. Currently their population runs into millions. Other smaller tribes like the Wanazaki, Wajita, Wakerewe of Tanzania; the Banyala, Samia, and Suba of Kenya and the Samia of Uganda are some of the most important ethnic groups that live on the shores of Lake Victoria. We have among the communities the Asians who, although constituting a minority, were brought by the colonialists to build the then Kenya-Uganda railway.

[4] http://www.buganda.com/crisis66.htm.

Fig. 5.3 Kabaka palace. **a** Main office, (**b**), the compound with Kabaka's flag **c** Kabaka's lake where he could go and bathe, and (**d**), remnant of Kabaka's vehicle when he fled during President Obote's take over in 1966

The group is among town dwellers and still controls major businesses in the Lake Victoria port towns and cities.

The earlier settlers brought with them distinct cultures into the basin. Through interactions, intermarriages amongst the different cultures became imminent. The cross transfers of communities were enhanced among the neighbourhood communities, through trade. The barter system amongst the communities became prominent as canoes and dhows transported goods. Trade and market centers were developed and later small towns were established in the Lake ports. This was the origin of the municipalities and cities that are now well established. The major towns include Kampala, Entebbe and Jinja in Uganda, with Kampala being the capital city of Uganda; Kisumu, Homa Bay in Kenya, with Kisumu being the third largest city in Kenya after Nairobi and Mombasa; and Mwanza, Musoma and Bukoba in Tanzania. These towns' populations run into millions. The peoples of the LVB share a relatively common history characterized by intensive interaction, extensive trading, intermarriage and welfare. In other cases, assimilation occurred, e.g., the Kenyan Suba were assimilated into Luo and now speak and practise Luo culture. Young Subas can hardly speak their

Fig. 5.4 Historical sites within the lake Victoria surrounding. **a** Path to Uganda's torture chambers during Idi Amin's regime, (**b**), (**c**) and (**d**), inside the torture chambers

native Suba language but Dholuo (Luo language).[5] In essence, the Subas are Luos for all intent and purpose. The regime of former president Moi attempted to divide the two communities through the 1988 census but the Subas and Luos remained a single entity. Furthermore, the communities within LVB also share the historic experience of external interventions in the area, such as the slave trade and colonialism.

In 1885, Tanzania, Rwanda, and Burundi were placed under the rule of the Imperial German Government. However, after the first World War, Tanzania was placed under the British mandate until the establishment of the Trusteeship system under the Charter of the United Nations, while Rwanda and Burundi came under Belgium rule. In 1887 Kenya was placed under the Imperial British East Africa Company, but the territory was transferred in 1895 to Britain. Likewise, the British established the Uganda Protectorate in 1894. However, after 1945, various nationalist movements emerged in this region, and began a sustained and successful effort to emancipate themselves from European domination. In 1961, Tanzania became independent followed by Uganda, Rwanda, and Burundi in 1962 while Kenya gained its

[5] https://www.standardmedia.co.ke/the-standard-insider/article/2001386038/how-abasuba-accepted-luo-love-and-lost-their-culture.

independence in 1963. The delay in Kenya's independence is argued to be due to Jaramogi Oginga Odinga's (a native Luo from Siaya County within Lake Victoria Basin) refusal to assent to the throne unless Jomo Kenyatta (Kenya's first president) was released from prison.

5.3.2 Impacts of Colonialism

The establishment of colonial rule over the people of the region brought with it resource management structures that removed the power from traditional leaders to the central governments of larger territories of Kenya, Uganda, and Tanganyika. This meant that people who had no interest in the Lake could be given the responsibility to manage resources such as fisheries. The ownership of the resources shifted with time. The population influx and lack of properly planned infrastructural development has brought many problems to the urban centers. These problems include non-functional sewerage systems, industrial pollution and shortage of water supply. These have become a serious headache for the citizens. County governments alone cannot address these problems. The citizens have to join them with 'own key' solutions through organized groups such as Non-Governmental Organizations (NGOs) and Community-Based Organizations (CBOs). Kenya changed its constitution in 2010 to allow for the devolved system and is proposing to increase county revenue from the current 15% to 35%. This, if properly managed, could see huge development within the Kenyan side of the LVB. However, rampant corruption and mismanagement of the resources that bedevilled the first county governments of Kisumu, Siaya, Migori, Homabay and Busia in the Kenyan side of LVB could not let this happen. Hopefully, the future county governments will learn from the mistakes of their predecessors. The current (2021) Kisumu county government under Professor Anyang Nyongo seems to have hit the right pedal.

5.4 GHA's Precious Lake: Benefits and Challenges

Of all the tropical's lakes, Lake Victoria stands tall as the greatest freshwater body. In the entire world, it comes only second to Lake Superior in the USA. Within its surrounding, it directly supports a population of more than 76 million East Africa's inhabitants (i.e., $\frac{1}{3}$ of the combined population) [5, 19], while globally it is the source of Tilapia (*Oreochromis niloticus*) and Nile Perch (*Lates niloticus*) [3]. So important is the Lake such that Egypt and Sudan entered the Nile treaties of 1929 and 1959 for exclusive use of its waters, see e.g., [7]. Just as oil is an important resource of the world, Lake Victoria and its basin plays a major role in the world such that it appears at the top of a number of lists as pointed out by [12]. Specific roles played by this Lake that are beneficial to the entire GHA region are [3, 6, 7, 13]:

- Agricultural (rain-fed as well as irrigated) potential. Its surface water could be used for irrigation whereas its modulating of the regional hydroclimate contributes to rainfall that is essential for rain-fed agriculture.
- It is a source of freshwater that supports livelihood of people living within its shores. Its waters is used for drinking, domestic use, in industries, for transportation and sports among others.
- It contributes to the river Nile, the lifeline of Egypt and Sudan.
- It is the habitat of Tilapia (*Oreochromis niloticus*) and Nile Perch (*Lates niloticus*), the two types of fish that are in great demand the world over together with other species of fish who reside in its waters. If properly managed through provision of infrastructure such as coolants, this resource could play a pivotal role in alleviating food insecurity.
- Its wetlands provide homage to various bird species, insects and thus preserving biodiversity.
- It helps fight poverty by creating employment in fish industry and transportation for people living within its environs.
- It acts as eco-tourism center.

The list could go on and on but we have chosen to mention only but a few.

Although the Lake and its basin boasts of such rich endowment, the fact of the matter is that it is a sick giant [4] (i.e., suffering from infestation of water hyacinths, pollution, siltation, sewerage discharge and poor management). This has necessitated the establishment of several measures to help address the challenges faced by this Lake. Such measures include the establishment of Lake Victoria Management Environmental Programme (LVEMP) and other institutions. The challenges facing the Lake Victoria Basin (LVB) include:

- Having as its dependants people who are faced with the health threat such as HIV/AIDs and currently (2021) COVID-19. With increased demands from the health sector, resources that would otherwise be used for the development, conservation and management of the Lake's resources are diverted towards the health sector. Moreover, a sick society would certainly be unproductive, especially when the able men are the ones expected to go fishing inside the lake.
- Being infested with the deadly water hyacinth, which is depriving it of oxygen and thus suffocating it, and posing threats to hydropower plants (see Fig. 5.5).
- Faced with people living in its surrounding languish in abject poverty although it is endowed with great resources [22].
- Consequences of decision and policies made in far parts of the world, i.e., ecological power, and by global economic structure [4].

In general, the Lake faces four critical and mutually reinforcing problems. The *first* is the *obsolete Nile treaties* that lock up its potential for irrigated agriculture. If these treaties could be done away with, and the lake used sustainably, then its contribution to irrigated agriculture could go a long way in alleviating food insecurity over the entire Greater Horn of Africa (GHA). *Second* is *productivity*. The most important commercial species caught is Nile Perch. A predatory species that has over time

Fig. 5.5 A huge floating island in Lake Victoria clogging a turbine in a hydroelectric power station. *Source* https://www.bbc.com/news/world-africa-52286296

eaten to near extinction many species endemic to the Lake that were once important sources of food and livelihood for the local communities. The scramble for the catch has become the source of conflict between fishers from the different countries. Overfishing is attributed to the use of inappropriate gear and insufficient enforcement capacity. Open access and inappropriate exploitation increases the fishing effort against the diminishing reproduction rate. In addition, territorial boundaries tend to concentrate the fishing effort on a limited area especially in Kenya. Authorities lack the resources nor the will to effectively police the Lake.

The *third* set of challenges are *environmental*, namely resource degradation emanating from catchment degradation and discharge of human, industrial and agricultural effluent [1, 5]. The *fourth* set of challenges is *market related* and includes price competitiveness, quality assurance and reliability of supply. The principal export product Nile Perch, is under intense price pressure from aquaculture, in particular the new tilapia varieties. While the fishery has made substantial progress in meeting European (EU) standards, it has not yet attained the eco-labelling certification that is increasingly required by the retailers [18].

Owing to these overwhelming evidence of the importance of Lake Victoria together with its emerging threats, a continuous monitoring of its status is inevitable. A comprehensive coverage of Lake Victoria's resources and related challenges are presented in Awange and Ong'ang'a [13]. Awange [6, 17] looks at the feasibility of monitoring this lake using the state of the art space-based techniques while Awange [7] weighs the Nile River Basin from space. While presenting the hydroclimate of Greater Horn of Africa, this book proposes the potential of irrigated agriculture, e.g., from Lake Victoria's surface water.

5.5 Fluctuations: Climatic or Anthropogenic Induced?

The lake's level has undergone fluctuations in the past. For example, with the receding of the lake waters in 2002–2006, acres of land that were lost to the floods of the 1960s were fast being reclaimed, creating sources of conflicts between man and wildlife. In some beaches, e.g., Usoma in Kenya, wetlands that were once breeding places for fish were dying up, leaving areas of land as playing fields for children and farmland. Ships were now forced to dock deep inside the lake, while the landing bays needed to be extended. Those who directly depended on the lake waters for domestic use were forced to go deeper into the lake to draw water, thus exposing women and children to water-borne diseases and risks of snakes and crocodiles. Water intakes that supplied major towns and cities had to be extended deeper into the lake, thus causing more financial burden to the municipalities that were already strained financially [23]. In 2020, the floods were back leading to speculations that the lake was retracing its 1960s level. Come 2019–2021, the rains came again and the lake's water level rose to a point where it was thought to be tracing its roots back to the 60s. This rise of the lake's level is reported by Khaki and Awange [17] to have been caused by the Indian Ocean Dipole (IOD) climate variability index.

With 80% of Lake Victoria water coming from direct rainfall, changes in the lake level are directly related to the variation in the water stored in its basin, which contributes around 20% in the form of river discharge [5, 6]. A decrease in stored basin water was therefore suspected to contribute to the drop in the lake level in 2002–2006. An analysis of the stored water in the Lake Victoria Basin (LVB) in relation to rainfall and evaporation was therefore necessary as a first diagnosis. This would provide water resource managers and planners with information on the state and changing trend of the stored water within the basin. Such basin scale observations could only be achieved through the use of satellites such as Gravity Recovery and Climate Experiment (GRACE) discussed in Chap. 3. Conventional methods for studying variations in stored water using, e.g., land surface and hydrological models could not diagnose the problem, see e.g., [23].

Having been motivated by the potential of the GRACE satellites, Awange et al. [23] undertook a satellite analysis of the entire LVB in an attempt to establish the cause of the decline in Lake Victoria's water levels. The GRACE and CHallenge Challenging Minisatellite Payload (CHAMP) satellites (Fig. 3.2 in Sect. 3.3.3) together with data from the Tropical Rainfall Measuring Mission (TRMM) satellite were employed in the analysis. Using 45 months of data spanning a period of 4 years (2002–2006), the GRACE satellite data were used to analyze the gravity field variation caused by changes in the stored waters within the lake basin. Figure 5.6 presents the annual variation of the geoid in the lake's basin during the high rain season months of March, April and May (MAM) for the period 2002–2006. The GRACE results indicated that the basin's total water storage dramatically decreased at a rate of 6.20 mm/month. These changes are expressed in equivalent water thickness (also known as total water storage (TWS)) in Fig. 5.7. For the period 2002–2006, the results indicate a general decline in the lake basin's water level at a rate of 1.83 km^3/month [23].

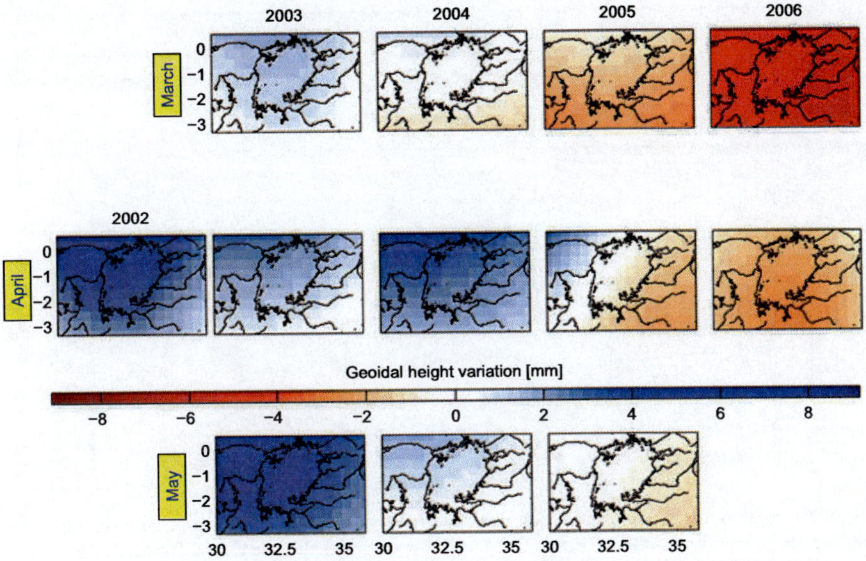

Fig. 5.6 Inter-annual comparison in the geoid during the high rainy season of MAM from 2002–2006. The figure indicates a decline in total water storage in the Lake Victoria Basin during this period. *Source* Awange et al. [23]

To validate the GRACE results, TRMM Level 3 monthly data for the same period of time were used to compute mean rainfall at a spatial resolution of $0.25^\circ \times 0.25^\circ$ (25×25 km), as shown in Fig. 5.8, from which the rainfall trends were analyzed (Fig. 5.9). To assess the effect of evaporation, Global Navigation Satellite System (GNSS) remote sensing data (59 CHAMP satellite occultations) for the period 2001 to 2006 were analyzed to define if tropopause warming took place [23]. The results indicated that the tropopause temperature fell in 2002 by about 3.9 K and increased by 2.2 K in 2003 and remained above the 189.5 K value of 2002. The tropopause heights showed a steady increase from a height of 16.72 m in 2001 and remained above that value reaching a maximum of 17.59 km in 2005, an increase in height by 0.87 m. Temperatures did not, therefore, increase drastically to cause massive evaporation. TRMM results indicated the rainfall over the basin (and directly over the lake) to have been stable during this period (see Figs. 5.8 and 5.9). Since rainfall over the period remained stable, and temperatures did not increase drastically to cause increased evaporation, the remaining major contributor during the period 2002–2006 was suspected to be discharge from the expanded Owen Falls (Nalubaale) dam. Awange et al. [23] concluded, thanks to the GRACE and GNSS satellites, that the fall in Lake Victoria's water level between 2002 and 2006, was due to human impact on the basin's environment (i.e., expanded dam) as opposed to natural factors.

In a related work, Swenson and Wahr [24] used satellite gravimetric and altimetric data to study trends in water storage and lake levels of multiple lakes in the Great

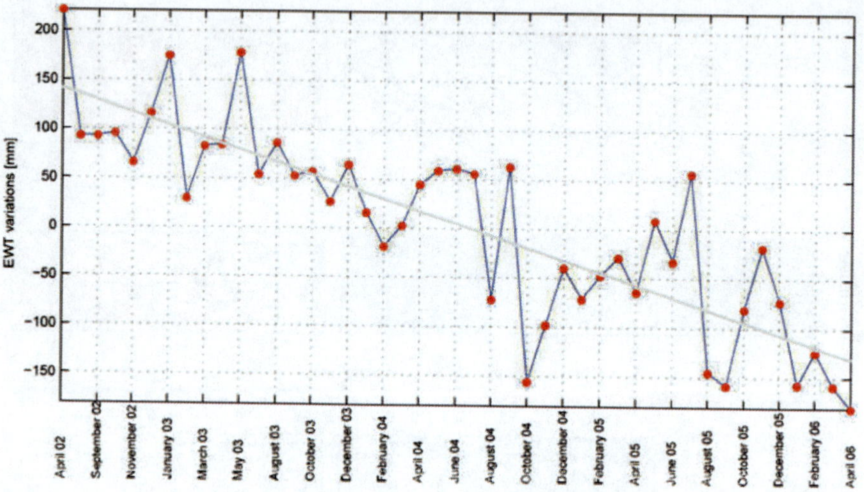

Fig. 5.7 Lake Victoria Basin total water storage (equivalent water thickness) changes between 2002–2006, as seen by the GRACE satellite. The figure indicates that the GRACE satellites observed a general decline in the lake's basin waters over this period (cf. Fig. 3.5 in Sect. 3.5.2 obtained from satellite altimetry). *Source* Awange et al. [23]

Rift Valley region of East Africa for the years 2003–2008. GRACE total water storage estimated by Swenson and Wahr [24] corroborated the findings of Awange et al. [23] that the lake's water level had declined by as much as 60 mm/year, while their altimetric data indicated that levels in some large lakes in the East African region dropped by as much as 1–2 m. Swenson and Wahr [24] concluded that the largest decline occurred in Lake Victoria and, like Awange et al. [23], attributed this to the role of human activities.

Both the findings of Awange et al. [23] and Swenson and Wahr [24] provide evidence that the GRACE satellites could be used to provide independent means of assessing the relative impacts of climate and human activities on the balance of stored water that does not depend on in-situ observations, such as dam discharge values, which may not be available to the public domain.

5.6 Concluding Remarks

Lake Victoria is a riparian resource shared by 3 East African countries: Kenya, Uganda and Tanzania. It serves as a source of water for irrigation, which if properly utilized, could significantly enhance GHA's food security. As argued previously, however, the caveats imposed by obsolete Nile treaties of the 1920s and 1960s have to be removed/modified and an agreed mechanism that supports equitable use of the lake's resources put in place. Furthermore, its modulation of the regions climate

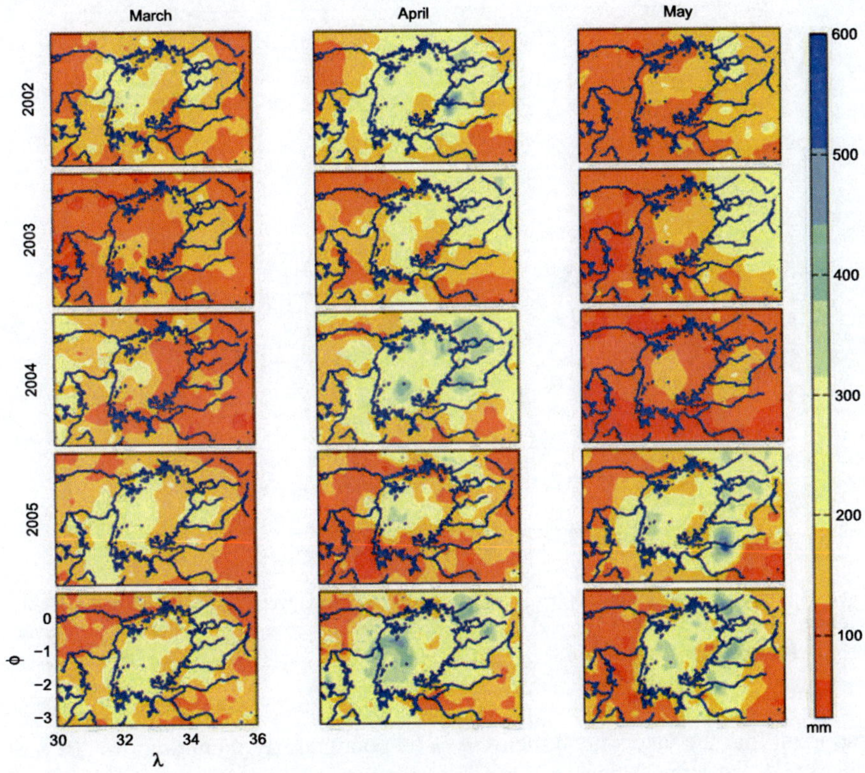

Fig. 5.8 Inter-annual comparison of rainfall over the Lake Victoria Basin during the high rainy season of MAM from 2002–2006 using TRMM results. *Source* Awange et al. [23]

leading to rainfall within its basin is significant for rain-fed agriculture. Moreover, it is vital for transport, domestic and livestock uses and a source of livelihood for over 42 million people living around the lake. Its fish (tilapia and Nile perch) are exported the world over. Its role as an indicator of environmental and climate change on long-term scales has been documented in [15]. The vital role of this lake and its surroundings as a source of food has been documented e.g. in Awange and Ong'ng'a [6, 7, 13]. The lake and its environment is under threat from environmental pollution and more recently from declining water levels.

This chapter has discussed the origin of Lake Victoria, its name and some elements of demographic features of the region. It is a region where the population growth rate is very high and average population density also high. These two phenomena have a direct bearing on the environment, leading to serious environmental degradation, soil erosion, pollution of the Lake and the decline in soil fertility. It is therefore imperative that measures be taken to address the issues concerning population pressure and environmental degradation. Whereas these measures are necessary, it is more

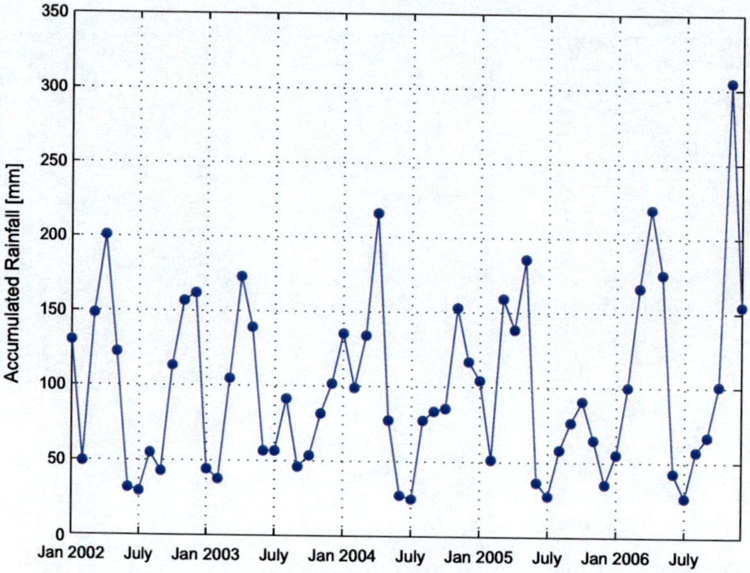

Fig. 5.9 Time series of rainfall 2002–2006 for the lake Victoria basin as observed by the TRMM satellite. The x-axis shows the annual cycle from January of a given year to January of the next year. *Source* Awange et al. [23]

important that the lake waters themselves be continuously monitored for the benefit of GHA at large. However due to its size, ground-based observation methods by themselves are insufficient hence the need of space-based methods discussed in Awange [6].

References

1. Morgan B, Awange JL, Saleem A, Hu K (2020) Understanding vegetation variability and their "hotspots" within Lake Victoria Basin (LVB: 2003–2018), 122. https://doi.org/10.1016/j.apgeog.2020.102238
2. Hickman GM, Dickins WHG, Woods E (1973) The lands and people of East Africa. Longman, Essex, UK
3. Aseto O, Ong'ang'a O (2003) Lake Victoria (Kenya) and its environs: resource. Opportunites and challenges. Africa Herald Publishing House, Kendu Bay, Kenya
4. Fuggle RF (2004) Lake Victoria: A case study of complex interrelationship. In: United Nations Environmental Program: Africa Environmental Outlook Case Studies
5. Awange JL, Saleem A, Sukhadiya RO, Ouma YO, Hu K (2019) Physical dynamics of Lake Victoria over the past 34 years (1984–2018): is the lake dying? Sci Total Environ 658:199–218
6. Awange JL (2021) Lake Victoria monitored from space. Springer Nature International
7. Awange JL (2021) Nile waters. Weighed from space. Springer Nature International
8. Johnson TC, Kelts K, Odada EO (2000) The Holocene History of Lake Victoria. Ambio, Vol 29 (1)

9. Ong'ang'a O (2002) Poverty and Wealth of fisherfolks in the Lake Victoria basin of Kenya. Africa Herald Publishing House, Kendu Bay, Kenya
10. Hughes RH, Hughes JS (1992) A directory of African Wildlife. With a chapter on Madagascar, IUCN/UNEP/WCMC
11. Balirwa SJ (1998) Lake Victoria wetlands and the ecology of the nile tilapia oreochromis niloticus. Linne. PhD dissertation, Balkema Publishers, Rotterdam, Netherlands
12. Pitcher TJ, Hart PJB (1995) The impact of species changes in African Lakes, Fish and Fisheries Series 18. Chapman and Hall, London, UK, p 601
13. Awange JL, Ong'ang'a O (2006) Lake Victoria-Ecology: resource of the Lake basin and environment. Springer, Berlin
14. Kull D (2006) Connections between recent water level drops in Lake Victoria, dam operations and drought. http://www.irn.org/programs/nile/pdf/060208vic.pdf. Accessed 25 Sep 2011
15. Nicholson SE (1998) Historical fluctuations of Lake Victoria and other lakes in the Northern Rift Valley of East Africa. In: Lehman JT (ed) Environmental change and response in East African lakes. Kluwer, Dordrecht, pp 7–35
16. Nicholson SE (1999) Historical and modern fluctuations of lakes Tanganyika and Rukwa and their relationship to rainfall variability. Clim Change 41:53–71. https://doi.org/10.1023/A:1005424619718
17. Khaki M, Awange J (2021) The 2019 2020 rise in Lake Victoria monitored from space: exploiting the state-of-the-art grace-fo and the newly released ERA-5 reanalysis products. Sensors 21(13):4304. https://doi.org/10.3390/s21134304
18. KLI (2004) Market based fisheries regulation strategy for Lake Victoria. A Report of the Regional Inception Workshop 6–8 June 2004, Lake Naivasha Country Club, Kenya, 50pp
19. Nile Basin Water Resources Atlas (2017). http://atlas.nilebasin.org/
20. Bremner J, Lopez-Carr D, Zvoleff A (2013) Using new methods and data to assess and address population, fertility, and environment links in the Lake Victoria Basin, population and the environment. Session 247. https://www.iussp.org/en/event/17/programme/paper/6136
21. Ehlin U (1997) Lake Victoria basin: natural resource under stress. Department of natural resource and environment. Sida, Stockholm
22. Aseto O, Ong'ang'a O, Awange JL (2003) Poverty. A challenge for the Lake Victoria basin. OSIENALA Series 5, Printed by Africa Herald Publishing House, Kendu Bay, Kenya
23. Awange JL, Sharifi M, Ogonda G, Wickert J, Grafarend EW, Omulo M (2008) The Falling Lake Victoria water level: GRACE. TRIMM and CHAMP satellite analysis. Water resource management. https://doi.org/10.1007/s11269-007-9191-y
24. Swenson S, Wahr J (2009) Monitoring the water balance of Lake Victoria, East Africa, from space. J Hydrol 370:163–176. https://doi.org/10.1016/j.jhydrol.2009.03.008

Chapter 6
GHA's Water Tower: Ethiopian Highlands

Ethiopia's hydrology plays a significant international role, being the headwaters of the Blue Nile Basin, where it contributes about 86% of the total annual flow of the Nile and also approximately 90% of inflow into Lake Turkana, a lake situated in the arid area of northern Kenya. In recent decades, however, extreme hydrological variability, seasonality, and anthropogenic factors have posed challenges to the region's water resource management. For example, due to the large and increasing population pressure, insufficient agricultural production, a low number of developed energy sources, and drought episodes, Ethiopia, which has almost 94% of its population depending on wood fuel, is constructing a major hydropower (Grand Ethiopian Renaissance Dam, GERD) and irrigation development schemes.—[90].

6.1 Summary

It is perplexing that a place can have plenty of water yet face perennial food insecurity. Such is the unfortunate tale of Ethiopia, the water tower of Greater Horn of Africa (GHA). Water towers are areas that generate high stream flow compared with other areas [89]. Within GHA, Ethiopian Highlands stand tall amongst its competitors and as such, understanding the spatio-temporal characteristics of its water storage changes is crucial for not only Ethiopia, a country that is facing a range of challenges in water management caused by anthropogenic impacts as well as climate variability but also for the overall management of the entire GHA's water resource. In addition to this, the scarcity of in-situ measurements of soil moisture and groundwater, combined with intrinsic "scale limitations" of traditional methods used in hydrological characterization are further limiting the ability to assess water resource distribution in the region. This chapter presents the work of [90] who applied remotely sensed and model data over Ethiopia for a period of 8 years (2003–2011) to (i) test the performance of models and remotely sensed data in modeling water resources distribution in un-gauged arid regions of Ethiopia, (ii) analyze the inter-annual and seasonal variability as well as changes in total water storage (TWS;

J. Awange, *Food Insecurity & Hydroclimate in Greater Horn of Africa*,
https://doi.org/10.1007/978-3-030-91002-0_6

surface, groundwater, soil moisture and vegetation) over Ethiopia, (iii) understand the relationship between TWS changes, rainfall, and soil moisture anomalies over the region, and (iv), identify the relationship between the characteristics of aquifers and TWS changes. The data used included (see Chap. 3 for a discussion); monthly gravity field data from the Gravity Recovery And Climate Experiment (GRACE) mission, rainfall data from the Tropical Rainfall Measuring Mission (TRMM), and soil moisture from the Global Land Data Assimilation System (GLDAS) model. The results showed that the western part and the north-eastern lowlands of Ethiopia experienced decrease in TWS water between 2003–2011, whereas all the other regions gained water during the same period. The impact of rainfall seasonality was also seen in the TWS changes. Applying the statistical method of Principal Component Analysis (PCA) to TWS, soil moisture and rainfall variations identified the dominant annual water variability in the western, north-western, northern, and central regions, and the dominant seasonal variability in the western, north-western, and the eastern regions. A correlation analysis between TWS and rainfall indicated a minimum time lag of zero to a maximum of six months, whereas no lag was noticeable between soil moisture anomalies and TWS changes. The delay response and correlation coefficient between rainfall and TWS appeared to be related to groundwater diffuse recharge mechanisms, revealing that most regions of Ethiopia received indirect recharge. The results further showed that the magnitude of TWS changes was higher in the western region and lower in the north-eastern region, and that the elevation influenced soil moisture as well as TWS.

6.2 Ethiopian Hydrogeological Regimes: Characterization

Ethiopia's hydrology plays a significant international role, being the headwaters of the Blue Nile Basin, where it contributes about 86% of the total annual flow of the Nile [52, 77] and also approximately 90% of inflow into Lake Turkana [32] a lake situated in the arid area of northern Kenya. In recent decades, however, extreme hydrological variability, seasonality, and anthropogenic factors are posing challenges to the region's water resource management. For example, due to the large and increasing population pressure, insufficient agricultural production, a low number of developed energy sources, and drought episodes, Ethiopia, which has almost 94% of its population depending on wood fuel, is planning major hydropower and irrigation development schemes [12, 83]. In addition to irrigation and hydroelectric dams, land degradation and changes in land cover in Ethiopia where forests are being converted to agricultural land are having impact on the Nile flow, see, e.g., [63, 74].

Characterizing water-storage in all its forms (surface, soil moisture, and groundwater) and their responses to the incoming (precipitation) and outgoing (evaporation and discharge) water masses, therefore, is of great importance in terms of understanding extreme changes in stored water triggered by natural and anthropogenic factors. This is of interest especially in areas where the seasonality of rainfall is strong and the welfare of the society relies on the availability of water, such as in Ethiopia. To quantify the stored water resource of a region, soil moisture and groundwater have been documented to play a significant role, e.g., [64, 65, 78]. For developing countries such as Ethiopia, however, in-situ networks of soil moisture and groundwater measurements are sparse. Therefore, these components are among the most difficult water budget parameters to obtain. To circumvent this shortfall, characterizing the hydrogeological regimes has been undertaken by considering their hydrogeological environments, see e.g., [11, 37, 43, 44, 88]. Most of the previous studies have concentrated on the climatic characteristics, e.g., [21, 58–60, 72, 73, 75, 82] and the use of environmental isotopes and hydrochemicals, e.g., [12] to trace the water availability in Ethiopia. For example, Berhane et al. [12] found the environmental isotopes and hydrochemicals approach to be the most effective tool for differentiating various forms of geochemical reaction, and to infer the environmental factors affecting groundwater quality and its flow in the region.

Besides concentrating on climatic characteristics, most of the studies above are also restricted to small scale hydrological characterizations, which do not reflect the large-scale water storage variability over Ethiopia. Moreover, isotopes and hydrochemical-based methods are costly, require skilled experts, and are often difficult to apply over large areas, which are not easy to venture into, particularly in developing countries such as Ethiopia. Furthermore, most of the hydrological studies over Ethiopia focused only on regional characterizations. Generalizing such local outputs to the whole of Ethiopia is extremely difficult due to its vast range of climatic and topographic conditions.

Total water storage (TWS; surface, groundwater, soil moisture and vegetation water), which is defined as the sum of all forms of water stored above and underneath the Earth's surface, is a key component of the terrestrial and global hydrological cycles that exerts important control over water, energy and biogeochemical fluxes, and thus plays a major role in the Earth's climate system [65, 66, 78, 80, 81, 99], see Sect. 3.3.3. The Gravity Recovery And Climate Experiment (GRACE) mission offers the possibility of remotely sensing global and regional TWS changes. Launched in 2002 as a joint project of the US and Germany, GRACE products have contributed enormously to the study of changes in total water storage. With the help of complementary data sets, GRACE-derived products offer the possibility of monitoring groundwater depletion in data poor regions of the world [29, 36, 64]. Although GRACE has the advantage of providing freely available data set capable of monitoring TWS changes at basin scales compared to isotopes and hydrochemical methods mentioned earlier, except for studies that have been done at continental or global scales [62, 70] and those in connection with the Nile Basin [4–9, 17, 51, 91–96], its products have not been applied specifically to the whole Ethiopian Basins at a local level. Within Ethiopia, for example, the application of GRACE products are reported, e.g., in Bonsor and Melesse et al. [17, 51], with respect to the study of the Blue Nile. For instance, Melesse et al. [51] presents the low and high flow characteristics of the Blue Nile River using wavelets and applies GRACE products to analyse moisture flux. One of the major challenges in applying the GRACE products to estimate TWS changes over Ethiopian Basins is the fact that only 2 out of the 12 basins (Abbay and Wabishebele) can barely fulfill the requirement for the smallest resolvable basin area of 200,000 km^2, see also [81]. Furthermore, Longuevergne et al. [49] have pointed out that an accurate estimate of TWS changes in small basins using GRACE-derived products requires a compromise between competing needs for noise suppression and spatial resolution. To overcome the spatial limitations, in [90], Ethiopia is divided into ten regions of equal sizes of $4° \times 4°$. For each region, TWS are computed from GRACE data using the approaches presented in Wahr et al. [84].

Awange et al. [90]'s contribution focus on the remotely-sensed TWS changes over Ethiopia using GRACE products. For the purpose of evaluating GRACE products, [90] also uses rainfall and soil moisture data based on products from the Tropical Rainfall Measuring Mission (TRMM) [45] and the Global Land Data Assimilation System (GLDAS) [66], respectively. The major aims of [90] are (i) understanding the response of Ethiopian aquifers to TWS changes from information obtained following the analysis of inter-annual (i.e., between years variability), and intra-annual (i.e., processes that occur on a time scale of less than one year, but more than one month) GRACE products, (ii) depicting the reaction of each region to hydrological input (rainfall) including any time delays, and (iii), identifying the dominant pattern of intra-annual, annual, and seasonal variability over Ethiopia by applying the statistical method of Principal Component Analysis (PCA) on GRACE-derived TWS, soil moisture, and rainfall patterns. Knowledge of the lag time is important for understanding the longest period over which the available stored groundwater can be sustainably exploited after rainy seasons to support irrigated agriculture. This chapter is organized as follows; in Sect. 6.3, the background of Ethiopian Highlands and its characteristics are discussed. Section 6.4 presents the remotely sensed/model products used and the employed analysis methods. The water tower is then characterized in Sect. 6.5 and finally, Sect. 6.6 summarizes the major findings of this chapter.

6.3 Ethiopian Highlands: Background

6.3.1 Location

Ethiopia, located between latitudes 3°15' N to 15°N and longitudes 33°E to 48°E, is a landlocked country bounded by Eritrea (North), South Sudan (South West), Sudan (North West), Kenya (South), Somalia (East) and Djibouti (North East), with a surface area of 1,127,127 km^2. Its altitude ranges from nearly 120 m below mean sea level in the Dallol depression to about 4620 m above mean sea level at Mount Ras Dashen. It contains three major physiographic regions that include the western highlands and its associated lowlands, the eastern highlands and its associated lowlands, and the Rift Valley in between them, running from north-east to south-west, separating the eastern and western highlands (Fig. 6.1). It is because of these physiographic influences on the drainage systems that Ethiopia is counted as the water tower of GHA, with twelve major basins; eight of which are river basins, one is a lake basin, and the remaining three are dry basins with no or insignificant out flow [31].

Ethiopia contributes to three major drainage systems (Fig. 6.1), the Mediterranean Sea drainage system (Abbay, Blue Nile, Baro-Akobo, Mereb and Tekeze), the Great East African Rift-Valley drainage system (Omo-Ghibe, Awash, Rift-Valley Lakes, Danakil and Aysha) and the Indian Ocean drainage system (Genale-Dawa, Wabishebelle and Ogaden). The groundwater resources of Ethiopia and their distribution vary depending on their respective geological, structural, and climatic conditions. The near-surface geological pattern that mainly govern the hydrogeological characteristics of Ethiopia constitutes the region's oldest basement rocks (Precambrian basement) (18%), Paleozoic and Mesozoic sedimentary rocks (25%), Tertiary volcanic (40%), and Quaternary sediments and volcanics (17%) [2]. It should be pointed out that there are also large areas with Tertiary sediments occurring mainly in the Rift Valley.

6.3.2 Climate

The climate of Ethiopia ranges from equatorial rainforest in the south and southwest, which is characterized by high rainfall and humidity; afro-alpine conditions on the summits of the Semien (western highlands) and Bale (eastern highlands) mountains, to the desert-like conditions of the Northeast, East and Southeast lowlands, e.g., [96]. The temperatures range from 60 °C at the Dallol depression, to freezing temperatures on the Mount Ras Dashen Plateau [54]. The mean annual rainfall varies from 3000 mm at Masha in the western highlands to barely 200 mm in the eastern lowlands [28, 68, 73]. Ethiopia experiences two rainfall seasons. The major rainy season (summer, regionally know as "Kiremt") extends from June to September, and accounts for nearly 60% [71], especially over the northern two thirds of the country. The minor rainy season (spring, regionally know as "Belg") usually begins in January/February and ends in April/May [20]. In general rainfall over Ethiopia is influenced by both local scale forcing mechanisms associated with the Ethiopian Highlands as well as heterogeneous land surface characteristics. In addition, the rainfall variability is significantly influenced by large (global) atmospheric circulation and sea surface temperatures. These large scale forcing mechanisms are normally expressed through El'Niño Southern Oscillation (ENSO) induced anomalies, Quasi-Biennial Oscillation (QBO), as well as west-east sea surface temperature gradients over the equatorial Indian Ocean (e.g., [15, 71, 94]). Table 6.1 shows the annual precipitation and potential evapotranspiration (ET) over the seven climatic zones over Ethiopia (Table 3 of Berhanu et al. [13]).

The common indicators of climate variability and changes are trends in precipitation and the maximum and minimum temperatures. Mengistu et al. [53] reported a consistent warming trend in the maximum and minimum temperatures over the past few decades in Ethiopia. However, rainfall shows a declining trend for the east, south and southwest parts of Ethiopia [73]. Additionally, for the central highlands, there is no evidence of trend in rainfall [19]. In general, there is no trend in the extremes of seasonal rainfall in Kiremt and Belg over Ethiopia (cf., [72]). However,

Table 6.1 Annual precipitation, average temperature, and potential evapotranspiration (ET). *Source* [90]

Climatic zones	Annual precipitation (mm)	Temperature °	Annual ET (mm)
Arid	<302	>27.5	2159
Semi-arid	302–350	27.5–21	1737–2159
Sub-moist	350–566	21–16	1431–1737
Moist	566–835	16–11	1124–1431
Sub-humid	835–1189	11–7.2	895–1124
Humid	1189–1711	<7.5	<895
Per-humid	>1711	–	–

Marshall et al. [50] have reported a gradual decline in evapotranspiration on the order of 5 mm/year over the coast of West Africa, the Sahel, and the western Ethiopian highlands, over a 31 years study period.

6.4 Satellite-Hydrological Model Products: Analysis

6.4.1 Gravity Recovery and Climate Experiment (GRACE)

GRACE (see Sect. 3.3.3), a US/German satellite gravimetry mission that tracks changes in the Earth's gravity field [81] has been the most commonly exploited satellite in the past decade for computing total water storage (TWS) changes. GRACE products have been used to quantify water storage changes at regional as well as global scales as reported for different regions, e.g., [9, 34, 36, 64]. The GRACE Release-05 (RL05) Level-2 (L2) dataset provided the processed time-variable gravity field products used in this chapter. The products are provided as sets of spherical harmonic coefficients (Stokes's coefficients) averaged over certain time periods (usually monthly). Spherical harmonic coefficients at high degrees are affected by correlated noise [46, 47] that need to be smoothed before being used for the hydrological analysis. These correlations can be reduced, using either an empirical method based on a polynomial fit, e.g. [79], or an a priori synthetic model of the observation geometry, e.g., [46]. In this regard, [90] employed the DDK3 [47] filtered monthly spherical harmonic coefficients dataset available from the official website of the International Center for Global Earth Models.[1] The suitability of the employed DDK filter is discussed, e.g., in Werth et al. [85]. The investigations of [90] covered the period from January 2003 to December 2011, and included 96 months of the German GeoForschungsZentrum (GFZ), Potsdam, products.

[1] (ICGEM) (http://icgem.gfz-potsdam.de/ICGEM/TimeSeries.html).

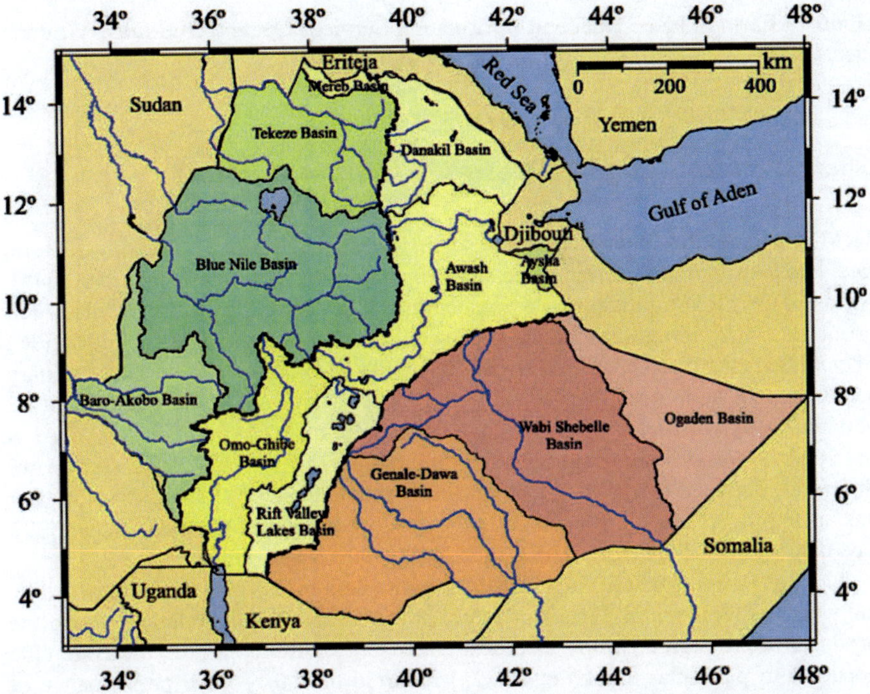

Fig. 6.1 Major rivers and basins in Ethiopia. The black dashed line represents the major water divides between the Mediterranean Sea Basin (West), the Rift Valley endorheic Basins (Center), and Indian Ocean Basin (East). *Source* Modified from Nyssen et al. [56]. The various colours mark out the separate 12 basins. The x-axis contain the longitudes (East) while the y-axis show the latitudes (North). *Source* [90]

6.4.2 Global Land Data Assimilation System (GLDAS)

GLDAS model is developed jointly by the National Aeronautics and Space Administration (NASA), the National Oceanic and Atmospheric Administration (NOAA), and the National Centers for Environmental Prediction (NCEP) [66]. It drives multiple offline (not coupled to the atmosphere) land surface models that integrate huge quantities of observation-based data. GLDAS executes its outputs globally at a relatively high spatial and temporal resolution, enabled by the Land Information System (LIS) [30]. The parameters in the GLDAS-monthly data fall into three main categories: water balance, energy balance, and forcing parameters. Water balance includes soil moisture parameters and other variables such as rainfall rate and surface runoff. To investigate the spatial and temporal variation of soil moisture over Ethiopia, the water balance monthly soil moisture data generated by NOAH LSM at a spatial resolution of $1^\circ \times 1^\circ$ is used. The data was obtained freely from NASA,

Goddard Earth Science Data and Information Services Center (GES DISC) (http://disc.sci.gsfc.nasa.gov/hydrology/dataholdings).

6.4.3 Tropical Rainfall Measuring Mission (TRMM)

TRMM, a joint USA/Japan satellite mission, is designed to survey the rain structure, rate, and distribution in tropical and subtropical regions (latitude range ±50°) [45]. Amongst the TRMM precipitation products are those derived from integrated rainfall estimates from various sensors on-board satellites such as Precipitation Radar (PR), Special Sensor Microwave Imager (SSM/I), infra-red (IR) data from geostationary satellites, and additional merged rain gauge observations [40, 48]. TRMM-3B42 version 7 [39], which was made available to the public on 22 May, 2012 was used. It has a spatial resolution of 0.25° × 0.25° and temporal resolution of 3-h. The major difference between the previous versions and the one used by [90] is that the later has an improved scheme for weighting the incorporated rain-gauge data [33, 39]. The data was obtained from NASA GES DISC (http://mirador.gsfc.nasa.gov/).

The motivation for selecting TRMM satellite data in this chapter is informed by the works of Dinku et al. [16, 23–25, 97, 98] in terms of the validation of satellite products in the region, which includes nearly half of Ethiopia (i.e., the Blue Nile portion). In particular, Dinku et al. [25] reported the fairly good performance of the TRMM- 3B42 rainfall product compared to locally available rain gauge data. It should be pointed out that although another product, TRMM-3B43, provides monthly data, [90] used the shorter daily TRMM-3B42 product mainly due to the fact that these have been validated in the region, see, e.g., [25].

6.4.4 Analysis Methods

As already stated in Sect. 3.3.3, GRACE products are limited in their usefulness to a spatial extent of greater than 200,000 km^2. On the one hand, only two of the Ethiopian Basins (Abbay and Wabishebele) can fulfill this criterion. On the other hand, Ethiopian catchments exhibit very different characteristics as they are climatically and environmentally extremely heterogeneous (see Sect. 6.3). In this respect, computing the TWS over Ethiopia as a whole might lead to wrong conclusions, since the basins' averages will follow the general variability of the area with the dominant signal. To overcome this problem, Ethiopia was subdivided into 4° × 4° (approximately 440 km × 440 km, i.e., 193,000 km^2) blocks that enable (i) the spatial extent required for the valid consideration of the GRACE products, and (ii), minimizing the errors that might result from generalizing products for Ethiopia as a whole. This way, ten regions are obtained, whose minimum/maximum latitude and longitude, and designation to be used in this chapter from now on are shown in Fig. 6.2. These subdivisions are made in such a way that larger physiographic regions of Ethiopia are represented as close as possible. Since GRACE data sets have already been smoothed using the DDK3 filter [47], the computation of GRACE-derived TWS is undertaken following the approach presented in Wahr et al. [84]. For all 10 regions (Fig. 6.2), TWS values were generated with an approximate monthly temporal resolution over the study period (2003–2011), where the average field from the study period (2003–2011) was removed from the monthly TWS. The missing data from June 2003 was filled using a simple linear interpolation. The average error of the derived TWS over the whole region of the study was estimated to be less than 5 mm of water columns. This error was estimated using the formal errors of the Stokes coefficients smoothed using DDK3 filter and converted to TWS cf. [84].

Monthly soil moisture down to 2 m beneath the surface has been computed from NOAA Land Surface Model (LSM) and water budget data obtained for four layers (0–10, 10–40, 40–100 and 100–200 cm), with spatial resolutions of $1° \times 1°$ and aggregated to a spatial domain of $4° \times 4°$. The average rainfall over each study region obtained from TRMM-3B42 was also rescaled to $4° \times 4°$ and a monthly temporal resolution. Since TRMM 3B42-V7 rainfall rate are 3-hourly averages centered at the middle of each 3-hour period (i.e., 0, 3, 6, 9, 12, 15, 18, and 21 h), they are converted to total daily rainfall by first multiplying each 3-hourly rainfall rate by 3 to get the total rainfall for each 3-h period, then added to obtain the desired monthly data, the sum of all the 3-hourly total rainfalls in within 24-h period is multiplied by the number of the days for a specific month. Both TRMM and GLDAS data were filtered using a Gaussian filter of 300 km radius. The radius of 300 km was selected to be consistent with the smoothing impact of the DDK3 filter applied to the GRACE products. This value has been tested, e.g., in Werth et al. [85] and was found to be applicable to nearly all river basins.

6.4.5 Groundwater Changes from GRACE and GLDAS

The most important variables to use when deriving groundwater storage changes (GW) from GRACE-TWS data are; (i) changes in surface water storage (SWS) associated with lakes and reservoirs, (ii) changes in soil moisture (SM), and (iii), changes in snow water equivalent (SWE). However, when compared to the land surface, only about 0.7% of Ethiopia is covered by water bodies [54]. Therefore, this chapter does not consider the contribution of surface water storage in computing TWS over Ethiopian regions since the areas of the existing water bodies and their fluctuations levels were quite small. The snow water equivalent was also not considered in the computation since its temporal variability over Ethiopia is negligible compared to the other contributors of TWS. Soil moisture anomalies were computed by removing the mean value over the extent of the study period (2003–2011) of the monthly soil moisture values. The monthly groundwater anomaly (ΔGWS) was then obtained by subtracting changes in soil moisture from TWS changes as:

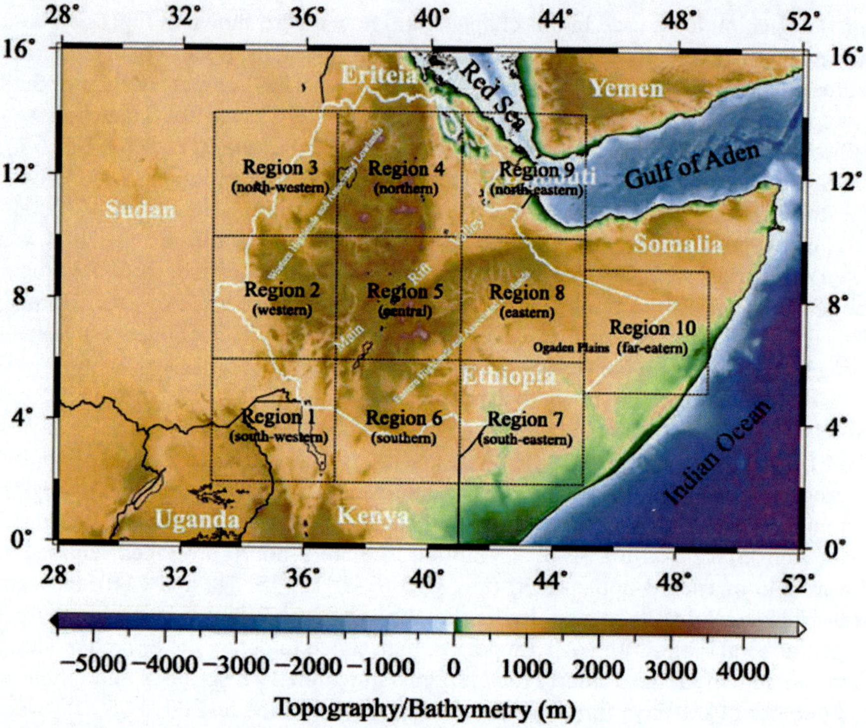

Fig. 6.2 Ethiopia is divided into ten study regions of 4° × 4° to meet the spatial extent requirement of the GRACE products of 200,000 km^2 and the physiographic regions. The x-axis contain the longitudes (East) while the y-axis show the latitudes (North). *Source* [90]

$$\Delta GWS = \Delta TWS - \Delta SM, \tag{6.1}$$

assuming that the surface water and snow are insignificant. This approach has been previously applied by Rodell et al. [67] in Northern India with the aim of estimating groundwater depletion.

In association with the GRACE-derived TWS estimates over the Ethiopian Highland, Bonsor et al. [17] used a recharge model for the Nile Basin, Zoomable Object Oriented Distributed Recharge Model (ZOODRM), to interpret the seasonal variation in TWS. They found that the simulated annual variation in groundwater storage and soil moisture accounts for 50–90% of the variations in the GRACE-derived TWS.

6.4.6 Total Water Storage Duration Curve (TDC) and Total Storage Deficit (TSD)

To simplify the information available for judging the characteristics of TWS, this chapter expresses the TWS changes as a single value through the use of the TWS duration curve (TDC). The TDC is derived through the principle of the Flow Duration Curve (FDC). The FDC is a frequency distribution formed from daily/monthly stream flow data and their exceedence probability [86]. In this approach, all the available data are first listed in descending order and given a rank (the maximum will take the value 1 and the minimum will take the last rank). Using the ranks m, the probability of exceedence (in percentage) is computed using the Weibull method [38, p. 23]:

$$\%Exceedence = \frac{m}{(N+1)} \times 100\%, \tag{6.2}$$

where m is rank and N is the sample size. TDC is obtained by plotting the TWS against its probability of exceedence. The slope STDC of the TDC is then calculated between the 33rd and 66th TWS percentiles using

$$S_{TDC} = \frac{TWS(33\%) - TWS(66\%)}{(66-33)}, \tag{6.3}$$

modified for the TWS case from Sawicz et al. [69] as this portion represents a relatively linear part of the TDC. This slope reflects the aquifer storage size and the aquifer through flow property.

To assess the anthropogenic effect on TWS through soil moisture, total storage deficit (TSD) in each regions is computed from GRACE derived TWS anomalies using the approach stated by Yirdaw et al. [87] as

$$TSD_i = \frac{TWS_i - Mean(TWS_i)}{Max(TWS_i) - Min(TWS_i)} \times 100\%, \tag{6.4}$$

where TSD_i is the Total Storage Deficit (%); TWS_j is the monthly total storage anomaly as derived from GRACE-measurements, Mean (TWS_i), Max (TWS_j), and Min (TWS_j) are the long-term mean, maximum, and minimum TWS, respectively, for each month j over the study period (2003–2011).

6.4.7 Statistical Analysis: Correlation and PCA

6.4.7.1 Correlation Analysis

To study the lag time of water storage variations (TWS and soil moisture) in response to rainfall, lagged correlation analyses at 95% confidence level were carried out between (a) TWS and rainfall, (b) soil moisture variation and rainfall, and (c), soil moisture anomalies and TWS.

6.4.7.2 Principal Component Analysis (PCA)

Principal Component Analysis (PCA) is a statistical technique that can be used to extract dominant uncorrelated patterns from spatio-temporal observations [61, 95]. In principle, PCA expands the derived TWS changes within the 10 defined regions in terms of new sets of orthogonal vectors known as empirical orthogonal functions (EOFs) associated with their uncorrelated temporal evolutions known as principal components (PCs) [34, 35]. The PCA method used in this chapter is based on the eigenvalue decomposition of the data-derived auto-covariance matrix [61]. Here, only the dominant components of TWS behavior over Ethiopia are shown. The less dominant components that most often correspond to lower correlations with overall TWS changes are not presented. To decide on the significant number of modes, North's rule of thumb, which states that if the sampling error of a particular eigenvalue is comparable to a nearby eigenvalue, then the sampling errors in the EOF will be comparable to the "nearby" EOF [55] was used (see [61] for details).

6.5 Hydrogeological Characterization

Figure 6.3 shows box plots used to highlight the overall patterns of response for the variations in water storage over Ethiopia computed from GRACE, soil moisture (from GLDAS), and rainfall (from TRMM) for the 10 sub-regions over the study period (2003–2011). The lower whisker of the box plot indicates the lowest observed value (sample minimum), the lower end of the box is the lower quartile (25%), the line across the box indicates the median, the upper end of the box specify the upper quartile (75%), and the upper whisker of the plot illustrates the highest observed value of the sample. The red crosses indicate outliers that represent cases that have values more than three times the height of the boxes. Overall, the results in Fig. 6.3 (top) show that the variation in TWS is greatest in the western highlands area (see Fig. 6.1), i.e., western (region 2), north-western (region 3), and northern (region 4) of Ethiopia, while the far eastern corner (region 10) shows relatively low variation. Highland dominated regions (Fig. 6.3, middle and bottom, regions 2, 3, 4 and 5) show strong variations in soil moisture and rainfall, while the lowland dominated segments display comparatively less variability (Fig. 6.3, middle and bottom).

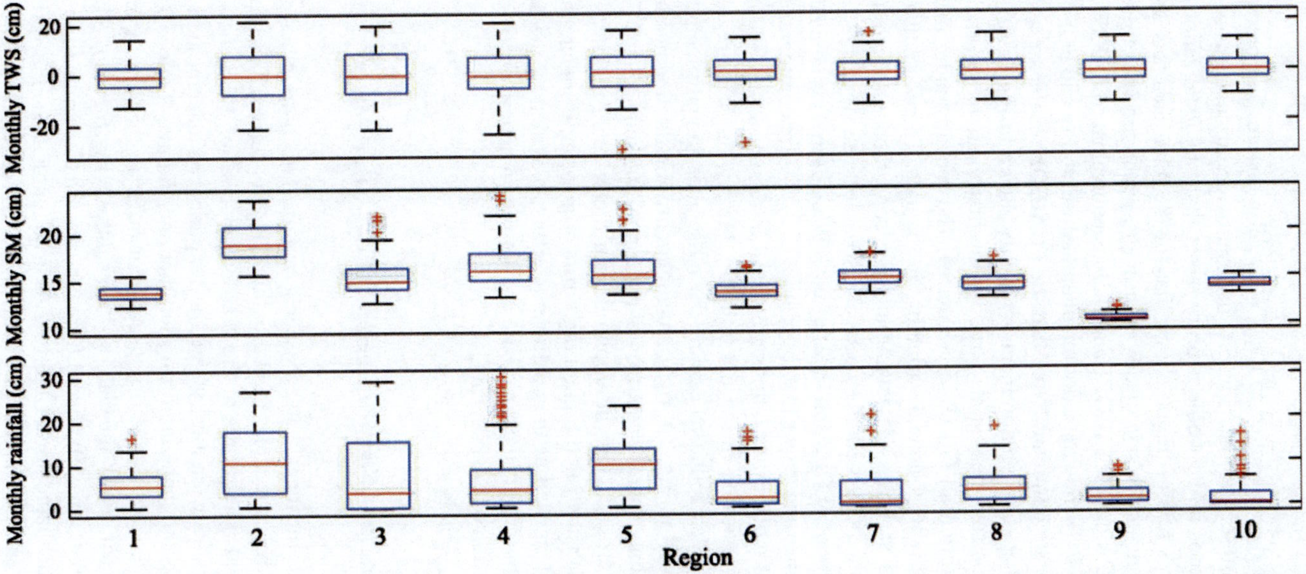

Fig. 6.3 Summary of TWS, soil moisture, and rainfall variability for each of the Ethiopian region shown in Fig. 6.2. The red crosses indicates outliers. TWS (top) is seen to be greatest in western (region 2), north-western (region 3) and northern (region 4) region and less so in the far east (region 10). Regions 2–5 (central) show a strong variation in soil moisture (middle) and rainfall (bottom). *Source* [90]

6.5.1 *Dominant Variability of Water-Storage over Ethiopia*

The PCA method was applied to the TWS, soil-moisture, and precipitation fields derived for the 10 sub-regions (Fig. 6.4) in order to identify patterns of simultaneous temporal variations in the whole country rather than any localization of the signal. Applying North's rule of thumb shows that for the three data sets above, only the first two components are statistically significant, which are shown in Fig. 6.4. The first mode of PCA on TWS (EOF1 and PC1) is equivalent to 65.83% of total TWS variability, with PC1 showing a dominant annual variability and EOF1 showing that regions 2–5 are the dominant. Note that the PCA/EOF results are somewhat close to the regions that show predominant variability shown in Fig. 6.3(a), as one could expect. The same statement is also true for the first mode of soil-moisture (67.4% of variance) and rainfall (70.1% of variance). The second modes of all three data sets extract mostly inter-annual variability in the data. There are also some annual and intra-annual variations detectable in the second modes, which are due to the incapability of the PCA method to perfectly separate such signals [35].

The second mode (EOF2 and PC2) of the TWS (Fig. 6.4) is equivalent to 14.9% of the total variance and shows that the regions 2, 3, 6–9 are dominant. The dominant inter-annual soil moisture variations are found in regions 1, 2, 5–8, and 10 (EOF2 and PC2 of soil-moisture in Fig. 6.4, middle). Finally, the same regions, i.e., 1, 2, 5–8, and 10 are also found to exhibit the most inter-annual precipitation (EOF2 and PC2 of rainfall in Fig. 6.4, right), again confirming the findings in Fig. 6.3a. It should be mentioned that the accuracy of the annual and semi-annual GRACE-TWS estimations are around 1 cm (in amplitude) for the selected regions. This has also been tested by considering the sampling errors originating from length-limited data sets [61, p. 199], as well as the coefficients errors, provided by the data producer, the GFZ Potsdam Center (the error-bars are not shown here in order to enhance the visual interpretation of the patterns).

6.5.2 *Annual and Seasonal Mean TWS Changes*

Figure 6.5 depicts the mean water mass lost or gained in cm/month aggregated over seasonal and annual time scales from monthly GRACE-derived TWS values over the 2003–2011 period. The results of the seasonal analysis of TWS show regions 1–5 having the same characteristics, where they experienced a decrease in TWS during winter and spring and gained it during summer and autumn. Regions 7 and 8 reflect the same characteristics as the first group, but differ in that their main water loss season is during winter and the water gain in summer (Fig. 6.5, top). The other three regions (region 6, 9, and 10) have different characteristics from all the others. The only characteristic that regions 6 and 10 have in common is that they gain water mass for 75% of the year (i.e., 3 seasons), while it is 50% for region 9. The results, from

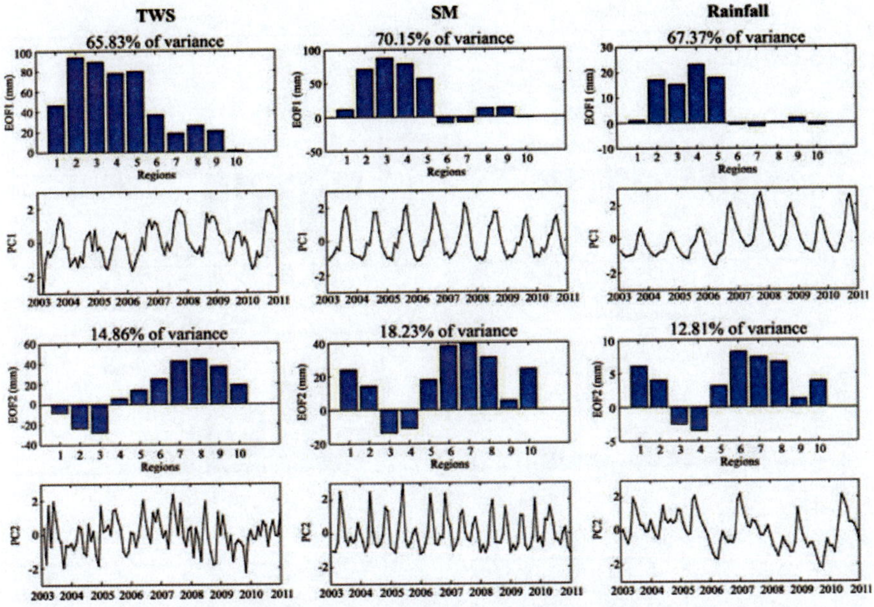

Fig. 6.4 Results of PCA, applied on the time series of 10 predefined regions; (left) TWS data from GRACE, (middle) soil moisture data from GLDAS (right) and, (right) rainfall data from TRMM. *Source* [90]

both GLDAS and GRACE products, confirm the fact that summer is the main water gaining season for Ethiopia corresponding to the main rainy season.

The annual mean indicates that mass has been gained in the northern (region 4), central and eastern (regions 5 and 8 respectively), southern (regions 6 and 7) and far eastern (region 10) regions, while the south-western (regions 1), western (region 2), north-western (region 3), and north-eastern (region 9) regions showed that mass has been lost (Fig. 6.5, bottom). The maximum mean annual water mass gains were recorded in central Ethiopia (region 5) during summer (Fig. 6.5), which may be attributed to three main factors. First, the larger part of the Rift Valley Basin lakes lies in this region (see Figs. 6.1 and 6.2), and could be exerting a greater surface water influence. The Ethiopian Rift Valley lakes' basin (Fig. 6.1), which lies in the main Ethiopian Rift, is a collection of cascade lakes, which might have underground connections with other lakes. However, regarding the second factor, considering the amount of surface water, the Rift Valley lakes' basin is known to have a closed watershed, and as such there is water going into the lakes, but have no surface water outlets, e.g., Lake Naivasha [5]. Thirdly, the region is again the starting point of four of the six Ethiopian international basins that include the Awash, Omo-Ghibe, Ganale-Dawa and Wabishebele.

The GRACE-derived TWS duration curve (TDC, see Eq. (2)) is presented in Fig. 6.6. From the probability of exceedence at the no loss/no gain point (i.e., point

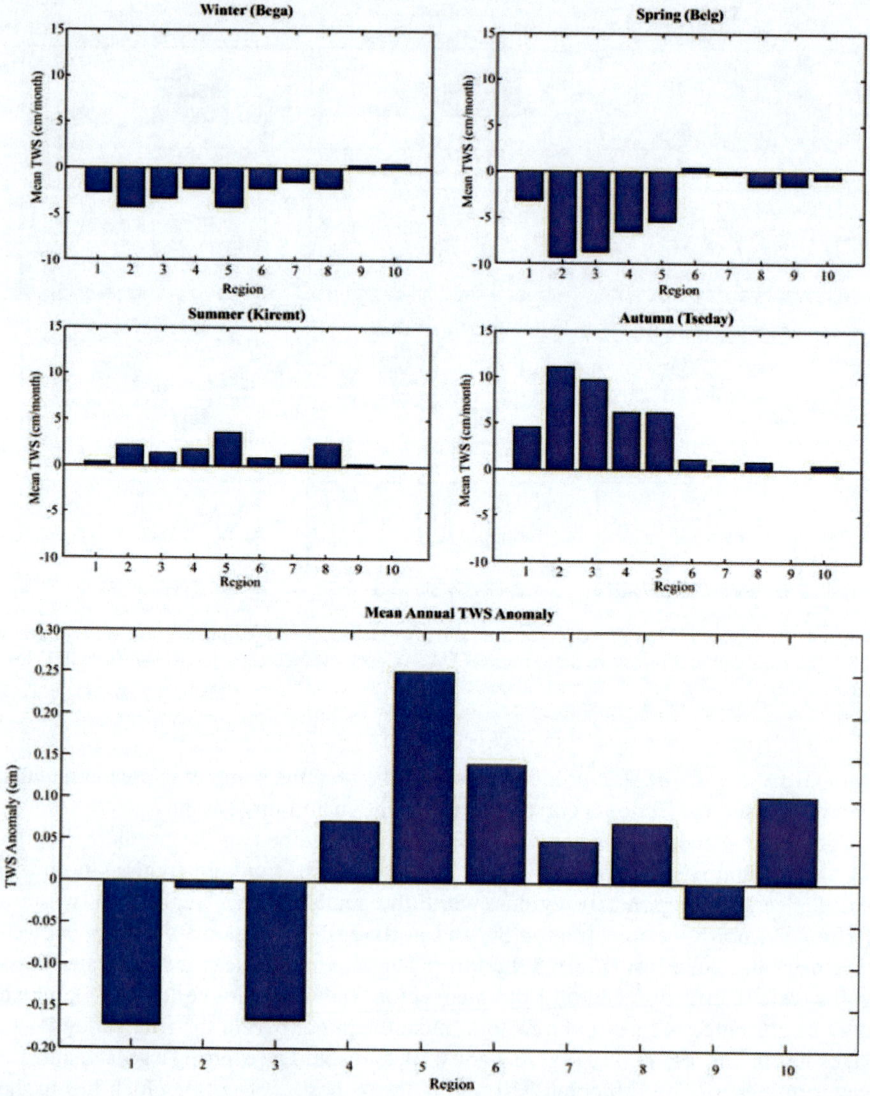

Fig. 6.5 Seasonal (top) and Annual (bottom) mean of TWS variation over Ethiopia derived from GRACE. *Source* [90]

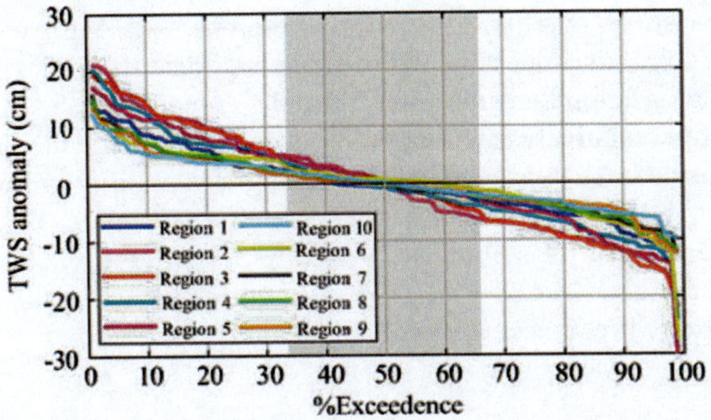

Fig. 6.6 Total Duration Curve (TDC) of the GRACE-based monthly TWS anomalies over the study period. The figure indicates that almost all regions gained water 45% of the time. Slope analysis of this figure show that over Ethiopia, TWS changes are high in the western region and low in the eastern region. The extreme fall at the end of the curves for regions 5 and 6 could be the consequence of the loss of a large amount of water in January and February 2003, which was the result of drought that propagated from the previous year. *Source* [90]

at which the TWS anomaly equals zero), it can be seen that almost all regions gained water mass only 45% of the time, and lost it 55% of the time. The TDC curves in Fig. 6.6 also allow us to reflect upon the probability of change in amount of TWS. For, example, the maximum water mass gain, which has a probability of occurrence of less than 1% over the 2003–2011 period, ranges from 12 cm/month in region 10 to 21 cm/month in region 2, while the maximum water loss with a percentage exceedence of about 99% ranges from 30 cm/month in region 5 to 10 cm/month in region 10.

To quantify the TWS changes into a single value, the slope of the TDC was calculated using Eq. (4). A high slope value means rapid response and small storage potential, while a low slope value means slow response to incoming water mass and large storage potential. For simplicity, we can group the regions based on their TDC slopes as, (i) $S_{TDC} > 30\%$ (only region 2), (ii) $20\% < S_{TDC} < 30\%$ (regions 3–5), and (iii), $10\% < S_{TDC} < 20\%$ (regions 1, 6–10). This indicates that over Ethiopia, the TWS changes are highest in the western region and lowest in the north-eastern region, i.e., consistent with the findings in Figs. 6.3 and 6.4.

6.5.3 Inter-annual Variation

Changes in TWS, soil moisture (SM), rainfall, and groundwater storage (GW) anomalies over the study period for each month have been compiled in Fig. 6.7. The year to year changes in TWS (Fig. 6.7a) shows that except for a few months, e.g., June (regions 1, 4 and 9), July (regions, 7 and 8) and September (region 6), there exist significant variability in TWS in all regions for each month. Figure 6.7b also show considerable variations in soil moisture for all months, where it is relatively strong during rainy seasons. The inter-annual plot of rainfall illustrated in Fig. 6.7c describes the considerable variation that exists in all months in regions 1 and 2, while regions 3–5, 8 and 9 show relatively low variability except during the major rainy season (June–September; JJAS). Additionally, there is a low variability in June–September for regions 6, 7 and 10, i.e., south east and east. The GW plots (Fig. 6.7d) present similar characteristics as the soil moisture and TWS. From Fig. 6.7, the variations of TWS, SM and GW anomalies are seen to be less correlated to rainfall for all regions with the exception of region 2 (western), see also Fig. 6.11.

6.5.4 Intra-annual Variation

Figure 6.8 shows the intra-annual variability of TWS, soil moisture, GW and rainfall developed from the mean of each month over the study period as opposed to the red lines in Fig. 6.7. From this, one can clearly see two distinct groups based on the behavior of their TWS anomaly values over the months. The first group includes regions 2–5, where the TWS and GW anomalies shows three similar parts; (i) the rising limb, which occurs from June to September, indicating the recovery period of TWS, (ii) the peak, which occurs in August, and (iii), the recession or falling limb that covers a longer period than the previous ones, i.e., starting from a point at which the peak is attained and extending up to June. The results indicate that, like the rainfall patterns, TWS shows seasonality, having seasons showing mass gain (JJA and some parts of SON), while in the other two seasons, the mass loss exceeds the gain in most regions of Ethiopia. This happens since the mentioned periods are the main rainy seasons in Ethiopia. The second group includes all the other regions with characteristics opposite to the first group. That is, they have short duration recession curves with very long recovery or rising limb period, attaining their maximum peaks in the SON season. Despite the fact that SON is the long rainy period in these regions, the length of the rising period might be attributed to the aquifer type and size, as well as recharge mechanism. Compared to the rainfall, and the response of GW and TWS, one can see that there appears to be very little fluctuation in soil moisture in all cases.

Examining now the soil moisture in greater detail (Fig. 6.9), one can visually identify extreme regions that include (i) the western part of Ethiopia (i.e., region 2), which has the highest soil moisture content of all the studied regions, (ii) the north-eastern (i.e., region 9), which has the lowest soil moisture content, and (iii), the intermediate regions, which include all the other areas. Apart from this, Ethiopia may be classified into three soil moisture variability regions. The first covers the western (region 2), north-western (region 3), northern (region 4) and central (region 5) sub-regions, which show little variability over the first half of the year, and then a significant rise in the second half, attaining their peaks in September. This is because the second half of the year is the main rainy season for most segments. The second soil moisture region (southern; region 6 and south-eastern; region 7) is characterized by

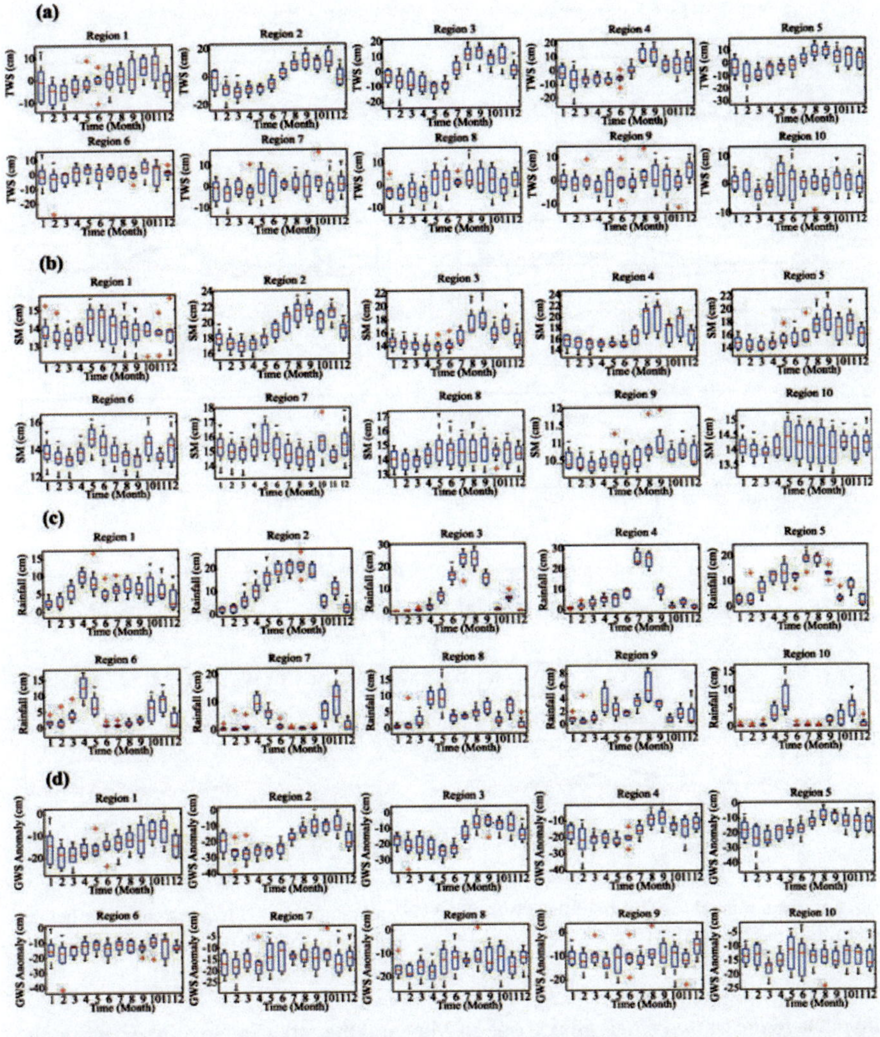

Fig. 6.7 Inter-annual variability of TWS (**a**), Soil Moisture (**b**), Rainfall (**c**) and GW anomaly **d** over Ethiopia. The variation in TWS (**a**), soil moisture (**b**), and GW (**d**) are unrelated to the rainfall (**c**), thus indicating a possible human influence on hydrological changes in Ethiopia. Note the difference in amplitude of the figures. Although (**a**) and (**d**) look similar, they differ considerably in amplitude. *Source* [90]

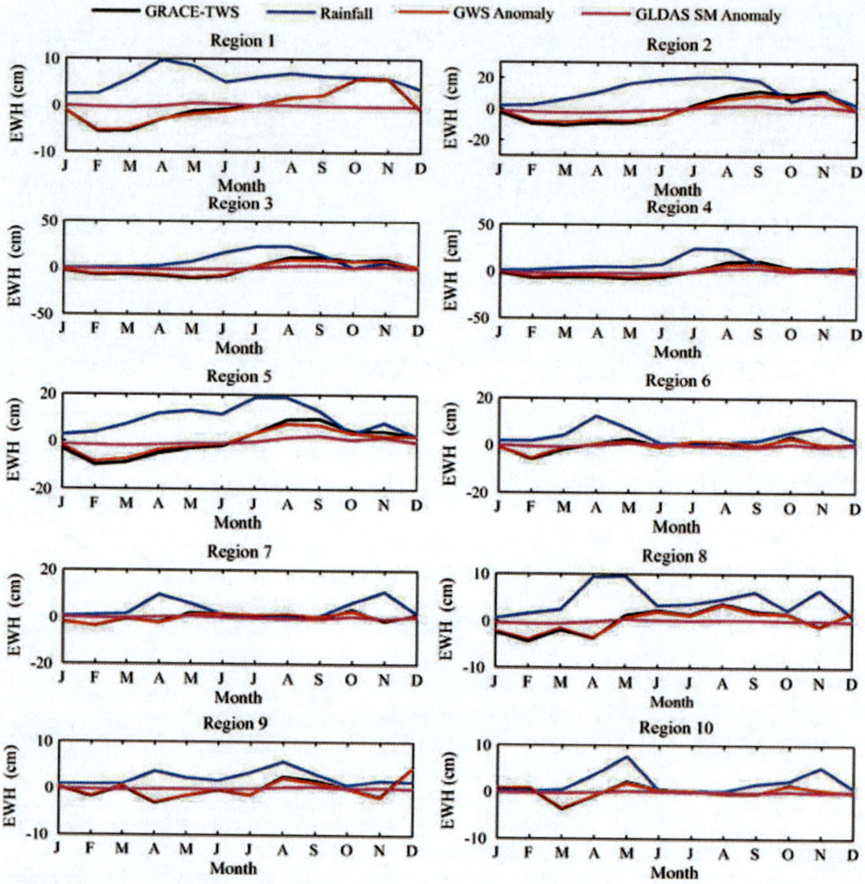

Fig. 6.8 Intra-annual variability of the TWS, SM, GW and rainfall EWH (equivalent water height) over Ethiopia. *Source* [90]

the occurrence of two small peaks, one in May and the other in November, while the third soil moisture region includes the south-western (region 1) and eastern (region 10), displaying very little variability in soil moisture throughout the year compared to others.

The soil moisture variation over Ethiopia bears a resemblance to the agro-ecological zonation described in Hurni [41]. Depending on the dominant agro-ecological zone in each region, three sub-divisions may exist. These are; (i) Dega/Woyna Dega consisting of regions 2, 4 and 5, (ii) the Bereha made up of regions 1, 6, 7, 8, 9 and 10, and (iii), the transitional zone Kolla/Dega, comprised only of region 3. The Dega/Woyna Dega group displays high soil moisture levels with strong seasonality, while the Bereha group shows little seasonality behavior as they receive little rainfall. Furthermore, Fig. 6.9 indicates two regions of comparatively extreme

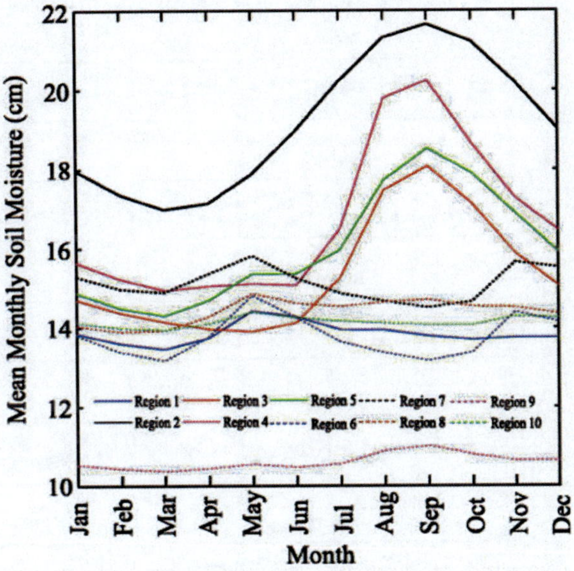

Fig. 6.9 Intra-annual variability of soil moisture over Ethiopia. *Source* [90]

wetness and dryness, i.e., regions 2 and 9, respectively. The possible reason for region 2's wetness is the fact that (i) this region receives rainfall almost throughout the year, comprising the area which obtains the highest annual rainfall amount in Ethiopia (see Sect. 6.3) and (ii), the land cover in the region is highly dominated by rainforest. In region 9, the Danakil depression region, which shows much lower soil moisture content, is known to be one of the driest regions of Ethiopia, where there is no significant river flow and only short period bimodal rainfall, very scarce vegetation cover, and high temperatures (see Sect. 6.3). Ertale, the area where volcanic magma can be seen at Earth's surface is also in this region, where high temperatures, combined with little rainfall, lead to the retention time of water in the soil to be very limited.

6.5.5 Correlation Between Different Data Sets

A correlation analysis between available data sets was made using the mean monthly data and is presented in Fig. 6.10 to study the time lag between rainfall and TWS, soil moisture and TWS, and soil moisture and rainfall. The analysis shows almost no lag between soil moisture and TWS in all regions, except regions 1 and 9, where lags of 5 and 1 month, respectively are seen. For TWS and rainfall, lags of 0 (region 10), 1 month (regions 4 and 7), 2 months (region 3), 3 months (regions 2, 5 and 8), 4 months (region 9), and 6 months (regions 1 and 6) have been observed, with strong correlations in regions 2–5. The lags between TWS and rainfall obtained from Fig. 6.10 are tabulated in Table 6.2. Larger lag periods are indicative of water-storages

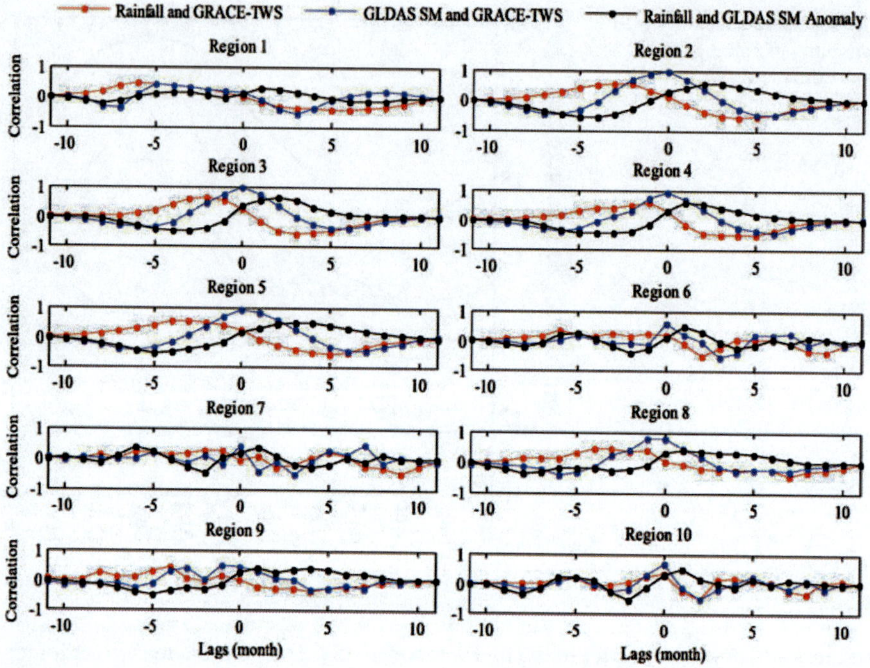

Fig. 6.10 Correlation between different data sets (TWS, soil moisture, and rainfall) at 95% confidence level. *Source* [90]

properties such as large groundwater reservoirs, indirect recharge mechanisms, or a combination of both, while smaller values are more likely to be attributed to direct recharge mechanisms, smaller groundwater reservoir or again their combinations.

6.5.6 Rainfall and Water Storage Changes: Relationship

From Figs. 6.7 and 6.8, one can see that there is little significant delineation between the plots of TWS and GW. It is, therefore, thought that the very large influence of TWS over Ethiopia comes from the groundwater storage. The relationship between rainfall and TWS is presented in Fig. 6.11. This is developed by accounting for the lag duration of TWS from rainfall. For example, in region 1, a lag of 6 months is used to assess how much of the rainfall observed in January contributed to the TWS in July. In all regions, except 2, 4 and 5, coefficients of determinant R^2 of less than 0.5 were obtained. For these regions, where high values of R^2 are obtained, i.e., the western (region 2), northern (region 4) and central (region 5), two deductions could be made, i.e., (i) either the regions are source of water (e.g., regions 2, 4 and 5),

Table 6.2 Recharge mechanisms, recharge rates, hydrogeology, aquifer flow, and storage type characteristics over different regions of Ethiopia (*Source* [1, 18]). The Table also includes the slope m of the TDC derived from GRACE-TWS for each region and the lag between GRACE-TWS and TRMM-rainfall. *Source* [90]

Region	Recharge mechanisms	Recharge rate (mm/yr)	GRACE-mean Annual TWS(mm/yr)	Hydrogeologic setting	Aquifer flow and Storage type	TDC slope (m)	Lag (months)
1	Recharge from wadi beds through local wadi floods	50–150	−26.64	Unconsolidated sediment	Inter-granular	−0.15	6
2	Direct diffuse recharge through soil and root zone. Recharge from wetland during high flood	150–250	−1.08	Volcanic rocks, metamorphic rocks and intrusives	Fracture	−0.344	3
3	Wadi bed through local wadi floods, Selective recharge from heavy rainfall	50–150	−19.92	Volcanic rocks, metamorphic rocks and intrusives	Fracture	−0.276	2
4	Selective recharges from heavy rain falls, direct diffuse+floods from highlands	50–150	8.64	Metamorphic rocks and intrusives	Fracture	−0.236	1
5	Selective recharges from heavy rain falls, Mountain block and mountain font	50–250	30.12	Metamorphic rocks and intrusives	Fracture	−0.221	3

(continued)

Table 6.2 (continued)

Region	Recharge mechanisms	Recharge rate (mm/yr)	GRACE-mean Annual TWS(mm/yr)	Hydrogeologic setting	Aquifer flow and Storage type	TDC slope (m)	Lag (months)
6	Selective recharges from heavy rainfalls wadi beds through local wadi floods	50–150	17.04	Metamorphic rocks and intrusives, consolidated	Fracture	−0.146	6
7	Selective recharges from heavy rainfalls wadi beds through local wadi floods	<50	6.00	Consolidated sediments	Inter-granular and Fracture	−0.136	1
8	Selective recharges from heavy rainfalls wadi beds through local wadi floods	<50–150	8.64	Consolidated and unconsolidated sediments	Inter-granular and Fracture	−0.138	3
9	Flood Water from the highlands	<50	−4.80	Metamorphic rocks and intrusives	Fracture	−0.116	4
10	Recharge from wadi beds through local wadi floods	<50	12.48	Consolidated sediments, unconsolidated Sediments	Fracture (Karst)	−0.138	0

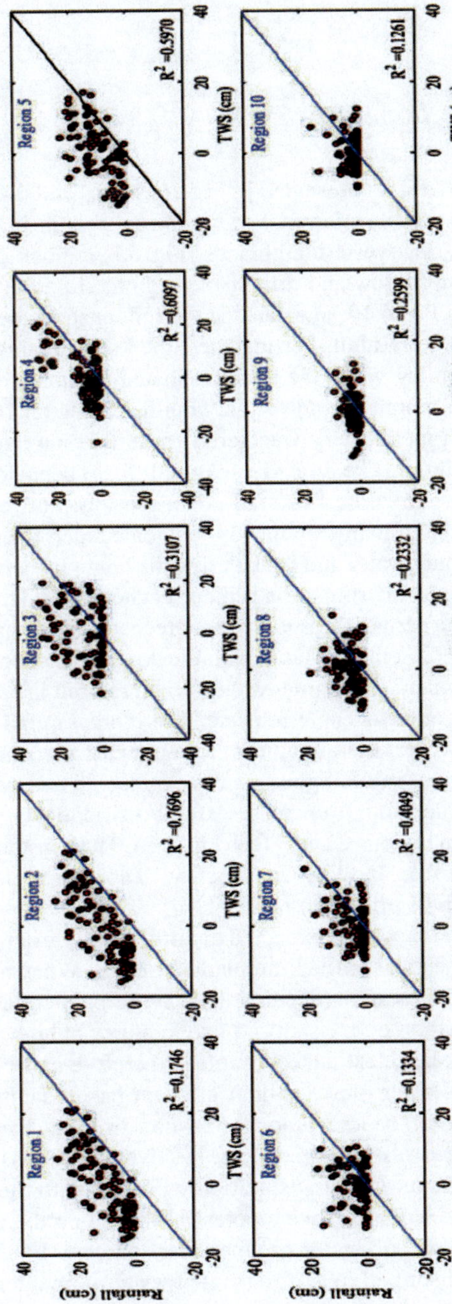

Fig. 6.11 Relationship between TRMM rainfall in January and GRACE TWS in July for each region at a lag of six months. This lag is chosen as an example to show the impact of the January rainfall on TWS in July. *Source* [90]

or (ii), the reliance of the groundwater storage variability on the rainfall pattern is minimal.

6.5.7 Topographic Impact on TWS

As one can see from the results presented in Figs. 6.3 (top) and 6.4, the influence of topography on the TWS may be inferred. From these figures, the highland dominated regions of Ethiopia, i.e., the western highlands (Fig. 6.2, regions 2–5) show greater variability in TWS than the lowland-dominated regions. In addition, from the lag correlation presented in Fig. 6.10, one can see the influence of the regime's hydrogeology on its response to rainfall. For instance, the karst dominated regions (i.e., 10 in Table 6.2) has zero lag while the unconsolidated sediment dominated region 1 has a lag of around 6 months. In the Krast dominated region 10, it is normal to get 0 lag because Krast means very weathered rocks that have many channels at the surface to absorb rainfall. The direct recharge (diffuse) behaviours are expected to be very obvious if the amount of rainfall is not too low. For region 1, it should be pointed out that although many studies directly calculated the cross-correlation between rainfall and groundwater and considering the lag to be between rainfall and groundwater recharge, such interpretation is incorrect because it ignores the fact that rainfall needs to go over a certain threshold before recharging groundwater [100]. So the lag of 6 months above just mathematically indicates the distance between the first rise of rainfall and the first rise of groundwater. When rainfall is below the recharge threshold, rainfall and groundwater is irrelevant. Therefore, rainfall in region 1 starts to rise in February, however, the amount is not enough to recharge groundwater. After March, the amount of rainfall goes over the threshold and groundwater starts to recharge. Although the rainfall decreases after April and May, region 1 keeps receiving the runoff from regions 2 and 5 (see Fig. 6.8). Thus, groundwater in region 1 keeps rising until October. In October, the rainy season in regions 2 and 5 also ends, no more runoff can be offered to region 1.

However, hydrological fluxes are expected to differ greatly from the outlet to high elevations for some of the areas in the Ethiopian highlands. Whereas precipitation is expected to increase with elevation (depending on aspect), potential evapotranspiration decreases with elevation due to reduced temperatures at higher elevations (cf., [13]). Based on water balance estimates, runoff is therefore expected to be greater at higher elevation than lower ones for high gradient basins facing the prevailing storm movement direction. The total amount of runoff from the basin will therefore depend on the hypsometric relationship, i.e., the relative proportions of drainage area at different elevations. Since evapotranspiration will vary with the type of vegetation, changes in land use will also have a potential impact on the amount and type of evaporation (from either surface or groundwater sources). In addition, the conversion of relatively undisturbed (e.g., forested) areas to more intensive agriculture will greatly increase surface runoff at the expense of groundwater recharge due to reduced infiltration capacity. It is therefore critical to assess the role of elevation in hydrological fluxes coupled with seasonal variability and changes under potential climate scenarios, see also Chap. 15 and [95].

6.5.8 Possible Human Influence on the Observed TWS

Ethiopia is a highly populated region that has seen a number of dams constructed (see e.g., [96]) and high levels of water withdrawal from groundwater and surface water reservoirs, which would impact upon the total water storage, e.g., [12, 83]. Large scale land-use and land-cover changes in Ethiopia due to agricultural expansion and heavy grazing have been reported, e.g., [14, 22]. In addition, Hurni and Nyssen et al. [41, 57] have reported that land degradation is a serious problem in Ethiopia. Ethiopia is an Agrarian country, where more than 85% of its population depend on irrigated agriculture, with production in certain regions being highly dependent on the available water resources (i.e., surface water and groundwater). The total area under irrigation in Ethiopia is about of 290,729 ha ($\sim$0.3% of the country's area), with about 2611 ha depending on groundwater irrigation [76]. The irrigation water demand in Ethiopia is estimated to be 40 km^3/year [27], with an estimated 2.6–6.5 km^3/year of groundwater potential [10]. The principal grain crops are wheat, barley (primarily cool-weather crops) and corn, sorghum, and millet (warm weather grain crops).

Since the GRACE products can resolve large areas to the tune of hundreds of km (i.e., 450 km $\times$ 450 km), the impact of anthropogenic activities at hill slope scales, such as those discussed in Descheemaeker et al. [22], on its signals will not be detectable. However, if such impacts occur over a wider scale that is of the same order as the resolution of GRACE, then they may be detectable.

TSD in percent (Eq. (4)) are computed from the GRACE-derived TWS anomalies and presented in Fig. 6.12. Depending upon their TSD patterns, Ethiopia can be divided into three different zones. In the first zone, we have regions 2–5, while regions 1 and 6 are in zone 2, with regions 7–10 in the third zone. For the three zones, TSD decreased from the beginning of 2003 until around mid-2006 for zones one and two, and January 2005 for the third zone. Then, for the first and third zones, it started to rise in 2006, while for the second group this was from around the beginning of 2007. The decrease in TSD starting from Jan-2003 indicates increasing dryness, which may be due to the propagation of the 2002 drought in Ethiopia [3, 94]. In regions where significant TSD has been seen from 2003 to 2006 (e.g., Zone 1 in Fig. 6.12), the main crops are teff, sorghum, maize, wheat and millet. Despite the decrease in TSD in the regions, there has been an increase in crop production area and yield [26], indicating on the one hand that such crops can be sustained under such water deficit conditions. However, on the other hand, this could indicate that the greater portion of this deficit in TWS could be due to groundwater storage depletion. This, however, remains an open question for future investigations.

In regions 1–3, and 9, over the study period, very significant water loss was identified (see Fig. 6.5, bottom), especially when compared to the recharge amount

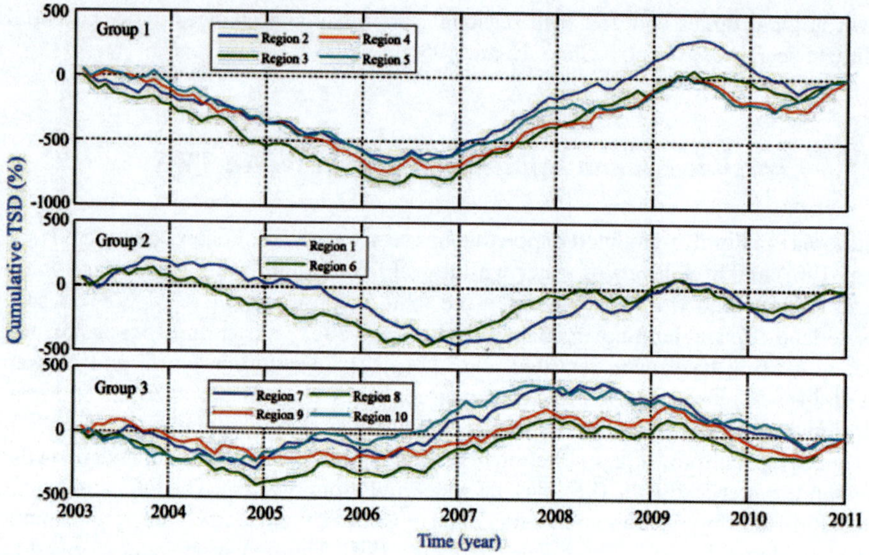

Fig. 6.12 Total Storage Deficit (TSD) derived for the main 10 regions defined over Ethiopia subdivided into different zones. The values are in percent (see Eq. (4))

per year, as stated in Table 6.2. This might be attributed in part to human exploitation and partly due to climate impacts discussed below.

From the developed TDC using 96 months of GRACE derived TWS data (Fig. 6.6), the lower end of the curve might be due to extreme water depletion cases, while the upper end results from high rainfall condition with long return periods. This is again an open question subject to future investigations. Nonetheless, to see the real variation of TWS, the slope of the TDC in the region where one can assume linearity is computed. If this slope is expressed in percentage, it varies between -14 and -35%. There are several factors that might affect the slope of the TDC, e.g., soil moisture, hydrogeological setting, aquifer flow, storage types, storage potential, recharge rates and recharge mechanisms. With the characteristics listed in Table 6.2, and their TDC slope, one notes that (i) region 2 has the smallest storage potential compared to the others, (ii) regions 3–5 have intermediate storage potentials, and (iii), all the other regions have relatively large storage potentials. From Fig. 6.11, although most regions have correlation coefficients of less than 0.5, Fig. 6.4 on the other hand show that certain regions gained TWS. This could indicate that a certain proportion of the TWS in the regions is contributed by subsurface inter flow and dominant indirect recharges.

6.5.9 Climate Impact on the Observed TWS

The extreme depletion in TWS and GW in certain regions during February 2003 (see PC1 of TWS in Fig. 6.4) could probably be due to the reported droughts in the year 2002 [3, 94]. This is further supported when they are considered together with the lag determined from TWS and rainfall correlation (Fig. 6.10). Rainfall over Ethiopia is known to be seasonal, having dry and rainy seasons [73]. The effect of this seasonality seems to be reflected in the TWS anomalies, having two distinct seasons on average, the water lose (winter, 21.3 mm/month and spring, 36.6 mm/month) and water gain (summer, 14.4 mm/month and autumn, 41.0 mm/month) seasons. PCA results showed that the annual peaks of rainfall are also very important and are highly correlated to the annual peaks of TWS, thus indicating that rainfall is the main driving factor for the rise in TWS levels in most parts of Ethiopia. Although climatic factors such as drought could be a contributor towards the fall in TWS levels, as pointed out earlier, human influences may also contribute.

The dominant hydrogeologic regimes, recharge mechanisms, recharge rates, aquifer flow and storage types for each region are presented, e.g., in [1, 18]. Here, the results obtained in relation to the properties of the regions described in the Table 6.2 are discussed. For the same aquifer flow and storage type, one can notice that there are different lag values, see e.g., regions 2 and 3 in Table 6.2, which could be attributed to the nature of the recharge mechanism in each region. Furthermore, regions that are dominated by the inter-granular aquifer flow and storage type are seen to have a longer lag periods (regions 1, and 6), while karst dominated regions show a direct response to rainfall (region 10).

In the north-west (region 3) a lag of 2 months is noted (Table 6.2). This shows that it takes 2 months from when it rains to the time when the impact on TWS is visible, hence making the transfer of water from one season to another impossible. This implies that for most parts of Ethiopia, the rainfall in the previous season might have an impact on the TWS of the current season at most, but cannot have further effect later. In regions 2, 5 and 8, the cross correlation between rainfall with TWS indicated a lag of 3 months (Table 6.2). Figure 6.11 also shows that there is a high correlation between rainfall and TWS in regions 2 and 5, indicating that the influence of rainfall on TWS in these regions is higher than subsurface inflow from adjacent regions. TWS in region 9 recorded a lag of 4 months from the rainfall (Table 6.2), although this region has low rainfall and also displays a poor correlation between rainfall and TWS anomaly (Fig. 6.11).

The mean annual rainfall for regions 1 and 6 have two small peaks, with a lag of 6 months obtained from the TWS/Rainfall cross correlation, implying that the autumn water gain is mainly from the rainfall received in winter. With the characteristics described in Table 6.2, these regions have the property of being able to transfer water mass gained from rainfall in a certain year to the next one. According to Kebede [42], 75% of Ethiopia's aquifers receive localized and indirect recharge. This implies that only 25% of Ethiopia's aquifers should exhibit a good correlation between TWS and rainfall. The R^2 results obtained from correlation analysis in Fig. 6.11 supports this assertion.

In the northern (region 4) and south eastern (region 7) regions, a lag of one month between TWS and rainfall was obtained from the cross correlation, implying that the aquifer responds to the rainfall after about one month. From the slopes of the TDC, region 4 has a lower storage potential than region 7, while the rainfall in region 4 is a bit higher than in region 7. One may therefore conclude that the chance that water can be transferred from a rainy to non-rainy season in such regions is very low. In this case, the rise or fall in TWS in certain season is mostly subject to the rainfall for that particular season.

6.6 Concluding Remarks

Time series of terrestrial water storage anomalies (TWS), soil moisture, and rainfall over Ethiopia were derived from GRACE, GLDAS, and TRMM data sets and products, respectively, and analyzed with the aim of better understanding the relationship between rainfall and groundwater variations over Ethiopia's aquifers. The inter-annual and seasonal variability of water storage, the relationship between water storage changes with rainfall and soil moisture, and the aquifers' characteristics and TWS anomalies have been derived over a period of 8 years from 2003 to 2011. The chapter indicated that:

(i) The western part and the north-eastern lowland of Ethiopia lost water during that period. From seasonal analysis, the seasons when water is gained are found to be summer and autumn, while the loss seasons are spring and winter for most parts of Ethiopia. Soil moisture and rainfall variations showed the dominant annual water variability in the western, north western, northern and central regions, and the dominant seasonal variability in the western and eastern regions.
(ii) Change in soil moisture was seen to have less influence over the total water storage, while groundwater storage has a dominant influence.
(iii) From a correlation analysis between TWS and rainfall, lags of 2–3 months for most regions were observed, indicating the capacity of a large portion of the Ethiopian groundwater storage (aquifers) to transfer water mass gained in certain periods to be less than 3 months. If validated at a finer spatial resolution, this could be vital information for water managers to know, that if an increase in TWS is required to back up water availability in dry seasons, one has to focus on increasing soil moisture and surface water storage rather than groundwater storages. The information could also be vital in land-use planning issues.

References

1. Abiye TA (2010) An overview of the transboundary aquifers in East Africa. J Afr Earth Sci 58:684–91. https://doi.org/10.1016/j.jafrearsci.2009.10.003 [Africa and the International Year of Planet Earth Promoting Earth sciences-based decision making in Africa 22nd Colloquium of African Geology]
2. Alemayehu T (2006) Groundwater occurrence in Ethiopia, Addis Ababa, Ethiopia: Addis Ababa University; 111 p. http://www.eah.org.et/docs/Ethiopian%20groundwater-Tamiru.pdf
3. Anderson S, Choularton R (2004) Retrospective analysis: 2002/3 crisis in Ethiopia: early warning and response, Technical Report The Regional Economic Development Services Office For East and Southern Africa (REDSO)
4. Awange J, Forootan E, Kuhn M, Kusche J, Heck B (2014) Water storage changes and climate variability within the Nile Basin between 2002–2011. Adv Water Res 73:1–15. https://doi.org/10.1016/j.advwatres.2014.06.010
5. Awange J, Forootan E, Kusche J, Kiema J, Omondi P, Heck B, Fleming K, Ohanya S, Gonçalves R (2013) Understanding the decline of water storage across the Ramser-Lake Naivasha using satellite-based methods. Adv Water Res 60:7–23. https://doi.org/10.1016/j.advwatres.2013.07.002
6. Awange JL (2012) Environmental monitoring. In: Environmental monitoring using GNSS environmental science and engineering. Springer, Berlin, Heidelberg, pp 1–12. https://doi.org/10.1007/978-3-540-88256-5_1
7. Awange JL, Anyah R, Agola N, Forootan E, Omondi P (2013) Potential impacts of climate and environmental change on the stored water of lake Victoria Basin and economic implications. Water Res Res 49:8160–73. https://doi.org/10.1002/2013WR014350
8. Awange JL, Kyalo Kiema JB (2013) Environmental monitoring and management, environmental science and engineering. Springer, Berlin, Heidelberg. https://doi.org/10.1007/978-3-642-34085-7_1
9. Awange JL, Sharifi MA, Ogonda G, Wickert J, Grafarend EW, Omulo MA (2008) The falling Lake Victoria water level: GRACE, TRIMM and CHAMP satellite analysis of the Lake Basin. Water Res Manag 22:775–96. https://doi.org/10.1007/s11269-007-9191-y
10. Awulachew S, Yilma A, Loulseged M, Loiskandl W, Ayana M, Alamirew T (2007) Water resources and irrigation development in Ethiopia. Working Paper 123 International Water Management Institute Colombo, Sri Lanka
11. Ayenew T, Demlie M, Wohnlich S (2008) Hydrogeological framework and occurrence of groundwater in the Ethiopian aquifers. J Afr Earth Sci 52:97–113. https://doi.org/10.1016/j.jafrearsci.2008.06.006
12. Berhane G, Martens K, Farrah NA, Walraevens K (2013) Water leakage investigation of micro-dam reservoirs in Mesozoic sedimentary sequences in Northern Ethiopia. J Afr Earth Sci 79:98–110. https://doi.org/10.1016/j.jafrearsci.2012.10.004
13. Berhanu B, Melesse AM, Seleshi Y (2013) GIS-based hydrological zones and soil geodatabase of Ethiopia. CATENA 104:21–31. https://doi.org/10.1016/j.catena.2012.12.007
14. Bewket W, Abebe S (2013) Land-use and land-cover change and its environmental implications in a tropical highland watershed, Ethiopia. Int J Environ Stud 70:126–39. https://doi.org/10.1080/00207233.2012.755765
15. Bewket W, Conway D (2007) A note on the temporal and spatial variability of rainfall in the drought-prone Amhara region of Ethiopia. Int J Climatol 27:1467–77. https://doi.org/10.1002/joc.1481
16. Beyene EG, Meissner B (2010) Spatio-temporal analyses of correlation between NOAA satellite RFE and weather stations' rainfall record in Ethiopia. Int J Appl Earth Obs Geoinf 12(Suppl 1):S69-75. https://doi.org/10.1016/j.jag.2009.09.006 [Supplement Issue on Remote Sensing for Africa—A Special Collection from the African Association for Remote Sensing of the Environment (AARSE)]

17. Bonsor HC, Mansour MM, MacDonald AM, Hughes AG, Hipkin RG, Bedada T (2010) Interpretation of grace data of the Nile Basin using a groundwater recharge model. Hydrol Earth Syst Sci Discuss 7:4501–33. https://doi.org/10.5194/hessd-7-4501-2010
18. Chernet T (1993) Hydrogeology of Ethiopia and water resources development. Ethiopia: Technical Report Ethiopian Institute of Geological Surveys, Ministry of Mines and Energy Addis Ababa
19. Cheung WH, Senay GB, Singh A (2008) Trends and spatial distribution of annual and seasonal rainfall in Ethiopia. Int J Climatol 28:1723–34. https://doi.org/10.1002/joc.1623
20. Chukalla A, Haile A, Schultz B (2013) Optimum irrigation and pond operation to move away from exclusively rainfed agriculture: the Boru Dodota spate irrigation scheme, Ethiopia. Irrig Sci 31:1091–102. https://doi.org/10.1007/s00271-012-0390-9
21. Conway D, Schipper ELF (2011) Adaptation to climate change in Africa: challenges and opportunities identified from Ethiopia. Global Environ Change 21:227–37. https://doi.org/10.1016/j.gloenvcha.2010.07.013
22. Descheemaeker K, Raes D, Nyssen J, Poesen J, Haile M, Deckers J (2009) Changes in water flows and water productivity upon vegetation regeneration on degraded hillslopes in northern Ethiopia: a water balance modelling exercise. Rangeland J 31:237–49. https://doi.org/10.1071/RJ09010
23. Dinku T, Ceccato P, Grover-Kopec E, Lemma M, Connor SJ, Ropelewski CF (2007) Validation of satellite rainfall products over East Africa's complex topography. Int J Remote Sens 28:1503–26. https://doi.org/10.1080/01431160600954688
24. Dinku T, Chidzambwa S, Ceccato P, Connor SJ, Ropelewski CF (2008) Validation of high-resolution satellite rainfall products over complex terrain. Int J Remote Sens 29:4097–110. https://doi.org/10.1080/01431160701772526
25. Dinku T, Pietro C, Keith C, Stephen JC (2010) Evaluating detection skills of satellite rainfall estimates over desert locust recession regions. J Appl Meteorol Climatol 49:1322–32. https://doi.org/10.1175/2010JAMC2281.1
26. Eberhardt M (2008) Cereal production in Ethiopia: a brief historical overview of land under cultivation and yields. http://users.ox.ac.uk/econstd/Brief_MEberhardt.pdf
27. Egziabher TG (2000) Regional development planning in Ethiopia: past experience, current initiatives and future prospects. East Africa Soc Sci Res Rev 16:65–94
28. Engida AN, Esteves M (2011) Characterization and disaggregation of daily rainfall in the Upper Blue Nile Basin in Ethiopia. J Hydrol 399:226–34. https://doi.org/10.1016/j.jhydrol.2011.01.001
29. Famiglietti JS, Rodell M (2013) Water in the balance. Science 340:1300–1. https://doi.org/10.1126/science.1236460
30. Fang H, Beaudoing HK, Rodell M, Teng WL, Vollmer BE (2009) Global Land data assimilation system (GLDAS) products, services and application from NASA hydrology data and information services center (HDISC). In: ASPRS 2009 annual conference, Baltimore, Maryland
31. FAO (2005) Irrigation in Africa in figures: AQUASTAT Survey—2005. Karen Frenken, FAO Water Reports number 29 Food and Agriculture Organization of the United Nations Rome
32. Ferguson A, Harbott B (1982) Geophysical, physical and chemical aspects of Lake Turkana. In: Hopson E (ed) Lake Turkana: a report of the findings of the Lake Turkana project 1972–1975. Overseas Development Administration, London, UK
33. Fleming K, Awange JL (2013) Comparing the version 7 TRMM 3B43 monthly precipitation product with the TRMM 3B43 version 6/6A and Bureau of Meteorology datasets for Australia. Aust Meteorol Oceanogr J 63:421–426. http://www.bom.gov.au/amm/docs/2013/fleming.pdf
34. Forootan E, Awange J, Kusche J, Heck B, Eicker A (2012) Independent patterns of water mass anomalies over Australia from satellite data and models. Remote Sens Environ 124:427–43. https://doi.org/10.1016/j.rse.2012.05.023
35. Forootan E, Kusche J (2012) Separation of global time-variable gravity signals into maximally independent components. J Geodesy 86:477–97. https://doi.org/10.1007/s00190-011-0532-5.10.1007/s00190-011-0532-5

36. Forootan E, Rietbroek R, Kusche J, Sharifi M, Awange J, Schmidt M, Omondi P, Famiglietti J (2014) Separation of large scale water storage patterns over Iran using GRACE, altimetry and hydrological data. Remote Sens Environ 140:580–95. https://doi.org/10.1016/j.rse.2013.09.025
37. Furi W, Razack M, Abiye TA, Kebede S, Legesse D (2012) Hydrochemical characterization of complex volcanic aquifers in a continental rifted zone: the Middle Awash Basin Ethiopia. Hydrogeol J 20:385–400. https://doi.org/10.1007/s10040-011-0807-1
38. Helsel D, Hirsch R (2002) Statistical Methods in Water Resources Techniques of Water Resources Investigations volume chapter A3 of Book 4. US Geological Survey
39. Huffman GJ, Bolvin DT (2012) TRMM and other data precipitation data set documentation (visited on 2012-03-23). ftp://meso-a.gsfc.nasa.gov/pub/trmmdocs/3B42_3B43_doc.pdf
40. Huffman GJ, Bolvin DT, Nelkin EJ, Wolff DB, Adler RF, Gu G, Hong Y, Bowman KP, Stocker EF (2007) The TRMM multisatellite precipitation analysis (TMPA): quasiglobal, multiyear, combined-sensor precipitation estimates at fine scales. J Hydrometeorol 8:38–55. https://doi.org/10.1175/JHM560.1
41. Hurni H (1988) Agroecological Belts of Ethiopia: explanatory notes on three maps at a scale of 1:1,000,000. Research Report Centre for Development and Environment University of Bern, Switzerland in association with The Ministry of Agriculture, Ethiopia Bern. Soil Conservation Research Programme Ethiopia
42. Kebede S (2013) Groundwater in Ethiopia: features, numbers and opportunities. Springer, Berlin, Heidelberg. [Springer Hydrogeology]
43. Kebede S, Travi Y, Alemayehu T, Ayenew T (2005) Groundwater recharge, circulation and geochemical evolution in the source region of the Blue Nile River Ethiopia. Appl Geochem 20:1658–76. https://doi.org/10.1016/j.apgeochem.2005.04.016
44. Kebede S, Travi Y, Asrat A, Alemayehu T, Ayenew T, Tessema Z (2008) Groundwater origin and flow along selected transects in Ethiopian Rift volcanic aquifers. Hydrogeol J 16:55–73. https://doi.org/10.1007/s10040-007-0210-0
45. Kummerow C, William B, Toshiaki K, James S, Simpson J (1998) The tropical rainfall measuring mission (TRMM) sensor package. J Atmos Oceanic Tech 15:809–17. https://doi.org/10.1175/1520-0426(1998)015
46. Kusche J (2007) Approximate decorrelation and non-isotropic smoothing of time variable GRACE-type gravity field models. J Geodesy 81:733–49. https://doi.org/10.1007/s00190-007-0143-3
47. Kusche J, Schmidt R, Petrovic S, Rietbroek R (2009) Decorrelated GRACE time-variable gravity solutions by GFZ, and their validation using a hydrological model. J Geodesy 83:903–13. https://doi.org/10.1007/s00190-009-0308-3
48. Liu Z, Ostrenga D, Teng W, Kempler S (2012) Tropical rainfall measuring mission (TRMM) precipitation data and services for research and applications. Bull Am Meteorol Soc 93:1317–25. https://doi.org/10.1175/BAMS-D-11-00152.1
49. Longuevergne L, Scanlon BR, Wilson CR (2010) GRACE Hydrological estimates for small basins: evaluating processing approaches on the high plains aquifer, USA. Water Resour Res 46. https://doi.org/10.1029/2009WR008564 [n/a-n/a]
50. Marshall M, Funk C, Michaelsen J (2012) Examining evapotranspiration trends in Africa. Clim Dyn 38:1849–65. https://doi.org/10.1007/s00382-012-1299-y
51. Melesse A, Abtew W, Dessalegne T, Wang X (2010) Low and high flow analyses and wavelet application for characterization of the Blue Nile River system. Hydrol Process 24:241–52. https://doi.org/10.1002/hyp.7312
52. Melesse AM (ed) (2011) Nile River Basin: hydrology, climate and water use. Springer, Netherlands
53. Mengistu D, Bewket W, Lal R (2013) Recent spatiotemporal temperature and rainfall variability and trends over the upper Blue Nile river basin, Ethiopia. Int J Climatol n/a–n/a. https://doi.org/10.1002/joc.3837
54. MoWE (2012) Ministry of Water & Energy of Federal Democratic Republic of Ethiopia. 2012 (visited on 2012-11-05). http://www.mowr.gov.et/index.php?pagenum=2.1&pagehgt=5495px

55. North GR, Bell TL, Cahalan RF, Moeng FJ (1982) Sampling errors in the estimation of empirical orthogonal functions. Mon Weather Rev 110:699–706. https://doi.org/10.1175/1520-0493110
56. Nyssen J, Clymans W, Descheemaeker K, Poesen J, Vandecasteele I, Vanmaercke M, Zenebe A, Van Camp M, Haile M, Haregeweyn N, Moeyersons J, Martens K, Gebreyohannes T, Deckers J, Walraevens K (2010) Impact of soil and water conservation measures on catchment hydrological response—a case in north Ethiopia. Hydrol Process 24:1880–95. https://doi.org/10.1002/hyp.7628
57. Nyssen J, Poesen J, Moeyersons J, Deckers J, Haile M, Lang A (2004) Human impact on the environment in the Ethiopian and Eritrean highlands—a state of the art. Earth-Sci Rev 64:273–320. https://doi.org/10.1016/S0012-8252(03)00078-3
58. Omondi P, Awange J, Ogallo L, Ininda J, Forootan E (2013) The influence of low frequency sea surface temperature modes on delineated decadal rainfall zones in Eastern Africa region. Adv Water Res 54:161–80. https://doi.org/10.1016/j.advwatres.2013.01.001
59. Omondi P, Awange J, Ogallo L, Okoola R, Forootan E (2012) Decadal rainfall variability modes in observed rainfall records over East Africa and their relations to historical sea surface temperature changes. J Hydrol 464–465:140–56. https://doi.org/10.1016/j.jhydrol.2012.07.003
60. Omondi PA, Awange JL, Forootan E, Ogallo LA, Barakiza R, Girmaw GB, Fesseha I, Kululetera V, Kilembe C, Mbati MM, Kilavi M, King'uyu SM, Omeny PA, Njogu A, Badr EM, Musa TA, Muchiri P, Bamanya D, Komutunga E (2014) Changes in temperature and precipitation extremes over the greater horn of Africa region from 1961 to 2010. Int J Climatol 34:1262–77. https://doi.org/10.1002/joc.3763
61. Preisendorfer R (1988) Principal component analysis in meteorology and oceanography. Elsevier
62. Reager JT, Famiglietti JS (2009) Global terrestrial water storage capacity and flood potential using grace. Geophys Res Lett 36. https://doi.org/10.1029/2009GL040826, [n/a-n/a]
63. Rientjes THM, Haile AT, Kebede E, Mannaerts CMM, Habib E, Steenhuis TS (2011) Changes in land cover, rainfall and stream flow in Upper Gilgel Abbay catchment, Blue Nile Basin—Ethiopia. Hydrol Earth Syst Sci 15:1979–89. https://doi.org/10.5194/hess-15-1979-2011
64. Rodell M, Chen J, Kato H, Famiglietti JS, Nigro J, Wilson CR (2007) Estimating groundwater storage changes in the Mississippi River Basin (USA) using GRACE. Hydrogeol J 15:159–66. https://doi.org/10.1007/s10040-006-0103-7
65. Rodell M, Famiglietti JS (2001) An analysis of terrestrial water storage variations in illinois with implications for the gravity recovery and climate experiment (grace). Water Resour Res 37:1327–39. https://doi.org/10.1029/2000WR900306
66. Rodell M, Houser PR, Jambor U, Gottschalck J, Mitchell K, Meng C-J et al (2004) The global land data assimilation system. Bull Am Meteorol Soc 85:381–94. https://doi.org/10.1175/BAMS-85-3-381
67. Rodell M, Velicogna I, Famiglietti JS (2009) Satellite-based estimates of groundwater depletion in India. Nature 460:999–1002. https://doi.org/10.1038/nature08238
68. Romilly TG, Gebremichael M (2011) Evaluation of satellite rainfall estimates over Ethiopian river basins. Hydrol Earth Syst Sci 15:1505–14. https://doi.org/10.5194/hess-15-1505-2011
69. Sawicz K, Wagener T, Sivapalan M, Troch PA, Carrillo G (2011) Catchment classification: empirical analysis of hydrologic similarity based on catchment function in the eastern USA. Hydrol Earth Syst Sci 15:2895–911. https://doi.org/10.5194/hess-15-2895-2011
70. Schmidt R, Flechtner F, Meyer U, Neumayer K-H, Dahle C, König R, Kusche J (2008) Hydrological signals observed by the GRACE satellites. Surv Geophys 29:319–34. https://doi.org/10.1007/s10712-008-9033-3
71. Segele ZT, Lamb PJ, Leslie LM (2009) Seasonal-to-interannual variability of Ethiopia/ Horn of Africa monsoon. Part I: associations of wavelet-filtered large-scale atmospheric circulation and global sea surface temperature. J Climate 22:3396–421. https://doi.org/10.1175/2008JCLI2859.1

72. Seleshi Y, Camberlin P (2006) Recent changes in dry spell and extreme rainfall events in Ethiopia. Theor Appl Climatol 83:181–91. https://doi.org/10.1007/s00704-005-0134-3.10.1007/s00704-005-0134-3
73. Seleshi Y, Zanke U (2004) Recent changes in rainfall and rainy days in Ethiopia. Int J Climatol 24:973–83
74. Senay GB, Asante K, Artan G (2009) Water balance dynamics in the Nile Basin. Hydrol Process 23:3675–81. https://doi.org/10.1002/hyp.7364
75. Shang H, Yan J, Gebremichael M, Ayalew SM (2011) Trend analysis of extreme precipitation in the Northwestern Highlands of Ethiopia with a case study of Debre Markos. Hydrol Earth Syst Sci 15:1937–44. https://doi.org/10.5194/hess-15-1937-2011
76. Siebert S, Henrich V, Frenken K, Burke J (2013) Update of the digital global map of irrigation areas (GMIA) to Version 5. Documentation Institute of Crop Science and Resource Conservation Rheinische Friedrich Wilhelms Universität Bonn and FAO Bonn, Germany and Rome, Italy
77. Sutcliffe J, Parks Y (1999) The hydrology of the Nile. Technical Report International Association of Hydrological Sciences Oxfordshire, UK. Special Publication No 5
78. Swenson S, Famiglietti J, Basara J, Wahr J (2008) Estimating profile soil moisture and groundwater variations using GRACE and Oklahoma Mesonet soil moisture data. Water Res Res44. https://doi.org/10.1029/2007WR006057 [n/a-n/a]
79. Swenson S, Wahr J (2006) Post-processing removal of correlated errors in GRACE data. Geophys Res Lett 33:L08402. https://doi.org/10.1029/2005GL025285
80. Syed TH, Famiglietti JS, Rodell M, Chen J, Wilson CR (2008) Analysis of terrestrial water storage changes from GRACE and GLDAS. Water Res Res 44. https://doi.org/10.1029/2006WR005779 [n/a-n/a]
81. Tapley BD, Bettadpur S, Ries JC, Thompson PF, Watkins MM (2004) GRACE measurements of mass variability in the Earth System. Science 305:503–5. https://doi.org/10.1126/science.1099192
82. Taye MT, Willems P (2012) Temporal variability of hydroclimatic extremes in the Blue Nile Basin. Water Res Res 48. https://doi.org/10.1029/2011WR011466 [n/a-n/a
83. Tesfagiorgis K, Gebreyohannes T, Smedt F, Moeyersons J, Hagos M, Nyssen J, Deckers J (2011) Evaluation of groundwater resources in the Geba Basin Ethiopia. Bull Eng Geology Environ 70:461–6. https://doi.org/10.1007/s10064-010-0338-3
84. Wahr J, Molenaar M, Bryan F (1998) Time variability of the Earth's gravity field: hydrological and oceanic effects and their possible detection using GRACE. J Geophys Res: Solid Earth 103:30205–29. https://doi.org/10.1029/98JB02844
85. Werth S, Güntner A, Schmidt R, Kusche J (2009) Evaluation of GRACE filter tools from a hydrological perspective. Geophys J Int 179:1499–515. https://doi.org/10.1111/j.1365-246X.2009.04355.x
86. Yadav M, Wagener T, Gupta H (2007) Regionalization of constraints on expected watershed response behavior for improved predictions in ungauged basins. Adv Water Res 30:1756–74. https://doi.org/10.1016/j.advwatres.2007.01.005
87. Yirdaw S, Snelgrove K, Agboma C (2008) GRACE satellite observations of terrestrial moisture changes for drought characterization in the Canadian prairie. J Hydrol 356:84–92. https://doi.org/10.1016/j.jhydrol.2008.04.004
88. Yitbarek A, Razack M, Ayenew T, Zemedagegnehu E, Azagegn T (2012) Hydrogeological and hydrochemical framework of Upper Awash River basin, Ethiopia: with special emphasis on inter-basins groundwater transfer between Blue Nile and Awash Rivers. J Afri Earth Sci 65:46–60. https://doi.org/10.1016/j.jafrearsci.2012.01.002
89. Nile Basin Water Resources Atlas (2017). http://atlas.nilebasin.org/
90. Awange JL, Gebremichael M, Forootan E, Wakbulcho G, Anyah R, Ferreira CG, Alemayehu T (2014) Characterization of Ethiopian mega hydrogeological regimes using GRACE, TRMM and GLDAS datasets. Adv Water Res 74:64–78. https://doi.org/10.1016/j.advwatres.2014.07.012

91. Awange J, Forootan E, Fleming K, and Odhiambo GO (2015) Dominant Patterns of Water Storage Changes in the Nile Basin During 2003–2013. In: Lakshmi V (ed) Remote sensing of the terrestrial water cycle. Wiley, NJ, pp 367–382. https://doi.org/10.1002/9781118872086.ch22
92. Khaki M, Awange J, Forootan E, Kuhn M (2018) Understanding the association between climate variability and the Nile's water level fluctuations and water storage changes during 1992–2016. Sci Total Environ 645:1509–1521. https://doi.org/10.1016/j.scitotenv.2018.07.212
93. Khaki M, Awange J, Forootan E, Kuhn M (2018b) Understanding the association between climate variability and the Nile's water level fluctuations and water storage changes during 1992–2016. Sci Total Environ 645:1509–1521. https://doi.org/10.1016/j.scitotenv.2018.07.212
94. Mpelasoka F, Awange JL, Zerihun A (2018) Influence of coupled ocean-atmosphere phenomena on the Greater Horn of Africa droughts and their implications. Science of the Total Environment, 610–611: 691–702. https://doi.org/10.1016/j.scitotenv.2017.08.109
95. Awange JL, Palancz B and Völgyesi L (2020) Hybrid imaging and visualization. employing machine learning with Mathematica—Python. Springer Nature International, Berlin. ISBN 978-3-030-26152-8, https://doi.org/10.1007/978-3-030-26153-5
96. Awange JL (2021) Nile Waters—Weighed from space. Springer Nature International
97. Awange JL, Ferreira VG, Khandu Andam-Akorful SA, Forootan E, Agutu NO, He XF (2015) Uncertainties in remotely-sensed precipitation data over Africa. Int J Climatol. https://doi.org/10.1002/joc.4346
98. Awange JL, Hu KX, Khaki M (2019) The newly merged satellite remotely sensed, gauge and reanalysis-based Multi-Source Weighted-Ensemble Precipitation: Evaluation over Australia and Africa (1981–2016). Sci Total Environ 670:448–465. https://doi.org/10.1016/j.scitotenv.2019.03.148
99. Agutu NO, Awange JL, Ndehedehe C, Mwaniki MW (2020) Consistency of agricultural drought characterization over Upper Greater Horn of Africa (1982-2013): Topographical, gauge density, and model forcing influence. Sci Total Environ 709, https://doi.org/10.1016/j.scitotenv.2019.135149
100. Hu K, Awange JL, Kun M, Saleem A (2019) Spatio-temporal groundwater variations associated with climatic and anthropogenic impacts in South-West Western Australia. Sci Total Environ 69. https://doi.org/10.1016/j.scitotenv.2019.133599

Part III
Extreme Climate: Drought

Chapter 7
Rainfall-SST Fluctuation: Predictability

Decadal scale fluctuations are crucial because they control water supplies, affect biota, and may modulate higher-frequency events such as floods and droughts. Furthermore, low frequency natural variability is important in global climate change issues because it may obscure human influences on hydrological variations.—[36].

7.1 Summary

Detailed knowledge of the long-term interface of climate and rainfall variability is essential for managing agricultural activities in Greater Horn of Africa countries. This chapter presents the work of [36] who investigated the space–time patterns of decadal rainfall variability modes over East Africa and their predictability potentials using Sea Surface Temperature (SST). The analysis includes observed rainfall data from 1920 to 2004 and global SSTs for the period 1950–2004. Simple correlation, trend and cyclical analyses, Principal Component Analysis (PCA) with VARIMAX rotation and Canonical Correlation Analysis (CCA) are employed. The results show decadal signals in filtered observed rainfall record with 10 years period during March–May (MAM) and October–December (OND) seasons. During June–August (JJA), however, cycles with 20 years period are common. Too much/little rainfall received in one or two years determines the general trend of the decadal mean rainfall. CCA results for MAM show significant positive correlations between the VARIMAX-PCA of SST and the canonical component time series over the central equatorial Indian Ocean. Positive loadings are spread over the coastal and Lake Victoria regions while negative loading are spread over the rest of the region with significant canonical correlation skills. For the JJA seasons, Atlantic SSTs has negative loadings centred on the tropical western Atlantic Ocean associated with the wet/dry regimes over western/eastern sectors. Although SST variability is known to contribute to extreme wet conditions, the wetter conditions in western parts of Eastern

J. Awange, *Food Insecurity & Hydroclimate in Greater Horn of Africa*,
https://doi.org/10.1007/978-3-030-91002-0_7

Africa during JJA have been shown to depend on three mechanisms (i) the incursions of moist Congo AirMass (CAM) associated with west-east migration of the Congo Branch of the walker circulation, (ii) for the drier conditions along the coast, this also depends on the SST anomalies over the Indian Ocean Dipole (IOD) region, and (iii), the influence of topography. The western region has higher elevations compared to the coastal regions and as such, the topography could be influencing the rainfall in favour of the western region compared to coastal regions. The highest canonical correlation skill between OND rainfall and the Pacific SSTs show that El'Niño/LaNiña phases are associated with wet/dry decades over the region.

7.2 Decadal Fluctuation in Climate System

Decadal variations of extreme climate impact negatively on agricultural production resulting into massive losses amongst the affected communities and thus deleterious effect on the economy of Greater Horn of Africa countries. Understanding the nature and causes of decadal fluctuations in climate system is an unresolved problem, partly because observed records are relatively short or sparse and because dynamical processes that operate on this time-scale have not firmly been understood. Over the region, much attention has been devoted to how and why precipitation varies in association with the El Niño-Southern Oscillation (ENSO) [15, 23, 33, 42] at diurnal, seasonal and inter-annual time-scales. The impacts of persistent decadal climate anomalies have far reaching socio-economic implications due to persistent climate stress that they would impose on the regional socio-economic systems.

Climate parameters have been observed in a global scale during the last several decades [7, 14, 57]. Examples of such variability include the North Atlantic Oscillations (NAO) phenomenon; drought in California, parts of Australia, or the Sahel and Eastern Africa. Their influences have been observed in lake level fluctuations and inter-annual rainfall records [3]. Impacts of such decadal variability of extreme climate events would generally require more challenging mitigation strategies. Mitigation and adaptation to any of the climate anomalies would depend on the magnitude and duration of the persistence of the anomalies. Mitigation and/or adaptation measures are likely to involve investment in infrastructure and changes in policy due to the potentially large magnitude of their effects.

Over Eastern Africa region, [24, 25, 27, 35, 50] have shown some evidence of decadal climate variability in the observed rainfall records. Using Climate Prediction Centre (CPC) Merged Analysis of Precipitation (CMAP) data and the Principle Component Analysis (PCA) method, [50] investigated variability on the October to December (OND) rainfall over Eastern Africa region based on the period 1961–2001. They found that the most dominant mode (EOF1 explaining about 29% of variance) corresponding to ENSO climate variability. They associated the second empirical orthogonal function (EOF2 explaining about 14% of variance) to decadal trend mode. From their results, the long-term rainfall variability was characterized by positive anomalies over the northern sector of Eastern Africa and opposite conditions over the southern sector.

Several studies in the region on inter-annual variations of East African rainfall and their possible linkages to global sea surface temperature (SST) changes have been undertaken, e.g., [15, 21, 23, 31, 37]. The main focus of these studies were especially on the relation between rainfall anomalies and SST perturbations over the equatorial Pacific and Indian Ocean basins, and to some extent, the Atlantic Ocean [12, 15, 22, 29, 33, 49]. These studies determine the dominant role played by the ENSO anomaly patterns in influencing the inter-annual variabilities of the equatorial East Africa rainfall [15, 34]. Note that the zonal temperature gradient over the equatorial Indian Ocean, often referred to as the Indian Ocean Dipole Mode (IOD) [47, 48], coupled with ENSO (i.e., IOD-ENSO) influence, have also been linked to some of the wettest periods in the region, such as rainfalls in 1961 and 2006 [8–10].

In this chapter, attempts are made to examine decadal trend mode in observed East Africa rainfall records and its possible linkage to decadal patterns of global SST records. The chapter's investigation extends the previous studies by considering a relatively longer period of rainfall data from 1920–2004, and global SSTs for the period 1950–2004 and studying their interactions within their overlapping periods. Use is made of advanced multivariate statistical analysis techniques such as VARIMAX-PCA and Canonical Correlation Analysis (CCA), which allow in-depth investigation of possible correlations between decadal SST and rainfall variations.

The remainder of the chapter is organized as follows: Sect. 7.3 briefly describes the data and analysis methods used. The trend results are presented in Sect. 7.4. Sections 7.5 and 7.6 summarize the main decadal patterns of rainfall and SST variabilities. The link between SSTs and decadal rainfall patterns over the region is discussed in Sect. 7.7. Section 7.8 gives a summary, the major findings and conclusions of the chapter.

7.3 Climate Products and Analysis Methods

In this analysis, monthly observed rainfall data obtained from IGAD Climate Prediction and Applications Centre (ICPAC), the Kenya Meteorological Department (KMD), Tanzania Meteorological Agency (TMA) and Uganda Meteorological Department (UMD) is employed. The observed monthly rainfall data used are from 38 stations (Fig. 7.1) unevenly distributed over East Africa [35]. Table 7.1 gives the station names, longitudes and latitudes while Fig. 7.1 presents the distribution of representative stations over the East African region. Also used in the study are reconstructed Reynolds SST data obtained from the United States (US) National Oceanic and Atmospheric Administration (NOAA) official website.[1] The data are archived as the optimum interpolation (OI), version 2, global SST values on 1° by 1° grid points.

[1] http://www.esrl.noaa.gov/psd/data/gridded/data.ncep.oisst.v2.html.

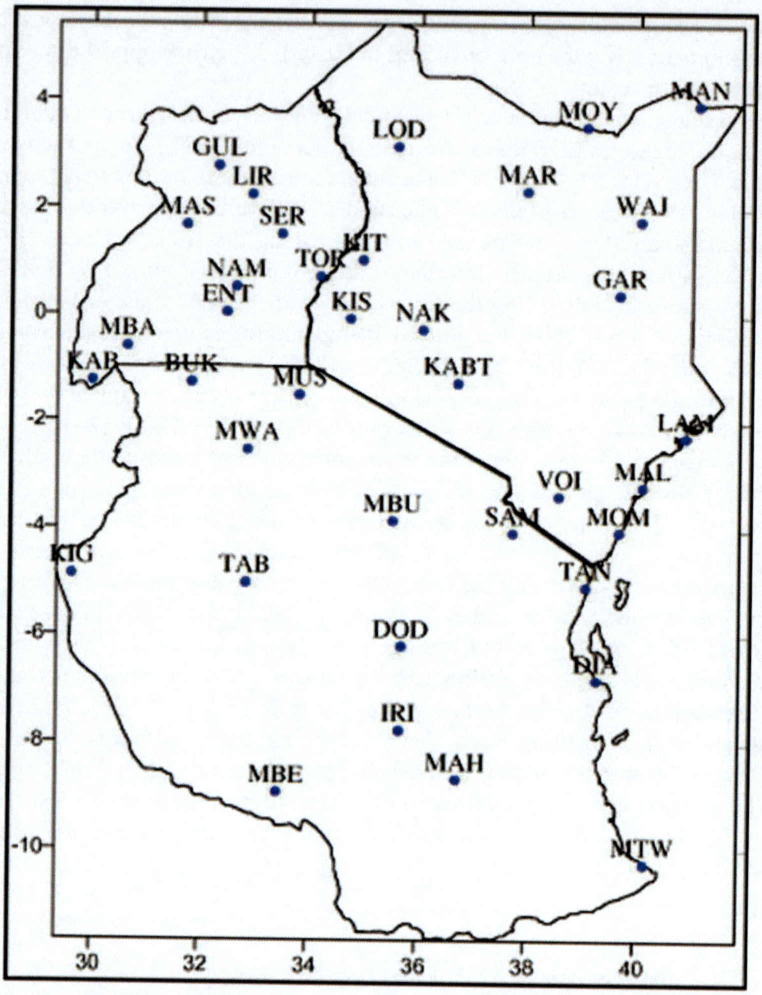

Fig. 7.1 Distribution of representative stations over the East African region. The x-axis represent the longitude while the y-axis represent the latitude. The full station names and their corresponding latitudes and longitudes are presented in Table 7.1. *Source* [36]

The SST values include *in-situ* and satellite SSTs observations plus those SSTs that are simulated by sea-ice cover. More on the SST data can be obtained from [43–45]. To link rainfall to SSTs, the correlation window for both data, i.e., 1950–2004 is employed.

Table 7.1 List and locations of stations used in the study (see Fig. 7.1 for the locations). *Note* the negative sign preceding Latitude implies "South". *Source* [36]

No.	Station name	Abbreviation	Latitude°	Longitude°
1	Lodwar	Lod	3.1	35.6
2	Kisumu	Kis	−0.1	34.7
3	Narok	Nar	−1.1	35.8
4	Dagoretti	Dag	−1.3	36.7
5	Mandera	Man	3.9	41.8
6	Wajir	Waj	1.7	40.0
7	Marsabit	Mar	2.3	37.9
8	Garissa	Gar	0.5	38.5
9	Moyale	Moy	3.5	39.0
10	Voi	Voi	−3.4	38.5
11	Malindi	Mal	−3.2	40.1
12	Mombasa	Mom	−4.	39.6
13	Lamu	Lam	−2.2	40.9
14	Bukoba	Buk	−1.2	31.4
15	Musoma	Mus	−1.3	33.4
16	Mwanza	Mwa	−2.2	32.5
17	Arusha	Aru	−3.2	36.3
18	Kigoma	Kig	−4.5	29.4
19	Same	Sam	−4.0	37.4
20	Tabora	Tab	−5.0	32.5
21	Tanga	Tan	−5.0	39.0
22	Mahenge	Mah	−8.6	36.7
23	Dodoma	Dod	−6.1	35.4
24	DIA	DIA	−6.5	39.1
25	Iringa	Iri	−7.3	35.4
26	Mbeya	Mbe	−8.5	33.2
27	Songea	Son	−10.4	35.3
28	Mtwara	Mtw	−10.2	40.1
29	Mbulu	Mbu	−3.8	35.5
30	Gulu	Gul	2.7	32.2
31	Lira	Lir	2.3	32.9
32	Soroti	Sor	1.7	33.6
33	Tororo	Tor	0.6	34.1
34	Entebbe	Ent	0.1	32.4
35	Masindi	Mas	1.6	31.7
36	Mbarara	Mba	−0.6	30.6
37	Kabete	Kabt	−1.2	36.7
38	Kabale	Kab	−1.4	29.2

7.3.1 Analysis Methods

Since the attention is primarily on the lower frequency (long wave-length) variabilities, a 9-point binomial coefficient filter is employed to smooth both the rainfall and SST time series so that all fluctuations of period shorter than 10 years are considerably suppressed. A Graphical method is then used to extract decadal trend modes while Mann–Kendall and the Spearman rank tests statistical methods that are based on rank statistics [16, 17, 56] are employed to test the significance of the observed trends.

The Principal Component Analysis (PCA) method is a statistical signal extraction technique based on diagonalization of the auto-covariance or auto-correlation matrix of a data set [4, 55]. In this chapter, the VARIMAX rotated version of PCA is applied to define dominant modes of variability of the low passed rainfall and SST series. The VARIMAX rotation is selected to improve the physical interpretation of the PCA modes and to derive more localized components, see e.g., [46]. To define the relationship between the dominant modes of decadal rainfall variabilities and SST variations in the global oceans, the Canonical Correlation Analysis (CCA) technique was adopted. Unlike PCA, CCA is a statistical technique that identifies a sequence of pairs of patterns in two multivariate data sets and constructs sets of transformed variables by projecting the original data onto these patterns [5, 6, 23, 53, 55]. CCA, therefore, can be regarded as a multivariate statistical technique that calculates linear combinations of a set of predictors that maximizes relationships, in a least square sense, to the similarly calculated linear combinations of a set of predictands.

The patterns are chosen such that new variables defined by projection of the two data sets onto these patterns exhibit maximum correlation but are uncorrelated with the projections of the data onto any of the other identified patterns. The superiority of CCA over other several techniques is its ability to operate on full fields of information and to objectively define the most highly related patterns of predictor and predictand [5, 15, 23, 31, 35].

CCA goes beyond the limitation of the simple correlation analysis by taking into consideration the full space and time dimensions of the fields analyzed and this is an exceptional skill capability of the technique. It also gives an extensive set of diagnostics that offer some insight into the physical base of the relationships used to form the predictions. The advantages of CCA include ability to operate on full fields of information and to objectively define the most highly related pattern of predictors and predictands. Its capability to define both the space and time evolution of the predictor dataset that best predicts an associated pattern of a predictand is efficient compared to simple correlation technique.

In this chapter, CCA is used to select pairs of spatial patterns of the two space/time dependent variable sets (VARIMAX-PCA of rainfall data and SSTs) such that the (time dependent) pattern amplitudes are optimally correlated. The strength and the sign of the corresponding patterns are described by the canonical correlation coordinates. Since the canonical series are normalized to unit variance, the canonical correlation patterns are expressed in the units of the variable they represent and indicate the "typical" strength of the mode of co-variation described by the patterns. The correlation between the canonical coordinates measures the degree of association between the canonical patterns of predictor and predictand variables [58].

A CCA transform pairs of original centred data vectors x' and y' into sets of new variables, called canonical variates, v_m and w_m, defined by the dot products

$$v_m = a_m^T x' = \sum_{i=1}^{I} a_{m,i} x', m = 1, ..., \min(I, J); \tag{7.1}$$

and

$$w_m = b_m^T y' = \sum_{j=1}^{J} b_{m,j} y', m = 1, ..., \min(I, J). \tag{7.2}$$

This construction of canonical variates is similar to that of the principal components u_m, in that each is a linear combination of (a sort of weighted average) of elements of the respective data vectors x' and y'. These vectors of weights, a_m and b_m, are called the canonical vectors. One data- and canonical-vector pair need not to have the same dimension as the other. Therefore, in Eq. 7.1, vectors x' and a_m each have I elements, while those of y' and b_m in Eq. 7.2 have J elements each. m is the number of canonical pairs, so called 'canonical variates' that can be extracted from the two data sets. In practice, m is derived as $m = min(I, J)$.

The canonical vectors a_m and b_m are the unique choices that result in the canonical variates having the properties

$$\mathrm{corr}[v_1, w_1] \geqslant \mathrm{corr}[v_2, w_2] \geqslant ... \geqslant \mathrm{corr}[v_m, w_m] \geqslant 0; \tag{7.3}$$

$$\mathrm{corr}[v_k, w_m] = \begin{cases} r_{c_m}, k = m \\ 0, k \neq m \end{cases} ; \tag{7.4}$$

and

$$\mathrm{Var}[v_m] = a_m^T [S_{x,x}] a_m = \mathrm{Var}[w_m] = b_m^T [S_{y,y}] b_m = 1. \tag{7.5}$$

Equation 7.3 shows that each of the m successive pairs of canonical variates exhibits no greater correlation than the previous pair. These correlations between the pairs of canonical are called the canonical correlations, r_C. Equation 7.4 states that each canonical variate is uncorrelated with all of the other canonical variates except its specific counterpart in the mth pair. Finally, Eq. 7.5 states that each canonical variate has unit variance.

7.4 Rainfall and Sea Surface Temperature Variability

Some examples of patterns of the decadal rainfall trend for both the smoothed and unsmoothed time series obtained are shown in Fig. 7.2a, b respectively associated with the long rainy season of MAM.

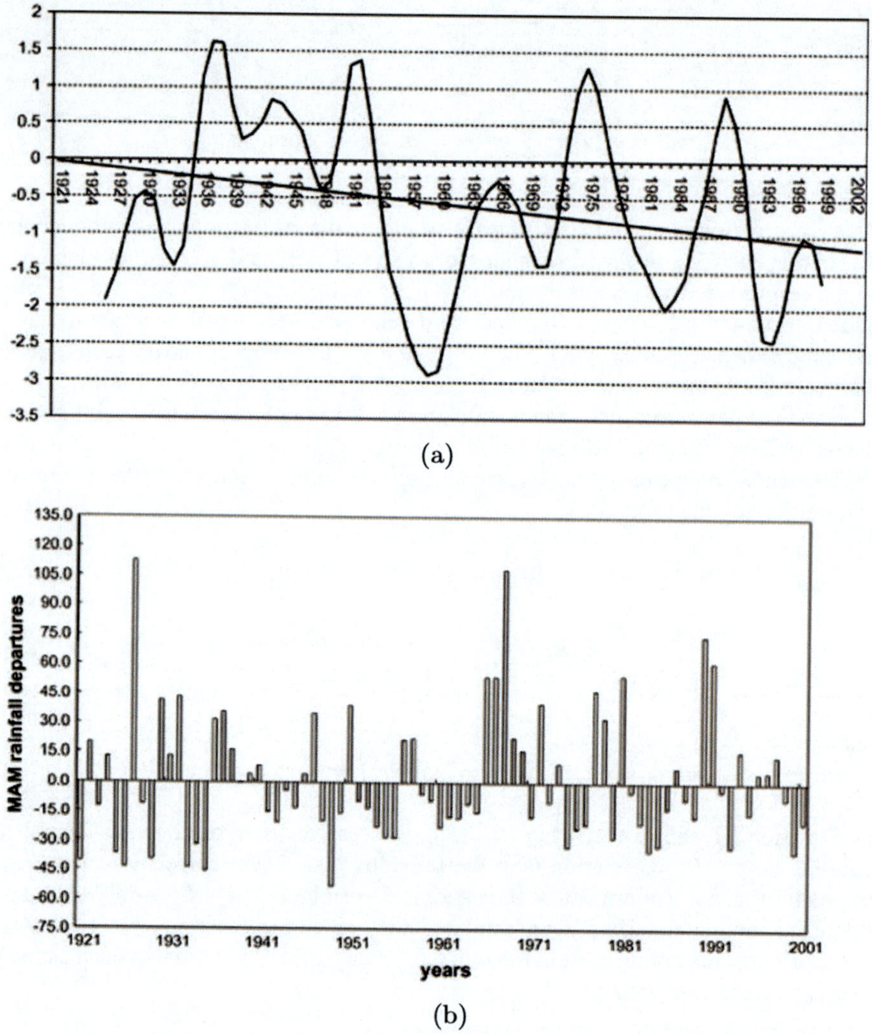

Fig. 7.2 **a** Smoothed inter-annual MAM rainfall anomalies (y-axis) for Voi in Kenya (i.e., station 10 in Table 7.1), and (**b**), unsmoothed inter-annual MAM rainfall anomalies for Voi in Kenya indicated in the y-axis. *Source* [36]

The ten year cycles are clearly discernible in the smoothed series. These modes are better illustrated when the time series of smoothed series are plotted as anomalies (Fig. 7.3a, b). The trend mode for the third short rainfall (dry) season associated with JJA rainfall shows that western and coastal parts of the region receive substantial amount of rainfall, which unlike MAM and OND seasons, are dominated by twenty years cycles (Fig. 7.3b). The major decadal signals observed from the graphs indicate that for the long rainy MAM seasons, the wet decades were 1921–1930, 1961–1970 and the late 1981–90 while the dry ones included 1931–1940, 1941–1950, 1951–1960, early parts of 1971–1980 and 1991–2000.

There are significant spatial variations in the observed decadal trend signals, with no noticeable decade with one specific dominant trend over the whole region. This could be attributed to the influence of regional and local factors including the existence of many large inland water bodies and complexity in the East Africa topography.

The short rainfall season of OND is the second major rain season for the region. The extreme events in one or two years within a decade influenced the general trend of the decadal mean rainfall. Example is the 1997/98 El Niño related floods that made 1991–2000 be a wet decade in most zones. The 1961/1970 decade was wet due to the heavy rainfall that was received over most parts of the region in 1961/1962 that resulted into the rise of Lake Victoria level by over 2.5 m [25, 38, 59].

The major decadal signals observed for OND seasons were wet decades of 1931–1941, 1951–1961, 1971–1981 and 1991–2001 while the dry decades included 1941–1951, 1961–1971, 1981–1991 and mid 1995–early 2001 (Fig. 7.3a). For the JJA seasons, the major decadal variability was relatively longer than ten years. The wet decades included 1951-mid 1961/1971 and mid 1981/1991-mid 1991/2001 while the dry ones included mid 1921–1951 and mid 1961/1971-mid 1981/1991 for the western parts of the region (Fig. 7.3b). The scenario was exactly opposite along the coastal region, i.e., the wet decades were generally 1921–1930, 1931–1940, 1961–1970, 1971–1980 while the dry decades were 1941–1950, 1951–1960, 1981–1990, 1991–2000. There was significant spatial variation in the observed trends (Fig. 7.4).

In order to establish whether the observed decadal trend and cyclic modes are significant, statistical tests on the differences amongst some decadal means and the Spearman rank were carried out [19]. A comparison of decadal means with the long-term seasonal rainfall means showed that the decades of 1921–1930 and 1961–1970 were generally wet while 1931–1940, 1951–1960 and 1991–2000 were generally dry during the long rainfall (MAM) seasons of the study period. In order to establish whether there was existence of any spatially coherent decadal differences, the spatial patterns of the various means were plotted in Fig. 7.4. Large scale wet/dry cases were, however, evident for a few specific years. Similar to the MAM seasons as already stated, this could have been due to the influence of regional and local factors, e.g., including the existence of many large inland water bodies and topographical complexity in the region [2, 20].

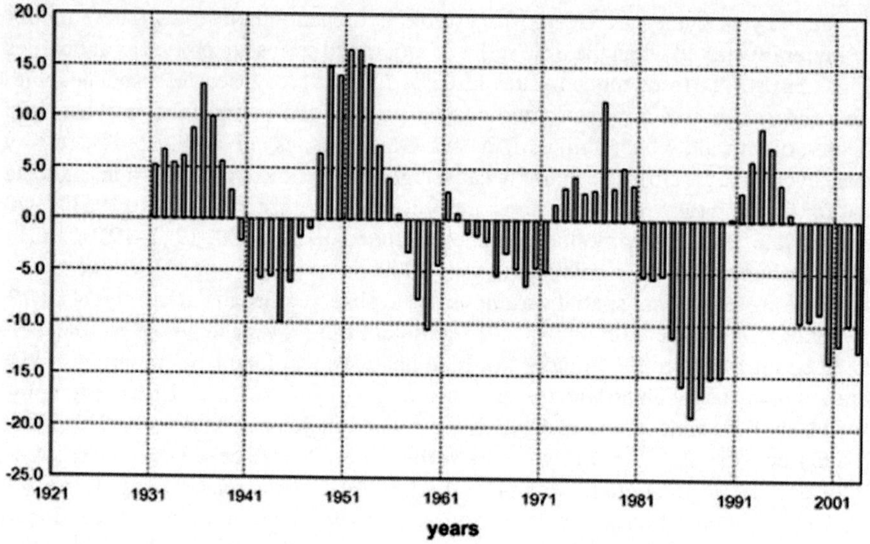

(a)

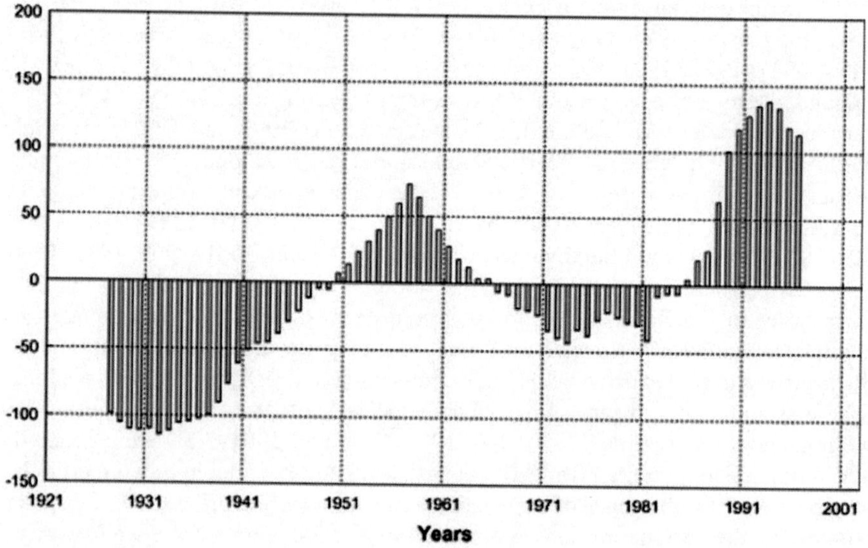

(b)

Fig. 7.3 **a** Graphical plot of the smoothed anomalies (y-axis) for OND decadal rainfall variability over Eastern Africa, and **(b)**, graphical plot of the smoothed anomalies (y-axis) for JJA decadal rainfall variability over western and coastal sub-regions. *Source* [36]

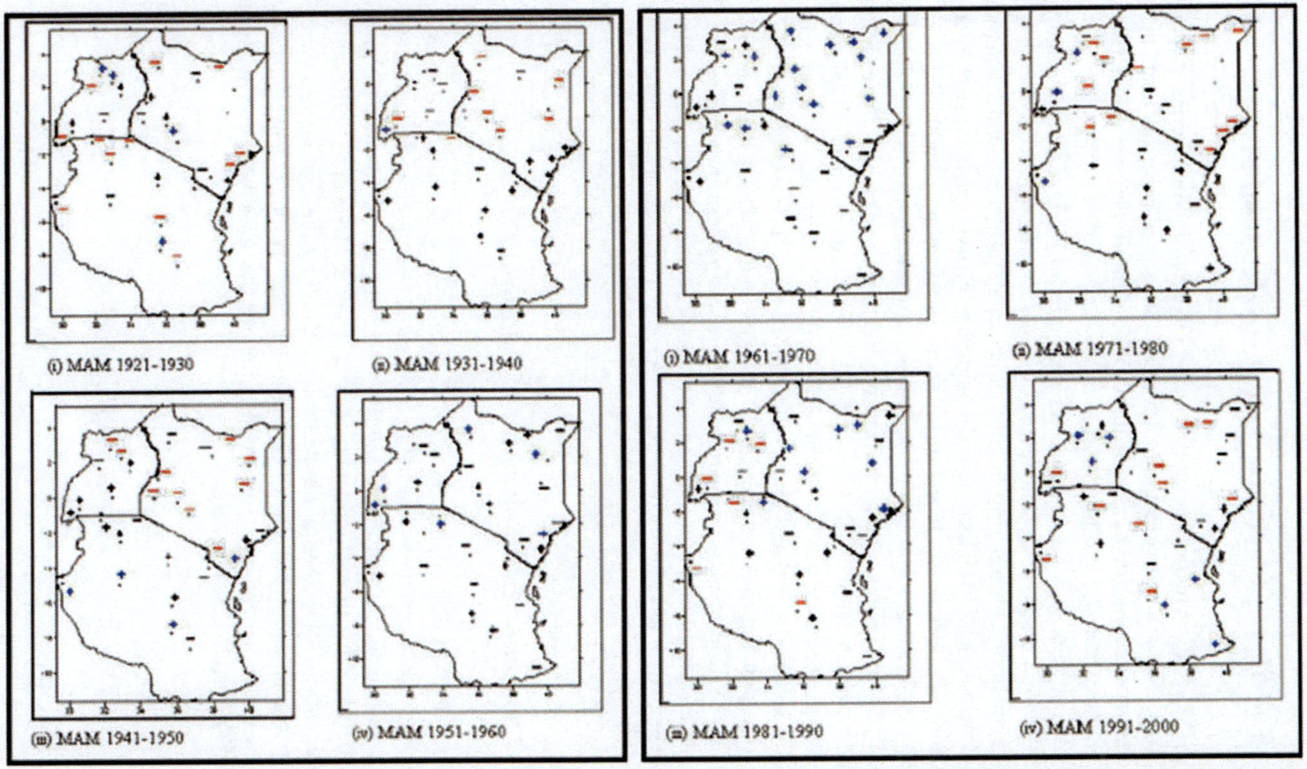

Fig. 7.4 Spatial distribution of the mean decadal rainfall for MAM. In the figure, wet means are indicated by (+) while dry cases are indicated by (−). Blue shows statistically significant increase in mean decadal rainfall, red (statistically significant decrease in mean decadal rainfall) and black (no significant increase/decrease in mean decadal rainfall). The size shows the magnitude of the increase/decrease. *Source* [36]

7.5 Rainfall Seasonal Variability

7.5.1 The Short October–December (OND) Rainfall

The VARIMAX-PCA method is applied on the OND rainfall time series. According to the Scree, Kaiser's criterion and North et al. [30] sampling errors tests, the first five modes accounting for 81.3% of total OND rainfall variance are statistically significant (see Table 7.2 and Fig. 7.5a).

For brevity, here, only the first 3 dominant modes of OND seasons are shown in Fig. 7.7. The first mode is extended nearly in all parts of the region except the south-eastern segment of Tanzania (Fig. 7.7a). Schreck et al. [50] also found a remarkably similar distribution of EOF loadings, although their analysis covered slightly a bigger domain. The variance in EOF1 of [50] was 28%, compared with 42.9% of the corresponding mode in Fig. 7.7. This difference may be attributed to the larger region covered by their analysis. The corresponding PC time series (Fig. 7.7a, right) indicates some consistence with ENSO variability (the so called 'cold ENSO' signal in [15, 33, 34]. According to PC1, the average cold ENSO events were pronounced in the decades 1980–1990 and this brought about general depressed rainfall in the region corresponding to the high peaks of the time series. The reverse condition is represented by PC2 in Fig. 7.7b with a dipole spatial pattern over the region. The third mode is related to the decadal trend mode [10], showing positive mode over the north and south of Lake Victoria and a decrease mainly over the eastern coastal regions (Fig. 7.7c).

7.5.2 The Long March–May (MAM) Rainfall

VARIMAX-PCA is also applied on MAM rainfall season. Results of the Scree and North et al. (1982) test show that the first six modes accounting for about 80% of variance are statistically significant (see Table 7.2 and Fig. 7.5b). Like in the previous section, only the first three dominant modes are shown. Mode one (EOF1 and PC1) depicts the north–south rainfall dipole brought about by the movement of the Sun from one hemisphere to the other, i.e., due to the Inter-Tropical Convergence Zone (ITCZ), while the second mode is related to the positive IOD mode and the decadal trend in EOF3 and PC3 (Fig. 7.8).

7.5.3 The Dry June–July (JJA) Rainfall

Implementing VARIMAX-PCA on the dry JJA seasons show that parts of the equatorial sector, covering northern Tanzania, western parts of East Africa and the coastal

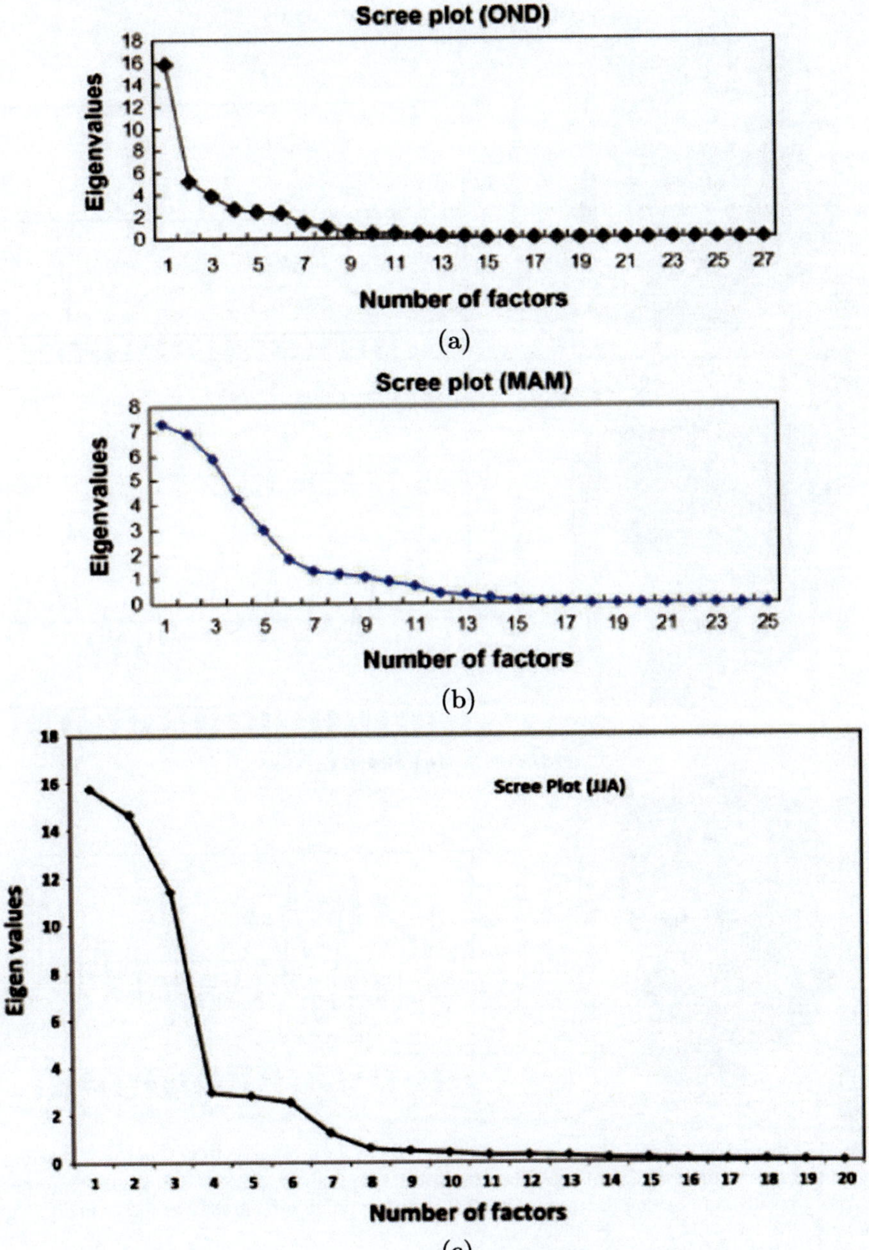

Fig. 7.5 **a** Scree's test selection of dominant principal components for OND rainfall seasons, **b** Scree's test selection of dominant principal components for MAM rainfall seasons, and (**c**), Scree's test selection of dominant principal components for June–July rainfall seasons. *Source* [36]

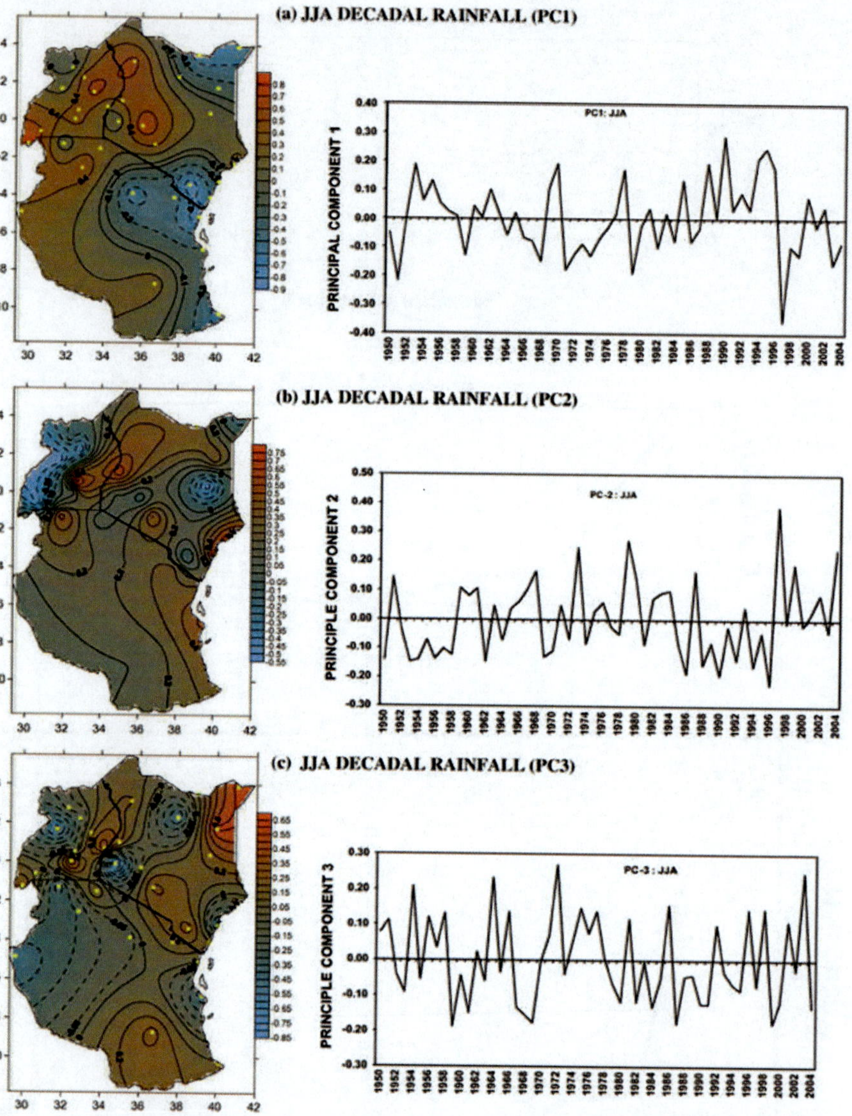

Fig. 7.6 Spatial and temporal patterns for EOF1 and PC1 (first row); EOF2 and PC2 (second row) and EOF3 and PC3 (third row) for JJA decadal rainfall. Dashed/solid contours represent negative/positive values; contour interval is 0.2. The yellow points refer to stations shown in Fig. 7.1. *Source* [36]

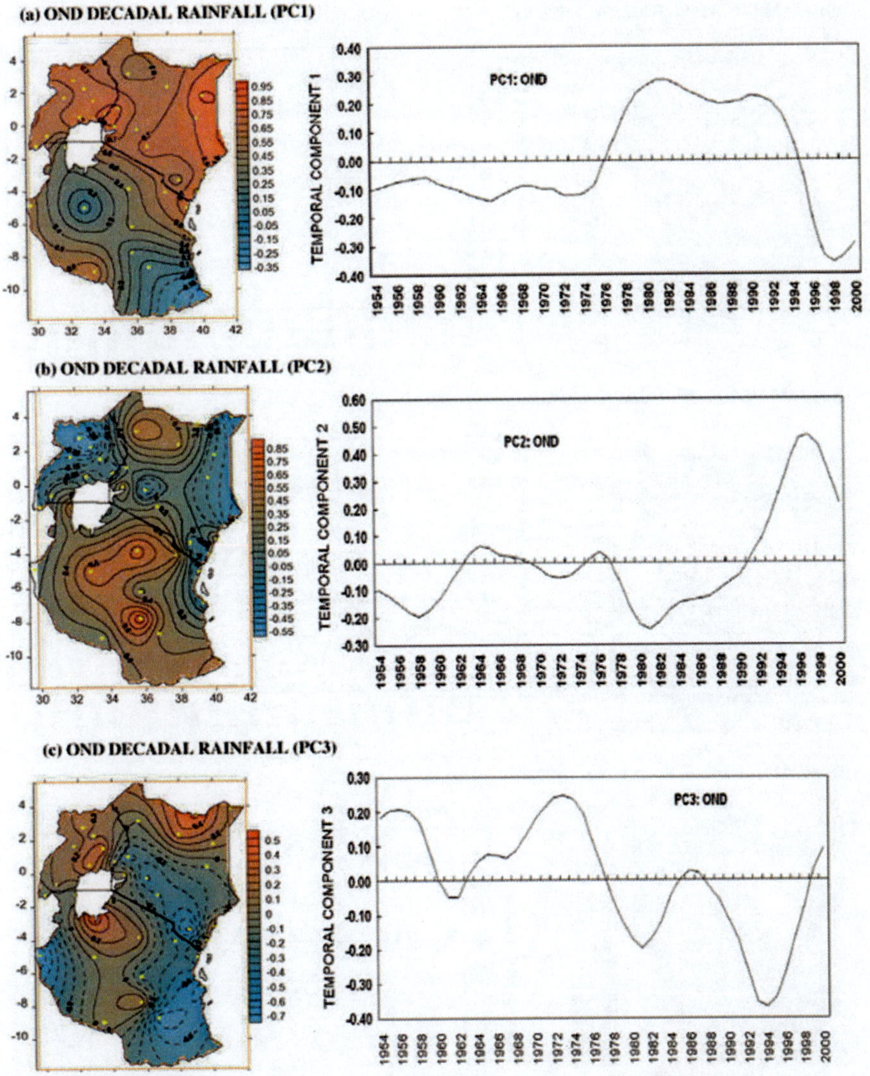

Fig. 7.7 Spatial and temporal patterns for EOF1 and PC1 (first row); EOF2 and PC2 (second row) and EOF3 and PC3 (third row) for OND decadal rainfall. Dashed/solid contours represent negative/positive values; contour interval is 0.2. The yellow points refer to stations shown in Fig. 7.1. *Source* [36]

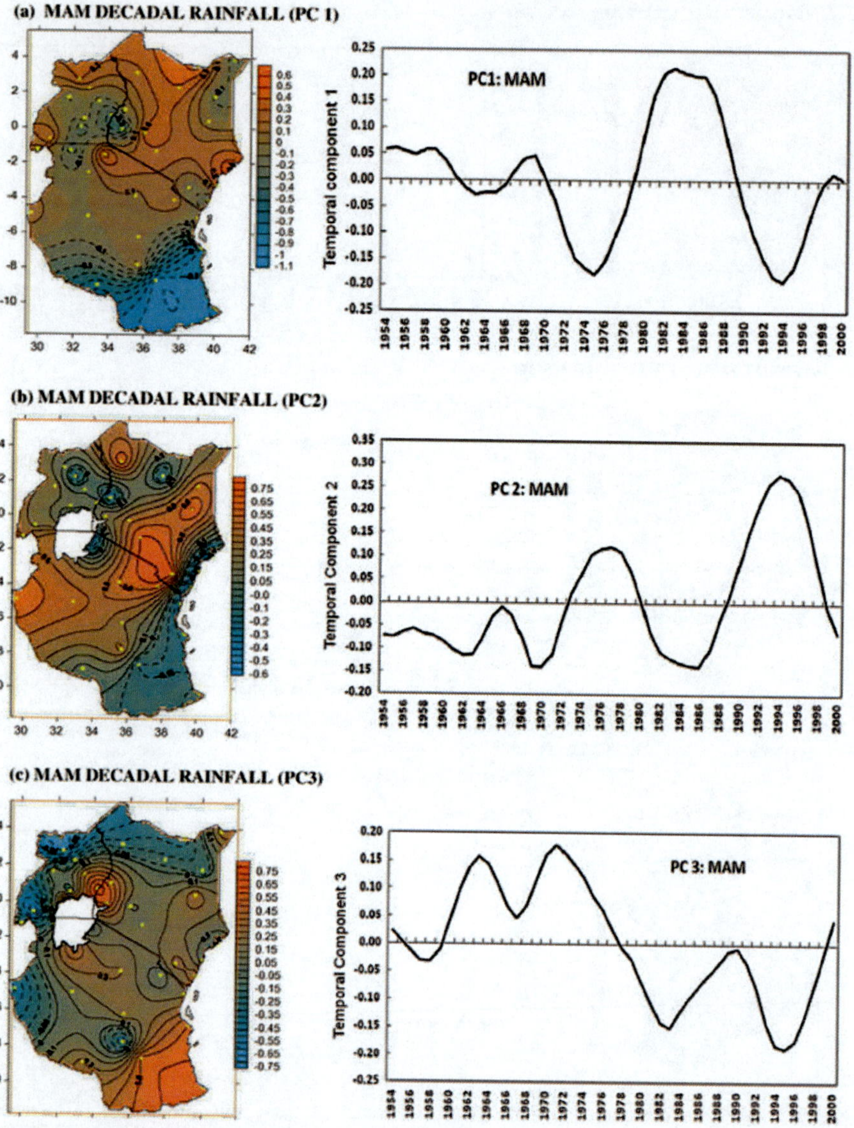

Fig. 7.8 Spatial and temporal patterns for EOF1 and PC1 (first row); EOF2 and PC2 (second row) and EOF3 and PC3 (third row) for MAM decadal rainfall. Dashed/solid contours represent negative/positive values; contour interval is 0.2. The yellow points refer to stations shown in Fig. 7.1. *Source* [36]

Table 7.2 Eigenvalues, variance and accumulated variance extracted by each mode of the decadal OND, MAM and JJA rainfall. *Source* [36]

Period	Factor	Eigenvalue	Variance extracted (%)	Cummulative variance (%)
OND	1	15.9	42.9	42.9
	2	5.3	14.2	57.1
	3	3.9	10.5	67.6
	4	2.7	7.2	74.8
	5	2.4	6.5	**81.3**
MAM	1	7.3	19.7	19.7
	2	6.9	18.6	38.3
	3	5.9	15.9	54.2
	4	4.2	11.4	65.6
	5	3.0	8.1	73.6
	6	2.3	6.3	**80.0**
JJA	1	15.8	28.6	28. 6
	2	14.6	26.6	55.2
	3	11.4	20.7	75.9
	4	2.9	5.4	81.3
	5	2.8	5.1	86.4
	6	2.5	4.6	91.0
	7	1.2	2.2	**93.2**

areas generally get rains (Fig. 7.7). Figure 7.5c and Table 7.2 indicate that seven PCA modes, accounting for about 93% of the total JJA variance, were significant.

Figure 7.6a displays the spatial pattern in which EOF 1 explains 28.6% of the total JJA variance. The distribution of the loadings is characterized by moisture incursion from the Congo air basin causing wetness in the western sector of the region [32]. The corresponding time series exhibits both strong inter-annual variability and low-frequency background variability. PC2 (Fig. 7.6b) show positive loadings in most parts of the study area except some pockets of the eastern and western segments. PC3 of JJA season show (Fig. 7.6c) both positive and negative loadings evenly distributed over the sectors that experience much rainfall during this season. All the time series of these PCs depicted strong decadal variability of rainfall. Most parts of the southern segments in Tanzania are dry during this season. The evolution of the background variability has positive trend in 1980/1990 decade, which reached its highest levels during the early 1998. There is indication of subsequent decline in the amplitude during the late 1990s and early 2000. Combined interpretation of the Regional-EOF1 distribution of loadings (Fig. 7.6a) and the corresponding time series suggests that the western sector of Eastern Africa had 1980/1990 decade wet while the southern sector was drier. The southern sector during this season is usually dry and thus could have resulted into the negative anomalies.

7.6 Sea Surface Seasonal Temperature Variability

This section presents the results of implementing the VARIMAX-PCA method on the SST records of some specific oceanic basins during OND, JJA and MAM seasons in order to compare the SST behaviours of these oceanic basins with the rainfall patterns over East Africa. The derived VARIMAX solutions are summarized in Table 7.3 and the derived spatial patterns are shown in Figs. 7.9–7.11. The first mode of VARIMAX-PCA derived from decadal SST of the Indian Ocean during OND seasons accounts for 38% of the total variance of SST. The positive loading is centred on the tropical equatorial Indian Ocean and the negative loading is located over the south-western Indian Ocean (Fig. 7.9a). The Indian Ocean EOF 1 for MAM (Fig. 7.9c), however, has a dipole structure like pattern of SSTs with positive centre near the Indo-Pacific area while negative centre located near the south-western Indian Ocean. The total variance accounted for this mode is 45.4% (Table 7.3).

Figure 7.10a shows the spatial patterns for Atlantic Ocean during the same OND seasons. Generally, the equatorial basin of Atlantic shows positive loadings while its northern sector exhibits negative loadings with the highest variance of 64.1% during OND season. It is noteworthy that the OND season had the first and second variances taking nearly all the total variances (Table 7.3).

Figure 7.10b indicates almost patterns in loading similar to Fig. 7.10a but with much stronger modes. The spatial patterns in Fig. 7.10c, d taking 32.5 and 39.7% respectively are generally similar to other patterns showing that the first dominant PC

Table 7.3 Percentage variance extracted by the first 4 Rotated Principal Components (RPCs) of decadal SST. *Source* [36]

		OND	DJF	MAM	JJA
Indian Ocean	PC1	38.0	40.5	45.4	35.2
	PC2	26.0	15.8	41.4	32.5
	PC3	20.1	15.7		24.9
	PC4	11.3	8.5		
Total variance		**95.4**	**80.5**	**89.8**	**92.6**
Atlantic Ocean	PC1	64.1	39.1	32.5	39.7
	PC2	34.9	30.6	31.5	34.9
	PC3		13.7	21.8	13.5
	PC4				
Total variance		**99.0**	**83.4**	**85.8**	**88.1**
Pacific Ocean	PC1	31.0	40.1	40.9	32.0
	PC2	23.4	30.0	28.2	23.7
	PC3	22.9	11.5	9.6	23.5
	PC4	10.3			
Total variance		**76.6**	**81.6**	**78.7**	**79.2**

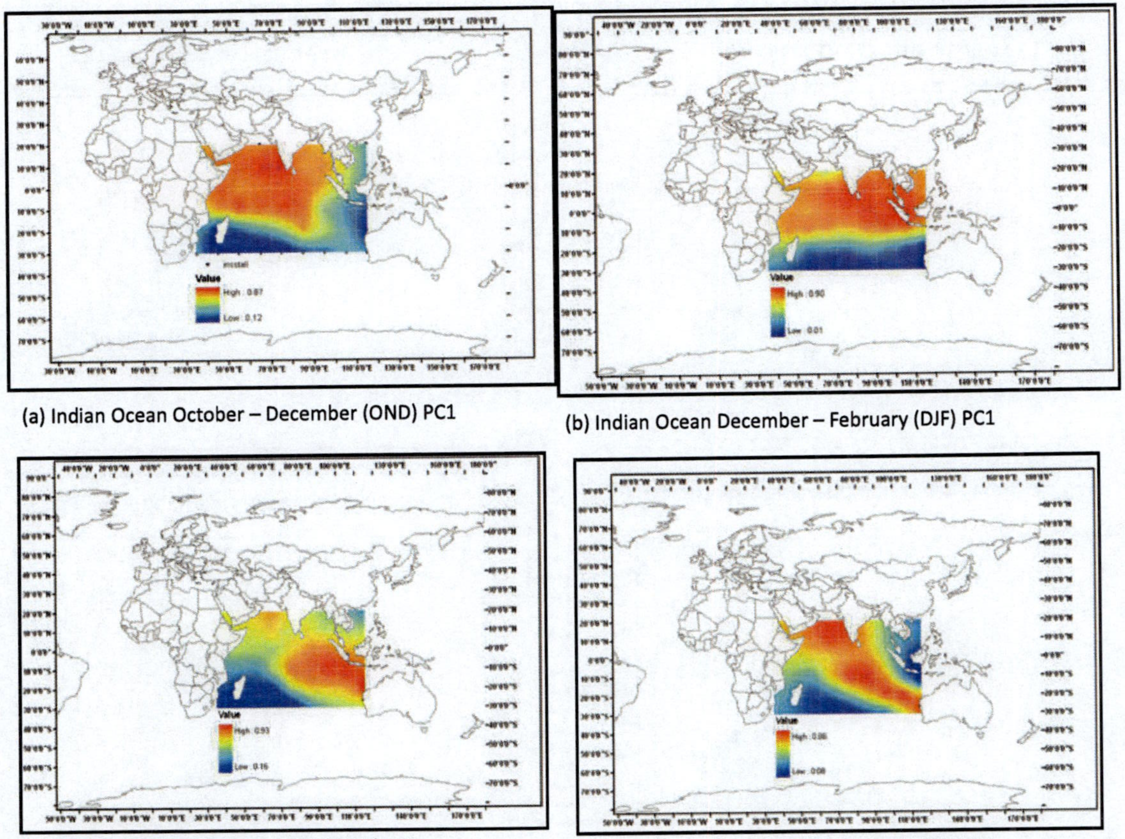

(a) Indian Ocean October – December (OND) PC1

(b) Indian Ocean December – February (DJF) PC1

(c) Indian Ocean March – May (MAM) PC1

(d) Indian Ocean June – August (JJA) PC1

Fig. 7.9 The spatial patterns of the first 9-term binomial coefficient filtered SST PCA modes for the Indian Ocean. The x-axis represent the longitude while the y-axis shows the latitude. *Source* [36]

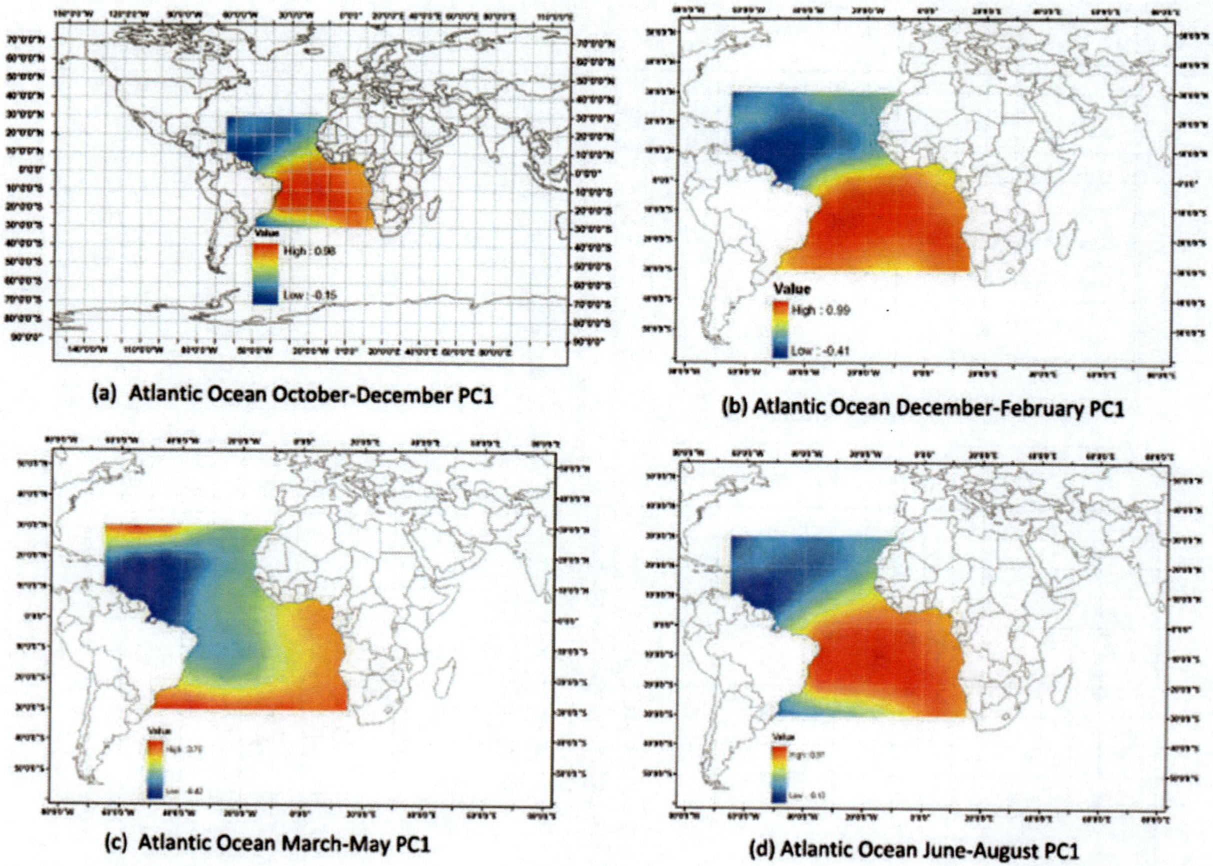

Fig. 7.10 The spatial patterns of the first 9-term binomial coefficient of the filtered SST PCA modes for the Atlantic Ocean. The x-axis represent the longitude while the y-axis shows the latitude. *Source* [36]

mode in all the four seasons is influenced by similar loading. This could be attributed to the strong and alternating north–south dipole pattern reflecting the known patterns of the Atlantic Tropical Dipole Oscillation [11]. In the Pacific Ocean, the first dominant mode for the four seasons seems to have positive/negative loadings over equatorial eastern/western ocean basin regions (Fig. 7.11) that seem to reflect the ENSO variability mode [15, 23, 34, 37]. This dipole structure like pattern of SSTs has negative centre near the Indo-Pacific area while the positive centre is located near the eastern equatorial Pacific Ocean.

The results seem to reflect the El'Niño/La'Niña variability mode [18, 51, 52].

7.7 Canonical Correlation Analysis (CCA)

In this section, the strength and the sign of the corresponding patterns are described by the canonical correlation coordinates. The CCA method takes into account analysis of the full space and time dimensions of the two fields (rainfall and SSTs), which make it superior comparing to a conventional correlation analysis.

7.7.1 CCA of the MAM Rainfall Seasons

The average December–February (DJF), JJA and MAM SSTs from the various ocean basins were independently correlated with MAM rainfall time series. MAM is the major rainfall season for the region and the skill of its predictability is still very low. Three significant modes were discernible for the Indian Ocean basin with DJF and MAM SSTs.

The canonical modes accounting for about 72 and 86.8% respectively of the total variance were selected as inputs into the CCA model. Figures 7.12 and 7.13 give examples of the CCA loading patterns for the DJF and MAM SSTs of the Indian Ocean respectively. An area of high significant positive correlation between the mentioned SSTs and the canonical component time series was evident over the central equatorial Indian Ocean (Fig. 7.12a). Similarly, there was significant correlation at most locations with positive loadings over the coastal and Lake Victoria regions and a negative loading over the rest of the region (Fig. 7.12b). The canonical correlation skill between rainfall and the predictor SST modes was about −0.79 (Fig. 7.12c). The canonical correlation score between MAM rainfall and the predictor SST modes was 0.72 and 0.96 for one and zero lags, respectively, Fig. 7.13c. The canonical scores of the pattern with warm SST in the Indian Ocean were increasing since the mid 1970s whereas the negative coupling was decreasing. Power et al. [39] in their analysis of decadal climate variability showed that decadal variability in Indian Ocean SST south of 40° is associated with rainfall variability over East Australia.

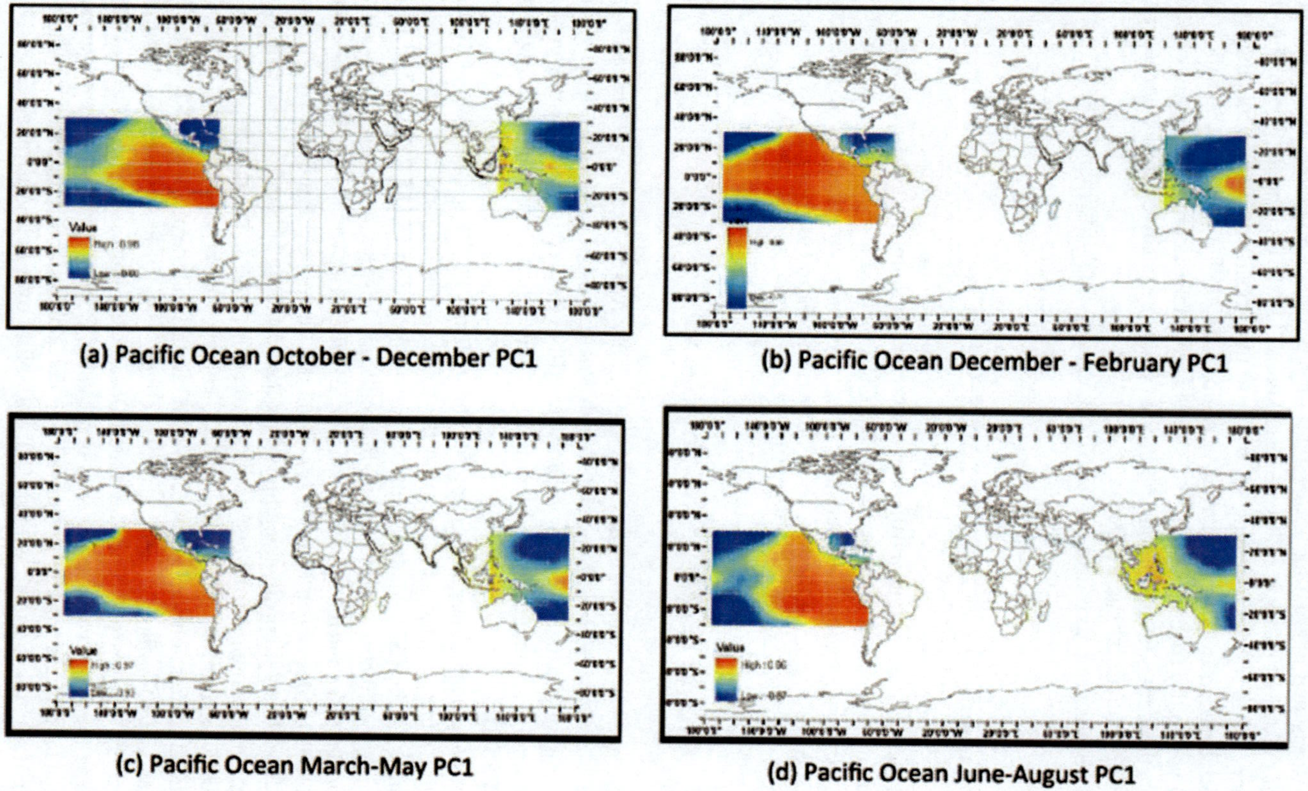

Fig. 7.11 The spatial patterns of the first 9-term binomial coefficient filtered SST PCA modes for the Pacific Ocean. The x-axis represent the longitude while the y-axis shows the latitude. *Source* [36]

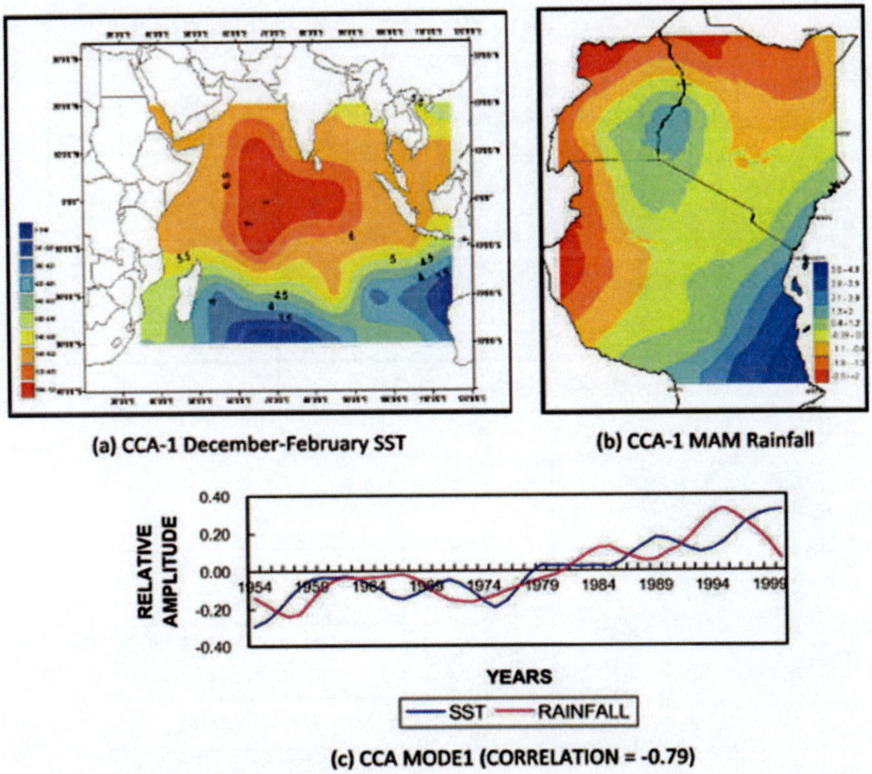

Fig. 7.12 The first spatial pattern pair for canonical correlation between decadal DJF of the Indian SST and MAM rainfall; **a** correlation between the predictor (SST) and the canonical vector (u); **b** correlation between the predictant (rainfall) and canonical vector (v), and; **c** normalized temporal functions (u and v) of the first CCA patterns for rainfall and SST. The x-axis represent the longitude while the y-axis shows the latitude. *Source* [36]

7.7.2 CCA of JJA Rainfall Season

The averages of MAM and JJA SSTs from the various ocean basins were independently correlated with JJA rainfall. Figures 7.14 and 7.15 depict loading patterns for the MAM and JJA SSTs of the Atlantic Ocean with JJA seasonal rainfall modes together with the corresponding temporal functions respectively. The negative loadings over the equatorial north-western and central Atlantic Ocean regions (Fig. 7.14a) are associated with the wet/dry regimes over western/eastern sectors of Eastern Africa (Fig. 7.14b).

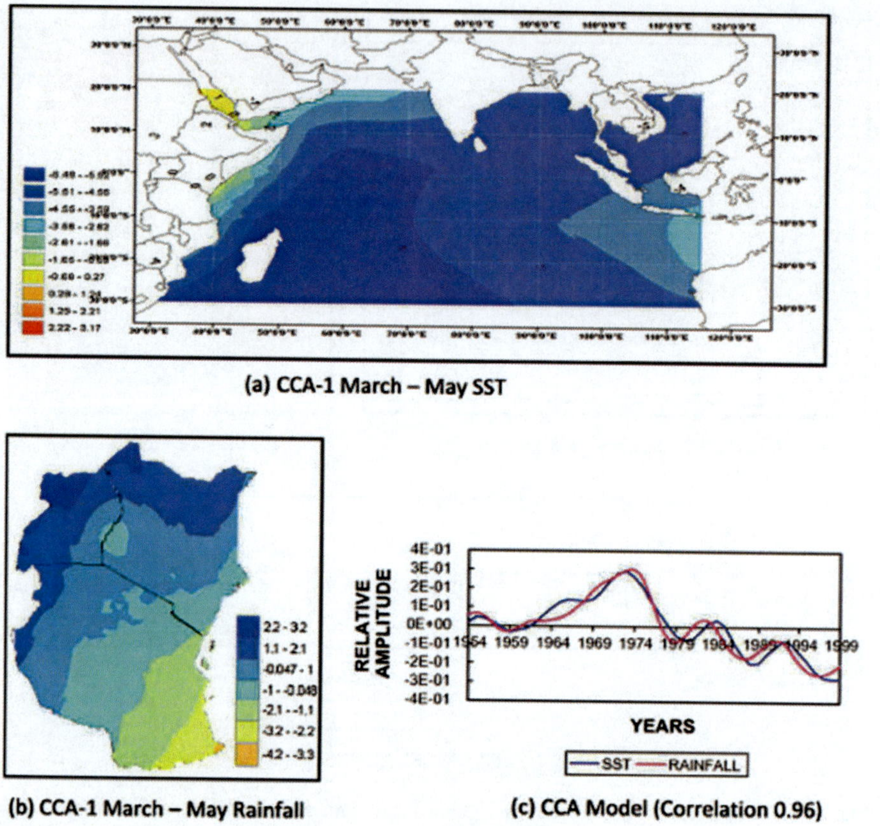

Fig. 7.13 The first spatial pattern pair for canonical correlation between decadal MAM of the Indian SST and MAM rainfall; **a** correlation between the predictor (SST) and the canonical vector (u); **b** correlation between the predictant (rainfall) and canonical vector (v) and; **c** normalized temporal functions (u and v) of the first CCA patterns for MAM of the Indian Ocean SST (lag zero). The x-axis represent the longitude while the y-axis shows the latitude. *Source* [36]

The canonical correlation score between rainfall and the predictor SST modes was 0.72 for lag one and 0.87 for zero lag. Lag zero that had maximum weights over the Atlantic Ocean basin and was positively correlated with JJA over the whole of western and coastal regions of Kenya together with Uganda (Fig. 7.15b). Similar results have been derived by previous studies, including those of [40, 41, 54] over the Indian Ocean and South African rainfall.

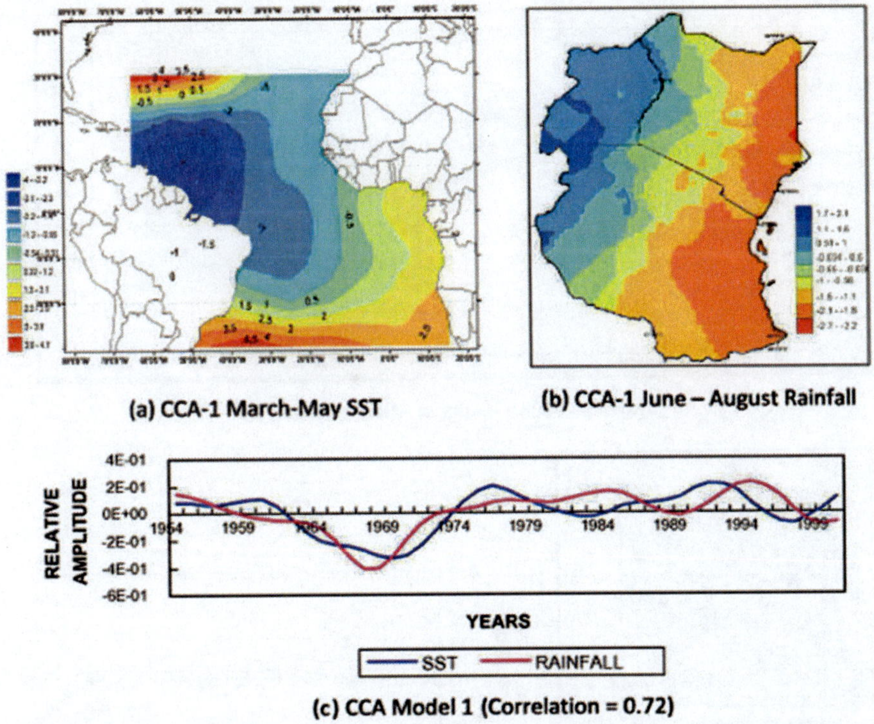

Fig. 7.14 The first spatial pattern pair for canonical correlation between decadal MAM Atlantic SST and JJA rainfall; **a** correlation between the predictor (SST) and the canonical vector (u); **b** correlation between the predictant (rainfall) and canonical vector (v) and; **c** normalized temporal functions (u and v) of the first CCA patterns for rainfall and SST. The x-axis represent the longitude while the y-axis shows the latitude. *Source* [36]

7.7.3 CCA of OND Rainfall Season

The average JJA and OND SSTs from the various ocean basins were independently correlated with OND rainfall. Figures 7.16 and 7.17 represent CCA loading patterns for JJA and OND of the Pacific Ocean SSTs and OND rainfall respectively. The highly negative loading over the equatorial eastern Pacific Ocean seems to be the major mode associated with the wetness in nearly the whole part of the region with pocket of dryness conditions over southern parts that generally have unimodal rainfall regimes. The canonical correlation skill between OND rainfall and the predictor SST modes was 0.88 at lag one showing stronger influence of eastern Pacific to the region. Figure 7.17 shows the CCA loading patterns for the OND rainfall and predictor of the Pacific Ocean SST modes at zero lag. The negative SST loading over the equatorial eastern Pacific Ocean is linked to the generally dry conditions in the region. The canonical correlation skill between OND decadal rainfall and the predictor SST

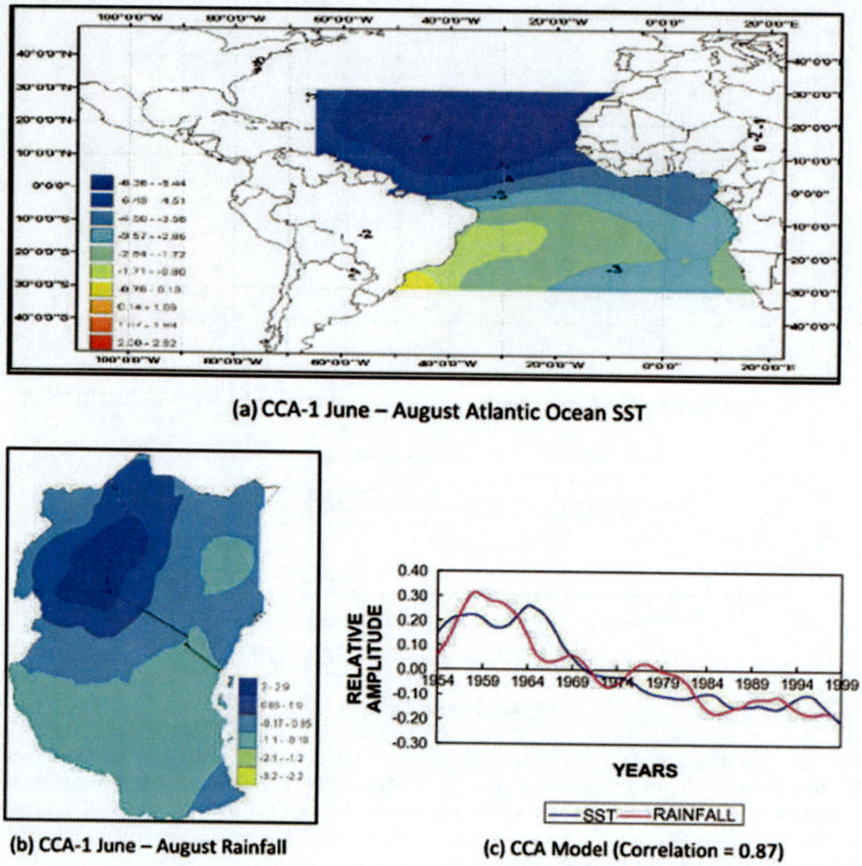

Fig. 7.15 The first spatial pattern pair for canonical correlation between decadal JJA Atlantic SST and JJA rainfall; **a** correlation between the predictor (SST) and the canonical vector (u); **b** correlation between the predictant (rainfall) and canonical vector (v) and; **c** normalized temporal functions (u and v) of the first CCA patterns for JJA SST (lag zero). The x-axis represent the longitude while the y-axis shows the latitude. *Source* [36]

modes is 0.97. Thus cold ENSO phase would be associated with depressed rainfall season over the whole region, while warm phase (El Niño) would be associated with enhanced decadal rainfall over most parts of the region.

7.8 Concluding Remarks

This chapter provided some evidence of decadal variability in the inter-annual patterns of East Africa rainfall. The MAM and OND seasonal rainfall are dominated by 10 year cycles of wet and dry phases, while the JJA season showed 20 year cycles of

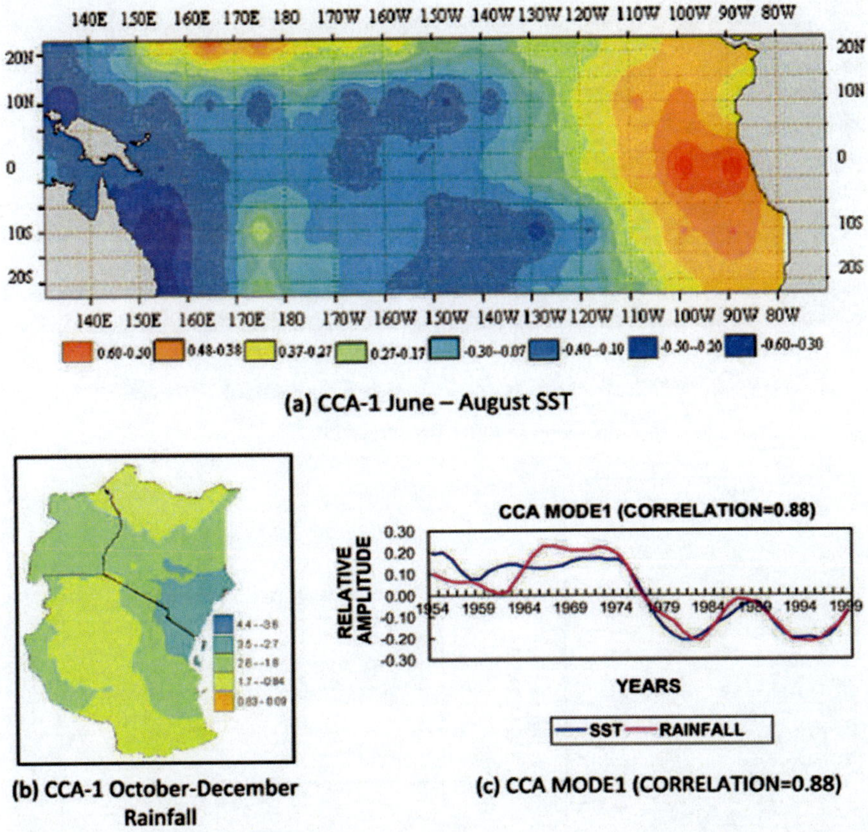

Fig. 7.16 The first spatial pattern pair for canonical correlation between decadal JJA Pacific SST and OND rainfall; **a** correlation between the predictor (SST) and the canonical vector (u); **b** correlation between the predictant (rainfall) and canonical vector (v) and; **c** normalized temporal functions (u and v) of the first CCA patterns for rainfall and SST. x-axis represent the longitude while the y-axis shows the latitude. *Source* [36]

wet and dry phases. Some teleconnections were also evident between the observed decadal rainfall variability patterns and SST variability modes over parts of the global oceans. The significant correlation between the rainfall and SSTs offers a useful indicator in predicting rainfall of the region at decadal time scale. Specifically, the chapter show that:

1. Trend analysis results showed that although no significant trend in the inter-annual patterns were discernible at many locations, too much or too little rainfall received in one or two years influenced the general trend of the decadal mean rainfall. Zones 1 and 8 in OND and MAM showed significant positive trends during this period of study, respectively. For the JJA season, when only the western and coastal parts of the region receive substantial amount of rainfall, no significant

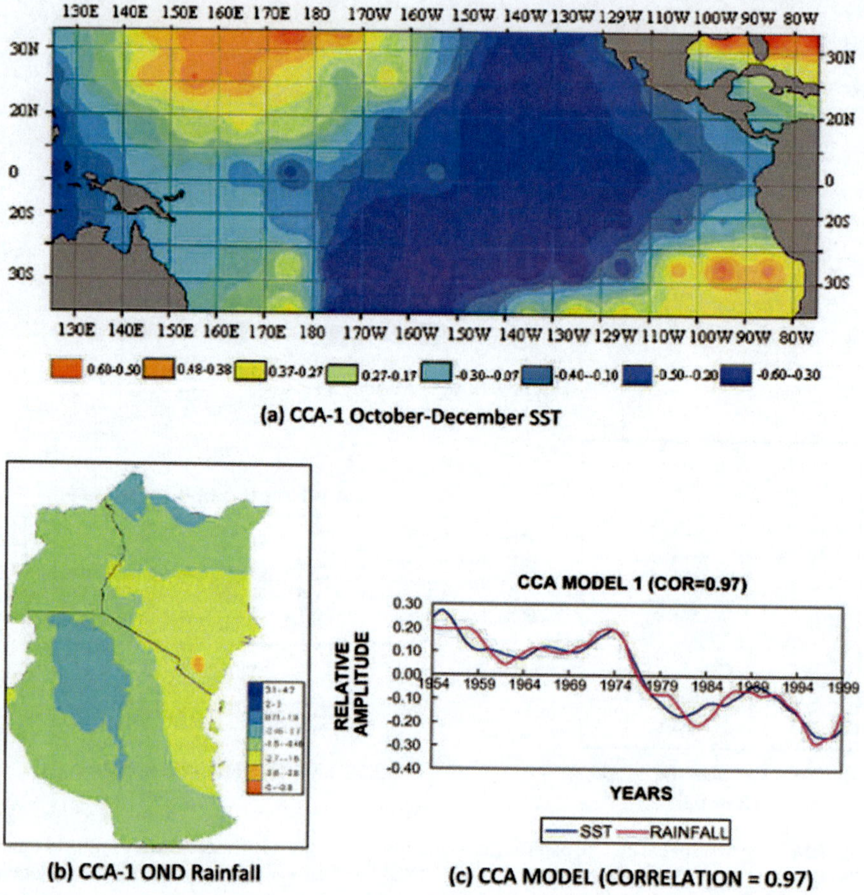

Fig. 7.17 The first spatial pattern pair for canonical correlation between decadal OND Pacific SST and OND rainfall; **a** correlation between the predictor (SST) and the canonical vector (u); **b** correlation between the predictant (rainfall) and canonical vector (v) and; **c** normalized temporal functions (u and v) of the first CCA patterns for OND SST (lag zero). x-axis represent the longitude while the y-axis shows the latitude. *Source* [36]

trends were observed although the decades after 1961 were wetter than before in these western regions but drier along the coastal regions. The wetness in the western region and dryness in the coastal region of Eastern Africa during JJA have been shown to depend on two mechanisms [13, 60, Professor Richard Anyah, Per. Comm. 2021] (i) The incursions of moist Congo AirMass (CAM) associated with west (highlands)-east (coast) migration of the Congo Branch of the walker circulation. Whenever there is a shift of walker circulation eastwards then most of western Equatorial Eastern Africa receives out of season (neither MAM nor OND) wet anomalies. This is also linked to interactions between Lake Victoria driven convection and the moist Congo AirMass, mostly reflected in above normal rainfall over LVB during JJA, and (ii), for drier conditions along the coast, which depends on the SST anomalies over the IOD region, a few studies have also shown that when the Congo walker circulation cell shifts eastward in response to SST anomalies over western Indian Ocean, areas of convection off the coast also shift. Apart from precipitation recycling over the lake, large scale moisture convergence/divergence over the highlands(west)/coast east are amplified by the SST gradients over western Indian Ocean and CAM, including sometimes cyclone activities of the southwest Indian Ocean (SWIO). Furthermore, the influence of topography should also be noted. The western region has high elevations compared to the coastal regions and as such, the topography could be influencing the rainfall in favour of the western region compared to coastal regions, see e.g., [1]. Nicholson [28] opine that the regionally integrated convergence in the lower troposphere, which though positive during the rainy season, is associated with coastal uplift and topographic uplift in the western highlands (see [26] for further discussion). No decade was observed to have the whole region dominated by one specific trend mode during the period of study except 1931–1940 and 1961–1970 during OND seasons.

2. Results from CCA, applied independently on the average DJF and MAM SSTs from the various ocean basins and MAM rainfall, show that three significant modes were discernible. One area of high significant positive correlation between the SSTs and the canonical component time series was evident over the central equatorial Indian Ocean. Similarly, there was significant correlation resulting into wet coastal and Lake Victoria regions with the rest of the region found to be dry.
3. The results from the average MAM and JJA SSTs correlated with JJA rainfall had negative loadings centred on the equatorial western and central Atlantic Ocean regions, which were associated with the wet/dry regimes over western/eastern sectors of the region. Linkages between the Atlantic Ocean basin and Eastern Africa during JJA are largely influenced by the space–time pattern of both zonal and meridional arms of the ITCZ.
4. The average JJA and OND SSTs with OND rainfall produced highly negative loading over the equatorial eastern Pacific Ocean that were associated with rainfall deficit in nearly the whole region but wet conditions over the southern parts of the region. The positive centre over the eastern equatorial Pacific Ocean, however, was associated with wet conditions in nearly all the region.

Finally, although not treated in this Chapter, it should be pointed out that changes in sea surface temperature would impact the marine ecosystem, which in turn will affect food security of the region. Of the GHA countries; Tanzania, Kenya, Somalia, Djibouti, Eritrea and Sudan have most of their populations depending on fishing as a source of food security since they boarder marine boundaries. Global warming resulting in SST change will therefore affect their marine ecosystems through, e.g., alteration of the migration and distributions patterns of various fish species, food webs, relationships between predators and prey, and the reproductive timing and success of species.[2] This will results not only on the loss of vital vitamin source but also of the economic income that can cushion the population against climate extreme shocks such as those of droughts and floods, which when they come, destroy crops (see Chap. 1).

References

1. Agutu NO, Awange JL, Ndehedehe C, Mwaniki MW (2020) Consistency of agricultural drought characterization over Upper Greater Horn of Africa (1982–2013): Topographical, gauge density, and model forcing influence. Sci Total Environ 709. https://doi.org/10.1016/j.scitotenv.2019.135149
2. Anyah RO (2005) Modelling the variability of the climate system over Lake Victoria basin. PhD. thesis, Department of Marine, Earth and Atmospheric Sciences, North Carolina University, USA
3. Awange JL, Ogalo L, Bae KH, Were P, Omondi P, Omute P, Mullo M (2008) Falling Lake Victoria water levels is climate a contributing factor? pp 1–20. https://doi.org/10.1007/s10584-008-9409-x
4. Awange JL, Palancz B, Völgyesi L (2020) Hybrid imaging and visualization. Employing machine learning with mathematica—python. Springer Nature International, Berlin
5. Barnett TP, Preisendorfer R (1987) Origins and levels of monthly and seasonal forecast skill for United States surface air temperatures determined by canonical correlation analysis. Mon Wea Rev 115:1825–1850
6. Barnston AG, He Y (1996) Skill of CCA forecasts of 3-month mean surface climate in Hawaii and Alaska. J Clim 9:2579–2605
7. Becker M, Llovel W, Cazenave A, Günter A, Crétaux J-F (2010) Recent hydrological behavior of the East African great lakes region inferred from GRACE, satellite altimetry and rainfall observations. Compt Rend Geosci 342(3):223–233. https://doi.org/10.1016/j.crte.2009.12.010
8. Black E (2004) The relationship between Indian Ocean sea-surface temperature and East African rainfall. Phil Trans R Soc A 2005(363):43–47. https://doi.org/10.1098/rsta.2004.1474
9. Black E, Slingo J, Sperber KR (2003) An observational study of the relationship between excessively strong short rains in the coastal East Africa and Indian Ocean SST. Mon Wea Rev 131:74–94
10. Bowden J, Semazzi FHM (2007) Empirical analysis of intraseasonal climate variability over the Greater Horn of Africa. J Clim 20(23):5715–5731
11. Chang P, Ji L, Li H (1997) A decadal climate variation in the tropical Atlantic Ocean from thermodynamic air-sea interactions. Nature 385:516–518
12. Goddard L, Graham NE (1999) The importance of the Indian Ocean for simulating rainfall anomalies over Eastern and Southern Africa. J Geophys Res 104:19099–19116
13. Finney DL, Marsham JH, Walker DP, Birch CE, Woodhams BJ, Jackson LS, Hardy S (2019) The effect of westerlies on East African rainfall and the associated role of tropical cyclones and the Madden-Julian Oscillation. Quart J R Meteorol Soc 146:647–664. https://doi.org/10.1002/qj.3698
14. Fogt Ryan L, Bromwich David H (2006) Decadal variability of the ENSO teleconnection to the high-latitude south pacific governed by coupling with the southern annular mode. J Clim 19:979–997. https://doi.org/10.1175/JCLI3671.1
15. Indeje M, Semazzi FHM, Ogallo LJ (2000) ENSO signals in East African rainfall and their prediction potentials. Int J Climatol 20:19–46
16. Kendall MG (1976) Time series. Charles Griffin, London

[2] http://www.oceanhealthindex.org/methodology/components/sea-surface-temperature.

17. Kendall MG, Stuart A (1961) Advanced theory of statistics. Charles Griffins, London
18. Mantua NJ, Hare SR, Zhang Y, Wallace JM, Francis RC (1997) A pacific interdecadal climate oscillation with impacts on salmon production. Bull Am Meteor Soc 78:1069–1079
19. Maritz JS (1981) Distribution-Free statistical methods. Chapman & Hall, 217 pp. ISBN 0-412-15940-6
20. Mukabana JR, Pielke RA (1996) Investigating the influence of synoptic scale winds and meso-scale circulations and diurnal weather patterns over Kenya using a mesoscale numerical model. Mon Wea Rev 124:224–243
21. Mutai CC (2003) The role of Indian Ocean SST on East African Short rains. In: Proceedings of the sixth Kenya meteorological society workshop on meteorological research, applications and services, Mombasa, Kenya, Sept 29–Oct 3, pp 55–59
22. Mutai CC, Ward MN (2000) East African rainfall and the tropical circulation/convection on intraseasonal to interannual time scales. J Clim 13:3915–3939
23. Mutemi JN (2003) Climate anomalies over eastern Africa associated with various ENSO evolution phases. PhD thesis, University of Nairobi, Kenya
24. Nicholson SE (1996) A review of climate dynamics and climate variability in Eastern Africa: the limnology climatology and paleoclimatology of the East Africa Lakes. Gordon and Breach, New York, pp 25–56
25. Nicholson SE (1998) Historical fluctuations of Lake Victoria and other Lakes in the Northern Rift valley of East Africa. In: Lehman JT (ed) Environmental change and response in East Africa Lakes. Kluwer Academic Publishers, Netherlands, pp 7–35
26. Nicholson SE, Funk C, Fink AH (2018) Rainfall over the African continent from the 19th through the 21st century. Global Planet Change Lett 165:114–127. https://doi.org/10.1016/j.gloplacha.2017.12.014
27. Nicholson SE (2000) The nature of rainfall variability over Africa on time scales of decades to millennia. Global Planet Change Lett 26:137–158
28. Nicholson SE (2017) Climate and climatic variability of rainfall over eastern Africa. Rev Geophys 55:590–635. https://doi.org/10.1002/2016RG000544
29. Nicholson SE, Entekhabi D (1987) Rainfall variability in equatorial and Southern Africa. Relationships with sea-surface temperature along the southwestern coast of Africa. J Clim Appl Meteorol 26:561–578
30. North GR, Bell TL, Cahalan RF (1982) Sampling errors in estimation of empirical orthogonal functions. A Meteorol Soc 110:699–706
31. Nyakwada W (2009) Predictability of East African seasonal rainfall with sea surface temperature gradient modes. PhD dissertation, Department of Meteorology University of Nairobi
32. Nyakwada W (2010) Predictability of east African seasonal rainfall with sea surface temperature gradient modes. PhD dissertation, Department of Meteorology University of Nairobi
33. Ogallo LJ (1988) Relationships between seasonal rainfall in East Africa and the Southern Oscillation. J Climatol 8:31–43
34. Ogallo LJ, Janowiak JE, Halpert MS (1988) Teleconnection between seasonal rainfall over Eastern Africa and Global Sea surface temperature anomalies. J Meteor Soc Jpn 66:807–822
35. Omondi PA (2005) Potential causes and predictability of the space—time patterns of the decadal rainfall variability modes over east Africa. MSc thesis, Department of Meteorology, University of Nairobi
36. Omondi P, Awange JL, Ogallo L, Okoola RA, Forootan E (2012) Decadal rainfall variability modes in observed rainfall records over East Africa and their relations to historical sea surface temperature changes. J Hydrol 464–465:140–156. https://doi.org/10.1016/j.jhydrol.2012.07.003
37. Owiti OZ (2005) Use of the Indian Ocean Dipole indices as predictor of east African rainfall anomalies. MSc thesis, Department Of Meteorology, University of Nairobi
38. Phoon SY, Shamseldin AY, Vairavamoorthy K (2004) Assessing impacts of climate change on Lake Victoria Basin, Africa. People-Centred approaches to water and environmental sanitation. In: 30th WEDC international conference, Vientiane, Lao PDR

39. Power S, Casey T, Folland C, Colman A, Mehta VM (1998) Interdecadal modulation of the impact of ENSO on Australia. Clim Dyn 15:319–324
40. Preston A (2005) Southern African rainfall variability and Indian Ocean Sea surface temperatures. An observational and modelling study. Oxford University Press, Cape Town
41. Reason CJC, Rouault M (2004) Links between the Antarctic Oscillation and winter rainfall over western South Africa. Geophys Res Lett 32(7):2005
42. Reason CJC, Allan RJ, Lindesay JA, Ansel TA (2000) ENSO and climate signals across the Indian Ocean basin in global context. Int J Climatol 20:1285–1327
43. Reynolds RW, Marsico DC (1993) An improved real-time global sea surface temperature analysis. J Clim 6:114–119
44. Reynolds RW, Smith TM (1994) Improved global sea surface temperature analyses using optimum interpolation. J Clim 7(929):948
45. Reynolds RW, Rayne NA, Smith TM, Stokes DC, Wang W (2002) An improved in-situ and satellite SST analysis for climate. J Clim 15:1609–1625
46. Richman MB (1986) Review article. Rotation of principal components. J Climat 6:293–335
47. Saji NH, Yamagata T (2003) Possible impacts of Indian Ocean dipole events on global climate. Clim Res 25(2):151–169
48. Saji NH, Yamagata T (2003) Structure of SST and surface wind variability during Indian Ocean dipole mode events: COADS observations. J Clim 16(16):2735–2751
49. Saji NH, Goswani BN, Vinayachandran PN, Yamagata T (1999) A dipole mode in the tropical Indian Ocean. Nature 401:360–363
50. Schreck JC, Semazzi FHM (2004) Variability of the recent climate of Eastern Africa. Int J Climatol 24:681–701
51. Tanimoto Y, Iwasaka N, Hanawa K, Toba Y (1993) Characteristic variations of sea surface temperature with multiple time scales in the North Pacific. J Clim 6:1153–1160
52. Trenberth KE, Hurrel JW (1994) Decadal atmosphere-ocean variations in the Pacific. Clim Dyn 9:303–319
53. Von Storch H, Zwiers F (1999) Statistical analysis in climate research. Cambridge University Press
54. Washington R, Preston A, Todd MC (2003) The role of Indian and Pacific Ocean SSTs in African rainfall variability. Bull Am Meteorol Soc 84(7):899–900
55. Wilks DS (1995) Statistical methods in the atmospheric sciences. Academic Press, pp 45–50, 399–402
56. WMO (1966) Climate change, WMO Tech Note No 79, Geneva. WMO
57. Lixin Wu, Zhengyu Liu (2005) North atlantic decadal variability: air-sea coupling, oceanic memory, and potential northern hemisphere resonance. J Clim 18:331–349. https://doi.org/10.1175/JCLI-3264.1
58. Xoplaki E, Gonzalez-Rouco J, Gyalistras J, Luterbacher D, Rickli R, Wanner H (2003) Interannual summer air temperature variability over Greece and its connection to the large-scale atmospheric circulation and Mediterranean SSTs 1950–1999. Clim Dyn 20:537–554
59. Yin X, Nicholson SE (1998) The water balance of Lake Victoria. J Hydrol Sci 43(5):789–811. Further reading. http://www.esrl.noaa.gov/psd/data/gridded/data.noaa.oisst.v2.html. Accessed Nov 2011
60. Zhao S, Cook KH (2021) Influence of Walker circulations on East African rainfall. Clim Dyn 56:2127–2147. https://doi.org/10.1007/s00382-020-05579-7

Chapter 8
Decadal Rainfall Variability: Link to Oceans

> The strong coupling between global basins and decadal rainfall can provide new prediction tools that are important for decadal rainfall prediction. This again can contribute to the formulation and improvement of prediction skills in numerical modeling and early warning several lead times ahead.—Omondi et al. [88]. This chapter seeks to answer the question whether there is any contribution/influence of anthropogenic effect on the low frequency SSTs?

8.1 Summary

Influence of low frequency global Sea Surface Temperatures (SSTs) modes on decadal rainfall modes over Greater Horn of Africa (GHA) region discussed in Chap. 7 is investigated in more details in this chapter. Fore-knowledge of rainfall distribution at decadal time scale in specific zones is critical for planning purposes. This chapter presents the work of Omondi et al., [88] where both rainfall and SST data that cover the period of 1950–2008 are subjected to a 'low-pass filter' in order to suppress the high frequency oscillations. VARIMAX-Rotated Principal Component Analysis (RPCA) is employed to delineate the region into decadal rainfall zones while Singular Value Decomposition (SVD) technique is used to examine the potential linkages of these zones to various areas of the tropical global oceans. Ten-year distinct decadal signals, significant at 95% confidence level, are dominant when observed *in-situ* rainfall time series are subjected to spectral analysis. The presence of variability at El Niño Southern Oscillation (ENSO)-related timescales, combined with influences in the 10–12 year and 16–20 year bands are also prevalent. Nine and seven homogeneous decadal rainfall zones for long rainfall season, i.e., March-May (MAM) and the short rainfall season, i.e., October-December (OND), respectively, are delineated. The third (dry) season of June–August (JJA), which is mainly expe-

J. Awange, *Food Insecurity & Hydroclimate in Greater Horn of Africa*,
https://doi.org/10.1007/978-3-030-91002-0_8

rienced in western and Coastal sub-regions has eight homogenous zones delineated. The forcing of decadal rainfall in the region is linked to the equatorial central Pacific Ocean, the tropical and South Atlantic Oceans, and the Southwest Indian Ocean. The high variability of these modes highlight the significant roles of all the global oceans in forcing decadal rainfall variability over the region.

8.2 Need for Decadal Climate Variability Information

Spatial and temporal variability of rainfall offers considerable challenges for assessing and understanding climate change and variability over Greater Horn of Africa (GHA) region. The region is already witnessing dire consequences of erratic climatic conditions that are likely to be associated with regional climate change [43]. The recurrences of floods and droughts have been associated with many socio-economic miseries. As pointed out in Chap. 1, this is a region with serious food insecurity and resources-based conflicts. IPCC [43] report provided clear evidences of climate change in the region with increased risks of climate extremes [36]. Unfortunately, the economies and livelihoods of the majority of the countries in the region still rely on rain dependent systems. Serious food insecurity, famines, and poverty among other miseries, are prevalent in the region. The rainfall pattern over the region is highly variable both in space and time because of the complexity of regional climates and the influence of regional geographic features, such as semi-arid lands, land cover variations, mountain chains, large lakes, land-sea contrasts, and the sea surface temperature (SST) changes of the surrounding Indian Ocean. Such diversity influences the distribution and statistics of rainfall extremes at all scales. This in turn may have a much greater impact on natural systems and human activities than mean precipitation [92].

Atmospheric motions are dominated by the gradients in temperature and the associated pressure gradients [10, 79]. The use of SST gradients can improve the representation of the driving forces of the general circulation. The SST gradients influence wind currents and the associated moisture transport [57]. Some of the SST gradients that have been documented to influence climate of the region and beyond include the Zonal SST gradients in the Indian Ocean associated with the Indian Ocean Dipole (IOD) [14, 25, 89, 106, 112, 132, 141]; the zonal SST gradient in the Pacific [15, 79] and the zonal and meridional SST gradients in the Atlantic Ocean [64, 77, 127] . The meridional and zonal SST gradients are linked through teleconnections with large-scale atmospheric circulation such as the Walker and Hadley cells [127]. The zonal and meridional SST gradients have a strong signature on the climate of the tropical areas [57, 128]. In the tropics, the zonal gradients are generally smaller than the meridional gradients but are more important in forcing low level tropical circulation and convergence than the meridional gradients [57].

There is particular interest in the coming decade, which represents a key planning horizon for infrastructure upgrades, insurance, energy policy, and business development. Societal needs for climate information on decadal time scales is continuing to grow in terms of its potential value and relevance as a driver in sector decision making. However, such information is currently lacking in the GHA region. Predictions and observationally-based analyses for decadal climate variability and change are therefore needed. A large timescale gap, especially the one that goes from one year to about 20 years, exists in climate predictions [43, 99, 116]. Long-term scale climate predictions over East African region has not been done with enough precision to adequately carry out assessment of impacts of climate anomalies at the sub-regional level. For interventions and planning purposes, climate information is required on average over 5–10 years. Only in the last decade has this received attention of the scientific community despite the relevance of this timescale to many societal and developmental applications, see e.g., [48, 94, 116]. These scientific and technical shortcomings in turn act as a major hurdle that limits the ability to design and justify investment decisions for the development and design of appropriate adaptation strategies.

Knowledge on decadal climate variability is popularly emerging as a new direction in climate science [26, 50, 60, 110] and [61], especially in light of increased demand for climate variability/change information. Numerous assessments of climate information user needs have identified this timescale as being important to infrastructure planners, water resource managers, and many others [48, 55, 61]. Decadal prediction lies between seasonal/interannual forecasting and longer term climate change projections, and further focuses on time-evolving regional climate conditions over the next 10–30 years. Perhaps the most striking finding is the linkage of decadal drought frequency in the Sahel to SST [58, 72, 75, 76, 142]. Other droughts on decadal-timescale, such as the "dust-bowl" in the Southern United States of America (USA) in the 1930s have been linked to variations in SST [9, 136, 137]. Nicholson [73] showed rainfall time series for 1901–1994 of three African regions displaying decadal rainfall variations. In South Africa, and some neighboring areas, an austral summer rainfall signal at near bi-decadal scales has been known for some time [122]. In the Pacific/Indian Ocean basins, there are strong signals of decadal variability associated with the Pacific Decadal Oscillation (PDO) with statistical links to the climate of the surrounding regions. For example, during the 20th century, El Niño- like phases of the PDO coincided with decades in which ENSOs impact on Australia was weak, whereas La Niña- like phases of the PDO coincided with decades in which ENSOs impact on Australia was strong [95] .

Over East Africa region of GHA, there is substantial evidence for decadal climate variability in the observed climate system [19, 68, 73, 85, 87, 109]. Decadal rainfall signals in East Africa have been linked to the Pacific Decadal Oscillation (PDO) [26]. Several parts of the region experience strong decadal signal during the short rains of OND season [19, 26, 85, 109]. Recently, [85, 86] provided some evidence of decadal variability in the interannual patterns of East Africa rainfall. They showed that some teleconnections were evident between the observed decadal rainfall and SST variability patterns over parts of the global oceans (sea also Chap. 7). They further found correlations between the rainfall and SSTs but no attempt was made to link regional homogenous zones to specific regions of global SSTs. Relatively, less attention has been directed at the influence of low frequency SST signals in forcing decadal rainfall in different sub-regions of East Africa.

The focus of the present chapter therefore, is to extends Chap. 7 by delineating the sub-regions of Eastern Africa based on decadal rainfall variability and investigate how these sub-regions are coupled to low frequency SST signals. Unlike in [85] where a simple correlation analysis was employed, this study applied the more sophisticated Singular Value Decomposition (SVD) technique in an attempt to link decadal rainfall to various global basins SSTs. The advantage of adding this step to the analysis is twofold: (i) It improves the performance of statistical methods with removing highly correlated observation in each region, which might add unrealistic weights to the zones with more stations and disturb the leading components towards the maximum variability of those zones, see e.g., [12], and (ii), it also groups together stations with common decadal rainfall characteristics into homogeneous zones, which considerably improves the interpretation of the results. Once the homogeneous regions are identified, investigations are again carried out for both spatial and temporal coupling over a longer period of time compared to the previous work where only temporal variability was considered and over a relatively shorter period of time.

This chapter is organized as follows: A brief discussion of the types of data used and methods of analysis are present in Sect. 8.3. The key results are discussed in Sect. 8.4, whereas summary and conclusions are provided in Sect. 8.5, which seeks to answer the question whether there is any contribution/influence of anthropogenic effect on the low frequency SSTs.

8.3 Climate Products and the Analysis Approach

8.3.1 Observed Climate Product

The rain gauge observations network over East Africa is quite sparse. Furthermore, the available data are also riddled with numerous gaps in both space and time. These limitations in the quantity and quality of in-situ observations impose substantial constraints on diagnostic studies of the regional climate (rainfall) variability. The observed rainfall data used in this chapter consist of monthly rainfall data (1920–2008) obtained from the Kenya Meteorological Department (KMD), Tanzania Meteorological Agency (TMA) and Uganda Meteorological Department (UMD). The in-situ rainfall stations used are the representative stations for 37 climatological zones used by the three countries in day-to-day operations [39]. These were derived from a total 136 records of rainfall stations scattered over East Africa through PCA analysis [42]. These climatological zones have been used by many authors in the region, e.g., [79, 85, 89–91] among others. A total of 37, unevenly distributed, gauge stations over East Africa were available for the analysis.

The sea surface temperature (SST) data was obtained from the National Center for Environmental Prediction/Climate Prediction Center (NCEP/CPC) for the period 1950–2008, which is distributed under the National Oceanic and Atmospheric Administration (NOAA)/Climate Diagnostic Center (CDC) Optimum Interpolation

(OI) SST Version 2 (OISSTv.2). The data are on 1.0° × 1.0° grid point resolution and is often known as optimal interpolation (OI) SST in literature following Reynolds and Smith [102]. The data set is a blend of insitu and satellite SSTs, but also includes those simulated by sea-ice cover [103, 113]. The data adjustment for biases was done using the method described by Reynolds [100] and Reynolds and Marsico [101]. Details of this methodology can be found in Reynolds and Smith [102]. All SST observed data within 30 °S–30 °N are discarded if the value is less than −2 °C or greater than 35 °C or if the SST anomaly lies outside ±3.5 times the climatological standard deviation.

SSTs have long memory and therefore have wide usage in most climate prediction models [67, 79, 89, 116, 117]. Several efforts have been made to improve the quality of SST records employed in climate prediction [115, 117]. Kanamitsu et al. [47], Smith and Reynolds [113, 115] and Smith et al. [117] among other authors, have discussed the details of these data. It is important to note that SST data to the south of 30 °S may not be of good quality as indicated in the study of Weare [131] and hence the restriction between 30 °S and 30 °N of the tropical global Oceans.

The methodology employed in the analysis for this chapter are briefly highlighted in the section that follows. However, in establishing the relationship between rainfall and temperature, a common window for both data sets is set as 1950–2008. Both data sets were low-pass filtered using a nine-term binomial coefficient filter or ten years moving average to remove all fluctuations equal to and less than 10 years. It is also noteworthy that when any time series is subjected to nine point Gaussian probability curve, four years of data are truncated from both ends of the series. The graphical presentations of these series include the plotting of the bar charts anomalies obtained when long-term mean decadal rainfall is subtracted from the filtered record and the ensuing series detrended.

8.3.2 Spectral Analysis

The spectrum of a time series is the distribution of variance of the series as a function of frequency. The object of spectral analysis is to estimate and study the spectrum. The spectrum contains no new information beyond that which is available in the Autocovariance Function but can be computed mathematically by transformation of the Autocovariance Function. But the spectrum and Autocovariance Function present the information on the variance of the time series from complementary viewpoints. The Autocovariance Function summarizes information in the time domain and the spectrum in the frequency domain. A simple model of a time series consisting of just a single sinusoidal component with superposed noise is given by Eq. 8.1, i.e.,

$$y_t = Rcos(\omega t + \phi) + Z_t = w_t + z_t, \tag{8.1}$$

where t is time, z_t is the random normal component in time t, w_t is the sinusoidal component; and R, ω and ϕ are the amplitude, angular frequency, and phase of the sinusoidal component respectively. Following Percival and Walden [93], the spectrum can be defined in terms of a more complicated model in which a time series X_t of length N consists of a linear combination of many sinusoids at fixed frequencies f_j and random amplitudes A_j and B_j:

$$X_t = \mu + \sum_{j=1}^{[N/2]} [A_j \cos(2\pi f_j t) + B_j \sin(2\pi f_j t)], \qquad t = 1, 2, ..., N \tag{8.2}$$

where μ is a constant term, the notation [N/2] refers to the greatest integer less than or equal to N/2, and the frequencies f_j are related to the sample size N by

$$f_j \equiv j/N, \qquad 1 \leqslant j \leqslant [N/2]. \tag{8.3}$$

The frequencies of the sinusoids are at intervals of 1/N and are the *Fourier frequencies*, or *standard frequencies*. The frequency f_j is *jth* standard frequency, which depends on the sample size. Using this model to define spectrum, the additional assumptions made is that the amplitudes A_j and B_j are random variables with expected values, i.e.,

$$E\{A_j\} = E\{B_j\} = 0 \tag{8.4}$$

and

$$E\{A_j^2\} = \{B_j^2\} = \sigma_j^2. \tag{8.5}$$

From the relationship between variance and amplitude of sinusoid, Eq. 8.5 implies that the variance associated with the *j*th standard frequency is σ_j^2. It is also assumed that the amplitudes associated with various standard frequencies are uncorrelated:

$$E\{A_j A_k\} = E\{B_j B_k\} = 0 \quad \text{for} \quad j \neq k \tag{8.6}$$

and

$$E\{A_j B_k\} = 0 \quad \text{for all} \quad j, k. \tag{8.7}$$

With the assumptions above, it can then be shown that the expected value of X_t is

$$E\{X_t\} = \mu, \tag{8.8}$$

the variance of X_t is

$$\sigma^2 = E\{(X_t - \mu)^2\} = \sum_{j=1}^{[N/2]} \sigma_j^2, \tag{8.9}$$

and the autocorrelation function of X_t is

$$\rho_k = \frac{\sum_{j=1}^{[N/2]} \sigma_j^2 \cos(2\pi f_j k)}{\sum_{j=1}^{[N/2]} \sigma_j^2}. \tag{8.10}$$

Equation 8.9 simply denotes that the variance of the series X_t is the sum of the variances associated with the sinusoidal components at the different standard frequencies. Thus, the variance of the series can be decomposed into components at the standard frequencies or simply expressed as a function of frequency.

For such model therefore, the spectrum can be defined as

$$S_j^2 \equiv \sigma_j^2, \qquad 1 \leqslant j \leqslant [N/2]. \tag{8.11}$$

A plot of S_j against frequencies f_j shows the variance contributed by the sinusoidal terms at each of the standard frequencies. From Eq. 8.9, the variance of X_t can then be expressed as the sum of the spectral components

$$\sigma^2 = \sum_{j=1}^{[N/2]} S_j. \tag{8.12}$$

The variance contributed at frequency f_j is the spectrum S_j at that frequency. The shape of the spectral values S_j plotted against f_j indicates which frequencies are most important to the variability of the time series.

Considering that σ_j^2 are spectral values, Eq. 8.10 gives an important relationship between the spectrum and the autocorrelation function. Here, the autocorrelation function is expressed as a cosine transform of the spectrum. Similarly the spectrum can be shown to be the Fourier transform of the autocorrelation function. The spectrum and autocorrelation function are therefore different characterizations of the same time series information. The autocorrelation function is a time-domain characterization and the spectrum is a frequency-domain characterization. From a practical point of view, the spectrum and autocorrelation function are complementary to each other. There is uncertainty in trying to understand such a process from a single sample when analyzing a spectrum. Spectral analysis is therefore concerned with estimating the unknown spectrum of the process from the data and with quantifying the relative importance of different frequency bands to the variance of the process.

Various methods have been developed to estimate the spectrum from an observed time series, e.g. [18, 24, 65, 93, 133]. Wiener–Khintchine theorem [134] says that for any stationary stochastic process with autocovariance function $\gamma(k)$, there exists a monotonically increasing function,

$$F(\omega),$$

such that

$$\gamma(k) = \int_0^{\pi} \cos \omega k d F(\omega). \tag{8.13}$$

Eq. 8.13 is called the spectral representation of the autocovariance function, and $F(\omega)$ is called the spectral distribution function. $F(\omega)$ has a direct physical interpretation: $F(\omega) =$ contribution to the variance of the series, which is accounted for by frequencies in the range $(0, \omega)$.

The range is restricted to angular frequencies $(0, \omega)$ since there is no variation at negative frequencies, so the lower limit on the range is 0. Further, if a process is measured at unit intervals, the highest possible frequency that can be studied corresponds to the wave that undergoes a complete cycle in two intervals, or an angular frequency of $\omega = \pi$. Thus, all the variation is accounted for by frequencies less than π:

$$F(\pi) = \text{var}(X_t) = \sigma_x^2. \tag{8.14}$$

The function $F(\omega)$ increases monotonically between $\omega = 0$ and $\omega = \pi$, and in this way $F(\omega)$ is similar to a cumulative distribution function (cdf). By scaling $F(\omega)$ by the variance, the normalized spectral distribution function given by Eq. 8.16 is obtained by

$$F^*(\omega) = F(\omega)/\sigma_x^2, \tag{8.15}$$

which gives the proportion of variance accounted for by frequencies in the range $\omega = 0$, and like a cdf, reaches a maximum of 1.0, since $F^*(\omega) = 1$.

The spectral distribution function gives the variance of the process at frequencies less than some frequency. By differentiating the spectral distribution function, the spectral density function, which gives the variance associated with each frequency is obtained

$$f(\omega) = \frac{dF(\omega)}{d\omega} \equiv \text{(power) spectral density function.} \tag{8.16}$$

A plot of the spectrum $f(\omega)$ against frequency ω is essentially a plot of the variance of a time series as a function of frequency. More precisely, a given value of $f(\omega)$ is a "variance per unit frequency", such that by integrating over some increment of $f(\omega)$ the variance associated with that range of frequencies is obtained.

In a graph of the spectrum, therefore, the area under the curve bounded by two frequencies represents the variance in that frequency range, and the total area underneath the curve represents the variance of the series. A peak in the spectrum represents relatively high variance in a frequency band centered on the peak.

In this analysis, normalized spectrum are plotted so as to isolate spectral properties of time series with greatly differing variances. The methods outlined in Koopmans [52] was adopted in testing for significance of spectral estimates. The upper and lower bounds of a confidence interval around each spectral estimate are computed using χ^2 distribution. Details of the computations of weights and spectral estimate can be obtained from Gottman [32] and Koopmans [52]. For a 95% confidence interval (CL), the critical values of χ^2 that cut off the bottom and top 5% are obtained from χ^2 distribution table. These are denoted as for the bottom 5% cutoff and for the top 5% cutoff critical values of χ^2 in Eq. (8.17). The spectral estimate for which the CL is set up is denoted by $s(f_i)$, where f_i is the specific frequency and $s(f_i)$ is the estimated power at that frequency. Thus the upper and lower bounds of a 95% CL for $s(f_i)$ are given by as:

$$\begin{aligned} \text{Lower bound of CL} &= s(f_i)/x^2_{0.95} \\ \text{Upper bound of CL} &= s(f_i)/x^2_{0.05} \end{aligned} \tag{8.17}$$

The width of the CL varies as a function of the size of $s(f_i)$, and so if $s(f_i)$ is plotted on a linear scale, a separate CL is set up for each spectral estimate. In this chapter, spectral analysis technique is employed to detect periodic or quasi-periodic fluctuations and further if the smoothed rainfall record is dominated by ten years cycle [44].

8.3.3 VARIMAX-Rotated Principal Component Analysis

Seasonal rainfall patterns over East Africa are known to be highly variable due to the existence of complex topography and large inland water bodies [2, 41, 67, 79, 80]. To prepare a spatially largest possible area for coherent rainfall having similar characteristics and associations with regional/global circulation parameters, homogenous zones are delineated. Spatial averages are more representative of the large-scale conditions than are data for individual stations [69]. The use of normalized regionally averaged series reduce two problems inherent in the analysis of rainfall in the sub-humid, tropical areas namely; the highly diverse means and variabilities and the randomness of the convective process reflected in individual station totals [41, 80]. Reduction of regional data and delineation of homogeneous climate zones is critical in studying mechanisms associated with these modes of variability. Regionalization and averaging of rainfall over large but homogeneous regions has the advantages of reducing meteorological noise in the data as well as minimizing the number of variables, which describe the regional climate variability [41, 67, 79, 80]. To effectively provide a better understanding of the physical processes responsible for decadal climate variability over the region, VARIMAX-Rotated PCA (RPCA) technique [41, 67, 79] is employed to isolate dominant modes of the decadal rainfall variability and cluster analysis on the rainfall stations network to group them into homogeneous decadal rainfall zones. The delineation of homogeneous zones is followed by identifying the stations with the largest correlation with the Principal Component (PC) time series associated with the first eigenvector of the decadal rainfall anomaly [11, 42].

The VARIMAX rotated version of PCA was thus applied to define dominant modes of variability of the low passed rainfall and SST series. The VARIMAX rotation is selected to improve the physical interpretation of the PCA modes and to derive more localized components [104]. PCA concept is therefore employed in this chapter to establish the dominant modes in both decadal rainfall and SST data. The spatial patterns of the dominant modes observed in the SST fields are thereafter used in the selection of SSTs modes from the major global basins for the model development. The use of PCA [5], however, requires the identification of the number of factors, which must be included in the solutions. A detailed discussion on selection rules are documented in e.g., [5, 23, 45, 46, 134].

8.3.4 Singular Value Decomposition

The purpose of Singular Value Decomposition (SVD) is to reduce a dataset containing a large number of values to a dataset containing significantly fewer values, but which still contains a large fraction of the variability present in the original data [124]. Often, the rainfall and SST data will exhibit large spatial correlations. SVD analysis results in a more compact representation of these correlations, especially with multivariate datasets and can provide insight into spatial and temporal variations exhibited in the data being analyzed.

In this chapter, SVD method aims to relate the two fields by decomposing their joint covariance matrix into singular values and two sets of paired-orthogonal vectors–one for each field. The covariance between the expansion coefficients of the leading pattern in each field is maximized. The singular values give the magnitude of the Squared Covariance Fraction (SCF) as accounted for by the various singular values [20, 125]. Thus, SVD is a better approach compared to simple correlation analysis between different components of climate systems.

The SVD method is used to further examine the detailed patterns of the dominant modes of decadal variations in rainfall and their coupling to the global SST changes. Detailed descriptions of SVD analysis can be found, e.g., in Bretherton et al. [20] and von Storch and Navarra [124]. Exploiting the spatial coherence of climate signals in data series with low signal to noise ratios facilitates their identification [59] while preserving spatial information and allowing the isolation of signals that might largely cancel in coarse spatial averaging (e.g., signals which largely involve dipole and quadrupole patterns). The usage of the SVD method, therefore, is similar to canonical correlation analysis (CCA) where two sets of orthogonal time series are produced, along with their corresponding spatial patterns. CCA aims to maximize correlation between variables [6], while SVD aims to maximize covariance between variables. The difference between CCA and SVD is that CCA identifies spatial patterns by maximizing the temporal correlation between two data fields, whereas SVD maximizes the temporal covariance between two data fields [140]. Barnston and Smith [7], Barnett and Preisendorfer [6] among others provide an in-depth analysis of CCA and its relation to SVD.

It should be noted that prior to SVD analysis, the series are subjected to a 'low-pass filter' in order to suppress the high frequency oscillations. The weights used are nine point Gaussian probability curve (0.01, 0.05, 0.12, 0.20, 0.24, 0.20, 0.12, 0.05, 0.01). The long-term-mean is subtracted and the time series further de-trended until the records clearly show no trends when linear regression technique and the 'Student's t' test are applied [27]. When trends in the data exist over time, the first structure often captures them. The purpose of this analysis is to find spatial and temporal correlations independent of trends and thus de-trending data before applying the analysis.

8.4 Decadal Rainfall Variability: Delineation and Linkage to SST

8.4.1 Delineation of East Africa into Climatic Zones

Table 8.1 gives a summary of the results obtained from rotated VARIMAX solutions. Using the Scree test; Kaiser's criterion and North et al. [78] sampling errors methods, nine PCA modes are found to be significant for the MAM season. These accounts

Table 8.1 VARIMAX-RPCA results of homogeneous group of stations extracted by each mode of the decadal MAM, OND and JJA seasonal rainfall. *Source* [88]

Period	Factor	Eigenvalue	Variance extracted (%)	Cumulative variance (%)
MAM	1	7.3	19.7	19.7
	2	6.9	18.6	38.3
	3	5.9	15.9	54.2
	4	4.2	11.4	65.6
	5	3.0	8.1	73.7
	6	2.3	6.4	80.1
	7	1.8	5.0	85.1
	8	1.2	3.3	88.4
	9	1.1	2.9	**91.3**
OND	1	15.87	42.9	42.9
	2	5.25	14.2	57.1
	3	3.87	10.5	67.6
	4	2.67	7.2	74.8
	5	2.40	6.5	81.3
	6	2.32	6.3	87.6
	7	1.38	3.7	**91.3**
JJA	1	7.9	24.1	24.1
	2	5.4	16.5	40.6
	3	4.3	13.0	53.6
	4	3.8	11.5	65.1
	5	3.1	9.5	74.6
	6	2.1	6.3	80.9
	7	1.6	4.9	85.8
	8	1.3	4.0	**89.8**

for 91.3% of total March–May decadal rainfall variance and thus is used for further analyses. Results from the second rainfall season of OND indicate seven modes accounting for 91.3% to be significant. During the Northern Hemisphere summer of June–August season, most of the region is dry except the western equatorial and coastal sectors influenced by Lake Victoria and Indian Ocean, respectively. Eight PCA modes accounting of 89.9% were significant during this season.

Figure 8.1 show 9, 7 and 8 unique climatologically homogeneous zones delineated for the MAM, OND and JJA seasons, respectively, using decadal rainfall records for the three countries. Stations with common characteristics are grouped together into homogeneous regions using spatial patterns of the dominant PCA modes. The PCA statistics (communality) at each individual zone is therefore used to identify a representative station for each zone, which is used for further analyses [11, 33, 39, 41, 80, 81, 83]. For example, correlation between the stations with highest communality in every cluster is computed and this helps in the mapping of stations into zones. Correlation coefficient values as high as 0.94 are observed between Dar-es-Salaam International Airport (DIA) and stations within zone 5 during OND season (Fig. 8.1b). To facilitate discussion of the regions, Fig. 8.2 presents the anomaly of decadal precipitation across some stations in each of the delineated homogenous regions. Determining the spatial form of these signals was the primary purpose of the PCA. The nine significant modes of variability passed both Scree and Kaiser's criterion tests but the eigen values were indistinguishable at 95% confidence level using North et al. [78] sampling error test. Figure 8.3 shows an example of ten years cycle dominant in most rainfall records that were subjected to spectral analysis. The spectral peaks are significant at 95% confidence level when both white and red noise hypotheses are tested. The ten years dominant cycles demonstrate the dominance of decadal variability in rainfall patterns over East Africa. There are, however, significant differences in the amplitudes and spectral bands of the spectral peaks from the various regions, which signifies some differences in the impacts of the decadal variability of various parts of East Africa due to the modification of the complex regional climate system.

Figure 8.4, however, depicts long-term linear trends towards negative values, which are detectable in most sub-regions. When linear regression technique and the 'Students t' test for testing if there is any significance in the trend are applied, there were no statistically significant trends detected. The results further show wet and dry decades recurring with approximately ten years cycle (e.g., Fig. 8.2) and sometimes extend over large areas of the region.

8.4.2 Links to Global SSTs

Table 8.2 provides the two main summary statistics of the SVD analysis obtained when the three global basins are separately analyzed with the three rainfall seasons of the region. These statistics provide a measure of the strength of the relationship between the two fields. The first statistic, the Squared Covariance Fraction (SCFk),

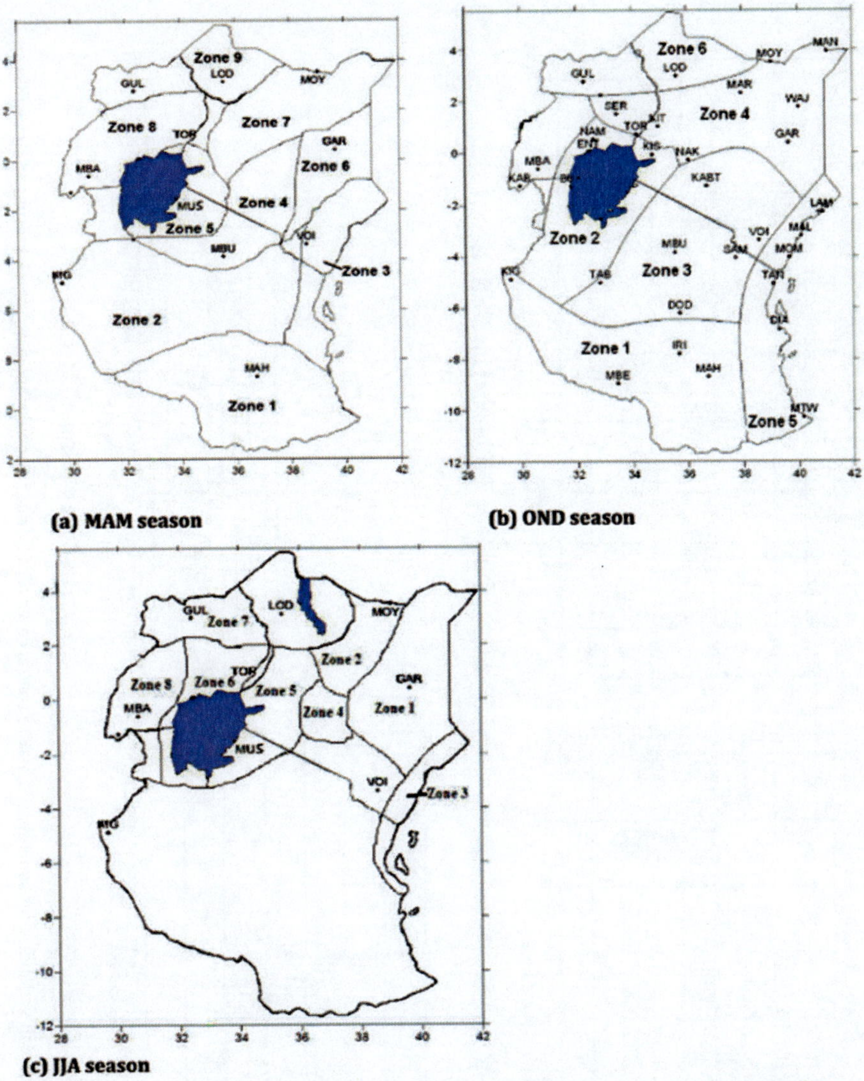

Fig. 8.1 Homogeneous climatic zones over East Africa obtained from combined PCA and simple correlation analyses using March–May (MAM), October–December (OND) and June–August (JJA) decadal rainfall variability. Blue areas in the figure represent water bodies of Lake Victoria and Turkana. *Source* [88]

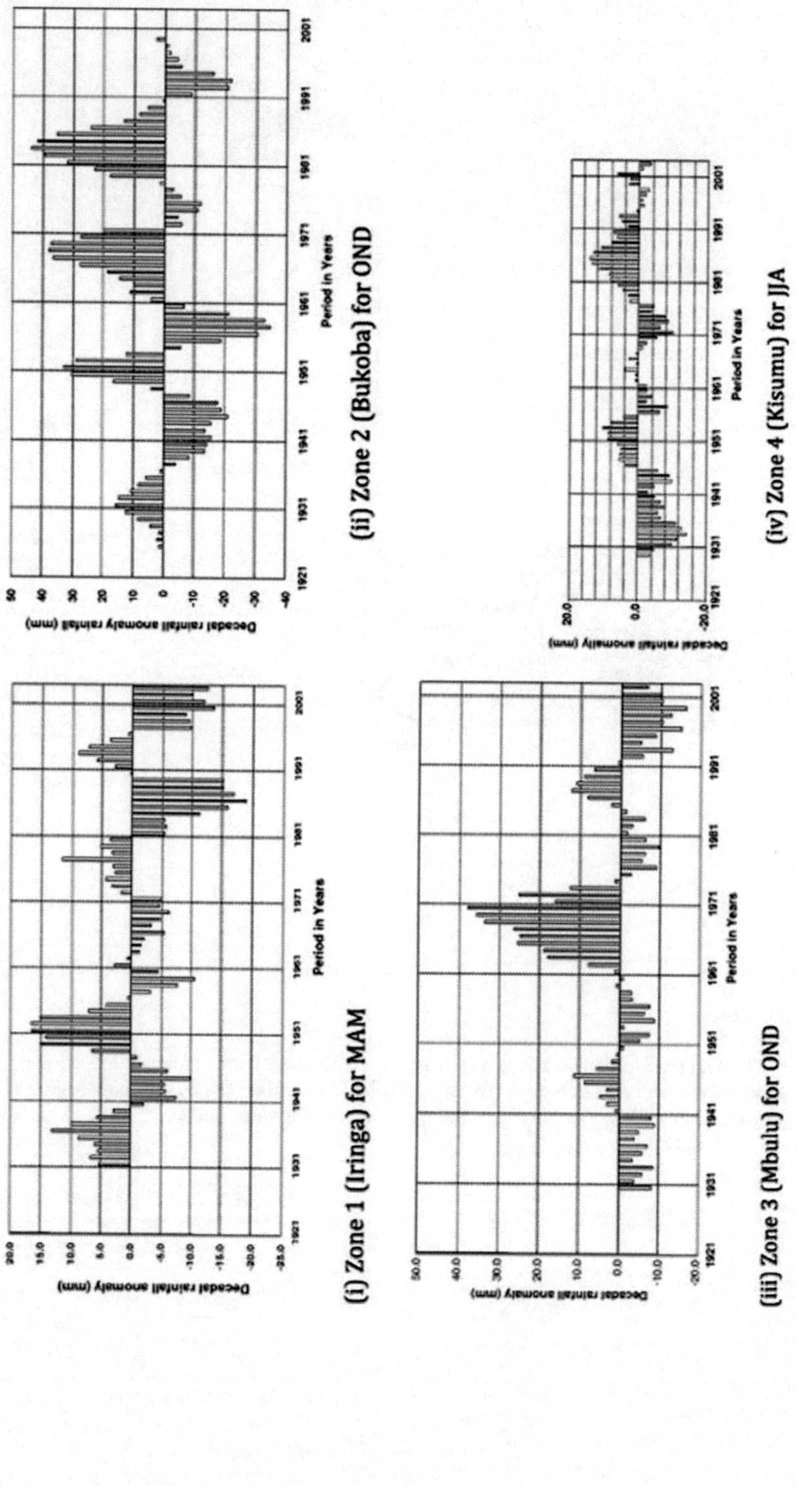

Fig. 8.2 Graphical plots of decadal rainfall variability for some representative stations over East Africa region during all the three seasons. *Source* [88]

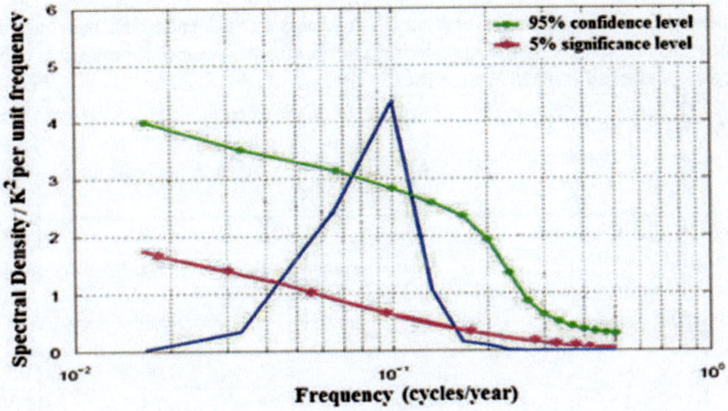

Fig. 8.3 Spectral analysis of the smoothed March–May seasonal rainfall for zone 8 as represented by Mbarara. *Source* [88]

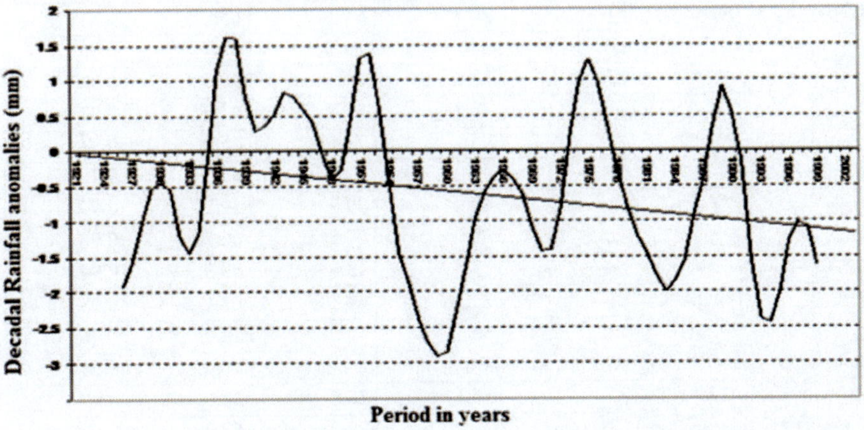

Fig. 8.4 Smoothed March–May rainfall anomalies depicting decadal variability with small percentages of the variance of the signals that are not statistically significant over Voi in Kenya. *Source* [88]

where k is the mode number, provides the percentage of the total squared covariance between the two fields explained by the SVD mode, and is proportional to the square of its singular value. This is a measure of the relative importance of the SVD mode in the relationship between the two fields. It should be noted that only the first three significant covariability modes of the squared covariance is shown in Table 8.2. The second statistic is the correlation coefficient (rk) between the two time series that represent the temporal variations of the mode in the two fields. It is a measure of the similarity between the time variations of the patterns of the two fields, or how strongly the two fields are related to each other with respect to time (significant at the 5% level) for the first three modes.

Table 8.2 Summary of some statistics from SVD analysis for decadal seasonal rainfall and the various global Ocean SSTs. r is the correlation coefficient between the expansion coefficient of SSTs and seasonal rainfall modes. *Source* [88]

Season	Oceans	Mode (k)	Square covariance fraction (SCF)	% of total mode covariance	Cumulative % covariance	Correlation (r)
MAM	INDIAN	1	20.5	49.9	49.9	0.73
		2	6.1	14.8	64.7	0.64
		3	4.1	9.9	74.6	0.38
	ATLANTIC	1	16.9	43.4	43.4	0.97
		2	9.0	23.1	66.5	0.98
		3	5.0	12.8	79.3	0.97
	PACIFIC	1	38.6	37.8	37.8	0.99
		2	25.5	25.0	62.8	0.98
		3	17.0	16.7	79.4	0.88
OND	INDIAN	1	26.7	65.2	65.2	0.87
		2	6.8	16.5	81.7	0.98
		3	3.6	8.8	90.5	0.94
	ATLANTIC	1	18.7	47.8	47.8	0.97
		2	9.0	23.1	71.0	0.94
		3	4.9	12.5	83.5	0.76
	PACIFIC	1	40.0	39.2	39.2	0.99
		2	28.0	27.5	66.7	0.96
		3	13.8	13.6	80.3	0.94
JJA	INDIAN	1	25.1	61.1	61.1	0.98
		2	6.9	16.8	78.0	0.96
		3	3.6	8.7	86.6	0.94
	ATLANTIC	1	15.1	38.6	38.6	0.99
		2	11.4	29.2	67.9	0.97
		3	5.5	14.1	82.0	0.88
	PACIFIC	1	43.0	42.1	42.1	0.99
		2	20.4	20.0	62.1	0.97
		3	15.6	15.3	77.4	0.85

Links between smoothed SST and decadal rainfall anomalies are investigated using SVD to establish the existence of covariance amongst East Africa rainfall and the specific global Ocean basin temperatures.

8.4.3 *The Three Oceans Versus MAM Modes*

Figure 8.5 shows the spatial and temporal patterns of the first SVD mode of MAM rainfall and Indian Ocean SST that accounted for 49.9% of the total square covariance. This mode is characterized by large pool of positive warming over western and northern Indian Ocean while negative loading are over south-eastern Indian Ocean (Fig. 8.5a). The corresponding precipitation pattern shows positive loadings over most parts of the region especially where MAM rains are dominant (Fig. 8.5b). It identifies the warming trend signal in the SSTs with strong positive loadings throughout the central Indian Ocean. A similar SST loading pattern has been reported by Smith and Reynolds [114]. The correlation coefficient of time series of the expansion coefficient of both SST and rainfall fields is 0.73 (Fig. 8.5c).

This first SVD mode is characterized by some east-west and north–south dipole patterns that are associated with much rainfall over most parts of the region. Some previous studies have shown strong positive SST anomalies over western Indian Ocean and negative SST anomalies off the Indonesia coast that has been referred to as the positive Indian Ocean Dipole (IOD) Mode [13, 14, 16, 17, 29, 35, 37, 63, 89, 106, 107, 112, 121, 132, 141]. The second SVD mode (SVD-2) that accounted for 14.8% variance has negative coherence loadings centered on the central equatorial Indian Ocean and positive loading over the south-western parts of the ocean basin (Fig. 8.6a). This mode is also characterized by negative loadings over the northern sector while positive loading is confined over the southern sector of the sub-region (Fig. 8.6b). It also showed stronger decadal trend mode (correlation coefficient, $r = 0.64$) in time series of expansion coefficient (Fig. 8.6c) compared to SVD-1 mode. Such a pattern in the Indian Ocean was observed in Nyakwada [79] and has been attributed to the mean seasonal SST pattern over the ocean when the overhead sun crosses the equator [70].

The third mode (SVD-3), which accounted for 9.9% of the variance showed positive loadings over the entire Indian Ocean with the highest loadings concentrated in the area between 20 °N and 15 °S (figure not shown). This pattern is associated with strong positive loadings for rainfall over the eastern sector and negative loadings over western sector of the region with strong decadal signal in the time series of expansion coefficients. It is worth noting that although this mode explains relatively low covariance, it was still able to display strong decadal variability for the MAM rainfall season [96–98]. The results from SVD analysis discussed above confirm the complexity of rainfall variability over East Africa during MAM rainfall season. SVD has however delineated three modes that could give more insight into the modes of decadal rainfall variability over the region. The dominance of the modes representing zonal SST variability in the Indian Ocean, which may be associated with the strong influence of this ocean on the climate of the region has been indicated and consistent with observation analyses by many other authors including Okoola [83], Ward and Folland [129], Goddard and Graham [31], Mutai [66], Nyakwada [79] among others.

Similarly, the influence of the Atlantic Ocean on MAM decadal rainfall was investigated using SVD technique. Moisture influx from the ocean is associated with

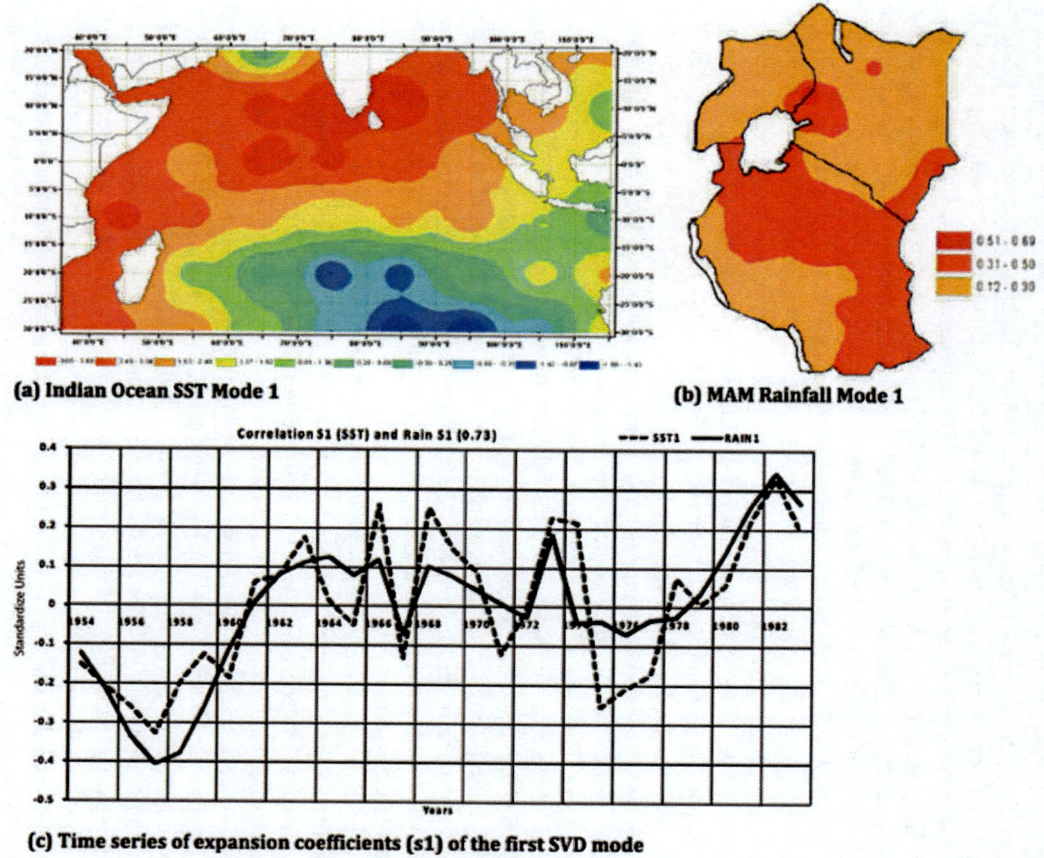

Fig. 8.5 Spatial patterns of the first SVD mode accounting for 49.9% of the covariance for **a** Indian Ocean SST, **b** MAM rainfall presented as homogeneous correlation maps **c** time series of expansion coefficients (s1) of the first SVD mode for March–May rainfall and Indian Ocean SST anomalies. x-axis represent the longitude while the y-axis shows the latitude. *Source* [88]

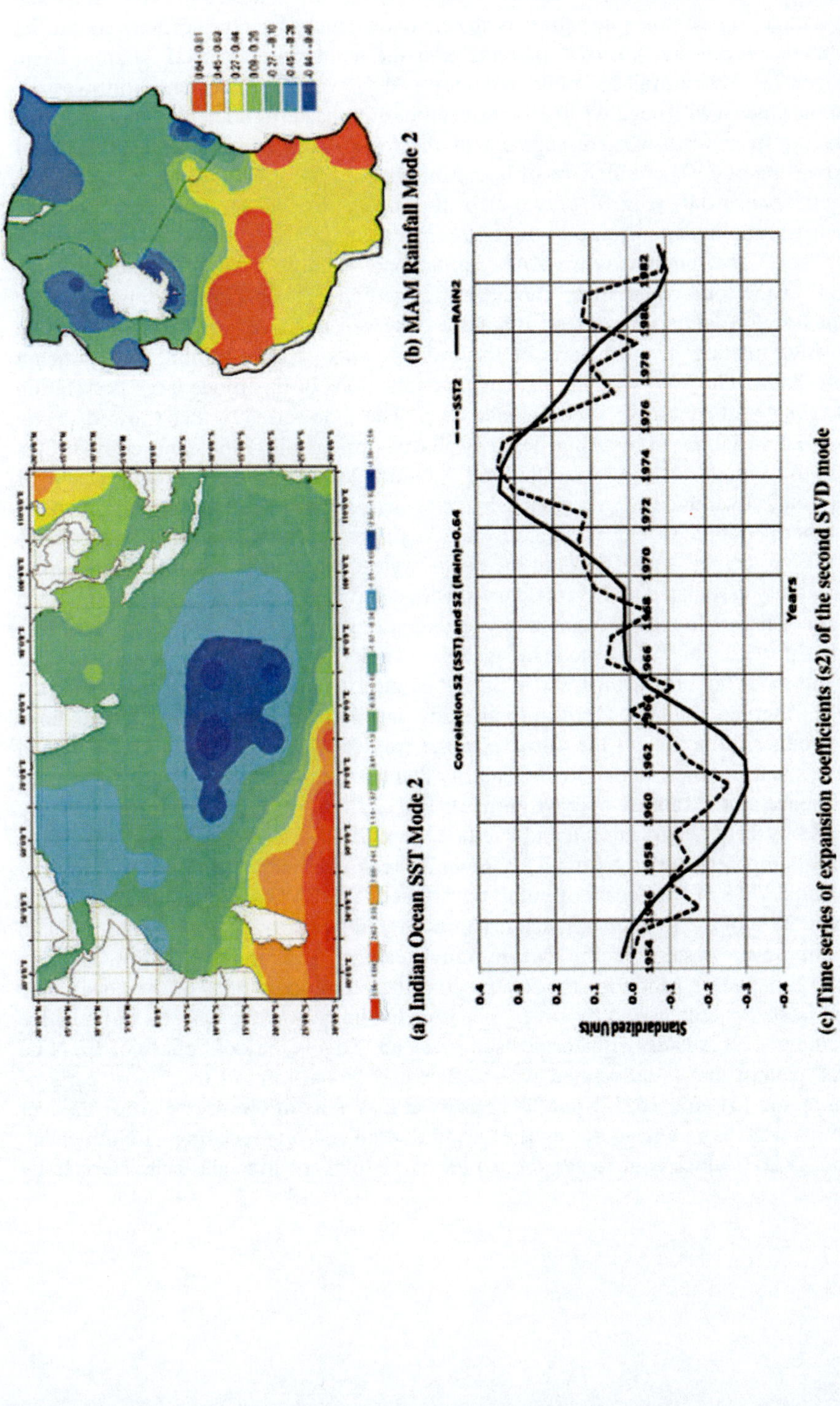

Fig. 8.6 Spatial patterns of the second SVD mode accounting for 14.8% of the covariance for **a** Indian Ocean SST, **b** MAM rainfall presented as homogeneous correlation maps **c** time series of expansion coefficients (s2) of the second SVD mode. x-axis represent the longitude while the y-axis shows the latitude. *Source* [88]

enhanced westerly circulation that also favours the incursions of moisture from the always wet tropical forests of Congo, Democratic Republic of Congo (DRC) and other central African countries. The SST variability in the Atlantic Ocean reaches its maximum in the period January to May [138]. The results of the analysis showed SVD-1 mode accounting for about 43% of the total covariance, and is characterized by a meridional dipole like pattern with negative loadings over the northern equatorial Atlantic Ocean and positive loadings over the southern equatorial Atlantic basin (Fig. 8.7a). The rainfall component of this mode showed a southeastern–northwestern dipole pattern with negative loadings over the south-eastern sub-sector while positive loading over north-western segment of the region (Fig. 8.7b). The time series of expansion of SVD coefficients of both rainfall and SST anomalies has a correlation coefficient of 0.97 (Fig. 8.7c) that also shows some trends and enhancement of the decadal amplitudes in some years (Fig. 8.7c).

The SVD-2 mode between MAM rainfall and Atlantic Ocean SST fields explains 23% of the total covariance (Table 8.2). Large coherent positive loadings over the equatorial Atlantic Ocean (Fig. 8.8a) are observed to be in association with large positive loading in most parts of the regions where MAM rainfall is significant (Fig. 8.8b). The time series of expansion coefficients of this mode has a correlation of 0.98 significant at 95% confidence level. The time series further showed strong decadal variation with strong positive phases during mid 1960s and early 1980s. The third mode for MAM rainfall and Atlantic Ocean SST that explains 12.8% of the total covariance is occasioned by inter-hemispheric positive loading over the southern Atlantic Ocean and negative loading over northern hemisphere with centre at (50 °W–35 °W, 4 °S–15 °N) of the basin (figure not shown). The regional rainfall variability associated to this mode has positive and negative loadings in the southern and northern sub-region respectively. This pattern was observed to be more or less opposite to that of SVD-2 mode in Fig. 8.8b. The time series of expansion coefficients of this mode has increasing trend in the 1960s and 1970s decades. Lindzen and Nigam [57] observed similar pattern in their study and indicated that such gradients have a stronger influence on the climate in the tropical regions. This mode displayed both zonal and meridional SST variability that has been observed to have significant influence on the rainfall over western Africa [130].

Analysis of MAM rainfall and Pacific Ocean SSTs showed a rather stronger coupling compared to other basins. This is manifested in the first three modes contributing to about 79% of the decadal rainfall covariance. The SVD-1 mode that contributed about 38% of the total covariance is characterized by warming over the eastern and cooling over western Pacific Ocean, reminiscent of El Niño SST pattern, see e.g., [28, 29]. SVD-2 Mode for the Pacific contributed to about 25% of the covariance and is also characterized by strong negative loadings over the eastern and positive loadings over western equatorial Pacific Ocean. This is the cold phase of La Niña SST pattern that is associated with deficient seasonal rainfall over most parts of the region [41, 67, 105]. Njau [74] observed that most of the severe droughts over Kenya were experienced during the MAM rainfall season preceding El-Niño event. The SVD-3, which contributes about 17% of the total covariance is characterized by

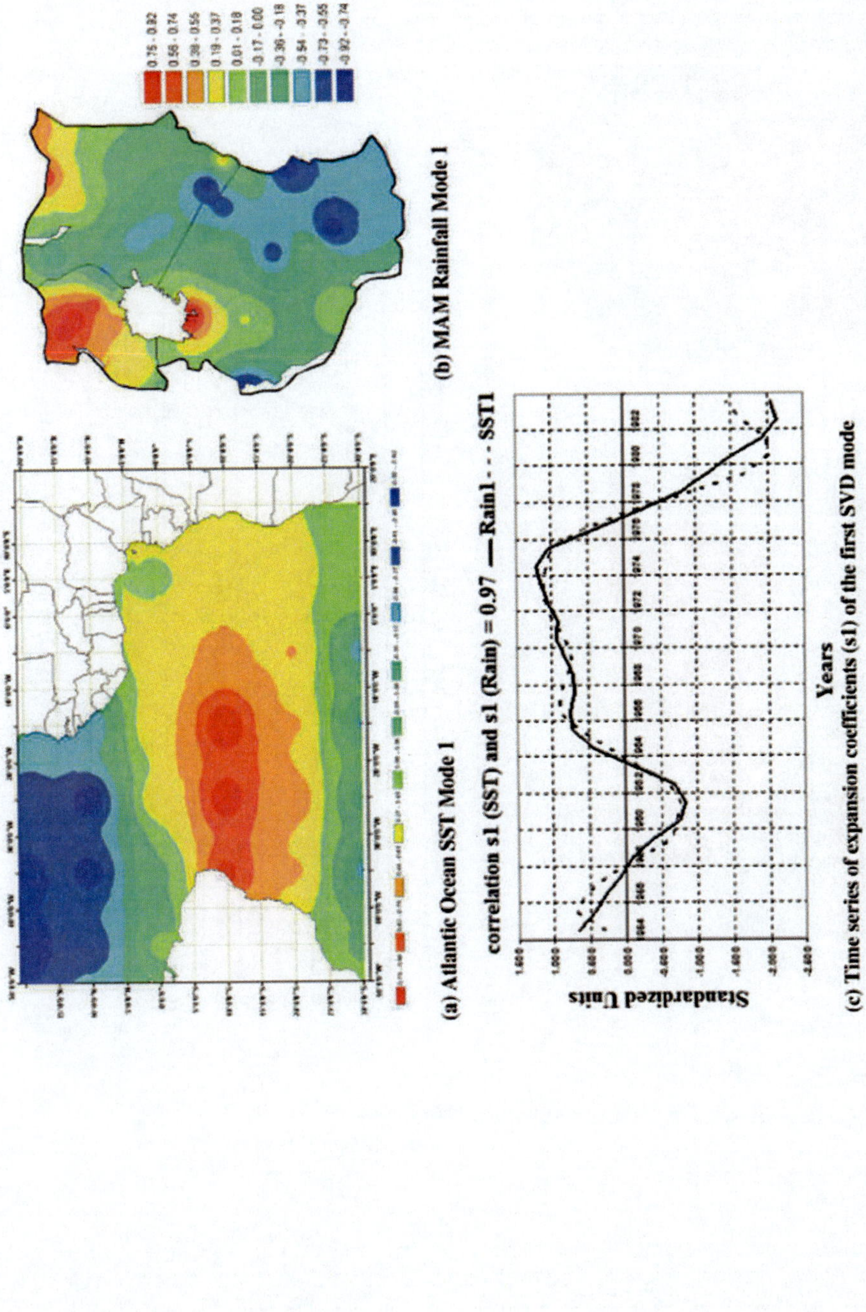

Fig. 8.7 Spatial patterns of the first SVD mode accounting for 43% of the covariance for **a** Atlantic Ocean SST, **b** MAM rainfall presented as homogeneous correlation maps **c** time series of expansion coefficients (s1) of the first SVD mode. x-axis represent the longitude while the y-axis shows the latitude. *Source* [88]

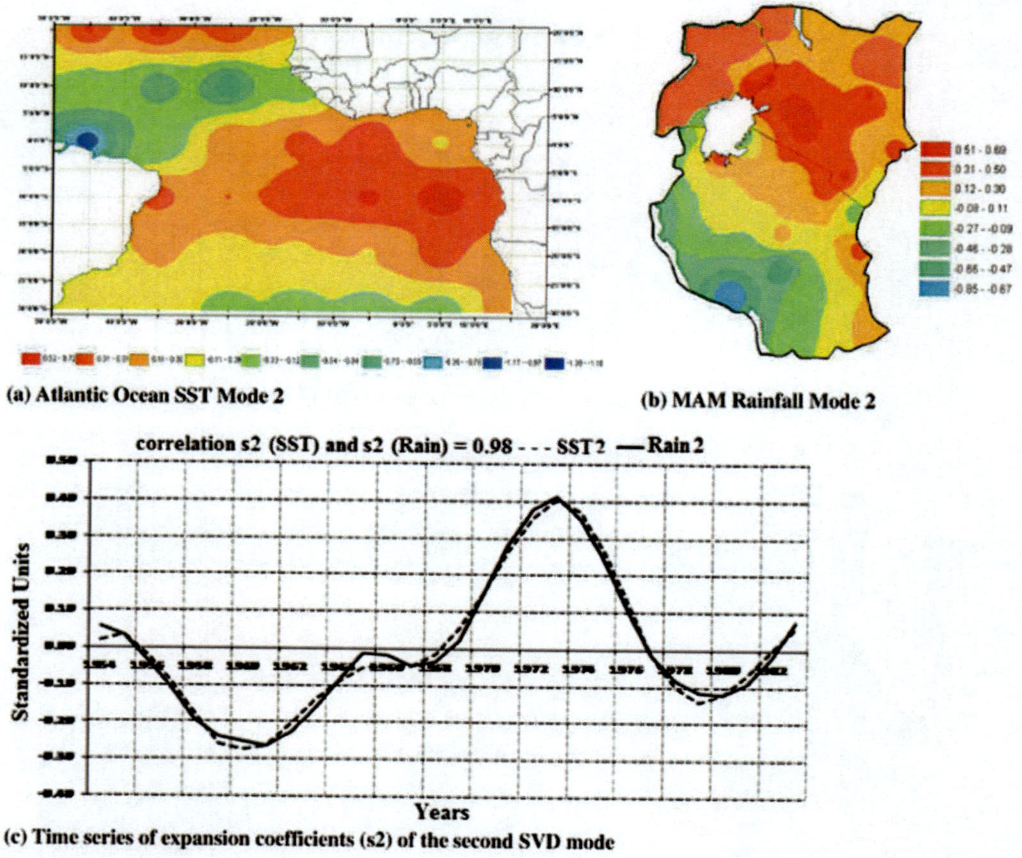

Fig. 8.8 Spatial patterns of the second SVD mode accounting for 23% of the covariance for **a** Atlantic Ocean SST, **b** MAM rainfall presented as homogeneous correlation maps **c** time series of expansion coefficients (s2) of the second SVD mode. x-axis represent the longitude while the y-axis shows the latitude. *Source* [88]

unique negative loadings over the eastern and western with positive loadings over the central equatorial Pacific Ocean.

8.4.4 The Three Oceans Versus OND Modes

The results obtained when the three global oceans were independently analyzed with OND decadal rainfall are now discussed. The results show stronger coupling between the three oceans and OND decadal rainfall compared to the MAM rainfall. The three leading SVD-coupled modes explain greater than 80% of the cumulative covariance compared to 75% for the MAM rainfall. Figure 8.9 shows the heterogeneous correlation patterns for the first mode in the SVD expansion for the Indian Ocean SST anomaly and OND decadal rainfall from 1954–1983 (30 years WMO climatology). Each map represents the correlation between the expansion coefficients of one field and the grid point anomaly values of the other field. The heterogeneous pattern for the first SVD mode (Fig. 8.9a) has two prominent features. The primary characteristic is the warming of the southeastern ocean as indicated by the positive loadings and cooling in southwestern ocean, mainly south of the equator. Secondly, this corresponds to negative loading patterns of rainfall over the southern sector and positive loading over the northern sector of the region. Mode-1 explains 65% of the total coupling covariance, which seems to represent the mean seasonal patterns of SST over the ocean consistent with the studies of Behera et al. [14], Schreck and Semazzi [109], Terray and Dominiak [118], and Tozuka et al. [121]. The time series of expansion of this mode showed a decreasing trend with correlation coefficient of 0.87 during the period of the study.

The heterogeneous pattern for the second SVD mode has spatial patterns exhibiting large coherent negative loading over the central equatorial Indian Ocean with small positive tongue over the Indo-Pacific Oceans (Fig. 8.10a). This signal is generally associated with negative loading patterns over most parts of the region especially over areas that receive substantial rainfall during OND season (Fig. 8.10b). This mode shows that the large-scale cooling of the ocean is associated with dry decades over the region. Nicholson and Kim [71], Harrison and Carson [34], Ihara et al. [40]; and Nyakwada [79] observed similar patterns over the Ocean in their respective studies. The time series of expansion coefficients of Mode-2 has correlation coefficient of 0.98 with significant decreasing decadal trend since 1960s.

Figure 8.11 shows the heterogeneous correlation patterns for the first mode in the SVD expansion for the Atlantic SST anomaly and OND rainfall (Table 8.2). The Atlantic Ocean is to the west of the region of East Africa and moisture influx from the ocean is associated with enhanced westerly circulation that also favours the incursions of moisture from the always wet tropical forests of Congo, DRC and other central African countries. The SST variability in the Atlantic Ocean reaches its maximum in the period January–May [138]. The first three SVD modes for Atlantic Ocean SST and October–December seasonal rainfall accounted for about 84% of the total square covariance. Mode-1 (47.8%) is characterized by a meridional dipole

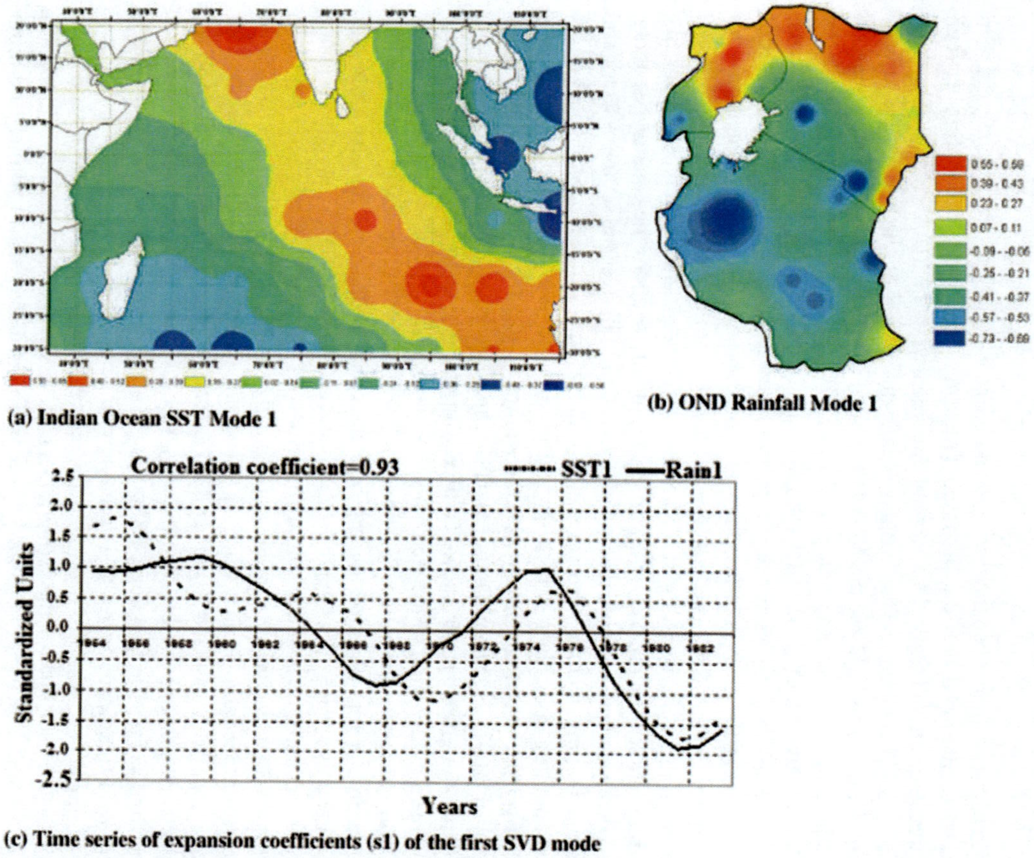

Fig. 8.9 Spatial patterns of the first SVD mode accounting for 65% of the covariance for **a** Indian Ocean SST, **b** OND rainfall presented as homogeneous correlation maps **c** time series of expansion coefficients (s1) of the first SVD mode. x-axis represent the longitude while the y-axis shows the latitude. *Source* [88]

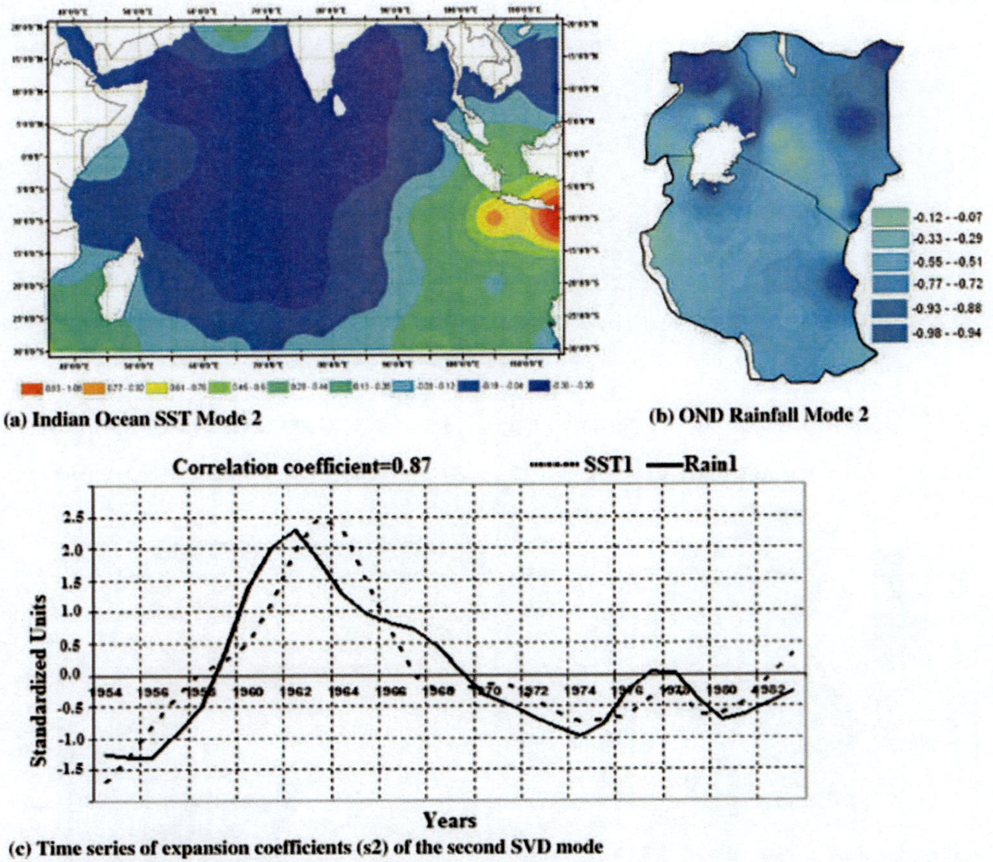

Fig. 8.10 Spatial patterns of the second SVD mode accounting for 16.5% of the covariance for **a** Indian Ocean SST, **b** OND rainfall presented as homogeneous correlation maps **c** time series of expansion coefficients (s2) of the second SVD mode. x-axis represent the longitude while the y-axis shows the latitude. *Source* [88]

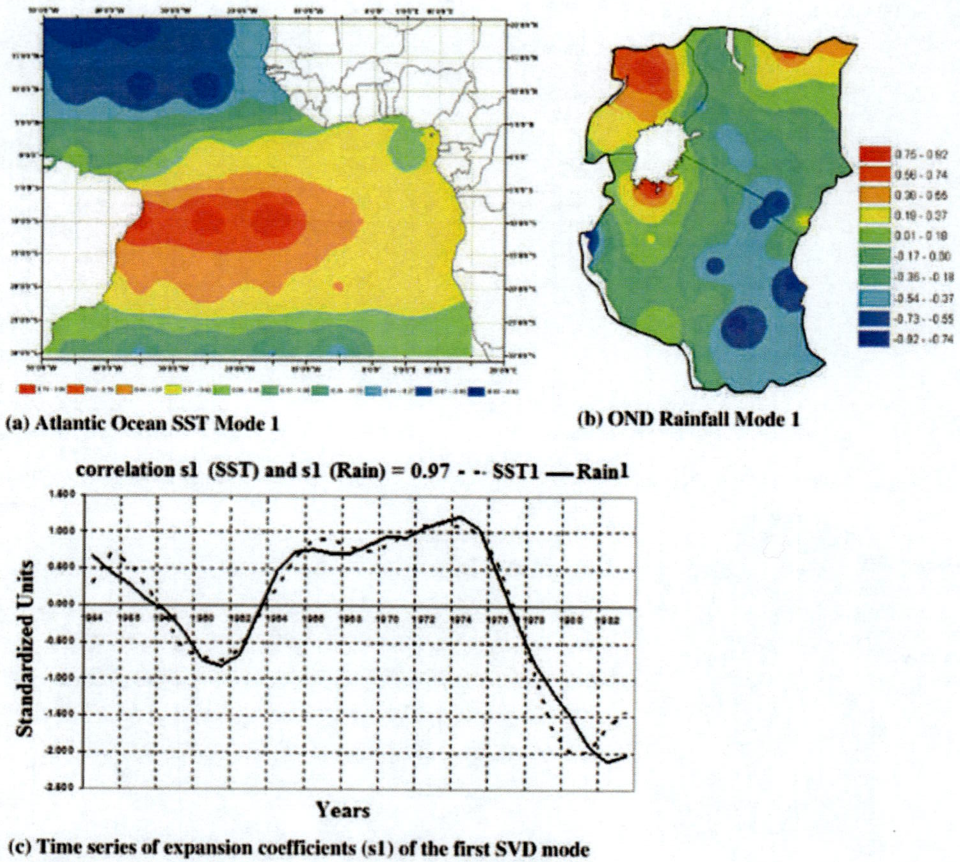

Fig. 8.11 Spatial patterns of the first SVD mode accounting for 47.8% of the covariance for **a** Atlantic Ocean SST, **b** OND rainfall presented as homogeneous correlation maps **c** time series of expansion coefficients (s1) of the first SVD mode. x-axis represent the longitude while the y-axis shows the latitude. *Source* [88]

like pattern with negative loadings over the northern equatorial Atlantic Ocean and positive loadings over the southern equatorial Atlantic basin (Fig. 8.11a). The OND rainfall component of this mode is characterized by south-eastern–north-western dipole pattern with negative and positive loadings respectively over the sub-sector (Fig. 8.11b). The time series of expansion of this mode has a correlation coefficient of 0.97 and also shows some enhanced decadal modes (Fig. 8.11c).

The spatial patterns observed for mode-2 seems to be opposite to that observed for SVD-1 mode for the same basin (Figure not shown). This shows that a change in the interhemispheric loading patterns in the Ocean causes corresponding change in the spatial and temporal regional rainfall loading patterns. For instance whenever there is negative loading in the northern Atlantic Ocean and positive in the south, the regional rainfall pattern is associated with positive loading in the northern sector and negative loading in the southern sector and vice versa. Similar changes are also true for the time series expansion coefficients of the two fields.

This suggests that the large scale changes in the global basins SSTs have significant influence on regional decadal climate variability. This emphasizes the importance of the ocean currents and the associated SST patterns in the study of climate variability and prediction. Strong relationship between ocean currents and climate has been established to be the major influence of regional climates [22, 49, 123]. Whenever there is enhancement of ENSO, this mode has always been manifested, and is also linked to the North Atlantic Oscillation (NAO) [56, 138].

The results on coupling between the Pacific Ocean and the October–December rainfall at decadal time scale over the region is summarized in Table 8.2 and also spatially shown by Fig. 8.12(a). Although Pacific is the furthest ocean from the region, it has some strong teleconnections to the East African climate, especially during ENSO and other years with large SST anomalies [41, 42, 51, 71, 82, 105, 109, 126, 130]. Circulation anomalies in Pacific Ocean have significant influence on Indian and Atlantic Oceans and the surrounding land areas [118, 135] including East African rainfall.

The first three modes for the Pacific SST and OND rainfall accounted for about 80% of the decadal rainfall covariance. The heterogeneous pattern for the first SVD mode that contributes about 39% of the total covariance mainly reveals the ENSO-like pattern in the central equatorial Pacific (figure not shown). The impact of this mode is observed to have similarity to Mode-1 of the Indian Ocean on East African rainfall (Fig. 8.5b). This may be due to close responses of the circulation amongst the three oceans [21]. It has been shown that ENSO is linked to the Indian Ocean variability through the modulation of walker circulation [53, 54, 108, 139].

The SSTs over the southern Indian Ocean during the December to February have been observed to influence climate shift in the Indian and Pacific Oceans and thus predictors of El Niño [118]. The SVD-2 Mode of the basin and OND rainfall (28% of covariance) is characterized by the La Niña-like SST pattern with large negative loadings over the eastern central equatorial Pacific Ocean and positive loading over the western equatorial Pacific Ocean (Fig. 8.12a). Such anomaly pattern is associated with negative loading over the region that causes depressed rainfall in the region (Fig. 8.12b). The time series exhibits strong decadal variability with a long-term

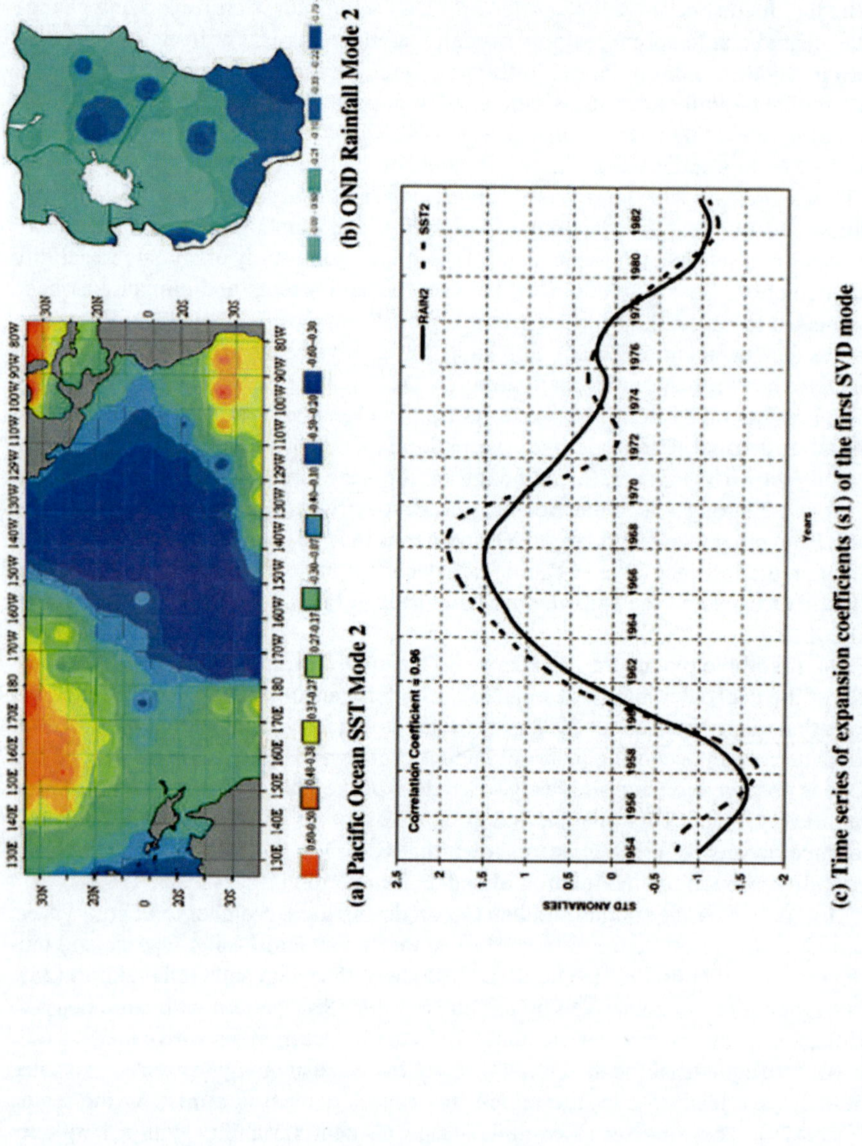

Fig. 8.12 Spatial patterns of the second SVD mode accounting for 28% of the covariance for **a** Pacific Ocean SST, **b** OND rainfall presented as homogeneous correlation maps **c** Time series of expansion coefficients (s2) of the second SVD mode. x-axis represent the longitude while the y-axis shows the latitude. *Source* [88]

change toward higher values in 1962–1970. The relative dry decade of 1960/1970 in the region noted by Omondi [84] is reflected in the Mode 2 time series.

8.4.5 The Three Oceans Versus JJA Modes

The June–August (JJA) season constitute the third rainy season when western and coastal parts of the region receive substantial amount of rainfall. Parts of the equatorial sector, covering northern Tanzania, western and coastal areas generally exhibit a trimodal rainfall regime centered on MAM, JJA and OND seasons. In Table 8.2, a summary of results for the three basins coupling with JJA decadal rainfall are given with the first three leading modes accounting for about 87% of the covariance. SVD Mode-1 is characterized by a zonal dipole mode with pools of negative and positive loadings observed over the western and eastern equatorial Indian Ocean (Fig. 8.13a) associated with depressed decadal rainfall over all parts of the region (Fig. 8.13b). Time series of expansion of the coefficients demonstrates strong decadal variations in both time series with expansion coefficient of 0.98 (Fig. 8.13c).

A rather interesting result presented in Fig. 8.14a for SVD Mode-1 when coupling between the JJA decadal rainfall and Atlantic Ocean accounted for 38.6% of the total covariance. An inter-hemispheric dipole with positive loading over the southern (0 °E–15 °E, 30 °S–15 °S) and the negative pole in the northern part of the tropical Atlantic Ocean (Fig. 8.14a) with strong decadal signal in time series of expansion coefficients (significant correlation of 0.99) was delineated (Fig. 8.14c). This pattern can therefore be linked to the possibly strong JJA decadal rainfall variability over the region (Fig. 8.14b). The second dominant SVD-2 explains 29.2% of the total JJA rainfall and SST covariance. Its spatial characteristic is opposite to that of OND rainfall and Atlantic Ocean (Fig. 8.11b) hence not shown in this study. There are, however, some differences in the magnitudes and spatial spread over some areas with high positive/negative loadings.

These results signify the importance of Atlantic Ocean modes on decadal rainfall over East Africa. The characteristics of the SVD modes are closely associated with some climate extremes affecting the region. These could be associated with the implications of these modes on the circulations, energy, and moisture induced by the ocean, and other inland rain generating systems. Barnston et al. [8] demonstrated that the time-space behaviour of the SST field alone influences the JJA seasonal rainfall over the region both on interannual and inter-decadal time-scales. Other similar results with JJAS season have been documented, e.g., by Gissila et al. [30], Segele and Lamb [111], Korecha and Barnston [51], and Zewdu et al. [143] among others.

The result of SVD mode one for the Pacific Ocean and JJA rainfall fields showed ENSO-like pattern discussed in the previous section. However, SVD-2 mode is characterized by large pool of positive SST anomalies over the central equatorial Pacific and negative SST anomalies in the equatorial western and eastern parts of the basin (Fig. 8.15a). The corresponding impact of this mode on regional rainfall

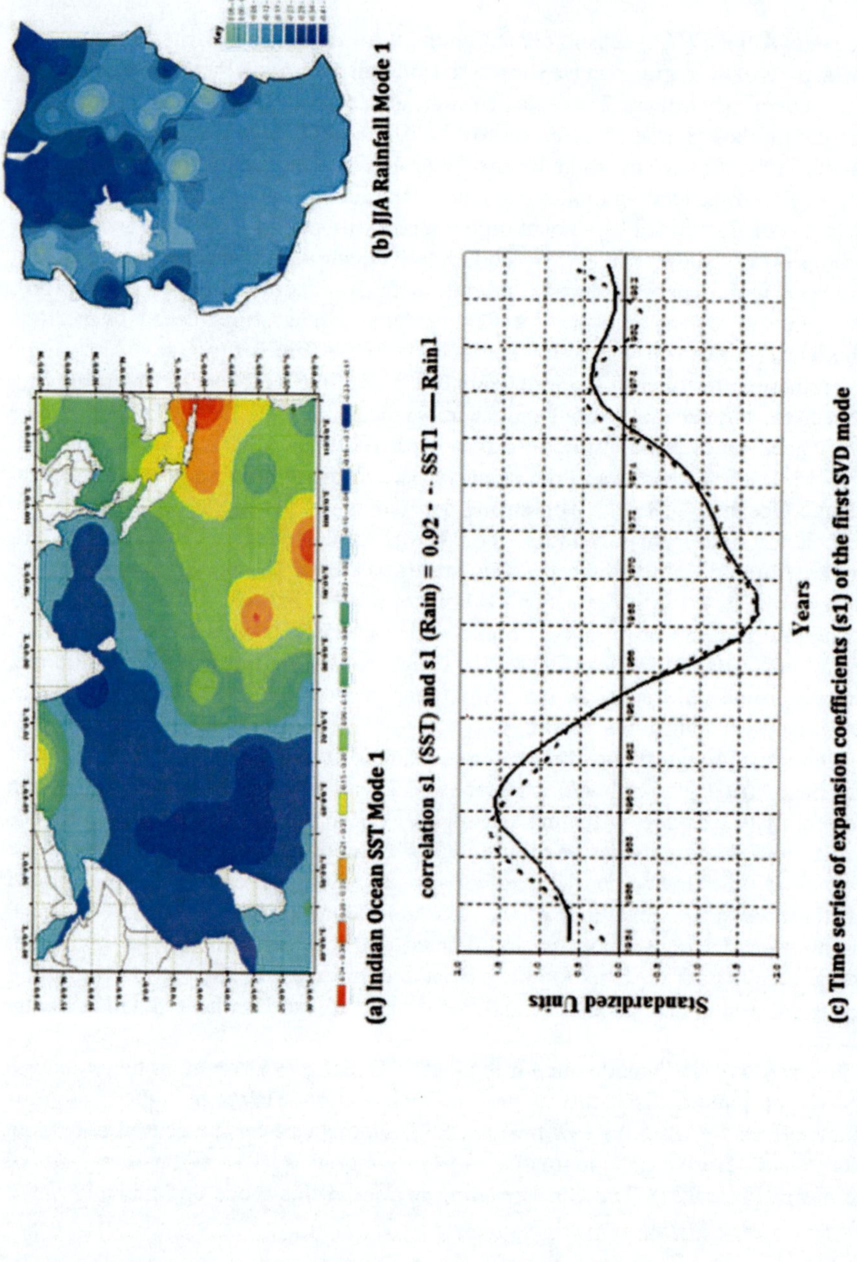

Fig. 8.13 Spatial patterns of the first SVD mode accounting for 61.1% of the covariance for **a** Indian Ocean SST, **b** JJA rainfall presented as homogeneous correlation maps **c** time series of expansion coefficients (s1) of the first SVD mode. x-axis represent the longitude while the y-axis shows the latitude. *Source* [88]

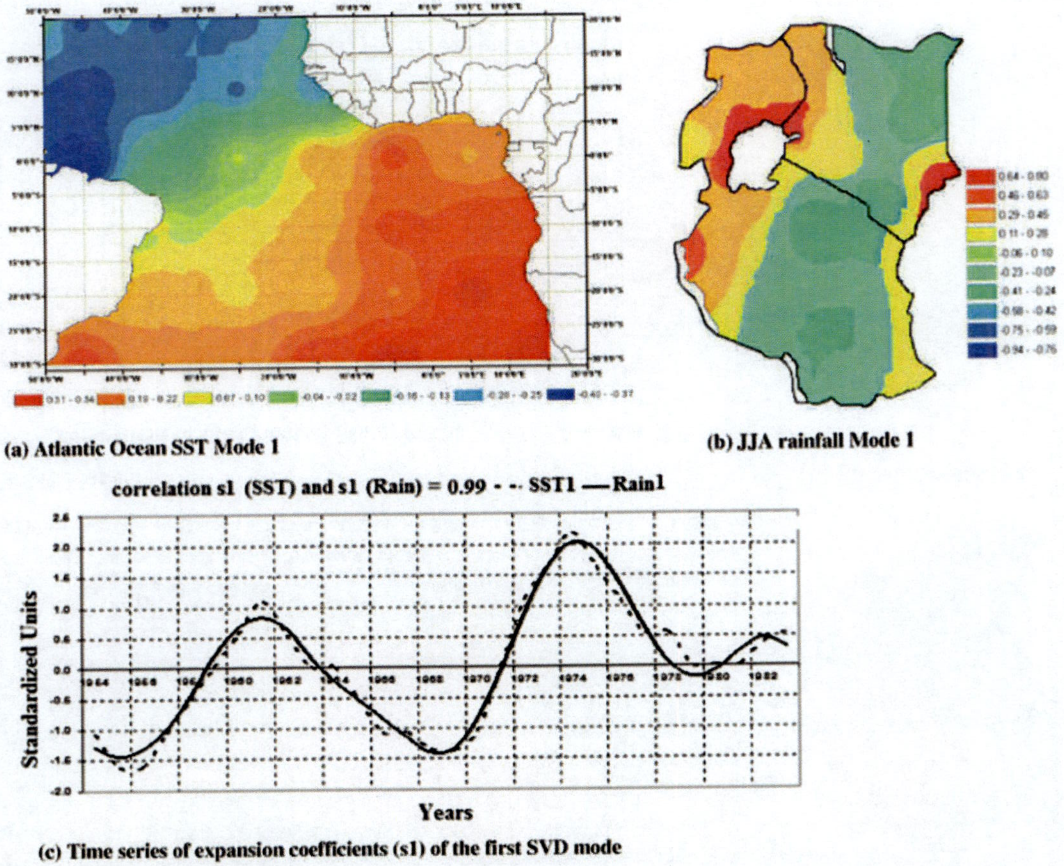

Fig. 8.14 Spatial patterns of the first SVD mode accounting for 38.6% of the covariance for **a** Atlantic Ocean SST, **b** JJA rainfall presented as homogeneous correlation maps **c** time series of expansion coefficients (s1) of the first SVD mode. x-axis represent the longitude while the y-axis shows the latitude. *Source* [88]

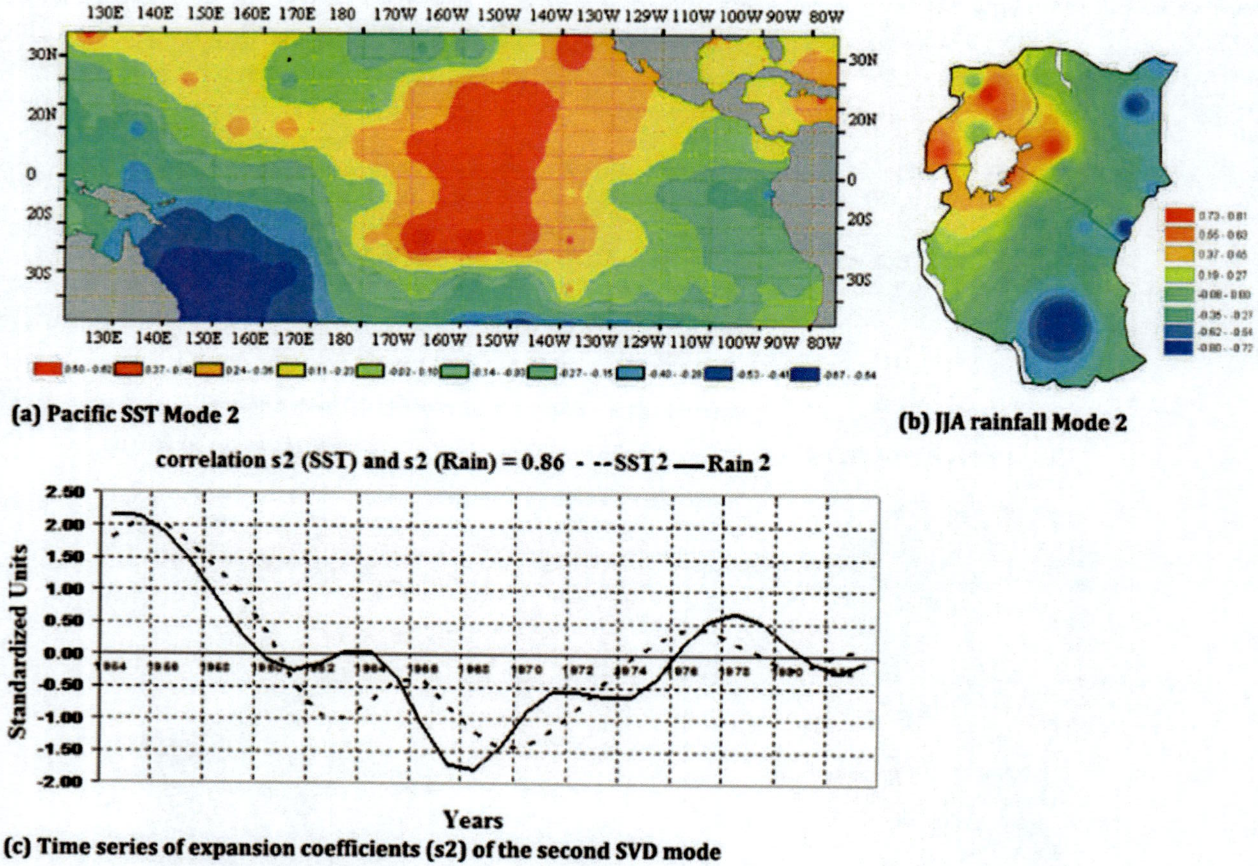

Fig. 8.15 Spatial patterns of the second SVD mode accounting for 20% of the covariance for **a** Pacific Ocean SST, **b** JJA rainfall presented as homogeneous correlation maps **c** time series of expansion coefficients (s2) of the second SVD mode. x-axis represent the longitude while the y-axis shows the latitude. *Source* [88]

is positive loading over western equatorial parts of the region and negative loading over the eastern highlands (Fig. 8.15b). Major warming events for this mode take place over the Central Pacific Ocean unlike the eastern/western Pacific of typical El'Niño/La'Niña evolution. Similar pattern in tropical Pacific SST variability mode was observed and referred to as trans-Niño index [119, 120] and later by Ashok et al. [3] who termed it 'El'Niño Modoki'. Cold SST anomalies (SSTA) are observed on both sides along the equator [3, 4, 38, 63, 119, 120] and this is associated with enhanced and depressed rainfall over western and eastern sectors respectively of East Africa region. El Niño Modoki involves ocean-atmosphere coupling processes, which include a unique tripolar sea level pressure pattern during the evolution, analogous to the Southern Oscillation in case of El'Niño [3, 4, 38, 63]. This anomalous warming event is different from conventional El Niño events but both take place in equatorial Pacific.

The results from SVD analyses have therefore demonstrated that although decadal rainfall variability modes are generally common at all locations, some strong signals are only restricted to specific areas. These signals are evident for all seasons over the entire region. It is a fact that some climate extremes experienced over the region such as floods of the early 1960s, 1997/98 and droughts of 1970s and early 1980s that extended over most parts of the region had decadal finger prints. The strong coupling between global basins and decadal rainfall can provide new prediction tools that are important for decadal rainfall prediction. This again can contribute to the formulation and improvement of prediction skills in numerical modeling and early warning several lead times ahead.

8.5 Decadal Rainfall Variability versus Food Security

Schreck and Semazzi [109] investigated the climate variability of GHA for the 1961–2001 'short rains' season of October through to December and established that the most dominant mode of variability corresponded to El'Niño–southern oscillation (ENSO) climate index while the second mode was associated with decadal variability. When the quasi decadal oscillation associated with ENSO and IOD are filtered, it is possible to identify the climate change/anthropogenic connection. Indeed, by isolating a separate eastern Africa trend mode in rainfall, [109] obtained a time evolution consistent with increasing warming of the global climate. Their results thus suggested a possible connection between the decadal variability mode and global warming (a product of anthropogenic activities, e.g., through CO_2 emissions).

Rainfall variability is one of the most important factors determining variability in agricultural production as it leads to crop failures, loss of livestock, and associated loss of income and even famine [1, 62]. Understanding of its variability at decadal time-scales is essential in addressing food insecurity of the GHA region as it could inform development of better predictive capacity at longer time scales that would considerably improve decision making and risk management in agriculture, thereby making important contribution toward sustainable development [62]. In this regards, the knowledge of extreme climate variations at decadal timescale gained from this chapter is useful for planning and decision making of long-term climate risk management strategies for sustainable socioeconomic development. This could relate to short-term decisions on the levels of inputs, types of crops, planting options, and marketing at the farm level as well as to the long-term design of land-use options at the policy level [62].

The primary focus of this chapter was to investigate the dominant spatial and temporal decadal rainfall variability modes over East Africa region of GHA and their linkages to decadal variability modes of the specific global oceans. The first investigation involved the use of rotated PCA (RPCA) and simple correlation analysis approaches to delineate East Africa region into zones with similar modes of decadal rainfall variability. This analysis was mainly to group together stations with common decadal rainfall characteristics into homogeneous zones. The stations highly correlated with each other were identified as representative stations for every zone used for the analyses in the chapter. Each observed rainfall record for the delineated representative station was thereafter subjected to spectral analysis to demonstrate the dominance of ten years cycles. Singular value decomposition (SVD) was then applied to evaluate the existence of covariance amongst regional rainfall and the individual global basins SSTs. VARIMAX-RPCA analysis delineated nine, seven and eight unique climatologically homogeneous zones for March–May (MAM), October-December (OND) and June–August (JJA) seasons, respectively, when smoothed rainfall records were used. It is also evident from the analysis that ten years cycle was dominant in most rainfall records with significant differences in the amplitudes and spectral bands of the spectral peaks from the various regions. This could be a reflection of the complex nature of the regional climate systems that include complex topography and large water bodies that include Lake Victoria. The results obtained from the teleconnections of the regional decadal rainfall variability patterns to the global SSTs using SVD analysis showed that:

1. Decadal variability of the SSTs in the Pacific, Atlantic and Indian Oceans all have significant influence on regional decadal rainfall variability.
2. Significant coupling of the SSTs and rainfall fields were observed with high values of Square covariance fraction (SCF) explained by the first three modes for each of the three oceans.
3. Generally, it was revealed that even when decadal variability of one of the oceans contributed most of the variance of decadal rainfall variability, the roles of the other two oceans were still very significant. This signifies the close interaction among the global oceans SSTs and the complex regional scale climate processes. The first three SVD modes of the basin SSTs and decadal rainfall accounted for over 75% of the total square covariance for all the seasons, with statistically significant expansion coefficients of time series of the SVD mode of the two fields' anomalies.
4. The SVD analysis for MAM rainfall season with Indian Ocean SST showed the first dominant mode accounting for 49.9% of the total square covariance and this was influenced mainly by the east/west zonal mode over the Indian Ocean.

5. Results for SVD analysis for OND rainfall season with Pacific Ocean SST showed the first dominant mode accounting 39.2% of the total square covariance. Decadal rainfall variability for this season was dominantly characterized by decadal variability modes of the ENSO phenomenon over the equatorial Pacific Ocean.
6. SVD analysis for the third rainfall season of June–August and the Atlantic Ocean SST showed the first dominant mode accounting for 38.6% of the total square covariance. The key modes of variability that seemed to drive the June–August rainfall significantly was the inter-hemispheric tropical Atlantic variability with meridional dipole pattern in the Atlantic Ocean. A unique tripolar mode of variability observed during this season was characterized by large pool of positive loading over the central equatorial Pacific and negative loadings in the equatorial western and eastern parts of the basin. This mode is associated with wet decades over western parts of the region and dry decades over the eastern sectors of the region.

This chapter provides means of using ocean variability modes to predict future decadal rainfall variability. Availability of timely and accurate climate information can provide useful tools for forward planning to reduce the vulnerability and risks of the hydro-meteorological hazards that currently are prevalent in the region. This lead time of the early warning is important for the general planning and management of the disaster risk reduction.

References

1. Allan RJ (2000) ENSO and climatic variability in the past 150 years. El'Niño and the Southern Oscillation: Multiscale variability and its impacts on natural ecosystems and society. In: Diaz HF, Markgraf V (eds). Cambridge University Press, pp 3–55
2. Anyah RO, Qiu W (2012) Characteristic 20th and 21st century precipitation and temperature patterns and changes over the greater horn of Africa. Int J Climatol 32:347–63
3. Ashok K, Behera SK, Rao SA, Weng H, Yamagata T (2007) El Niño Modoki and its possible teleconnection. J Geophys Res 112:C11007. https://doi.org/10.1029/2006JC003798
4. Ashok K, Iizuka S, Rao SA, Saji NH, Lee WJ (2009) Processes and boreal summer impacts of the 2004 El Niño Modoki: an AGCM study. Geophys Res Lett 36:L04703. https://doi.org/10.1029/2008GL036313
5. Awange JL, Palancz B, Völgyesi L (2020) Hybrid imaging and visualization. Employing machine learning with Mathematica—Python. Springer Nature International, Berlin
6. Barnett TP, Preisendorfer R (1987) Origins and levels of monthly and seasonal forecasts skill for United States surface air temperatures determined by canonical correlation analysis. Mon Weather Rev 115:1825–50
7. Barnston AG, Smith TM (1996) Specification and prediction of global surface temperature and precipitation from global SST using CCA. J Clim 9:2660–97
8. Barnston AG, Thiao W, Kumar V (1996) Long-lead forecasts of seasonal precipitation in Africa using CCA. Weather Forecast 11:506–20
9. Bark DL (1978) History of American droughts. In: Rosenberg NJ (ed). North American droughts. A selected symposia series. Westview Press, Boulder Colorado, p 9–23
10. Barry RG, Chorley RJ (1968) Atmosphere, weather and climate, 4th edn. Methuen, London, New York, p 407

11. Basalirwa CPK (1991) Raingauge network designs for Uganda; PhD thesis, University of Nairobi, Kenya
12. Basalirwa CPK (1995) Delineation of Uganda into climatological rainfall zones using the method of principal component analysis. Int J Climatol 15:1161–77
13. Becker M, Llovel W, Cayenave A, Günter A, Crétaux J-F (2010) Recent hydrological behavior of the East African great lakes region inferred from GRACE, satellite altimetry and rainfall observations. C R Geosci 342:223–33. https://doi.org/10.1016/j.crte.2009.12.010
14. Behera SK, Luo JJ, Masson S, Delecluse P, Gualdi S, Navarra A (2005) Paramount impact of the Indian Ocean dipole on the East African short rains: ACGCM study. J Clim 18:41–54
15. Bjerkness J (1969) Atmospheric teleconnections from the equatorial Pacific. Mon Weather Rev 97:163–72
16. Black E, Slingo J, Sperber KR (2003) An observational study of the relationship between excessively strong short rains in the coastal East Africa and Indian Ocean SST. Mon Weather Rev 131:74–94
17. Black E (2005) 2004: the relationship between Indian Ocean sea-surface temperature and East African rainfall. Philos Trans R Soc A 363:43–7. https://doi.org/10.1098/rsta.2004.1474
18. Bloomfield P (1976) [Fourier analysis of time series: an introduction]. Wiley, pp 256–799 [Smoothed periodogram method of spectral analysis]
19. Bowden J, Semazzi FHM, Anyah R, Schreck C (2004) Intraseasonal to decadal variability of the greater horn of Africa. North Carolina State University, Raleigh, North Carolina
20. Bretherton CS, Smith C, Wallace JM (1992) An intercomparison of methods for finding coupled patterns in climate data. J Clim 5:541–60
21. Burroughs WJ (1999) The climate revealed. Mitchell Beazley. An Imprint of Octopus Publishing Group Ltd, 2–4 Heron Quays, London, p 129
22. Cai W, Cowan T (2007) Trends in the southern Hemisphere circulation in IPCC AR4 models over 1950–1999: Ozone depletion versus greenhouse forcing. J Clim 20(4):681–93
23. Cattel RB (1966) The scree test for the number of factors. Multivar Behav Res 1:245–59
24. Chatfield C (2004) The analysis of time series, an introduction, 6th ed, New York: Chapman & Hall/CRC. A readable presentation of time series in frequency and time domains—a good introductory text
25. Clark CO, Webster PJ, Cole JE (2003) Interdecadal variability of the relationship between the Indian Ocean zonal mode and East African coastal rainfall anomalies. J Clim 16:548–54
26. CLIVAR VACS (2007) Decadal variability issues in Africa
27. Cohen SJ (1983) Classification for 500 mb height anomalies using obliquely rotated principal components. J Appl Meteorol 22(12):1975–88
28. Forootan E, Awange JL, Kusche J, Heck B, Eicker A (2012) Independent patterns of water mass anomalies over Australia from satellite data and models. J Remote Sens Environ. 10.1016/j.rse.2012.05.023
29. Garci'a-Garci'a D, Ummenhofer CC, Zlotnicki V (2011) Australian water mass variations from GRACE data linked to Indo-Pacific climate variability. Remote Sens Environ 115:2175–83. https://doi.org/10.1016/j.rse.2011.04.007
30. Gissila T, Black E, Grimes DIF, Slingo JM (2004) Seasonal forecasting of Ethiopian summer rains. Int J Climatol 24:1345–58
31. Goddard L, Graham NE (1999) Importance of the Indian Ocean for simulating rainfall anomalies over eastern and southern Africa. J Geophys Res 104(D16):19099–116
32. Gottman JM (1981) Time-series analysis. A comprehensive introduction for social scientists. Cambridge University Press, Cambridge, pp 400–430
33. Gregory S (1975) On the delimitation of regional patterns of recent climatic fluctuations. Mon Weather Rev 30:276–87
34. Harrison DE, Carson M (2007) Is the world ocean warming? Upper ocean temperature trends: 1950–2000. J Phys Oceanogr 37(2):174–87
35. Hastenrath S, Polzin D (2003) Circulation mechanisms of climate anomalies in the equatorial Indian Ocean. Meteorol Z 12:81–93

36. Hasternrath S, Polzin D (2004) Dynamics of the surface wind field over the equatorial Indian Ocean. Q J R Meteor Soc 130:503–517
37. Huang B, Shukla J (2007) Mechanisms for the interannual variability in the tropical Indian Ocean. Part I: the role of remote forcing from the tropical Pacific. J Clim 20:2917–2936
38. Hye-Mi Kim, Webster PJ, Curry JA (2009) Impact of shifting patterns of Pacific Ocean warming on North Atlantic tropical. Science 325(5936):77–80
39. ICPAC (1999) Homogeneous climatological zoning DMC lecture notes, pp 29–43 [chapter 3]
40. Ihara C, Kushnir Y, Cane MA (2008) Warming trend of the Indian Ocean SST and Indian Ocean dipole from 1880 to 2004. J Clim 21:2035–46
41. Indeje M (2000) Prediction and numerical simulation of regional climate of equatorial eastern Africa. PhD thesis, North Carolina State University, USA, p 327
42. Indeje M, Semazzi FHM, Ogallo LJ (2000) ENSO signals in East African rainfall seasons. Int J Climatol 20:19–46
43. Solomon S, Qin D, Manning M, Chen Z, Marquis M, Averyt KB, Tignor M, Miller HL (eds) (2007) Climate change 2007: the physical science basis. Contribution of working group I to the fourth assessment report (AR4) of the intergovernmental panel on climate change. IPCC, summary for policymakers. Cambridge University Press, Cambridge, UK, New York, NY, USA
44. Jenkins GM, Watts DG (1968) Spectral analysis and its applications. Holden-Day, CA, USA
45. Kaiser HF (1958) The varimax criterion for analytic rotation in factor analysis. Psychometrika 23:187–200
46. Kaiser HF (1959) Computer program for varimax rotation in factor analysis. Educ Physchol Meas 19:1413–20
47. Kanamitsu M, Ebisuki W, Wollen J, Yang SK, Hnilo JJ, Fiorino M, Potter GL (2002) NCEP—DOE AMIP—II reanalysis, R-2. Bull Am Meteorol Soc 83(11):1631–43
48. Keenlyside N, Latif M, Junclaus J, Kornblueh L, Roeckner E (2008) Advancing decadal climate scale prediction in the North Atlantic. Nature 453:84–8
49. Keller K, Deutssch C, Hall MG, Bradford DF (2007) Early detection of changes in the north Atlantic meridional overturning circulation: implications for the design of ocean observation systems. J Clim 20(2):145–57
50. Knutson TR, Tuleya RE (2004) Impact of CO_2-induced warming on simulated hurricane intensity and precipitation: sensitivity to the choice of climate model and convective parameterization. J Clim 17:3477–95
51. Korecha D, Barnston AG (2007) Predictability of June to September rainfall in Ethiopia. Mon Weather Rev 135:628–50
52. Koopmans LH (1995) The spectral analysis of time series. Probability and mathematical statistics, 2nd ed, vol 22
53. Kug JS, Kirtman BP, Kang IS (2006) Interactive feedback between ENSO and the Indian Ocean in an interactive ensemble coupled model. J Clim 19:6371–81
54. Kug JS, Kang IS (2006) Interactive feedback between the Indian Ocean and ENSO. J Clim 19:1784–801
55. Lee TCK, Zwiers F, Zhang X, Tsao M (2006) Evidence of decadal climate prediction skill resulting from changes in anthropogenic forcing. J Clim 19:5305–18
56. Li S, Robinson WA, Hoerling MP, Weickmann KM (2007) Dynamics of the extratropical response to a tropical Atlantic SST anomaly. J Clim 20(3):560–74
57. Lindzen RS, Nigam S (1987) On the role of sea surface temperature gradients in forcing low-level winds and convergence in the tropics. J Atmos Sci 44:2418–36
58. L'Hote Y, Mahe G, Some B, Triboulet JP (1896 to 2002) Analysis of a Sahelian annual rainfall index from; the drought continues. Hydrol Sci J—J Des Sci Hydrol (47):563–572
59. Mann EM, Park J (1996) Joint spatiotemporal modes of surface temperature and sea level pressure variability in the Northern during the last century. J Clim 9:2137–62
60. Meehl GA, Hibbard KA (2007) A strategy for climate change stabilization experiments with AOGCMs and ESMs. WCRP informal report no 3/2007, ICPO publication no112, IGBP report no 57, World climate research programme, Geneva, p 35

61. Meehl GA, Goddard L, Murphy J, Stouffer, Boer G, Danabasoglu, Dixon K, Giorgetta MA, Greene A, Hawkins E, Hegerl G, Karoly D, Keenlyside N, Kimoto M, Kirtman B, Navarra A, Pulwarty R, Smith D, Stammer D, Stockdale T (2009) Decadal prediction: can it be skillful? Bull Am Meteorol Soc 345–67
62. Meinke H, de Voil P, Hammer GL, Power S, Allan R, Stone RC, Folland C, Potgieter A (2005) Rainfall variability of decadal and longer time scales: signal or noise? J Clim 18:89–90. https://doi.org/10.1175/JCLI-3263.1
63. Meyers G, McIntosh P, Pigot L, Pook M (2007) The years of El'Niño, La Niña, and interactions with the tropical Indian Ocean. J Clim 20:2872–80
64. Moura AD, Shukla J (1981) On the dynamics of droughts in northeast Brazil: observations, theory, and numerical experiment with a general circulation model. J Atmos Sci 38:2375–653
65. Muhati FD, Ininda JM, Opijah FJ (2007) Relationship between ENSO parameters and the trends and periodic fluctuations in East African rainfall. J Kenya Meteorol Soc 1(1):20–43
66. Mutai CC (2003) The role of Indian Ocean SST on East African Short rains. In: Proceedings of the sixth Kenya meteor. Social workshop on meteor. Research, applications and services, Mombasa, Kenya, 29 Sept to 3 Oct 2003, pp 55–59
67. Mutemi JN (2003) Climate anomalies over eastern Africa associated with various ENSO evolution phases. PhD thesis, University of Nairobi, Kenya
68. Muthama NJ, Manene MM, Ndetei CJ (2008) Simulation of decadal precipitation over Nairobi in Kenya. J Sci 13:43–54
69. Nicholson SE, Entekhabi D (1986) The quasi-periodic behavior of rainfall variability in Africa and its relationship to the Southern Oscillation. Arch Meteorol Geophys Bioclimatol Ser A 34:311–48
70. Nicholson SE, Entekhabi D (1987) Rainfall variability in equatorial and Southern Africa: relationships with sea surface temperature along the south-western coast of Africa. J Clim Appl Meteorol 26:561–78
71. Nicholson SE, Kim J (1997) The relationship of the El'Niño-Southern Oscillation to African rainfall. Int J Climatol 17:117–35
72. Nicholson SE (2000) Land surface processes and Sahel climate. Rev Geophys 38:117–39
73. Nicholson SE (2000) The nature of rainfall variability over Africa on time scales of decades to millennia. Global Planet Change 26:137–58
74. Njau LN (2006) Diagnostics and predictability of East Africa rainfall with tropospheric circulation parameters. PhD dissertation, Department of Meteorology University of Nairobi, 170pp
75. Ndehedehe CE, Awange JL, Agutu NO, Okwuash O (2018) Changes in hydro-meteorological conditions over tropical West Africa (1980–2015) and links to global climate. Global Planet Change. https://doi.org/10.1016/j.gloplacha.2018.01.020
76. Ndehedehe CE, Awange JL, Kuhn M, Agutu NO, Fukuda Y (2017) Climate teleconnections influence on West Africa's terrestrial water storage. Hydrol Processes 31(18):3206–3224. https://doi.org/10.1002/hyp.11237
77. Nobre P, Shukla J (1996) Variations of sea surface temperature, wind stress, and rainfall over the Tropical Atlantic and South America. J Clim 9:2464–79
78. North GR, Bell TL, Cohalan RF (1982) Sampling errors in estimation of empirical orthogonal functions. Mon Wea Rev 110:699–706
79. Nyakwada W (2009) Predictability of East African seasonal rainfall with sea surface temperature gradient modes. PhD dissertation, Department of Met: University of Nairobi
80. Ogallo LJ (1988) Relationships between seasonal rainfall in East Africa and the Southern Oscillation. J Climatol 8:31–43
81. Ogallo LJ (1989) The spatial and temporal patterns of the East African seasonal rainfall derived from principal component analysis. J Clim 9:145–67
82. Ogallo LJ, Suleiman KA (1987) Rainfall characteristics in East Africa during the El'Niño years. In: Proceedings of the first technical conferences on meteorological research in Eastern and Southern Africa, Nairobi, Kenya, 6–9 Jan 1987, pp 76–80

83. Okoola REA (1996) Space-time characteristics of the ITCZ over equatorial Eastern Africa during anomalous rainfall years. PhD thesis, Department of Meteorology, University of Nairobi, Kenya
84. Omondi PA (2005) Potential causes and predictability of the space—time patterns of the decadal rainfall variability modes over east Africa. MSc, thesis, Department of Meteorology, University of Nairobi
85. Omondi PA, Ogallo LA, Okoola RE (2009) Decadal rainfall variability modes in observed rainfall records over East Africa and their predictability using sea surface temperature. J Kenya Meteorol Soc 3:37–54
86. Omondi PA, Awange JL, Ogallo LA, Okoola RA, Forootan E (2012) Decadal rainfall variability modes in observed rainfall records over East Africa and their relations to historical sea surface temperature changes. J Hydrol 464–465:140–56
87. Omondi PA, Ogallo LA, Anyah RO, Muthama JM, Ininda J (2012) Linkages between global sea surface temperatures and decadal rainfall variability over Eastern Africa region. Int J Climatol. http://dx.doi.org/10.1002/joc.3578
88. Omondi P, Awange JL, Ogallo L, Ininda J, Forootan E (2013) The influence of low frequency sea surface temperature modes on delineated decadal rainfall zones in Eastern Africa region. Adv Water Res 54:161–180. https://doi.org/10.1016/j.advwatres.2013.01.001
89. Owiti OZ (2005) Use of the Indian Ocean dipole indices as predictor East African rainfall anomalies. MSc thesis, Department Of Meteorology, University of Nairobi, Kenya
90. Owiti Z, Zhu W (2012) Spatial distribution of rainfall seasonality over East Africa. J Geogr Region Plan 5(15):409–21. https://doi.org/10.5897/JGRP12.02
91. Owiti Z, Ogallo L, Mutemi J (2008) Linkages between the Indian Ocean dipole and East African seasonal rainfall anomalies. J Kenya Meteorol Soc 2(1):3–17
92. Parry ML, Canziani OF, Palutikof JP, van der Linden PJ, Hanson CE (2007) Contribution of working group II to the fourth assessment report of the intergovernmental panel on climate change. Cambridge University Press, Cambridge, UK, New York, NY, USA, p 2007
93. Percival DB, Walden AT (1993) Spectral analysis for physical applications. Cambridge University Press [Definition of spectrum in context of a sinusoidal model, Eqs (1.6)–(1.16)]
94. Pohlmann H, Jungclaus JH, Köhl A, Stammer D, Marotzke J. Initializing decadal climate predictions with the GECCO oceanic synthesis: effects on the North Atlantic. J Clim 3926–3938 https://doi.org/10.1175/2009JCLI2535.1
95. Power S, Folland C, Colman A, Mehta V (1999) Inter-decadal modulation of the impact of ENSO on Australia. Clim Dyn 15:319–24
96. Rao SA, Yamagata T (2004) Abrupt termination of Indian Ocean dipole events in response to intraseasonal disturbances. Geophys Res Lett 31:L19306. https://doi.org/10.1029/2004GL020842
97. Rao SA, Behera SK (2005) Subsurface influence on SST in the tropical Indian Ocean: structure and interannual variability. Dyn Atmos Ocean 39:103–35
98. Rao SA, Masson S, Luo JJ, Behera SK, Yamagata T (2007) Termination of Indian Ocean dipole events in a general circulation model. J Clim 20:3018–35. https://doi.org/10.1175/JCLI4164.1
99. Reason CJC, Landman W, Tennant W (2006) Seasonal to decadal prediction of Southern African climate and its links with variability of the Atlantic Ocean. Am Meteorol Soc. https://doi.org/10.1175/BAMS-87-7-941
100. Reynolds RW (1988) A real-time global sea surface temperature analysis. J Clim 1:75–86
101. Reynolds RW, Marsico DC (1993) An improved real-time global sea surface temperature analysis. J Clim 6:114–9
102. Reynolds RW, Smith TM (1994) Improved global sea surface temperature analyses using optimum interpolation. J Clim 7:929–48
103. Reynolds RW, Rayne NA, Smith TM, Stokes DC, Wang W (2002) An improved in-situ and satellite SST analysis for climate. J Clim 15:1609–25
104. Richman MB (1986) Rotation of principal components. Review article. J Clim 6:293–335

105. Ropelewski CF, Halpert MS (1987) Global and regional scale precipitation patterns associated with the El-Nino/Southern Oscillation. Mon Weather Rev 115:1606–26
106. Saji NH, Goswami BN, Vinayachandran PN, Yamagata T (1999) A dipole mode in the tropical Indian Ocean. Nature 401:360–3
107. Saji NH, Yamagata T (2003) Possible impacts of Indian Ocean dipole events on global climate. Clim Res 25:151–69
108. Saji NH, Yamagata T (2003) Interference of teleconnection patterns generated from the tropical Indian and Pacific Oceans. Clim Res 25:151–69
109. Schreck JC, Semazzi FHM (2004) Variability of the recent climate of Eastern Africa. Int J Climatol 24:681–701. https://doi.org/10.1002/joc.1019
110. Seager R, Kushnir Y, Ting M, Cane M, Naik N, Miller J (2008) Would advance knowledge of 1930s SSTs have allowed prediction of the dust bowl drought? J Clim 21:3261–81
111. Segele TZ, Lamb PJ (2005) Characterization and variability of Kiremt rainy season over Ethiopia. Meteorol Atmos Phys 89:153–80
112. Singhrattna N, Rajagopalan B, Clark M, Kumar KK (2005) Seasonal forecasting of Thailand summer monsoon rainfall. Int J Climatol 25:649–64
113. Smith TM, Reynolds RW (2002) Bias corrections for historic sea surface temperatures based on marine air temperatures. J Clim 15:73–87
114. Smith TM, Reynolds RW (2003) Extended reconstruction of global sea surface temperatures based on COADS data (1854–1997). J Clim 16:1495–510
115. Smith TM, Reynolds RW (2004) Improved extended reconstruction of SST (1854–1997). J Clim 17:2466–77
116. Smith DM, Cusack S, Colman AW, Folland CK, Harris GR, Murphy JM (2007) Improved surface temperature prediction for the coming decade from a global climate model. Science 317:796–9
117. Smith TM, Reynolds RW, Peterson TC, Lawrimore J (2008) Improvements to NOAA's historical merged land-ocean surface temperature analysis (1880–2006). J Clim 21:2283–96
118. Terray P, Dominiak S (2005) Indian Ocean Sea surface temperature and El'Niño Southern Oscillation. J Clim 18:1351–68
119. Trenberth KE, Stepaniak DP (2001) Indices of El Niño evolution. J Clim 14:1697–701
120. Trenberth KE, Caron JM, Stepaniak DP, Worley S (2002) The evolution of ENSO and global atmospheric temperatures. J Geophys Res 107:4065. https://doi.org/10.1029/2000JD000298
121. Tozuka T, Luo JJ, Mason S, Yamagata T (2007) Decadal modulations of the Indian Ocean dipole in the SINTEX-F1 coupled GCM. J Clim 20:2881–94
122. Tyson PD, Preston-Whyte RA (2000) The weather and climate of southern Africa. Oxford University Press, Southern Africa, p 396
123. Valsala VK, Ikeda M (2007) Pathways and effects of the Indonesian throughflow water in the Indian Ocean using particle trajectory and tracers in an OGCM. J Clim 20(13):2994–3017
124. von Storch H, Navarra A (eds) (1995) Analyses of climate variability—applications of statistical techniques. Springer
125. Wallace JM, Smith C, Bretherton CS (1992) Singular value decomposition of wintertime sea surface temperature and 500-mb height anomalies. J Clim 5:561–76
126. Wang G, Eltahir EAB (1999) Use of ENSO information in medium- and long-range forecasting of the Nile in floods. J Clim 12:1726–37
127. Wang C (2002) Atlantic climate variability and its associated atmospheric circulation cells. J Clim 15:1516–36
128. Ward MN (1998) Diagnosis and short-lead time prediction of summer rainfall in tropical North Africa at inter-annual and multi-decadal Timescales. J Clim 1:3167–91
129. Ward MN, Folland CK (1991) Prediction of seasonal rainfall in the north nordeste of Brazil using eigenvectors of sea surface temperature. J Climatol 11:711–43
130. Wassila MT, Barnston AG, Kumar V (1999) Prediction of African rainfall on the seasonal timescale. J Geophys Res 104(D24):31589–97
131. Weare BC (1977) Empirical orthogonal analysis of Atlantic ocean surface temperatures. Quart J R Meteorol Soc 103:467–78

132. Webster PJ, Moore AM, Loschnigg JP, Leben RR (1999) Coupled ocean—atmosphere dynamics in the Indian Ocean during 1997–1998. Nature 401:356–9
133. Wilks Daniel S (1995) Statistical methods in the atmospheric sciences. Academic Press, New York, p 467p
134. Wilks DS (2006) Statistical methods in atmospheric sciences. In: International geophysics series, 2nd ed, vol 91. Elsevier, Academic Press, UK, pp 45–50, 399–402, 627pp
135. Wolter K (1987) The Southern Oscillation in surface circulation and climate over the tropical Atlantic, Eastern Pacific and Indian Ocean as captured by cluster analysis. J Appl Meteorol 26:540–58
136. Woodhouse CA, Overpeck JT (1998) 2000 years of drought variability in the central United States. Bull Am Meteorol Soc 79:2693–714
137. Worster D (1979) Dust bowl: the Southern great plains in the 1930s. Oxford University Press, New York, p 277
138. Wu L, He F, Liu Z, Li C (2007) Atmospheric teleconnections of tropical Atlantic variability: interhemispheric, tropical-extratropical, and cross-basin interactions. J Clim 20(5):856–70
139. Xie S-P, Annamalai H, Schott FA, McCreary JP Jr (2002) Structure and mechanisms of South Indian Ocean climate variability. J Clim 15:864–78
140. Xinhua C, Timothy JD (1995) Orthogonal rotation of spatial patterns derived from singular value decomposition analysis. J Clim 8:2631–43
141. Yu L, Rienecker MM (2000) Indian Ocean warming of 1997–1998. J Geophys Res 105:16923–16939
142. Zhang R, Delworth TL (2006) Impact of Atlantic multidecadal oscillations on India/Sahel rainfall and Atlantic hurricanes. Geophys Res Lett 33:L17712. https://doi.org/10.1029/2006GL026267
143. Zewdu TS, Lamb PJ, Leslie LM (2009) Large-scale atmospheric circulation and global sea surface temperature associations with Horn of Africa June-September rainfall. Int J Climatol 29:1075–100

Chapter 9
Extreme Temperatures and Precipitation

Extreme events not only cause property damage, injury, hunger, loss of life and threaten the existence of some species, but also drive changes in natural and human systems much more than average climate. Impacts of extreme climate change and variability lead to human suffering, particularly for the poor as witnessed, for example, during the prevalent droughts and floods in the GHA region. Overall, the future occurrence of warm days and nights are projected to be more frequent in the entire GHA, while the occurrence of cold night events is likely to decrease. The overall precipitation in the region decreased between 2002 and 2007 as seen from water storage changes. [57]. *Of interest in this Chapter is the contradiction in which temperature and evaporation increases but precipitation decreases leads to the question on "what actually causes the decline in total precipitation over GHA when there is an increase in warm days and nights"? Furthermore, the chapter seeks to understand how the decline in total precipitation and total water availability has negatively impacted agriculture over GHA on the one hand, and what measures have been established to overcome this, and whether they been effective on the other hand considering the fact that much of the agriculture in GHA is rain fed. Its overall effect on human health is also discussed.*

9.1 Summary

Special reports on climate extremes have shown evidences of changes in the patterns of climate extremes at global, regional and local scales. Understanding the characteristics of climate extremes at regional and local levels is critical not only for the development of preparedness and early warning systems, but is also fundamental in the development of any adaptation strategies. There is still very limited knowledge regarding the past, present and future patterns of climate extremes in the Greater Horn of Africa (GHA). This Chapter presents the work of [57], which was supported by the World Bank Global Facility for Disaster Reduction and Recovery (WB-GFDRR) and implemented by the World Meteorological Organization. It was organized in terms of three workshops with three main objectives; (i) analysis of daily rainfall and temperature extremes for ten countries in the GHA region using observed in-situ data

J. Awange, *Food Insecurity & Hydroclimate in Greater Horn of Africa*,
https://doi.org/10.1007/978-3-030-91002-0_9

running from 1971 to 2006, (ii) assessing whether the United Kingdom Met-office and Hadley centre Providing REgional Climates for Impact Studies (UK-PRECIS) modelling system can provide realistic representation of the past and present climate extremes as observed by available in-situ data, and (iii), studying the future regional climate extremes under different scenarios to further assess the expected changes in climate extremes. In the work of [57], use is made of the outputs of these workshops and also include post-workshop analyses to assess the changes of climate extremes within the GHA. The results show a significant decrease in total precipitation in wet days greater than 1 mm and increasing warm extremes, particularly at night, while cold extremes are decreasing. Considering a combination of geophysical models and satellite gravimetry observations from the Gravity Recovery and Climate Experiment (GRACE) mission in the frame of GRACE daily Kalman-smoothing models, for the years 2002 to 2010, a decline in total water storage variations over the GHA is observed.

9.2 Temperatures and Precipitation: Background

Extreme events not only cause property damage, injury, hunger, loss of life and threaten the existence of some species, see e.g. [22], but also drive changes in natural and human systems much more than average climate [58, 61, 72]. Impacts of extreme climate change and variability lead to human suffering, particularly for the poor as witnessed, for example, during the prevalent droughts and floods in the GHA region [49]. Analysis of changes in extreme climate events, therefore, is important due to the potentially high social, economic and ecological impacts of such events, e.g., [7, 13]. Limited availability of long records of daily climate data in some parts of the world, including the GHA, hampers efforts to analyse the impacts of climate variability/change on the frequency and severity of climate extremes [26]. Long records of daily climate data for the GHA region are discussed by [14, 18, 53–55, 81]. Availability of long-term high quality data in the region is hampered by inadequate monitoring networks; gaps in the records; a general decline of number of stations; chronic under-funding; differences in processing and quality control; and differences in data policies [81]. It is evident that the GHA countries share pronounced climatic trends [72] and variability and are vulnerable to extreme climatic conditions, e.g., [5, 22, 54, 55, 64].

It is increasingly becoming apparent that behind the ongoing research and debate on climate change, many parts of Africa are already witnessing dire consequences of erratic climatic conditions [5, 67] that are likely associated with regional climatic changes [30, 31]. This is expected to pose unprecedented challenges to most African economies that are significantly hinged on a predominantly rain-fed agriculture. Further challenges lie in understanding low frequency multi-decadal and centennial climate variability in the vastness and uniqueness of the complex African terrain and climate systems over eastern Africa [54, 55, 72].

In recent years over the GHA region, particularly in Kenya, Ethiopia and Somali, climate related extremes have been the dominant trigger of natural disasters. The region has recently witnessed frequent episodes of both excessive, e.g. [6, 35, 42, 73] and deficient rainfall, e.g. [34, 35, 43, 54, 55]. Consequently, there has been an upsurge in interest both from the scientific communities, e.g. [52, 56] and policy makers to the risk of increased extreme climatic events. A recent climate analysis for East Africa was conducted by [20] examining air temperature trends at 60 stations across Kenya. After spatially interpolating the station-based data, [20] reports finding a statistically significant upward trend in minimum temperature (Tn) in the Kenyan Highlands region. Omumbo et al. [56] found evidence of a warming trend in observed maximum, minimum and mean temperatures at Kericho (Kenya) during the period 1979–2009 using gold standard meteorological observations. An upward trend of $\sim$0.2 °C/decade was observed in all three temperature variables ($p < 0.01$). Mean temperature variations in Kericho were associated with large-scale climate variations including tropical SST ($r = 0.5$; $p < 0.01$). Other authors have also studied malaria resurgence using extreme climatic events over the region, e.g. [36, 59] among others. Analysis of rainfall, minimum and maximum temperature records showed increasing trends in annual minimum and maximum temperatures from 1951 to 2002 (0.4 and 0.2 °C/decade, respectively) but decreasing trend in rainfall from 1901 to 1950, 1951 to 2002 and 1901–2002 [21]. In analysis of rainfall seasonality index (SI), precipitation concentration index (PCI) and Modified Fournier Index (MFI), [23] found no statistically significant trends in observed rainfall for the hyper-arid region of Sudan during the common data period of 1945–2007.

Some studies using Global Climate Models (GCMs) have shown that changes in climate over the region are expected in a global warming scenario [5, 40, 66]. These are likely to include changes in the intensity, duration, and frequency of droughts and floods, heat waves, etc., and will have serious implications on agriculture, human health, as well as human activities. It is well documented that some parts of the GHA region are perennially prone to droughts and floods [34, 42, 43, 49, 53].

Although there is a lack of long records of daily data for extreme climate change detection, international collaboration is significantly improving the situation, culminating in an analysis, see, e.g. [4] that provided a near-global perspective on changing climate extremes for the Intergovernmental Panel on Climate Change (IPCC) Fourth Assessment Report [40]. Yet quantifiable information, describing how weather and climate extremes are changing over the GHA region has been unavailable. In preparation for the Intergovernmental Panel on Climate Change (IPCC) Fifth Assessment Report, a major effort by the Expert Team on Climate Change Detection and Indices (ETCCDI) was undertaken to analyse how extremes are changing over as much of the world as possible [4]. This included intensive international collaboration on data exchange and analysis, and, where data were not available, holding regional climate change workshops to generate information on extremes [4]. This global assessment initiative has greatly benefited from the contributions from a series of workshops [60] coordinated by ETCCDI which is jointly sponsored by the WMO Commission for Climatology (CCl), the Joint Commission for Oceanography and Marine Meteorology (JCOMM), as well as the Research Programme on Climate Variability and Predictability (CLIVAR). The ETCCDI workshops sought to bring participants together from countries within a data sparse region to fill in data gaps and to provide capacity building.

The availability of daily observation data is steadily improving and has led to the development of gridded regional [37] and global datasets [16]. For some areas such as the GHA region where daily data availability is still relatively poor, assembling and research using this data would be useful in decision making and policy formulation on the climate sensitive sectors of the region.

To address the issue of data shortage in GHA, therefore, and in line with global standard practice, e.g. [4], this chapter reports on the work of [57] that was intended to help fill the data gap in the region by assembling the necessary climate observations. Further, the understanding and use of regional climate downscaling tool known as PRECIS would help GHA countries design adaptation policies and reduce climate associated risks.

This chapter is divided into four parts namely; (1) the assessment of the adequacy of regional climate observations and trends for adaptation purposes, (2) the assessment of the adequacy and reliability of available model based climate projections for adaptation needs, (3) the assessment of the expected changes in climate extremes needed to assist in developing effective adaptation and climate risk management strategies, and (4), the analysis of changes in total water storage for the period 2002–2010 in order to assess vulnerability of GHA region to climatic variations over that period. The chapter is inherently regional in nature since, to be useful, regional climate models need observational support from as wide a region as possible. Climate does not recognize national boundaries, so in order to analyse and validate models, it is necessary to take a regional approach.

The chapter is organized as follows; in Sect. 9.3, hydroclimate products and the analysis methods are presented. Section 9.4 presents the temperature and precipitation trends. The chapter is then concluded in Sect. 9.7 by (i) looking at the contradiction in which temperature and evaporation increases but precipitation decreases, (ii) discussing the impacts of the decline in total precipitation and total water availability on agriculture over GHA, (iii) presents the measures that have been established to overcome this, and whether they been effective considering the fact that much of the agriculture in GHA is rain fed, and (iv), considers the overall effect on human health in GHA.

9.3 Hydroclimate Products and the Analysis Methods

9.3.1 Station Data and Quality Control

It is important to note that three workshops were organized to aid with the collection of data from the member countries that would hitherto not be possible when organized by an individual country. The workshop participants brought along with them selection of their best quality digitized daily temperature and precipitation series (Fig. 9.1).

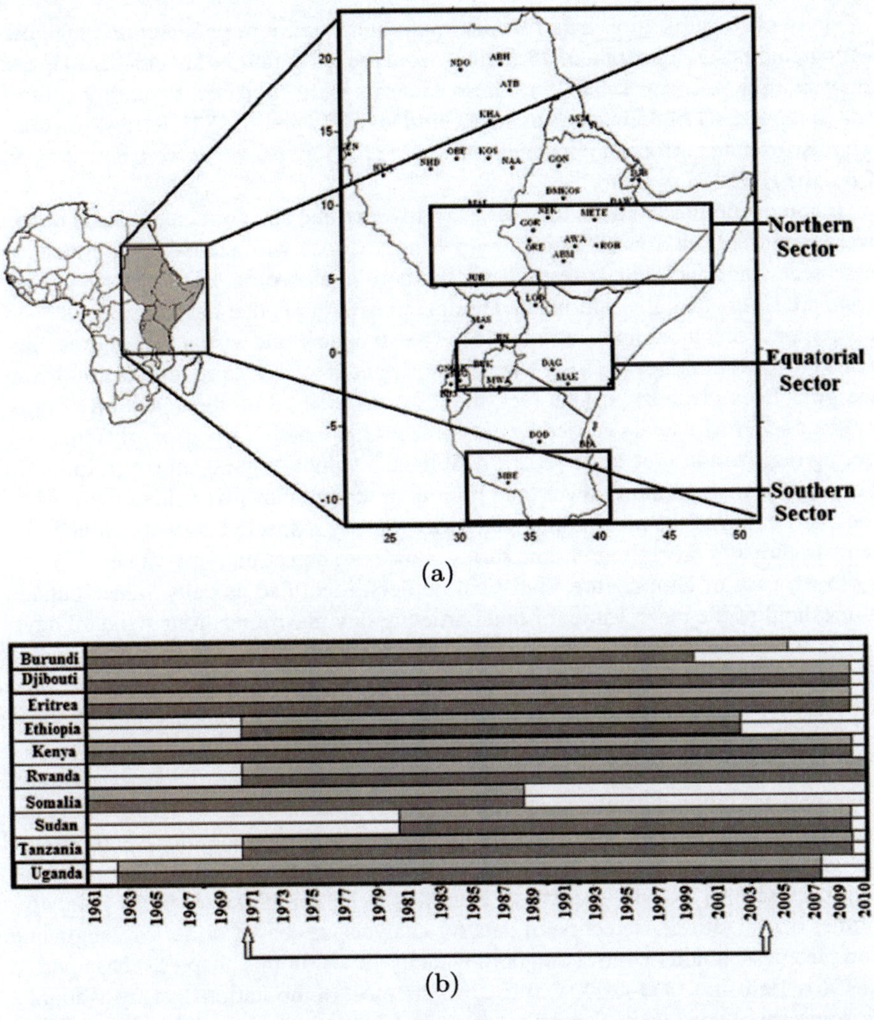

Fig. 9.1 **a** Location of the Greater Horn of Africa region in Africa together with stations used in the study (see Table 9.1 for full names, the values of latitude and longitude for each station), and **b**, available data for the time series used in the analysis. White indicates no data; upper box indicate rainfall while the lower one indicates temperature. The arrows show the common years available for all the countries. *Source* [57]

Daily observed station data for maximum and minimum temperatures together with precipitation for a total of 73 stations from the ten countries are employed in the analysis (Fig. 9.1a and Table 9.1). These data sets were subjected to quality control and homogeneity of their series using RClimDex software [61, 84]. RHtest software was also used to perform homogeneity tests [2, 3]. These packages were downloaded from the ETCCDI website.[1]

Inhomogeneities in all the time series were identified and corrected. Linear trends were identified via a least-squares regression analysis with statistical significance assessed using a two-tailed t-test. These 'standard' meteorological observations were compared with spatially interpolated temperature datasets that have been developed for regional or global applications [56]. The statistical and visual procedures contained in the RClimDex package were complemented by the required tests, following the guidelines given in [15]. In fact, the tests are focused on the detection of non-systematic errors usually caused by data processing, which happens most frequently during digitization procedure [2, 3]. Ambiguous values (e.g. negative precipitation or maximum temperature lower than minimum temperature) were identified. Also, the distribution of the precipitation data was visually inspected, as were plots of the temperature and precipitation time series in order to detect outlying values.

In the case of temperature, statistical outliers, identified as daily values outside a threshold of the mean value for that particular day plus/minus four standard deviations were also flagged. The suspicious data were validated, set to missing value or corrected on the basis of subjective inspection of partial time series for the adjacent days at the same period with other years and by spatial comparison with those available from close neighbouring stations.

Once the data passed the quality control checks, they were evaluated for homogeneity. The station data underwent homogeneity testing using the RHtest software package [2, 3], which helps in identifying step changes in a time series by comparing the goodness of fit of a two-phase regression model with that of a linear trend for the entire series [76, 78, 79]. RHtest is used to help identify series break points for further investigations. Selection of data for analysis was based on series' length and completeness, quality control and homogeneity [2, 3]. In this chapter, a base period of 1961–1990 and 1971–2000 was used where most of the station data are available. In some cases such as in Somalia, Sudan and Rwanda; a shorter base period was used; for example, the data supplied for Gikongoro in Rwanda began in 1967, a base period of 1967–1996 was used. To be included in the analysis, time series need to be sampled for at least 30 years and contain fewer than 10% of missing/rejected values. The reference period of 1971–2000 was chosen to maximize the number of stations with available data for calculation of the percentile-based indices. Even with this maximization, only 8 countries with a total 58 of the 73 original stations had long enough homogeneous periods to be included in the analysis, and not all the indices were calculated for all the stations. Figure 9.1a shows the location of stations and Fig. 9.1b the data availability period for these stations.

[1] http://cccma.seos.uvic.ca/ETCCDI/.

Table 9.1 List of stations (see Fig. 9.1 for their exact locations). Note that the data for Somalia are missing. *Source* [57]

No.	Station	STN	Code	Longitude	Latitude	Period
1	ABU HAMAD	ABH	62640	33.3	19.5	1960–2002
2	ARBA MINCH	ABM	63500	37.7	6.1	1986–2007
3	ASMARA	ASM	63021	38.9	15.3	1960–2009
4	ATBARA	ATB	62680	34.0	17.7	1960–2000
5	AWASSA	AWA	63460	38.5	7.1	1971–2007
6	BUJUMBURA	BUJ	64390	29.3	−3.3	1970–2010
7	BUKOBA	BUK	63729	31.8	−1.3	1960–1991
8	GIKONGORO	GIK		30.1	−1.6	1967–2010
9	GISENYI	GNY		29.3	−1.7	1971–2010
10	KAMEMBE	KMBE		28.9	−2.5	1971–2010
11	METE	METE		39.9	8.87	1983–2007
12	NAAMA	NAA		34.08	12.44	1963–2000
13	NAHOUD	NHD		28.26	12.42	1960–2000
14	OBIED	OBE		30.14	13.1	1960–2000
15	ORE	ORE		35.53	8.15	1951–2007
16	DAGORETTI	DAG	63741	36.8	−1.3	1960–1991
17	DAR.I.AIRP.	DIA	63894	39.2	−6.9	1971–2009
18	DEBRE MARCOS	DMKOS	63334	37.7	10.4	1970–2003
19	DIRE DAWA	DDAWA	63471	41.9	9.6	1970–2003
20	DJIBOUTI	DJB	63125	43.2	11.6	1979–2010
21	DODOMA	DOD	63862	35.8	−6.2	1971–2009
22	GENEINA	GEN	62770	22.5	13.5	1960–2001
23	GONDAR	GON	63331	37.4	12.5	1970–2003
24	GORE	GOR	63403	35.6	8.2	1970–2003
25	JINJA	JIN	63682	33.2	0.5	1970–2003
26	JUBA	JUB	62941	31.4	4.5	1960–2000
27	KHARTOUM	KHA	62721	32.6	15.6	1960–2000
28	KIGALI	KIG	64387	30.1	−2.0	1971–2010
29	KOSTI	KOS	62772	32.4	13.1	1960–2000
30	LODWAR	LOD	63612	35.6	3.1	1960–1991
31	MAKINDU	MAK	63766	37.8	−2.3	1960–1991
32	MALAKAL	MAL	62840	31.7	9.6	1960–2000
33	MASINDI	MAS	63654	31.7	1.7	1960–1991
34	MBEYA	MBE	63932	33.5	−8.9	1970–2009
35	MWANZA	MWA	63756	32.9	−2.5	1960–1991
36	NDONGOLA	NDO	62650	30.5	19.2	1960–2000
37	NEKEMTE	NEK	63340	36.6	9.1	1952–2007
38	NYALA	NYA	62790	24.9	12.1	1960–2000
39	ROBE	ROB	63474	40.0	7.1	1983–2007
40	SOMALIA	SOM				

Of the ten countries that had participants in the workshop, Somalia and Sudan had data time series falling outside the 1971–2000 period (Fig. 9.1b). Since many of the stations had data problems prior to 1970, the analysis was limited to the period 1971–2009. For percentile-based indices (e.g. the number of days exceeding the 90th percentile of minimum temperature), the methodology uses bootstrapping for calculating the baseline period values in order to avoid discontinuities in the indices time series at the beginning or end of the base period following the approach by [83].

The PRECIS modelling system was applied over the region to develop high-resolution climate scenarios. It was driven with initial and lateral boundary conditions using the UK Met Office Hadley Center Regional Climate Model (HadRM3P), which is a high-resolution atmospheric component of the Hadley Centre coupled ocean-atmosphere GCM-HadCM3, with a resolution of 1.875° in longitude and 1.25° in latitude. Details of the Global Climate Models (GCMs) used in verification of the PRECIS Regional Climate Model (RCM) can be obtained in the study of [68].

9.3.2 *Gravity Recovery and Climate Experiment (GRACE)*

Global and regional water cycles are related to different phenomena within the Earth system, including variations in atmosphere, hydrosphere, ice cover and land surface in various ways, while ranging from sub-seasonal and inter-annual to decadal and secular interactions. These all make it difficult to develop realistic models to simulate or predict water variations.

The gravity field of the Earth and its temporal variations, measured globally by GRACE (a joint US–German satellite mission launched in March 2002, see Sect. 3.3.3; however, are interrelated to total water storage (TWS) variations of the Earth [74]. This offers the unique opportunity to detect spatio-temporal variations of water within the Earth system from space [69, 70].

The GRACE mission consisted of two identical spacecrafts flying approximately 220 km apart in the same near-polar orbit of about 450 km [69, 70]. The main observable is the distance between the two satellites, measured using a microwave ranging system. Additional tracking information is provided by Global Positioning System (GPS) receivers on board each of the spacecraft and Satellite Laser Ranging (SLR) reflector. On-board accelerometers sense non-conservative forces such as atmospheric drag and solar radiation pressure. Time-variable gravity field solutions are obtained by the exploitation of GRACE observation data over certain time intervals, namely daily to monthly gravity field solutions. The solutions are computed in terms of spherical harmonics (SHs), see details in Sect. 3.3.3. GRACE time-variable products have been frequently used to study water variations and their relations to climate change, as documented, for example, by [8–10, 12, 29, 62].

This chapter made use of the ITG-GRACE2010 daily solutions [46], which were computed up to the SHs of degree and order 40, covering October 2002 to September 2009. For computation of such daily fields, the WaterGAP global hydrology model (WGHM), the atmospheric model European Centre for Medium-Range Weather

Forecasts (ECMWF), and the ocean circulation model (OMCT) have been used, within the framework of a Kalman-smoother procedure, to derive the temporal correlations while using GRACE observations as the main information for computation of the gravity models. Details of computations can be found in the study of [45].

The priority of using daily solutions for studying water variations when it is compared to those of monthly solutions is that it allows to recover fast gravity field (its equivalent water) variations as detailed as possible with reasonable temporal resolution. For studying TWS variation over the GHA, first, 2588 daily gravity solutions covering October 2002 to September 2009 were downloaded from the official website of the Astronomical Physical und Mathematical Geodesy (APMG) group at Bonn University.[2] These fields were then used to generate the global TWS values according to the approach of [74] and the boundary of the GHA as shown in Fig. 9.1a was extracted from the fields. Finally, spatially averaged TWS along with their corresponding cumulated TWS variations were computed (see Fig. 9.6b). Daily GRACE products, in this study, have been used to examine the impact of climate variability on total water storage variations for recent years (2002–2009) when the station data sets were not available. In fact, TWS variability reflected in GRACE data indicates the prolonged impact of decadal climate change over the region. Discussions on the use of GRACE-TWS for drought monitoring, for instance, can be found in the study of [38].

9.3.3 Extreme Climate Analysis: Trend and Indices

All stations from different countries were analysed. Trends for individual stations were calculated by adapting Sen's [65] slope estimator. This method has been applied in other similar works describing extreme indices [2, 3, 17, 83] and also adapted to climatological data by [82] in a study of annual temperatures over Canada and by [80] in their analysis of extreme wave heights over the Northern Hemisphere. Trends are significant at the 5% level when results are ± 1.96 standard deviations from the median trend. At least 70% of annual data are required to be non-missing to calculate a trend and refer to trends as being significant if they were determined to be statistically significant at the 5% level. To avoid biased estimates, station level trends were not calculated for series with excessive missing values.

A total of 27 indices, based upon recommendations of the ETCCDI, were calculated using RClimDex [16, 17]. Many of the indices use locally defined thresholds, making it easier to compare results over a wide region. The indices were primarily based on station level thresholds calculated over a base period, such as the 90th percentile of minimum temperature. These thresholds were determined for each day of the year using data from that day and two days on either side of it over the course of the base period. Table 9.2 lists the indices presented in this chapter while detailed

[2] http://www.igg.uni-bonn.de/apmg/index.php?id=itg-grace2010.

Table 9.2 List of stations (see Fig. 9.1 for their exact locations). Full definitions are available from the ETCCDI website http://cccma.seos.uvic.ca/ETCCDI/. Note that the data for Somalia are missing. *Source* [57]

ID	Indicator name	Description	Units
FD*	Frost days	Count of days where TN (daily minimum temperature)<0 °C	d
SU	Summer days	Count of days where TX (daily maximum temperature)>25 °C	d
ID	Icing days	Count of days where TX<0 °C	d
TR	Tropical nights	Count of days where TN>20 °C	d
GSL*	Growing season length	Annual count of days between first span of at least six days	d
		where TG (daily mean temperature)>5 °C and first span in	
		second half of the year of at least six days where TG<5 °C	
TXx		Monthly maximum value of daily maximum temperature	°C
TNx		Monthly maximum value of daily minimum temperature	°C
TXn		Monthly minimum value of daily maximum temperature	°C
TNn		Monthly minimum value of daily minimum temperature	°C
TN10p	Cold nights	Count of days where TN<10th percentile	%
TX10p	Cold day-times	Count of days where TX<10th percentile	%
TN90p*	Warm nights	Count of days where TN>90th percentile	%
TX90p	Warm day-times	Count of days where TX>90th percentile	%
WSDI*	Warm spell duration index	Count of days in a span of at least six days where TX>90th percentile	%
CSDI	Cold spell duration index	Count of days in a span of at least six days where TN>10th percentile	d
DTR	Diurnal temperature range	Mean difference between TX and TN (°C)	
RX1day	maximum 1-d Precipitation	Highest precipitation amount in 1-d period	mm
RX5day*	Maximum 5-d Precipitation	Highest precipitation amount in 5-d period	mm
SDII*	Simple daily intensity index	Mean precipitation amount on a wet day	mm
R10mm*	Heavy precipitation days	Count of days where RR (daily precipitation amount)≥10 mm	d
R20mm	Very heavy precipitation days	Count of days where RR≥20 mm	d
Rnnmm		Count of days where RR ≥ user-defined threshold in mm	d
CDD*	Consecutive dry days	Maximum length of dry spell (RR < 1 mm)	d
CWD	Consecutive wet days	Maximum length of wet spell (RR ≥ 1 mm)	d
R95p TOT*		Precipitation due to very wet days (>95th percentile)	mm
R99pTOT		Precipitation due to extremely wet days (>99th percentile)	mm
PRCPTOT		Total precipitation in wet days (>1 mm)	mm

descriptions of the indices and the exact formulae for calculating them are available on the ETCCDI web page.[3]

All the indices are essentially anomalies from the same base period. However, some precipitation indices could potentially be dominated by those stations with the greatest precipitation, as those stations may see precipitation vary from year to year by more than the total annual precipitation at stations with the least total precipitation [3]. To determine whether this was the case for the stations analysed, precipitation indices were also calculated by first standardizing the indices (dividing by the index's standard deviation). As a comparison of both approaches revealed similar shape and trends, the standardized indices were not used and the results are provided through the analysis of the simple anomaly series.

9.3.4 Modelling of Extreme Rainfall and Temperature

Using the PRECIS regional climate modelling system, [57] analysed the distribution of extremes of temperature and precipitation in GHA in the recent past (1961–1990) and in a future (2071–2100) climate under the IPCC SRES A2 and B2 emissions scenarios [40], Chap. 19, p. 18. Rainfall and temperature were simulated by PRECIS RCM and compared with observations (in areas in which data were available) for the period 1961–1990. The main task was to evaluate the simulations of climate (particularly precipitation) of regional climate model by comparing them with available data, and thereby assessing the uncertainty associated with future climate predictions. The UK-Met Office high resolution PRECIS model runs were compared with available data to assess its performance. The model was used to generate future climate projections to demonstrate how they can be used and interpreted at the national level. Emphasis was placed on determining the extremes, trend and the variance explained by each model. The purpose of the simulation of regional climate was to examine and compare statistics relevant to the region in observations extremes and regional circulation model. This further demonstrated the value of climate observations and regional models for decision making, to provide advice on model performance and limitations, and to improve capabilities across the region for using climate data records and model projections.

9.3.5 Advanced Statistical Analysis of GRACE's Water Storage Products

Independent component analysis (ICA) is a higher-order statistical technique, which can be viewed as an extension to the commonly used principle component analysis (PCA) [11, 27, 28]. Using an ICA algorithm, the input data (spatio-temporal

[3] http://cccma.seos.uvic.ca/ETCCDI/.

observations) are assumed to consist of a linear mixture of unknown source signals, which cannot be directly measured. By incorporating higher order statistical information contained in the data in the decomposition procedure, ICA extracts statistically independent components that reflect spatial and temporal manifestations of physical processes hidden in the data [33, 47].

Generally, there are two alternative ways to implement ICA on a temporal sequence of gridded datasets in which either temporally independent components or spatially independent time series can be estimated. The methods are respectively called temporal ICA and spatial ICA, see e.g. [27, 29] for details. Omondi et al., [57]'s work presented in this chapter made use of temporal ICA method, since the scope of their study was to extract the temporal behaviour of TWS changes for the period of October 2002 to September 2009 in order to further investigate the impacts of rainfall after the long-term study of 1970–2000.

9.4 Temperature and Precipitation Trends

9.4.1 Trends in Temperature Indices

Trends for the temperature indices for some selected countries are shown in Table 9.3 in comparison to global and other regional indices. The two countries, that is, Ethiopia and Kenya were used for comparison purposes since their data coverage was relatively good and trends calculated included similar window period of 1971–2003. The warm extremes were increasing while the cold extremes were decreasing, thus these series clearly indicated significant warming. Individual stations show most spatial coherence in the TN90p index, that is, frequency of nights warmer than the 90th percentile. Nearly half of the available stations indicated a significant increase in this index over the period 1971–2004. Sample time series for the percentile-based temperature indices are shown in Fig. 9.2 for Asmara in Eritrea. The frequencies of warm days and nights, relative to the base period 1961–1990, increased strongly between 1961 and 1990, with a large increase in the number of nights per year exceeding the 90th percentile threshold. There were also large reductions in the frequency of cold nights and cold days over the 49 years. The warmest day and night of the year is warming at a rate approximately comparable to the global average.

In general, over the entire region, the frequency of warm days and warm nights has increased, and the frequency of cold days and cold nights has decreased. This agrees with the results from other studies that have analysed these trends across different parts of the world [17, 19, 32, 44]. However, the results for the absolute temperature indices (TXx, TNx, etc.) defined for the entire region were sensitive to the large variability in these indices across the region. The percentile indices (e.g. TN90p) are more robust across large regions because they accounted for the influence of local climate effects. There has been a significant increase in the absolute annual maximum of both daily maximum and minimum temperatures, again in common

Table 9.3 Regional trends in temperature indices. The trends for the globe area from [4, 17] are based on the time period 1955–2003. A trend significant at the 5% level is marked with bold font. *Source* [57]

Index	Guinea	Central Africa	Zimbabwe	Global	Kenya	Ethiopia	Units
Warmest day	**0.14**	**0.25**	0.15	**0.21**	**0.35**	0.11	°C/decade
Warmest night	**0.17**	**0.21**	0.10	**0.30**	0.17	**0.33**	°C/decade
Coldest day	**0.23**	**0.13**	0.00	0.37	0.02	0.10	°C/decade
Coldest night	0.04	**0.23**	0.02	**0.71**	**0.21**	**0.32**	°C/decade
DTR	0.12	0.00	0.11	-0.08	0.22	0.61	°C/decade
Cold Night frequency	−0.21	**−1.71**	**−1.24**	**−1.26**	**−1.10**	**−1.23**	% of days in a year/decade
Cold Day frequency	**−2.15**	**−1.22**	**−1.05**	**−0.62**	**−1.6**	**−1.0**	% of days in a year/decade
Warm night frequency	**1.19**	**3.24**	0.71	**1.58**	**1.44**	**2.14**	% of days in a year/decade
Warm day frequency	**1.56**	**2.87**	**1.86**	**0.89**	**1.07**	0.65	% of days in a year/decade

with the global picture [17, 19, 32, 44]. The coldest day and night of the year is warming slower than the global average, although planetary trend for the coldest day is not significant.

9.4.2 *Trends in Precipitation Indices*

The map in Fig. 9.3b depicts total precipitation in wet days (>1 mm) (PRCPTOT), and is an example indicating relative lack of spatial coherence for precipitation in the region. The number of stations with significant negative or positive trends is low. Sample graphs of trends, calculated for various precipitation indices are shown in Figs. 9.3a, 9.4a, b, while Table 9.4 compares similar trends at regional and global scales [4, 17] at similar window period. Western Lake Victoria, southern Sudan and western Ethiopia generally show significant decreases in total precipitation (e.g., see Fig. 9.3b). For Asmara (Fig. 9.3a) and Djibouti (Figure not shown), there are sharp drops in the total annual precipitation time series around 2000–2010. Likely associated with the decrease in total precipitation, the length of the maximum number of consecutive dry days is increasing in Asmara and Djibouti, while the length of the

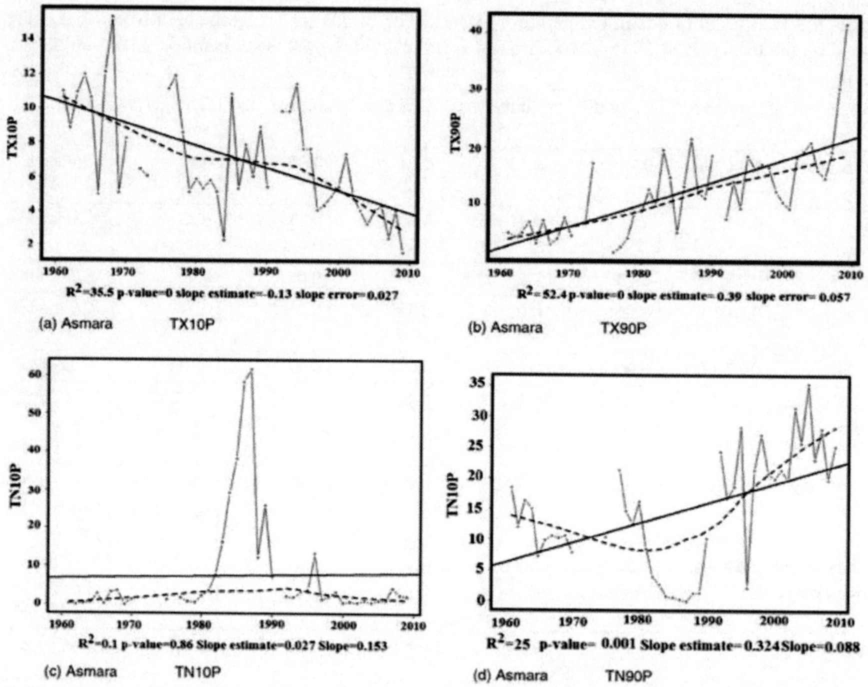

Fig. 9.2 Time series 1971–2010 for **a** cold days (TX10P), **b** warm days (TX90P), **c** cold nights (TN10P) and **d** warm nights (TN90P) (units: %). Bold line indicates ordinary least squares fit for Asmara, Eritrea. *Source* [57]

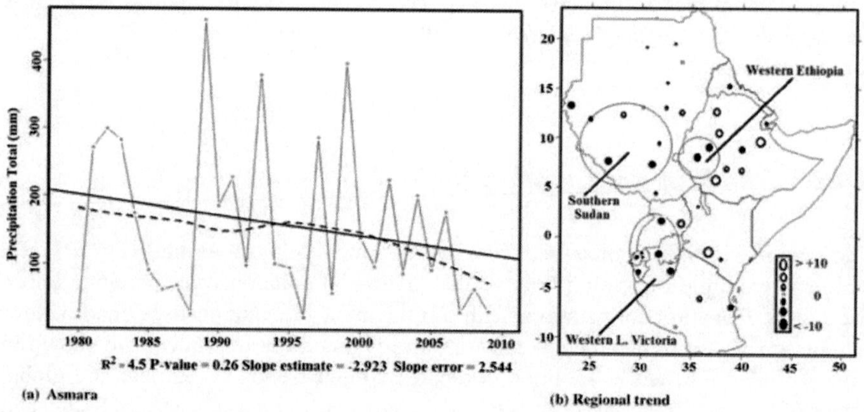

Fig. 9.3 Precipitation total (PRCPTOT). Individual station's time series and regional trend **a** Individual time series 1980–2010 for Asmara in Eritrea, and **b** regionally averaged station trends. Positive (negative) trends are shown in circles. Large (small) circles indicate significant (insignificant) trends. *Source* [57]

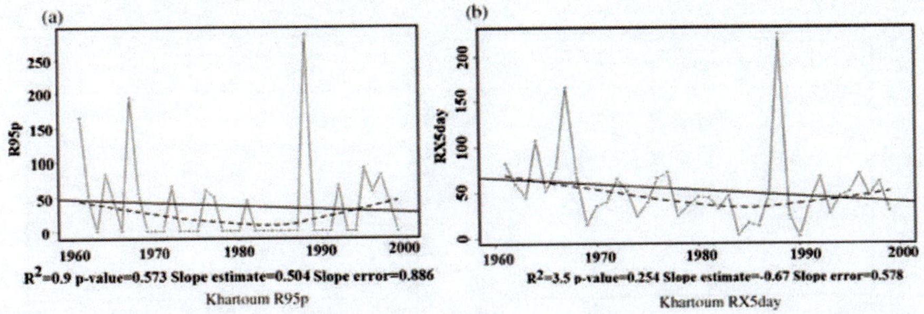

Fig. 9.4 Time series 1961–2000 for **a** contribution from very wet days (R95p, units: mm), and **b** annual maximum 5-d precipitation amounts (RX5day, units: mm) for Khartoum. Bold lines indicate ordinary least squares fit. *Source* [57]

Table 9.4 Regional trends in precipitation indices. The trends for the globe area from [4, 17] are based on the time period 1971–2003. A trend significant at the 5% level is marked with bold font. *Source* [57]

Index	Indian Ocean	Himalayas	Indo-Pacific	Global	Northern sector	Equatorial sector	Southern sector	Units
PRCPTOT	**81.84**	**41.77**	−2.86	5.91	−2.92	−0.85	10.3	mm/decade
SDII	**1.05**	**1.55**	**0.25**	0.05	−0.81	−0.89	−0.13	mm/d/decade
CDD	0.66	2.61	−1.01	**−1.19**	0.37	0.32	0.45	d/decade
CWD	0.10	−0.24	−0.13	**−0.07**	−0.05	−0.50	−0.07	d/decade
RX1day	1.12	1.70	−1.12	0.26	0.48	−0.33	−0.72	mm/decade
RX5day	5.96	16.39	0.90	**0.73**	0.67	-0.67	0.12	mm/decade
R10mm	2.09	0.00	−0.14	**0.03**	−0.29	−0.28	0.35	d/decade
R20mm	1.26	0.53	−0.04	0.06	−0.01	−0.04	0.14	d/decade
R95p	22.66	**82.30**	12.24	**4.68**	12.9	−0.50	1.79	mm/decade
R99p	−12.61	32.39	4.98	**3.38**	51.1	−2.30	8.05	mm/decade

maximum number of consecutive wet days shows a significant decrease (figure not shown). The Simple Daily Intensity Index (SDII), which takes into account the number of days with rainfall greater than or equal to 1 mm shows no significant changes. In general, decreasing trend in total precipitation in wet days (>1 mm) is observed in the north western sector (western Ethiopia and southern Sudan), and equatorial sector around Lake Victoria, while much of Ethiopia had significant positive increase (Fig. 9.3b). The precipitation due to very wet days greater than 95th percentile (R95p) index (Fig. 9.4a) indicate that the annual amount of precipitation contributed on days exceeding the long-term 95th percentile has decreased from about 50 mm to around 30 mm in Khartoum, but this change is non-significant (Fig. 9.4a). The highest precipitation amount in 5-d period or maximum 5-d precipitation (RX5day) index (Fig. 9.4b) depicts significant reduction over the 40-year period.

Similar time series of R95p for the southern sector is represented by Dodoma (Fig. 9.5a). There are increases in R95p over the southern sector, a marginal decrease

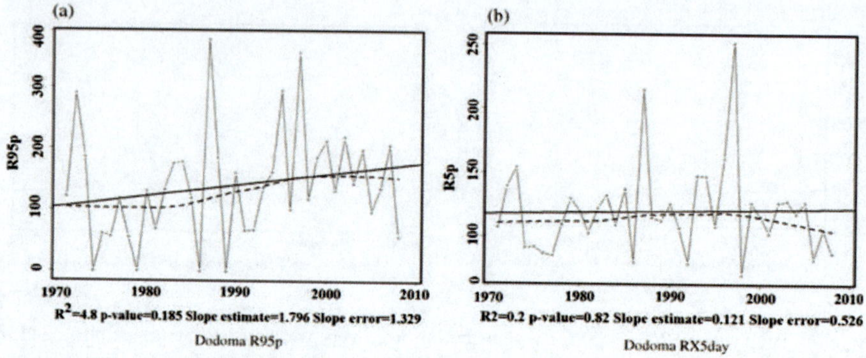

Fig. 9.5 Time series 1970–2010 for **a** contribution from very wet days (R95p, units: mm), and **b** annual maximum 5-d precipitation amounts (RX5day, units: mm) for Dodoma. Bold lines indicate ordinary least squares fit. *Source* [57]

over the equatorial sector, and a decrease over the northern sub-region. Between them, only the change within the southern sector is statistically significant. Table 9.4 lists the regional trends for the precipitation indices and also the global trends. The same problems exist with defining some of the precipitation indices across the whole region that applied to the absolute temperature indices, and indices defined relative to a local climatology (e.g. percentile based) are preferable for comparing across such a large region.

Compared to the temperature indices, there are fewer significant trends in the precipitation indices. In contrast to the other sub-regions, the northern sector has decreasing trends in all precipitation indices, apart from the consecutive dry day index, suggesting a consistent change towards drier conditions. However, it must be emphasized that these trends are non-significant. Over the region as a whole, the precipitation trends are mixed [30, 31]. This does not parallel the global results of [17] indicating consistent trends towards wetter conditions across nearly all of the indices, although it should be noted that analysis was for a different time period (1971–2005) and had only limited coverage of the tropics.

9.4.3 Relationship Between Precipitation and TWS Changes

Regarding the precipitation results, it was clear that the overall precipitation over the GHA is declining. To support the precipitation results of the last 7 years (2002–2009), daily TWS products were used as described in Sect. 9.3.2. The goal was to see whether the total water availability of the region was affected by climate variations or not. As a matter of fact, TWS tells quite more sophisticated story of water variations over the GHA region by providing information on daily precipitation minus evaporation minus run-off over the region. The results of spatially averaged TWS over the GHA

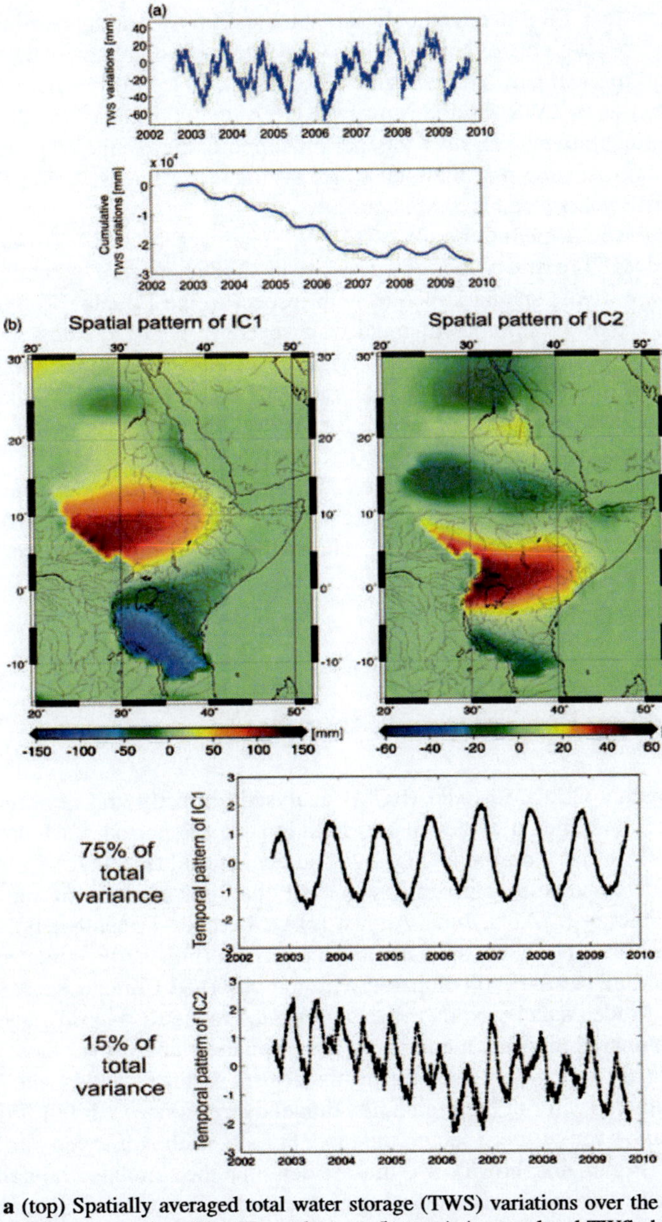

Fig. 9.6 **a** (top) Spatially averaged total water storage (TWS) variations over the GHA, derived from daily Kalman-smoother GRACE products, **a** (bottom) Accumulated TWS changes over the GHA. **b** Implementing the temporal ICA method over 2588 d of TWS maps over the GHA, computed from Kalman-smoother daily GRACE products provided by the APMG group at Bonn University. Independent patterns are ordered according to the variance they represent. One can reconstruct each mode of TWS variability by multiplying the spatial patterns with their corresponding temporal components. *Source* [57]

(Fig. 9.6) show that TWS declined between 2002 and 2007. An increase in TWS in 2007 is associated with the El Niño southern oscillation phenomenon, which usually brings rainfall to most parts of the region [39, 41, 50, 51]. This has been followed again by a decline in TWS variations over the GHA up to the end of the 2002–2009 period. The cumulative TWS over this period supports the results of precipitation (e.g., Fig. 9.3a), showing that the total water availability decreased over the GHA during the last 7 years (see Fig. 9.6b, bottom).

Applying the ICA method on GRACE-TWS data over the GHA shows that the first ICA mode (IC1) extracts 75% of variability in TWS changes. The spatial pattern of IC1 shows a dipole spatial structure with respect to the Equator. The temporal pattern of IC1 shows a dominant annual water cycle over GHA. The second ICA mode (IC2) corresponds to 15% of the variance, while the temporal pattern shows a summation of a long-term trend and an inter-annual variability. Considering the spatial pattern, therefore, a declining rate for the regions of the Lake Victoria Basin and the surrounding lakes are found. The declining pattern of TWS is also extended [see e.g., [8]] up to the south of Sudan (c.f., Fig. 9.3b). In contrast, over the tropical regions as well as Ethiopia a slight increase of TWS during 2002–2010 is seen (c.f., Fig. 9.3b extended [see e.g., [85]]). The spatial and temporal patterns of IC2, therefore, confirm the results of precipitation, which was derived for the long-term period during 1970–2000.

9.5 Modelling Precipitation Extremes

PRECIS Regional Climate Model (RCM) analysed correctly and reproduced the mean seasonal and annual cycles of precipitation for the period 1961–1990 over the southern (Fig. 9.7a), northern (Fig. 9.7b) and equatorial (Fig. 9.7c) sectors. The mean surface temperature climatology for all the four seasons of the region, that is, March–May (MAM), June–August (JJA), October–December (OND) and December–February (DJF) are spatially and temporally (Fig. 9.7d) simulated for the baseline period of 1961–1990 compared with the observed Climate Research Unit (CRU) data (gridded data based on the set aggregated to the RCM grid). The results show that, compared to CRU, the model underestimates rainfall over most parts of eastern highlands during OND while over the central sector (around Lake Victoria area) and southern sector of the region, the model overestimates rainfall (Fig. 9.7a). On the contrary, over northern sector, the model produces the observed rainfall reasonably well (Figure not shown). It is thus evident that the simulated rainfall by the PRECIS model is fairly consistent with the observed values over most parts of the study area. Results from the inter-annual variability of PRECIS simulated surface air temperature showed that during JJA season, some sections of the region recorded the lowest temperature and the warmest temperatures during the DJF season (Fig. 9.7b). The model results agree reasonably well with the observed patterns in terms of the spatial location of the extreme maximum temperatures.

The trends calculated for projected precipitation indices for the period 2010–2040 are shown in Fig. 9.8. Only central Uganda represented by Mbarara, Masindi

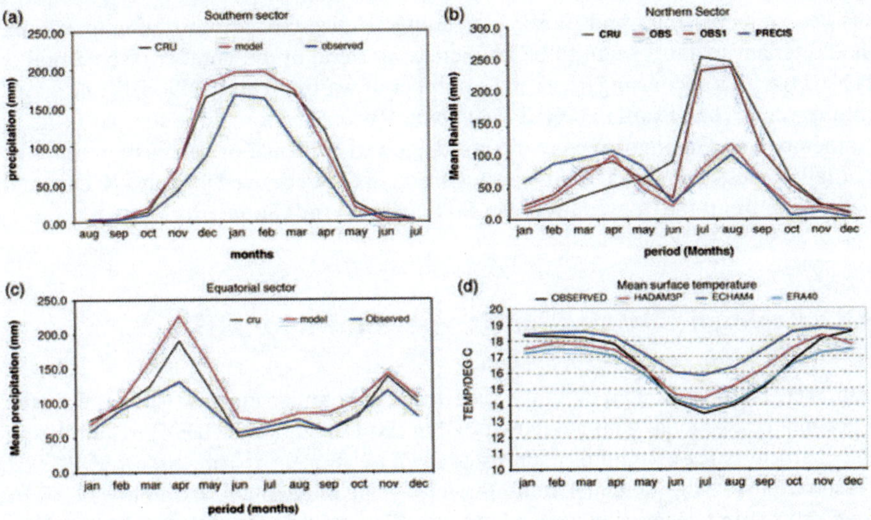

Fig. 9.7 Simulated and observed mean climate cycles for some stations located in various sectors of the region. *Source* [57]

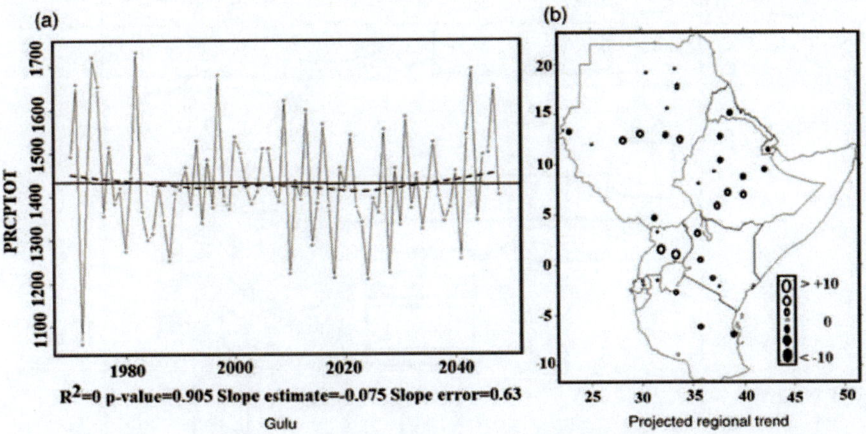

Fig. 9.8 Projected Precipitation Total (PRCPTOT in mm). Individual station time series and regional trend **a** Individual time series 2010–2040 for Gulu in Uganda **b** Regionally averaged station trends. Positive (negative) trends are shown in circles. Large (small) circles indicate significant (non-significant) trends. *Source* [57]

and Jinja show significant increases in total precipitation. Sample time series for Gulu is shown by Fig. 9.8a and regional projected trends in Fig. 9.8(b). Total annual rainfall (PRCPTOT) showed a decreasing trend for many of the stations over Sudan except for the Khartoum that had significant positive trend. Generally, consecutive wet days (CWD) showed a decreasing trend while the consecutive dry days (CDD) showed an increasing trend. While temperature indices vary from station to station, the dominant features seem to be an increasing trend in the number of cold nights (TN10p) with a decreasing trend in the number of warm nights (TN90p). Time series for projected Total Rainfall (PRCPTOT) over Rwanda, where three stations (Kigali, Kamembe and Gikongoro) were analysed showed evidence of decreasing trends in rainfall. Results show general increasing trend in Consecutive Dry Days (CDD) and decreasing trend for Consecutive Wet Days (CWD) for Gikongoro.

9.6 Regional Climate Models: Assessment for GHA

Endris et al. [24] evaluated the ability of 10 regional climate models (RCMs) from the Coordinated Regional Climate Downscaling Experiment (CORDEX) in simulating the characteristics of rainfall patterns over GHA (Fig. 9.9). They assessed RCMs' seasonal climatology, annual rainfall cycles, and interannual variability of GHA over three homogeneous subregions (North Eastern Africa-NEA (region 4), East

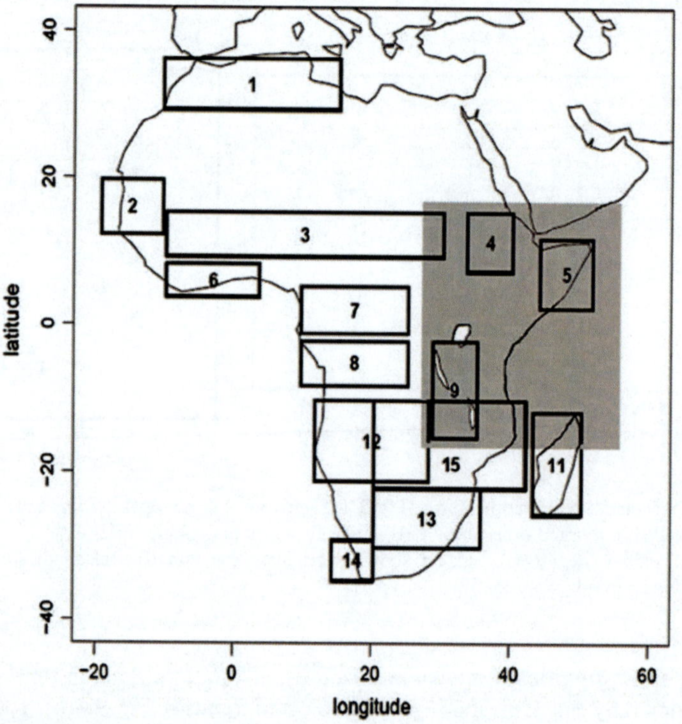

Fig. 9.9 Three GHA subregions (represented by region 4 (North Eastern Africa-NEA), 5 (East Eastern Africa-EEA), and 9 (South Eastern Africa-SEA) that were utilized by [24] to assess 10 regional climate models (RCMs). *Source* [25]

Eastern Africa-EEA (region 5), and South Eastern Africa-SEA (region 9)) against a number of observational datasets. The ability of the RCMs in simulating large-scale global climate forcing signals was further assessed by compositing the El Niño–Southern Oscillation (ENSO) and Indian Ocean dipole (IOD) events [24]. Endris et al. [24] found that most RCMs reasonably simulated the main features of the rainfall climatology over the three subregions and also reproduced the majority of the documented regional responses to ENSO and IOD climate variability indices. Significant biases in individual models were noticed depending on the subregion and season. Endris et al. [24] found that the ensemble mean had better agreement with observation than individual models, and demonstrated that the multimodel ensemble mean simulated GHA's rainfall adequately and is therefore usable in assessing future climate projections over GHA.

9.7 Concluding Remarks

A set of daily station observations from countries in the GHA region were for the first time compiled and analysed to enable assessment of changes in climate extremes over the region. Most stations showed decrease of total precipitation in wet days greater than 1 mm (PRCPTOT) as well as heavy rainy days (R10 mm), maximum 1-d precipitation (Rx1day), maximum 5-d precipitation (Rx5day), heavy precipitation days (R10 mm) and warm spell duration (WSDI). Index for warm days (Tx90P) showed increasing trend, while the index for cool days (TX10P) showed decreasing trend. The TXn index (monthly minimum value of daily maximum temperature) showed an increasing trend in most parts of the region. The TNx index (monthly maximum value of daily minimum temperature) showed an increasing trend over most parts of the region, and more significantly in the north-eastern and southern sectors of the region; indicating that there is a general increasing trend of warm nights for most of the stations in the region. Increasing trend in warm nights is indicative of significant night time warming.

Thus, increasing trends for warm nights were the most spatially coherent index consistent with the results of other regional workshops [19, 44] and the global analysis [4, 17]. Less spatial coherence trends in precipitation indices across the region and fewer trends that are locally significant when compared with the temperature indices are observed. In the few cases where statistically significant trends in precipitation indices are identified for regions and sub-regions, there is generally a trend towards wetter conditions consistent with the global results of [4].

In summary, the chapter aimed at:

(i) Assessing the adequacy of regional climate observations and trends for adaptation purposes. In this regard, the chapter found that there is inadequate in-situ data for an individual country analysis but provided sufficient ground for regional level climate analysis. The results further showed increasing/ decreasing trend in warm/cold extremes. Furthermore, frequencies of warm days and nights increased strongly, with a large increase in the number of nights per year exceeding the 90th percentile threshold between 1961 and 1990. On the contrary, precipitation patterns are mixed with fewer significant trends except significant decrease in total precipitation in wet days greater than 1mm across the whole region.

(ii) Assessing the adequacy and reliability of available model based climate projections for adaptation needs. To this end, the chapter found that the simulated climate is fairly consistent with the observed values over most parts of GHA.

(iii)) Assessing the expected changes in climate extremes needed to assist in developing effective adaptation and climate risk management strategies. Here, the chapter established that, generally, the model projected decreasing/increasing trend in consecutive wet/dry days with variations in temperature indices from one station to another. The dominant features in the region seem to be an increasing trend in the number of cold nights with a decreasing trend in the number of warm nights. Increasing trends for warm nights were the most spatially coherent index consistent with the results of other regions of the globe.

(iv) Assessing changes in the total water storage for the period 2002–2010. Here, the chapter established a decline in total water availability over the GHA region during the 2002–2009 period. The findings, therefore, showed that increasing trends in both night and day temperatures had the most spatially coherent indices. The model simulates well the spatial distribution of extreme temperature and rainfall events when compared with present climate observations, with temperature simulation being more realistic across the region. Overall, the future occurrence of warm days and nights are projected to be more frequent in the entire GHA, while the occurrence of cold night events is likely to decrease. The overall precipitation in the region decreased between 2002 and 2007 from the TWS products. The decline in total precipitation and total available water in GHA would not only negatively impact on the rain-fed agriculture but also impact on pastoralists who rely on surface water as well as irrigated agriculture that depends on groundwater (see Chaps. 1 and 13). Some of the measures put in place to overcome this are discussed in Sect. 1.3. Although these measures are normally put in place, sometimes they do not yield fruit. For example, in order to strengthen Kenya's resilience to drought and climate change, in 2016, the European Union funded the WaTER programme in Kenya focused on protecting the "Water Towers" or high-elevation forests, which acts as natural reservoirs from which Kenya draws its water supply [1]. As noble as the program was, it was suspended due to the country's forceful evictions of local communities from the forest [1]. In this case, the forceful eviction resulted in health related issues, e.g., exposure to mosquitoes and lack of safe drinking water for those evicted.

(v) For the 10 RCMs assessed over GHA, ensemble mean had better agreement with observation than individual models. Moreover, multimodel ensemble mean simulated GHA's rainfall adequately and as such, is recommended for assessing future climate projections over GHA.

The contradiction above in which temperature and evaporation increases but precipitation decreases leads to the question on "*what actually causes the decline in total precipitation over GHA when there is an increase in warm days and nights*"? The answer to this question is not simple. A two-fold deduction could be made as follows. *First* (Professor Richard Anyah, Pers. Comm), this contradictions between positive temperatures and evaporation on the one hand, and negative precipitation on the other hand has been a subject of a number of debates and studies, e.g., [63], leading F. H. H. Semazzi to come up with what is called the "*East Africa rainfall paradox*". The Paradox indicates that all global model projections have consistently shown that GHA and East Africa to be having increase in precipitation yet for the past decades, the rainfall amount has been decreasing, see e.g., [48]. This defies some of the climatological theories on relationships between water vapor and temperature (Clausius-Clapeyron). Two theories have so far been put forward to try and explain this 'paradox', i.e., (i) although the rainfall intensity over parts of GHA have increased, the rainfall season has shortened especially for the (MAM) [71], and (ii), the impact of aerosols on rainfall in East Africa, which tend to suppress rainfall during certain days despite the overall increase in temperature and evaporation [75, 77].

Second (Associate Professor Freddie Mpelasoka, Pers. Comm), rainfall/precipitation is more of moisture advection than local evaporation. Therefore, changes in precipitation will be a result of changes in the general circulation (synoptic features) conducive to moisture advection for rainfall generation for a given location. Regarding the effect of temperature rise, is the increase of moisture holding capacity before precipitation can occur, which can be interpreted in terms of extreme rainfall events. So it's not true that evaporation due to temperature rise results in increased precipitation because of local evaporation. Air masses which can produce rainfall over the GHA are well documented.

References

1. Abshir S (2020) Climate Change and Security in the Horn of Africa: Can Europe help to reduce the risks? https://www.eip.org/wp-content/uploads/2020/10/csen_policy_paper_climate_change_and_security_in_the_horn_of_africa.pdf [Accessed 10/09/2021]
2. Aguilar E, Peterson TC, Ramírez Obando P, Frutos R, Retana JA, Solera M, Soley J, González García I, Araujo RM, Rosa Santos A, Valle VE, Brunet M, Aguilar L, lvarez LA, Bautista M, Castañón C, Herrera L, Ruano E, Sinay JJ, Sánchez E, Hernández Oviedo GI, Obed F, Salgado JE, Vázquez JL, Baca M, Gutiérrez M, Centella C, Espinosa J, Martínez D, Olmedo B, Ojeda Espinoza CE, Núñez R, Haylock M, Benavides H, Mayorga R. (2005) Changes in precipitation and temperature extremes in Central America and northern South America, 1961–2003. J Geophys Res 110:D23107. https://doi.org/10.1029/2005JD006119
3. Aguilar E, Aziz Barry A, Brunet M, Ekang L, Fernandes A, Massoukina M, Mbah J, Mhanda A, do Nascimento DJ, Peterson TC, Thamba Umba O, Tomou M, Zhang X (2009) Changes in temperature and precipitation extremes in western central Africa, Guinea Conakry, and Zimbabwe, 1955–2006. J Geophys Res 114:D02115. https://doi.org/10.1029/2008JD011010

4. Alexander LV, Zhang X, Peterson TC, Caesar J, Gleason B, Tank AMGK, Haylock M, Collins D, Trewin B, Rahimzadeh F, Tagipour A, Kumar KR, Revadekar J, Griffiths G, Vincent L, Stephenson DB, Burn J, Aguilar E, Brunet M, Taylor M, New M, Zhai P, Rusticucci M, Vazquez-Aguirre JL (2006) Global observed changes in daily climate extremes of temperature and precipitation. J Geophys Res 111:D05109. https://doi.org/10.1029/2005JD006290
5. Anyah RO, Qiu W (2011) Characteristic 20th and 21st century precipitation and temperature patterns and changes over the Greater Horn of Africa. Int J Climatol 32:347–363. https://doi.org/10.1002/joc.2270
6. Anyah RO, Semazzi FH (2006) Climate variability over the Greater Horn of Africa based on NCAR AGCM ensemble. Theor Appl Climatol 86:39–62
7. Arnell NW (2004) Climate change and global water resources: SRES emissions and socio-economic scenarios. Global Environ Change 14(1):31–52. https://doi.org/10.1016/j.gloenvcha.2003.10.006
8. Awange J, Sharifi M, Ogonda G, Wickert J, Grafarend E, Omulo M (2008) The falling Lake Victoria water level: GRACE, TRIMM and CHAMP satellite analysis of the lake basin. Water Res Manag 22(7):775–796. https://doi.org/10.1007/s11269-007-9191
9. Awange J, Sharifi M, Baur O, Keller W, Featherstone W, Kuhn M (2009) GRACE hydrological monitoring of Australia: current limitations and future prospects. J Spat Sci 54(1):23–36. https://doi.org/10.1080/14498596.2009.9635164
10. Awange J, Fleming KM, Kuhn M, Featherstone WE, Heck B, Anjasmara I (2011) On the suitability of the 4° × 4° GRACE mascon solutions for remote sensing Australian hydrology. Remote Sens Environ 115(3):864–875. https://doi.org/10.1016/j.rse.2010.11.014
11. Awange JL, Palancz B, Völgyesi L (2020) Hybrid imaging and visualization. employing machine learning with mathematica - python. Springer Nature International, Berlin. 978-3-030-26152-8, https://doi.org/10.1007/978-3-030-26153-5
12. Becker M, Llovel W, Cazenave A, Günter A, Crétaux J-F (2010) Recent hydrological behavior of the East African great lakes region inferred from GRACE, satellite altimetry and rainfall observations. CR Geosci 342(3):223–233. https://doi.org/10.1016/j.crte.2009.12.010
13. Bohle HG, Downing TE, Watts MJ. 1994. Climate change and social vulnerability: toward a sociology and geography of food insecurity. Global EnvironChange 4(1): 37–48. https://doi.org/10.1016/0959-3780(94)90020-5
14. Brant L, Ileana B, George N, Leila K, Carvalho MV, Gabriel B, Senay DA, Leroux S, Funk C (2012) Seasonality of African precipitation from 1996 to 2009. J Clim 25(12):4304–4322. https://doi.org/10.1175/JCLI-D-11-00157.1
15. Brunet M, Saladié O, Jones P, Aguilar E, Moberg A, Lister D, Walther A, Almarza C (2008) A case-study/guidance on the development of long-term daily adjusted temperature datasets, WCDMP-66. World Meteorol. Org, Geneva, p 46
16. Caesar J, Alexander L, Vose R (2006) Large-scale changes in observed daily maximum and minimum temperatures - creation and analysis of a new gridded dataset. J Geophys Res 111:D05101. https://doi.org/10.1029/2005JD006280
17. Caesar JA, Alexander LV, Trewin B, Tsering K, Sorany L, Vuniyayawa V, Keosavang N, Shimana A, Htay MM, Karmacharya J, Jayasinghearachchi DA, Sakkamart J, Soares E, Hung LT, Thuong LT, Hue CT, Dung NTT, Hung PV, Cuong HD, Cuong NM, Sirabaha S (2010) Changes in temperature and precipitation extremes over the Indo-Pacific region from 1971 to 2005. Int J Climatol 31(6):791–801. https://doi.org/10.1002/joc.2118
18. Camberlin P, Philippon N (2002) The East African March-May rainy season, its teleconnections and predictability over the 1968–1997 period. J Clim 15:1002–1019
19. Choi G, Collins D, Ren G, Trewin B, Baldi M, Fukuda Y, Afzaal M, Pianmana T, Gomboluudev P, Huong PT, Lias N, Kwon WT, Boo KO, Cha YM, Zhou Y (2009) Changes in means and extreme events of temperature and precipitation in the Asia-Pacific Network region, 1955–2007. Int J Climatol 29:1906–1925
20. Christy JR, Norris WB, McNider RT (2009) Surface temperature variations in East Africa and possible causes. J Clim 2009(22):3342–3356

21. Conway D, Mould C, Bewket W (2004) Over one century of rainfall and temperature observations in Addis Ababa, Ethiopia. Int J Climatol 24:77–91
22. Downing TE (1991) Vulnerability to hunger in Africa: a climate change perspective. Glob Environ Chang 1(5):365–380. https://doi.org/10.1016/0959-3780(91)90003-C
23. Elagib NA (2010) Changing rainfall, seasonality and erosivity in the hyper-arid zone of Sudan. Land Degrad Dev 22(6):505–512. https://doi.org/10.1002/ldr.1023
24. Endris H, Omondi P, Jain S, Chang'a L, Lennard C, Hewitson B, Awange J, Ketiem P, Dosio A, Nikulin G et al (2013) Assessment of the performance of CORDEX Regional Climate Models in Simulating Eastern Africa Rainfall. J Clim 26:8453–8475. https://doi.org/10.1175/JCLI-D-12-00708.1
25. Favre A, Stone D, Cerezo R, Philippon N, Abiodun B (2011) Diagnostic of monthly rainfall from CORDEX simulations over Africa: Focus on the annual cycles. In: Proceedings of international conference on the coordinated regional climate downscaling experiment–CORDEX, Trieste, Italy, World Climate Research Program. http://indico.ictp.it/indico/getFile.py/access?resId53&materialId51&confId5a10131
26. Folland CK, Karl TR, Christy JR, Clarke RA, Gruza GV, Jouzel J, Mann ME, Oerlemans J, Salinger MJ, Wang SW (2001) Observed climate variability and change. Climate Change 2001. The Scientific Basis - Contribution of Working Group I to the Third Assessment Report of the Intergovernmental Panel on Climate Change. Cambridge University Press, New York, pp 99–181
27. Forootan E, Kusche J (2012) Separation of global time-variable gravity signals into maximally independent components. J Geodesy 86(7):477–497. https://doi.org/10.1007/s00190-011-0532-5
28. Forootan E, Kusche J (2013) Separation of deterministic signals, using independent component analysis (ICA). Stud Geophys Geod 57(1):17–26. https://doi.org/10.1007/s11200-012-0718-1
29. Forootan E, Awange J, Kusche J, Heck B, Eicker A (2012) Independent patterns of water mass anomalies over Australia from satellite data and models. J Remote Sens Environ 124:427–443. https://doi.org/10.1016/j.rse.2012.05.023
30. Funk C, Dettinger MD, Michaelsen JC, Verdin JP, Brown ME, Barlow M, Hoell A (2008) Warming of the Indian Ocean threatens eastern and southern African food security but could be mitigated by agricultural development. Proc Natl Acad Sci 105(32):11081–11086
31. Funk C, Michaelsen J, Marshall MT (2012) Mapping recent decadal climate variations in precipitation and temperature across Eastern Africa and the Sahel. In: Wardlow BD, Anderson MC, Verdin JP (eds) Remote sensing of drought: innovative monitoring approaches. Taylor and Francis, CRC Press, Boca Raton, FL
32. Griffiths GM, Chambers LE, Haylock MR, Manton MJ, Nicholls N, Baek HJ, Choi Y, Della Marta PM, Gosai A, Iga N, Lata R, Laurent V, Maitrepierre L, Nakamigawa H, Ouprasitwong N, Solofa D, Tahani L, Thuy DT, Tibig L, Trewin B, Vediapan K, Zhai P (2005) Change in mean temperature as a predictor of extreme temperature change in the Asia-Pacific region. Int J Climatol 25:1301–1330
33. Hannachi A, Unkel S, Trendafilov NT, Jolliffe IT (2009) Independent component analysis of climate data: a new look at EOF rotation. J Clim 22:2797–2812. https://doi.org/10.1175/2008JCLI2571.1
34. Hastenrath S, Polzin D, Mutai C (2007) Diagnosing the 2005 drought in equatorial East Africa. J Clim 29:4628–4637
35. Hastenrath S, Polzin D, Mutai C (2010) Diagnosing the droughts and floods in equatorial East Africa during boreal autumn 2005–08. J Clim 23:813–817
36. Hay SI, Rogers DJ, Randolph SE, Stern DI, Cox J, Shanks GD, Snow RW (2002) Hot topic or hot air? Climate change and malaria resurgence in East African highlands. Trends Parasitol 18:530–534
37. Haylock MR, Hofstra N, Klein Tank AMG, Klok EJ, Jones PD, New M (2008) A European daily high-resolution gridded data set of surface temperature and precipitation for 1950–2006. J Geophys Res 113:D20119. https://doi.org/10.1029/2008JD010201

38. Houborg R, Rodell M, Li B, Reichle R, Zaitchik BF. 2012. Drought indicators based on model-assimilated Gravity Recovery and Climate Experiment (GRACE) terrestrial water storage observations. Water Res Res 48(7). https://doi.org/10.1029/2011WR011291
39. Indeje M, Semazzi FHM, Ogallo LJ (2000) ENSO signals in East African rainfall and their prediction potentials. Int J Climatol 20:19–46
40. IPCC (2007) Climate change 2007: the physical science basis. In: Solomon S, Qin D, Manning M, Chen Z, Marquis M, Averyt KB, Tignor M, Miller HL (eds) Contribution of working group I to the fourth assessment report of the intergovernmental panel on climate change. Cambridge University Press, Cambridge, NY
41. Janowiak J (1988) An investigation of interannual rainfall variability in Africa
42. Kijazi AL, Reason CJC (2009) Analysis of the 2006 floods over northern Tanzania. Int J Climatol 29:955–970
43. Kijazi AL, Reason CJC (2009) Analysis of the 1998 to 2005 drought over the northeastern highlands of Tanzania. Climate Res 38:209–223
44. Klein Tank AMG, Peterson TC, Quadir DA, Dorji S, Zou X, Tang H, Santhosh K, Joshi UR, Jaswal AK, Kolli RK, Sikder A, Deshpande NR, Revadekar JV, Yeleuova K, Vandasheva S, Faleyeva M, Gomboluudev P, Budhathoki KP, Hussain A, Afzaal M, Chandrapala L, Anvar H, Amanmurad D, Asanova VS, Jones PD, New MG, Spektorman T. 2006. Changes in daily temperature and precipitation extremes in central and south Asia. J Geophys Res-Atmos 111(D16105). https://doi.org/10.1029/2005JD006316
45. Kurtenbach E (2011) Entwicklung eines Kalman-Filters zur Bestimmung kurzzeitiger Variationen des Erdschwerefeldes aus Daten der Satellitenmission GRACE. PhD dissertation, Bonn University, Germany
46. Kurtenbach E, Mayer-Gürr T, Eicker A (2009) Deriving daily snapshots of the Earth's gravity field from GRACE L1B data using Kalman filtering. Geophys Res Lett 36:L17102. https://doi.org/10.1029/2009GL039564
47. Lotsch A, Friedl MA, Pinzón J (2003) Spatio–temporal deconvolution of NDVI image sequences using independent component analysis. IEEE Trans Geosci Remote Sens 41(12)
48. Lyon B, Vigaud N (2017) Unraveling East Africa's climate paradox. In: Climate extremes. Wiley Inc, Hoboken, NJ, pp 265–281. https://doi.org/10.1002/9781119068020.ch16
49. Lyon B, DeWitt DG (2012) A recent and abrupt decline in the East African long rains. Geophys Res Lett 39(2):L02702. 10.L029/2011GL050337
50. Mutemi JN (2003) Climate Anomalies Over Eastern Africa Associated with Various ENSO Evolution Phases. PhD thesis, University of Nairobi, Kenya
51. Ogallo LJ, Janowiak JE, Halpert MS (1988) Teleconnection between seasonal rainfall over Eastern Africa and Global Sea surface temperature anomalies. J Meteorol Soc Jpn 66:807–822
52. van Oldenborgh GJ, Philip SY, Collins M (2005) El Niñno in a changing climate: a multi-model study. Ocean Sci 1:81–95
53. Omondi PA (2011) Agricultural drought indices. In: Proceedings of an expert meeting, 2–4 June, 2010, Murcia, Spain, WMO, AGM- 11WMO/TD No. 1572, WAOB-2011, pp 106–112
54. Omondi P, Awange JL, Ogallo LA, Okoola RA, Forootan E (2012) Decadal rainfall variability modes in observed rainfall records over East Africa and their relations to historical sea surface temperature changes. J Hydrol 464–465:140–156. https://doi.org/10.1016/j.jhydrol.2012.07.003
55. Omondi P, Ogallo LA, Anyah R, Muthama JM, Ininda J (2012) Linkagesbetween global sea surface temperatures and decadal rainfall variability over Eastern Africa region. Int J Climatol 33(8):2082–2104. https://doi.org/10.1002/joc.3578
56. Omumbo JA, Lyon B, Samuel MW, Connor SJ, Madeleine CT (2011) Raised temperatures over the Kericho tea estates: revisiting the climate in the East African highlands malaria debate Malaria Journal 10: 12 http://www.malariajournal.com/content/10/1/12
57. Omondi P, Awange JL et al (2014) Changes in temperature and precipitation extremes over the Greater Horn of Africa region from 1961 to 2010. Int J Climatol 34:1262–1277. https://doi.org/10.1002/joc.3763

58. Parmesan C, Root TL, Willig MR (2000) Impacts of extreme weather and climate on terrestrial biota. Bull Am Meteorol Soc 81:443–450. https://doi.org/10.1175/1520-0477(2000)081<0443:IOEWAC>2.3.CO;2
59. Pascal M, Ahumada JA, Chaves LF, Rodo X, Bouma M (2006) Malaria resurgence in the East African highlands: temperature trends revisited. PNAS 103(15):5829–5834. https://doi.org/10.1073/pnas.0508929103
60. Peterson TC, Manton MJ (2008) Monitoring changes in climate extremes: a tale of international collaboration. Bull Am Meteor Soc 89:1266–1271
61. Peterson TC, Easterling DR, Karl TR, Groisman P, Nicholls N, Plummer N, Torok S, Auer I, Bohm R, Gullett D, Vincent L, Heino R, Tuomenvirta H, Mestre O, Szentimrey T, Salinger J, Forland EJ, Hanssen-Bauer I, Alexandersson H, Jones P, Parker D (1998) Homogeneity adjustments of in-situ atmospheric climate data: a review. Int J Climatol 18:1493–1517
62. Ramillien G, Cazenave A, Brunau O (2004) Global time variations of hydrological signals from GRACE satellite gravimetry. Geophys J Int 158(3):813–826. https://doi.org/10.1111/j.1365-246X.2004.02328.x
63. Rowell DP, Booth BBB, Nicholson SE, Good P (2015) Reconciling Past and Future Rainfall Trends over East Africa. J Climate 28(24):9768–9788. Retrieved Sep 11, 2021, from https://journals.ametsoc.org/view/journals/clim/28/24/jcli-d-15-0140.1.xml
64. Schreck CJ, Semazzi FHM (2004) Variability of the recent climate of eastern Africa. Int J Climatol 24:681–701
65. Sen PK (1968) Estimates of regression coefficient based on Kendalls Tau. J Am Stat Assoc 63:1379–1389. https://doi.org/10.2307/2285891
66. Shongwe ME, van Oldenborgh GJ, van den Hurk BJJM, van Aalst MK (2010) Projected changes in mean and extreme precipitation in Africa under global warming. Part II: East Africa. J Clim 22:3819–3837
67. Shongwe ME, van Oldenborgh GJ, van den Hurk B, van Aalst M (2011) Projected changes in mean and extreme precipitation in Africa under global warming. Part II: East Africa. J Clim 24:3718–3733
68. Simon W, Hassell D, Hein D, Jones R, Taylor R (2004) Installing and using the hadley centre regional climate modelling system, PRECIS, Version 1.1 . Met Office Hadley Centre, Exeter
69. Tapley B, Bettadpur S, Ries J, Thompson P, Watkins M (2004) GRACE measurements of mass variability in the Earth system. Science 305:503–505. https://doi.org/10.1126/science.1099192
70. Tapley B, Bettadpur S, Watkins M, Reigber C (2004) The gravity recovery and climate experiment: mission overview and early results. Geophys Res Lett 31:L09607. https://doi.org/10.1029/2004GL019920
71. Thiery W, Davin E, Seneviratne S et al (2016) Hazardous thunderstorm intensification over Lake Victoria. Nat Commun 7:12786. https://doi.org/10.1038/ncomms12786 [accessed 11/09/2021]
72. Tierney JE, Smerdon JE, Anchukaitis KJ, Seager R (2013) Multidecadal variability in East African hydroclimate controlled by the Indian Ocean. Nature (London) 493(7432):389–392
73. Viste E, Korecha D, Sorteberg A (2012) Recent drought and precipitation tendencies in Ethiopia. Theore Appl Climatol 112(3–4):535–551
74. Wahr J, Molenaar M, Bryan F (1998) Time variability of the Earth's gravity field: hydrological and oceanic effects and their possible detection using GRACE. J Geophys Res 103(B12):30205–30229. https://doi.org/10.1029/98JB02844
75. Wainwright CM, Marsham JH, Keane RJ et al. (2019) Eastern African Paradox' rainfall decline due to shorter not less intense Long Rains. npj Clim Atmos Sci 2:34. https://doi.org/10.1038/s41612-019-0091-7 [accessed 11/09/2021]
76. Wang XL (2003) Comments on "Detection of undocumented change points: a revision of the two-phase regression model." J Clim 16:3383–3385
77. Walker DP, Marsham JH, Birch CE, Scaife AA, Finney DL (2020) Common mechanism for interannual and decadal variability in the East African Long Rains. Geophys Res Lett 47:e2020GL089182. https://doi.org/10.1029/2020GL089182
78. Wang XLL (2008) Accounting for autocorrelation in detecting mean shifts in climate data series using the penalized maximal t or F test. J Appl Meteorol Climatol 47:2423–2444. https://doi.org/10.1175/2008JAMC1741.1

79. Wang XLL (2008) Penalized maximal F test for detecting undocumented mean shift without trend change. J Atmos Oceanic Tech 25:368–384. https://doi.org/10.1175/2007JTECHA982.1
80. Wang XL, Swail VR (2001) Changes of extreme wave heights in Northern Hemisphere oceans and related atmospheric circulation regimes. J Climate 14:2204–2221. https://doi.org/10.1175/1520-0442(2001)014<2204:COEWHI>2.0.CO;2
81. WMO (2003) Report of the GCOS/GTOS/HWRP expert meeting on hydrological data for global studies, WMO/TD–No. 1156
82. Zhang XB, Vincent LA, Hogg WD, Niitso A (2000) Temperature and precipitation trends in Canada during the 20th century. Atmosp-Ocean 38:395–429
83. Zhang X, Aguilar E, Sensoy S, Melkonyan H, Tagiyeva U, Ahmed N, Kutaladze N, Rahimzadeh F, Taghipour A, Hantosh TH, Albert P, Semawi M, Ali MK, Al-Shabibi SHM, Al-Oulan Z, Taha Zatari T, Al Dean Khelet IK, Hamoud S, Ramazan Sagir R, Demircan M, Eken M, Adiguzel M, Lisa Alexander L, Peterson TC, Wallis T (2005) Trends in Middle East climate extreme indices from 1950 to 2003. J Geophys Res 110:D22104. https://doi.org/10.1029/2005JD006181
84. Zhang X, Zwiers FW, Hegerl G (2009) The influences of data precision on the calculation of temperature percentile indices. Int J Climatol 29:321–327. https://doi.org/10.1002/joc.1738
85. Awange JL, Gebremichael M, Forootan E, Wakbulcho G, Anyah R, Ferreira CG, Alemayehu T (2014) Characterization of Ethiopian mega hydrogeological regimes using GRACE, TRMM and GLDAS datasets. Adv. Water Resour. 74: 64–78. https://doi.org/10.1016/j.advwatres.2014.07.012

Chapter 10
GHA Droughts: Coupled Ocean-Atmosphere Phenomena

> For the GHA region, rather than responding successfully to the frequent recurrent droughts that afflict the region, the communities are invariably devastated by famine crisis, instabilities in national economies and political tensions. For example, the Ethiopian "biblical" famines of 1973–74 and 1984–85 left about 200,000 and 400,000, people dead, respectively, with the former disaster resulting in the overthrow of Emperor Haile Selassie. The latter contributed to the end of the Marxist regime of Mengistu Haile Mariam.—Mpelasoka et al., [31].

10.1 Summary

Drought-like humanitarian crises in the Greater Horn of Africa (GHA) are increasing despite recent progress in drought monitoring and prediction efforts. Notwithstanding these efforts, there remain challenges stemming from uncertainty in drought prediction, and the inflexibility and limited buffering capacity of the recurrently impacted systems. The complexity of the interactions of ENSO, IOD, IPO and NAO, arguably remains the main source of uncertainty in drought prediction. To develop practical drought risk parameters that potentially can guide investment strategies and risk-informed planning, this chapter presents the work of [31] who quantified drought characteristics that underpin drought impacts management. Drought characteristics that include probability of drought-year occurrences, durations, areal-extent and their trends over 11 decades (1903–2012) were derived from the Standardized Precipitation Index (SPI). Transient probability of drought-year occurrences, modelled on Beta distribution, across the region ranges from 10 to 40%, although most fall within 20–30%. For more than half of the drought events, durations of up to 4, 7, 14 and 24 months for the 3-, 6-, 12- and 24 month timescales were evident, while 1 out of 10 events persisted for up to 18 months for the short timescales, and up to 36 months or more for the long timescales. Apparently, only drought areal-extent showed statistically significant trends of up to 3%, 1%, 3.7%, 2.4%, 0.7%, −0.3%

J. Awange, *Food Insecurity & Hydroclimate in Greater Horn of Africa*,
https://doi.org/10.1007/978-3-030-91002-0_10

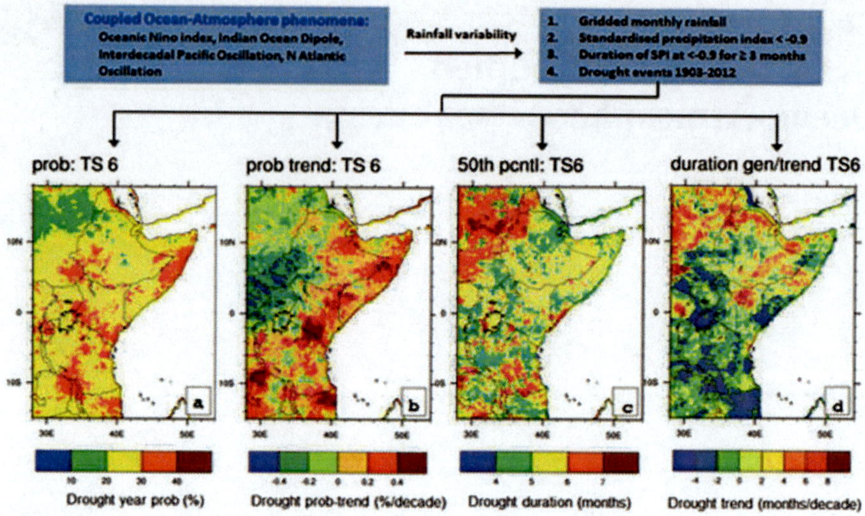

Fig. 10.1 Graphical abstract of Mpelesoka et al. [31]

and −0.6% per decade over Sudan, Eritrea, Ethiopia, Somalia, Kenya, Uganda and Tanzania, respectively. Since there is no evidence of significant changes in drought characteristics, the peculiarity of drought-like crises in the GHA can be attributed (at least in part) to unaccounted for systematic rainfall reduction. This highlights the importance of distinguishing drought impacts from those associated with new levels of aridity. In principle, drought is a temporary phenomenon while aridity is permanent, a difference that managers and decision-makers should be more aware. The outcome of this Chapter are graphically summarized in Fig. 10.1.

10.2 Frequently Recurring GHA's Droughts: Challenges

The Greater Horn of Africa (GHA) shares many common experiences with the rest of the world when it comes to impacts of drought. However, the impacts of drought in some Member States of the GHA region often appear to have more adverse effects on sustainable development than elsewhere in similar developing countries [48]. For example, for the southern Africa region, drought in Botswana is regarded as a learning opportunity to improve/achieve water security [46]. But for the GHA region, rather than responding successfully to the frequent recurrent droughts that afflict the region, the communities are invariably devastated by famine crisis, instabilities in national economies and political tensions. For example, the Ethiopian "biblical" famines of 1973–74 and 1984–85 left about 200,000 and 400,000, people dead, respectively, with the former disaster resulting in the overthrow of Emperor Haile Selassie [2, 19]. The latter contributed to the end of the Marxist regime of Mengistu Haile Mariam [25].

Nicholson [35] describes the relatively recent drought related crisis in the GHA region that prevailed during much of the period 2008–2011 triggering extreme food shortages and massive migration. In 2017, parts of GHA were in the midst of a major drought [17]. The most affected areas included most of Somalia, south-eastern Ethiopia, north-eastern and coastal Kenya, and northern Uganda, where Somalia and parts of Kenya faced severe famine. It is increasingly alarming that being in dire need of food assistance in the GHA is becoming a permanent feature of the region. Almost every year, including 2014, 2015, 2016, 2017 and 2021 famine headlines appear in the news as drought related crisis [11].

While droughts may continue to be a major problem, for some regions of the GHA, other factors such as armed conflicts and international politics, are invariably responsible for propelling a situation of economic hardship caused by droughts (see Chap. 1 as well as [10]). For example, on a long-term basis, environmental degradation, poor water resource management and poor governance are important compounding causes of severe drought impacts in the region [45]. More importantly, aridity reconstruction studies [44] show that the region is increasingly becoming drier. This systematic persistent decline in rainfall, particularly during much of the region's primary rainy season (March-April-May) is evident in the last 30 years' rainfall record [27, 52]. Whether this decline trend is associated with internal multi-decadal climate variability due to changes in the tropical Pacific [27, 54] (see also Chaps. 7 and 8) or anthropogenically driven warming in the Indian Ocean or western Pacific region [26], its impacts are yet to be distinguished from those associated with droughts. The impact of changes in aridity levels is often hardly distinguished from that of droughts. Drought is a recurrent feature of climate variability that occurs in virtually all climate regimes, and is different from aridity, which is a rather permanent feature [32]. Similarly, as underlying impoverishment of population increases, it is increasingly more difficult to distinguish between humanitarian crises triggered by drought impact and those stemming from chronic poverty [10].

Indeed, there are still many challenges in monitoring and prediction capabilities, as well as a perspective of the current understanding of drought and key research gaps. For example, almost all drought studies reiterate the influence of ENSO phenomenon on drought occurrences, with respect to drought prediction. However, there are other important ocean-atmosphere phenomenon. Such as the Indian Ocean Dipole (IOD), Inter-decadal Pacific Oscillation (IPO) and the North Atlantic Oscillation (NAO). The main challenge is to account for the interactions of different systems of climate variability and their teleconnections [7, 8, 33, 43]. The overall influence of climate variability drivers depends on their concurrent modes [3]. Hence, this is the main source of uncertainty in predictions of climatic extremes including droughts particularly, when they are solely based on ENSO phenomenon [24]. In addition to limited prediction skill, possibly the lack of flexibility in the impacted systems for the GHA underscore the effect of prediction. For example, in 2017, in Somalia and coastal Kenya cropping lands, 70–100% crop failure was registered [17]. Livestock mortality has been particularly devastating amongst small ruminants with mortality rate ranging from 25 to 75% in the cross border areas of Somalia-Kenya-Ethiopia. This is happening regardless of early warnings by the Inter-Governmental Authority on Development, potentially meant to elicit early actions (preparedness and mitigation measures). Apparently, moving from crisis to risk management in the GHA requires planning that places more weight on risk assessment and the development and implementation of mitigation actions and programs as suggested in [50].

This chapter has three main objectives: (1) quantification of the variation of influence among climate variability drivers across the region, (2) quantification of drought characteristics that underpin drought impacts management, and (3), examination of consistence of current increase in drought-like crises with trends in drought characteristics over the GHA region; The analyses include; (i) relating drought occurrences with climate variability drivers, (ii) modelling of transient probability of drought-year occurrences, (iii) determining drought duration, (iv) determining drought areal-extent, and (v), examining trends in rainfall.

10.3 Centennial Precipitation and SST Products

Monthly rainfall for the 1901–2013 period on a 0.1° × 0.1° grid across the GHA region are drawn from the Centennial Trends Greater Horn of Africa precipitation dataset [12]. The CenTrends data set provides a reasonably complete and accurate gridded precipitation products.

Sea Surface Temperatures (SSTs) drawn from the National Centers for Environmental Prediction (NCEP)/National Center for Atmospheric Research (NCAR) reanalysis dataset [23] for the 1948–2013 period are used to derive indices of four major climate variability drivers that include the Oceanic Niño Index (ONI), which represents the El'Niño Southern Oscillation (ENSO), Indian Ocean Dipole (IOD), Inter-decadal Pacific Oscillation (IPO), and the Northern Atlantic Oscillation (NAO).

10.4 Drought Characterization Approach

10.4.1 Identification of Drought Events

Monthly rainfall time series are transformed into the Standardized Precipitation Index (SPI), developed by Mckee et al., [30] to capture rainfall variability from which occurrences of drought events are unveiled for the 1903 April–2013 March period. SPI is a probability index that gives a good representation of rainfall variability, quantifying abnormal wetness and dryness levels. Mathematically, SPI is based on the cumulative probability of a given rainfall event occurring at a location. The historic rainfall data of the location is fitted in to a gamma distribution, which fits the precipitation distribution quite well. The cumulative probability gamma function subsequently transforms into a standard normal random variable. For comparison purposes, the World Meteorological Organization (WMO) recommends the use of SPI in monitoring of dry spells [53].

Since much of rainfall is experienced in short rainy seasons and most of it often concentrates in a few heavy falls, small shifts in the large-scale weather patterns at different timescales significantly alter the amount and/or the distribution of rainfall. Therefore, SPI is used at four timescales to facilitate the interpretation and relevance of rainfall anomalies to different systems. For this chapter, focus is placed on the 3-month, 6-month, 12-month and 24-month timescales, which are generally relevant to a range of agricultural and hydrological systems. For a given timescale, a drought event begins any time when the SPI is continuously less than negative 0.9 for at least 3 months, and ends when the SPI becomes greater than negative 0.9. This value is the threshold for dry conditions in the SPI classification, which is equivalent to 0.18 cumulative probability [30].

10.4.2 Modelling the Probability of Drought-Year Occurrences

Since the occurrences of a drought-year (derived from the SPI time series) are statistically independent in that there is no contribution of antecedent conditions, the modelling of transient probabilities of drought-year occurrences is rather straightforward. The series of hits and misses of drought-year occurrences form proportions that can be represented by a Binomial or Geometric distribution with parameter p [20, 49]. However, in practice these portions exhibit extra variations that cannot be explained by a simple Binomial or Geometric distribution because the parameter p does not remain constant in the course of time [37, 51]. For the parameter p to assume a continuous distribution in the parameter space $0 < p < 1$, the best way is to use the Beta distribution [13].

Beta distribution is a natural conjugate prior distribution in the Bayesian sense (i.e., evidence about the true state) and represents all possible values of unknown probabilities. Suppose these probabilities constitute a continuous random variable that follow a Beta distribution with parameters α and β, where $0 < \alpha < 1$ and $0 < \beta < 1$; α and β are chosen to reflect any existing information/belief, then the probability density function of takes the form of Eq. 10.1 [9],

$$f(x|\alpha, \beta) = \frac{x^{\alpha-1}(1-x)^{(\beta-1)}}{B(\alpha, \beta)}, \quad 0 < x < 1, \tag{10.1}$$

where $B(\alpha, \beta) = \frac{\Gamma(\alpha)\Gamma(\beta)}{\Gamma(\alpha+\beta)}$ is the beta function (acting as a normalizing constant) and $\Gamma(\alpha)$ is the gamma function:

$$\Gamma(\alpha) = \int_0^\infty x^{\alpha-1} e^{-x} dx. \tag{10.2}$$

The mean and variance of the Beta random variable x are

$$\mu = \frac{\alpha}{\alpha + \beta}, \tag{10.3}$$

and

$$\sigma^2 = \frac{\alpha\beta}{(\alpha + \beta)^2(\alpha + \beta + 1)}, \tag{10.4}$$

respectively.

Intuitively, the Beta distribution is particularly attractive for modelling the dynamics of the probability of drought-year occurrences, as more information become available. If h and m are numbers of hits and misses of drought-year occurrences respectively, at the ith time step, while the mathematics for proving the updating procedure is a bit involved, the operation is very simple. The Beta distribution takes the form of Eq. 10.5 [9].

$$Beta(\alpha_i, \beta_i) = Beta(\alpha_{i-1} + h, \beta_{i-1} + m). \tag{10.5}$$

For each grid-cell, the initial parameters α, β are estimated from a prior probability based on the proportion count of years in drought to the total number of years.

10.4.3 Coupled Ocean-Atmosphere Phenomena Influencing Drought Occurrences

The calculations of climate variability indices that related to four major coupled ocean-atmosphere phenomena of influence for the GHA adopted the climatology of 1976–2005 (i.e., 30-year period centered 1990). This is consistent with the current IPCC baseline for climate change assessment.

1. The ONI is calculated as running 3-month mean SST anomaly for the region covered by the Niño 3.4 index (i.e., 5°N–5°S, 120°–170°W) [34].
2. The IOD is the difference in SST between two areas (hence a dipole) – a western pole in the Arabian Sea (10°S–10°N, 50°–70°E) and an eastern pole in the eastern Indian Ocean south of Indonesia (10°S–22°N, 90°–110°E). The index is discussed in [40].
3. The Inter-decadal Pacific Oscillation (IPO) is an oceanographic/meteorological phenomenon. The index is based on the difference between the SST averaged over the central equatorial Pacific and the average of the SST in the Northwest and Southwest Pacific [15]. The regions used to calculate the index are: Region 1 (25°N–45°N, 140°E–145°W); Region 2 (10°S–10°N, 170°E–90°W); and Region 3 (50°S–15°S, 150°E–160°W).

4. The NAO is a large scale seesaw in atmospheric mass between the subtropical high (38.7°N, 9.1°W) and the polar low (65.4°N, 25°W). The index was defined by [21]. The index and its main characteristics are widely discussed in [36, 38].

Subsequently, the correlation analysis between individual indices was carried out on grid-cell basis, across the GHA region.

10.5 Influence of Climate Variability Drivers

10.5.1 Response to Drought Drivers Across GHA

The correlation of the probability of drought-year occurrence with individual indices of climate variability drivers are exhibited in Fig. 10.2 for ONI, IOD, IPO and NAO. The figure shows that the association of drought with the drivers of climate variability is not uniform across the GHA region, and more importantly the association can be of opposite direction. Positive and negative correlations with ONI indicate when El Niño and La Niña induce droughts, respectively. Similarly, the positive and negative correlations with IOD, IPO and NAO indicate when positive and negative modes induce drought conditions, respectively. Given the temporal (i.e., annual) level of association, the magnitudes of the correlation coefficients are arguably not expected to be high, rather, good enough to display the spatial patterns of the association direction. Figure 10.3 provides a comparison between the respective climate variability drivers in GHA.

Details of the areal extent of positive and negative associations of drought-year occurrences at 3-, 6-, 12- and 24-month time-scales with individual drivers are given in Table 10.1 for the seven GHA Member States. The proportions of the area for each Member State indicate that El Niño driven droughts (positive ONI correlation) affect 19% of the area of Eritrea, prominently for the 3-month time-scale drought; also 19% for Ethiopia (6-month time-scale); 14–40% for Somalia, and 7–43% for Kenya across all drought time-scales. On the other hand, the La Niña driven droughts (negative ONI correlation) affect 7–15% of Sudan across the four drought time-scales; 4% and 8% for Kenya, at 3-, and 12-month drought time-scales. The La Niña driven droughts affect Uganda the most (30–60% of the area) across all drought time-scales; followed by Tanzania (about 10–38%).

The area proportion exhibiting IOD positive correlation with drought occurrence is highest for Eritrea, Ethiopia, and Sudan (74–96% across all time-scales). This influence decreases to 14–40% over Somalia, 12–29% for Kenya, 6–37% over Uganda and less than 6% for Tanzania. Conversely, the negative IOD correlation is demonstrated over 26% of Somalia, 34% for Kenya, 31% and 64% for Uganda and Tanzania, respectively. The IPO positive correlation with drought occurrences in Somalia, Kenya, Uganda and Tanzania covers up to 85% of the area. On the other hand, IPO is negatively correlated with drought occurrences in Eritrea, Ethiopia and Sudan, over 25–88% of the area across the 4 drought time-scales. The proportion of area

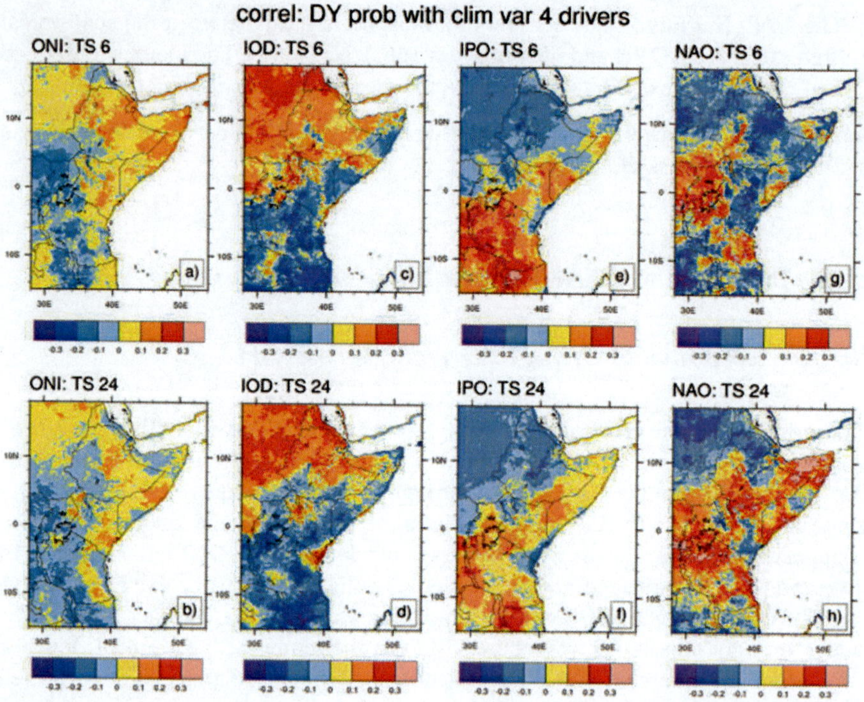

Fig. 10.2 Correlation coefficients of drought-year probability with ONI, IOD, IPO and NAO for 6- and 24-month drought timescales: panels [**a**, **c**, **e**, **g**] and [**b**, **d**, **f**, **h**], respectively. Positive and negative correlations with ONI indicate when El Niño and La Niña induce droughts, respectively. Similarly, the positive and negative correlations with IOD, IPO and NAO indicate when positive and negative modes induce drought conditions, respectively. The IOD and IPO for TS6 have exactly opposite spatial patterns regionally (IOD is positively correlated in the north and negatively correlated in the south, whereas the IPO is exact opposite). *Source* Mpelesoka et al. [31]

coverage of positive NAO for Uganda is 67–90%, Tanzania (30–63%), Kenya (27–90%), Somalia (12–63%), Sudan (13–23%), and only 10% over Ethiopia for 24-month timescale. On the other hand, the negative NAO correlation covers 26–46% of Eritrea, Sudan (10–30%), Somalia (25–35%), and 11–24% for Kenya, Uganda and Tanzania.

10.5.2 Reliability of ENSO in Drought Prediction

Almost all drought studies reiterate the influence of ENSO phenomenon on drought occurrences, with respect to drought prediction. However, surprisingly, many of them do not go beyond correlation analysis [4, 14], that merely suggests the potential teleconnection without desired explicit relationships. The main challenge is to account

Table 10.1 Areal extent (%) of regions where drought years at 3-, 6-, 12- and 24-month time-scales are positively or negatively correlated (coefficient magnitude greater than 0.1) with individual drivers of rainfall variability, namely the Oceanic Niño Index (ONI), Indian Ocean Dipole (IOD), Inter-decadal Pacific Oscillation (IPO) and Northern Atlantic Oscillation (NAO). *Source* [31]

Region	Drought time-scale	ONI	Proportion (%) under positive correlation IOD	IPO	NAO	ONI	Proportion (%) under negative correlation IOD	IPO	NAO
Eritrea	3	18.92	72.34	0.00	5.69	0.54	0.29	24.70	41.06
	6	8.98	94.39	0.00	6.45	2.14	0.00	45.55	26.31
	12	4.56	86.04	0.00	0.66	0.47	0.00	46.01	46.01
	24	2.46	96.39	0.00	1.31	2.91	0.00	60.71	42.06
Ethiopia	3	8.30	91.86	0.27	2.82	0.11	0.27	87.43	11.10
	6	18.82	82.46	0.51	2.12	0.09	0.39	86.79	7.75
	12	3.74	88.49	0.69	7.56	0.86	0.69	86.90	7.22
	24	2.05	87.43	2.08	10.45	0.11	1.62	33.78	6.34
Sudan	3	1.47	75.14	7.95	13.81	12.21	9.57	62.64	24.31
	6	0.62	89.02	0.00	21.69	14.74	0.00	79.21	30.18
	12	0.00	74.39	0.00	23.28	12.07	0.43	68.96	10.51
	24	0.00	72.58	0.00	15.75	7.32	3.30	67.04	20.72
Somalia	3	27.57	39.93	25.53	11.72	0.00	26.30	3.54	24.66
	6	40.42	24.00	25.98	12.51	0.02	25.68	3.58	34.58
	12	21.51	14.04	19.85	22.14	0.78	22.99	1.69	13.12
	24	14.35	15.63	16.40	62.54	0.10	27.56	0.12	1.93
Kenya	3	43.15	13.77	27.53	12.74	4.24	65.46	2.66	61.66
	6	15.26	12.28	26.00	7.83	1.35	23.92	17.83	41.36
	12	6.68	13.67	15.63	68.69	7.56	19.71	5.95	14.83
	24	20.24	29.42	16.07	27.64	1.19	25.48	38.33	11.35
Uganda	3	0.31	13.13	31.93	67.61	31.39	43.41	1.02	5.04
	6	0.03	37.20	13.31	89.69	60.72	2.37	7.13	0.03
	12	0.00	3.60	29.83	77.79	31.24	36.31	3.90	0.00
	24	3.84	5.87	30.87	66.99	29.63	41.46	1.74	3.65
Tanzania	3	1.34	5.74	72.72	63.49	37.24	52.72	0.62	11.00
	6	0.17	2.45	84.57	37.87	16.98	66.14	0.74	22.13
	12	0.26	3.85	74.74	30.43	8.96	75.84	2.94	24.36
	24	2.70	2.66	34.27	62.96	15.36	63.26	1.94	1.82

for the interactions of different systems of climate variability and their teleconnections. The overall influence of climate variability drivers depends on their concurrent modes [3]. Therefore, this is the main source of uncertainty in the prediction of climatic extremes including droughts when the prediction is solely based on ENSO phenomenon [24].

Although ENSO is generally acknowledged to play a major role in triggering worldwide extreme climatic events, it is evident in the observations and models that

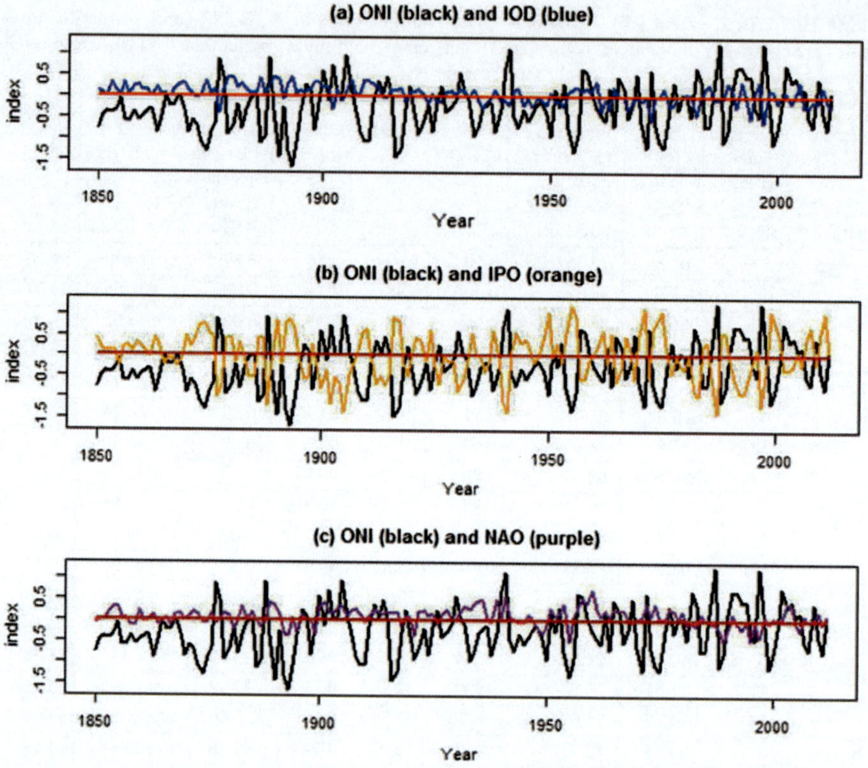

Fig. 10.3 Comparison of variability of annual mean of ONI with **a** IOD, **b** IPO and **c** NAO for the period 1850–2012. Complex interactions manifest between modes of these 4 major climate variability drivers. For example, ENSO extends its influence on modes of IOD and NAO, which in turn feed back onto ENSO [22]. The interactions between pairs of modes can alter their strength, periodicity, seasonality, and ultimately their predictability. *Source* Mpelesoka et al. [31]

not all droughts are ENSO driven [6]. This is demonstrated in Fig. 10.4 where proportions of drought-year occurrences in ENSO years to the total number of drought-year occurrences over the GHA region are shown for the 1950–2011 period. Here, an ENSO year refers to a year in which ENSO is either in El Niño or La Niña mode, as distinguished from a normal/neutral year [5]. Generally, ENSO-driven droughts are dominant over Sudan, Eritrea, most of Ethiopia except the south-eastern areas, Somalia, western and south-western Tanzania. However, the highest proportions of about 70% of drought-year occurrences are demonstrated at 3- and 6-month timescales (DTS3 and DTS6 (Fig. 10.4), respectively) and generally much lower proportions (<30%) at 12- and 24-month timescales (DTS12 and DTS24, respectively).

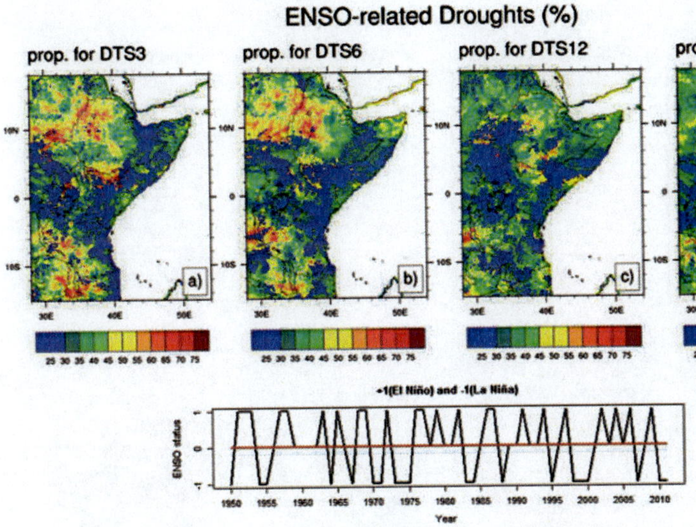

Fig. 10.4 Proportion (%) of ENSO-related drought-year occurrences to the total drought-year occurrences for the 1950–2012 period. It is evident in the observations that not all droughts are ENSO driven. The highest proportions of about 70% of drought-year occurrences are demonstrated at 3- and 6-month, and generally much lower proportions (<30%) at 12- and 24-month timescales, respectively. *Source* Mpelesoka et al. [31]

10.6 GHA's Drought Characteristics

Drought characteristics derived here include drought frequency (in terms of probability of drought-year), duration, and areal-extent of drought events. The importance of these elements are based on the view that duration (persistence) is a major determinant factor of drought severity (i.e., extent of impact devastation), in that once drought is declared, the impacted system gets incapacitated. Therefore, severity becomes a function of drought duration and at state or institutional level, the drought areal-extent exacerbates the challenge in terms of mitigation.

10.6.1 Probability of Drought-Year Occurrences

Long-term probability of drought-year occurrences across the GHA region range from 20 to 30%, except for isolated areas, which show probabilities of 10–20% and 30–40% on the low and high sides, respectively. The probabilities of drought-year occurrences are similar for all drought time-scales for most of the areas as shown in Fig. 10.5 for the 6- and 24-month drought timescales in panels (a) and (e), respectively. However, patches of relatively higher probabilities become more pronounced

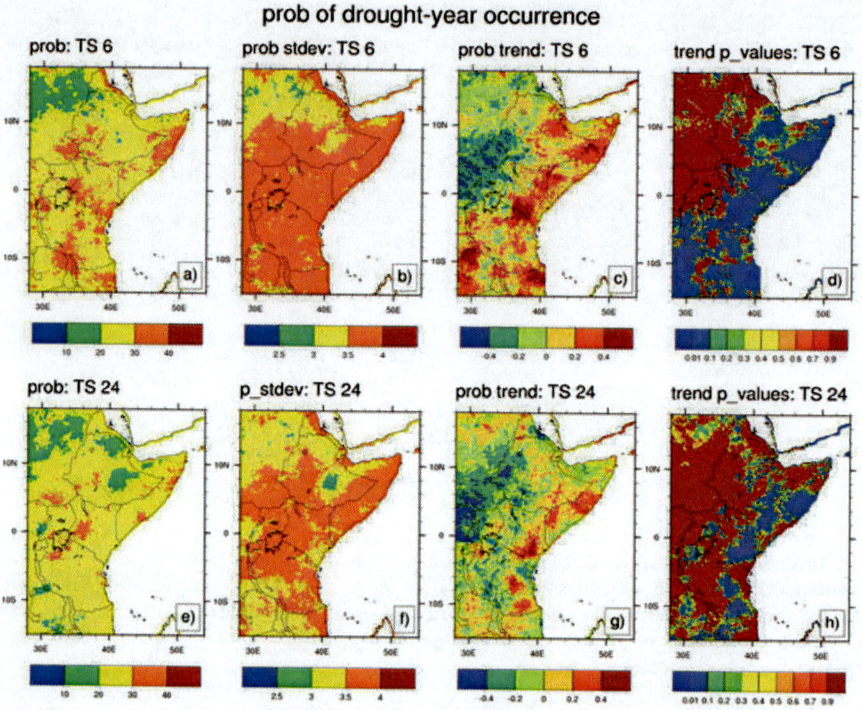

Fig. 10.5 Long-term drought-year probability (%), standard deviation (%), trend (% per decade) and their p-values for 6- and 24-month drought time scales: panels **a–d** and **e–h**, respectively. It is evident that the long-term probability of drought-year occurrences across the GHA region range from 20 to 30%, except for isolated areas, which exhibit probabilities as low as 10% and up to 40% on the lower and higher cases, respectively. This reaffirms drought to be an intrinsic part of the GHA landscape. *Source* Mpelesoka et al. [31]

for the 3- and 12-month (not shown) time-scales. Generally, there is consistency in drought-year occurrences, as indicated by the small range of probability variance (Fig. 10.5, panels (b) and (f)).

Nevertheless, the long-term trends in the probability of drought-year occurrences (Fig. 10.5, panels (c) and (g) in conjunction with (d) and (h)), respectively, suggest a general increase of the probability of drought-year occurrences in the north-eastern and some parts to the south of the region. For the drought-year at short timescale, significant increase is evident over north-eastern Ethiopia, Somalia, Kenya and some parts of Tanzania. The increases become more pronounced for droughts at the 3- and 12-month (not shown) timescales, with increases in the probability of occurrences of up to +0.4% per decade. Elsewhere, there is a general tendency of decrease in the probability of drought-year occurrences up to −0.2% per decade on the average. However, the decreases are not as significant as for the increase trends where applicable.

10.6.2 Duration of Drought Events

Drought events exhibit duration of about 4, 5–7, 6–14 and 8–24 months for the 3-, 6-, 12- and 24-month timescales, respectively for half of the time as shown by the 50th percentile. On the other hand, the 90th percentile durations suggest that 1 out of 10 droughts at short timescales can persist for 3 – 18 months. At higher (12- and 24-month) timescales, drought persistence exhibit longer duration of up to 36 and 72 months, respectively. Figure 10.6 demonstrates the 50th and 90th percentiles of drought durations for 6- and 24-month timescales in panels (a) and (b); and (e) and (f), respectively.

The results show a tendency of decreasing trend in drought duration of up to 2 months/decade over most of Uganda and central Tanzania (extending to the north-eastern and south-western areas) for droughts at 3- and 6-month timescales. For droughts at 12-month timescales, the decreasing trend becomes widespread over Tanzania, eastern Kenya, southern and northern Somalia and south-western Ethiopia. For most of the northern sector of the GHA (Sudan, Eritrea, Ethiopia, Somalia and northern Kenya) an increasing trend in drought duration is evident for the 3- and 6-month timescales. On the contrary, for the droughts at 24-month timescale, the entire region exhibits steep positive trends in drought duration (up to a few years per decade), except over the south-western Tanzania. This pattern is demonstrated in Fig. 10.6 panels (c) and (g) for the 6- and 24-month timescales.

However, the respective p-values from a statistical test of non-zero trend significance ($\alpha = 0.05$) shown in panels (d) and (h), suggest that these trends in duration of drought events are not statistically significant. This is consistent with results of other studies worldwide [1, 28], also as reported in the IPCC special report (SREX), that there is no clear evidence of trends in the observed drought characteristics [18].

10.6.3 Drought Areal-Extent

Another important element for drought risk assessment is the areal-extent (% of total land area) experiencing drought at a time. Figures 10.7 and 10.8, show that there is remarkable variability in drought areal-extent over years. For example, around the period 1910–1925, the GHA region experienced droughts covering up to 80% of the area, and only about 20% for the 1930–1960. Then after the1960s, drought areal-extent is shown to be on the rise, even beyond the 1910–1925 levels in the last decades. However, there is a general increasing trend representing an overall increase of 22% in the region over the course of the 20th Century and the first decade of the 21st Century. For some of the individual Member States, the proportion almost doubled, while others shrunk considerably, as detailed below.

However, due to the differences in responses to drought drivers across the region, (discussed in Sect. 10.5.1), the drought areal-extents among Member States of the GHA region also differ. Increasing trends of 0.26–0.34% per annum over Sudan;

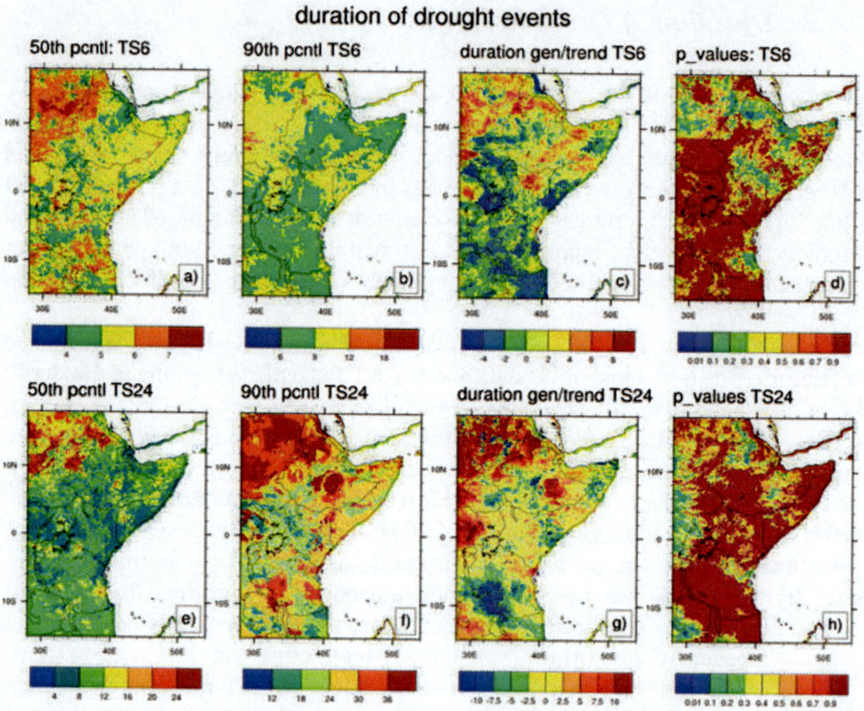

Fig. 10.6 The 50th and 90th percentiles of drought durations for 6- and 24-month timescales in panels **a** and **b**; and **e** and **f**, respectively. Long-term trend in general duration (months/decade) of droughts at 6- and 24-month time-scales in panels **c** and **g**, with their respective p-values from a statistical test of non-zero trend significance at 0.05 alpha (panels (d) and (h). The duration of more than half of the drought events is up to 7 and 24 months for the 6- and 24-month timescales, respectively. On the other hand, the 1 out of 10 drought events, can persist for up to 18 months (6-month timescales). Droughts at 24-month timescale, tend to persist much longer (≥36 months). Nevertheless, there is no significant trend in drought duration. *Source* Mpelesoka et al. [31]

+0.10% for Eritrea; +(0.34–0.40)% over Ethiopia; +(0.13–0.35)% for Somalia; +(0.04–0.10)% over Kenya; −0.10 to +0.04% for Uganda; and −0.13 to +0.01% over Tanzania are demonstrated. In the aspect of drought areal-extent, therefore, the threat is real for the GHA region in general, except for Tanzania and Uganda to a lesser extent where decreasing trends are evident for drought timescales greater than six months.

Increasing trends in drought areal-extent have serious implications that include worsening food security and water availability, straining the resilience of communities across the region. In Somalia alone, the number of people needing assistance rose from five million in September 2015 to over 6.2 million in February 2017 (more than half the country's population). This included a drastic increase in those facing "crisis" and "emergency" situations, from 1.1 million six months earlier to a projected 3 million between February and June 2017 [47]. Similar situation can be said

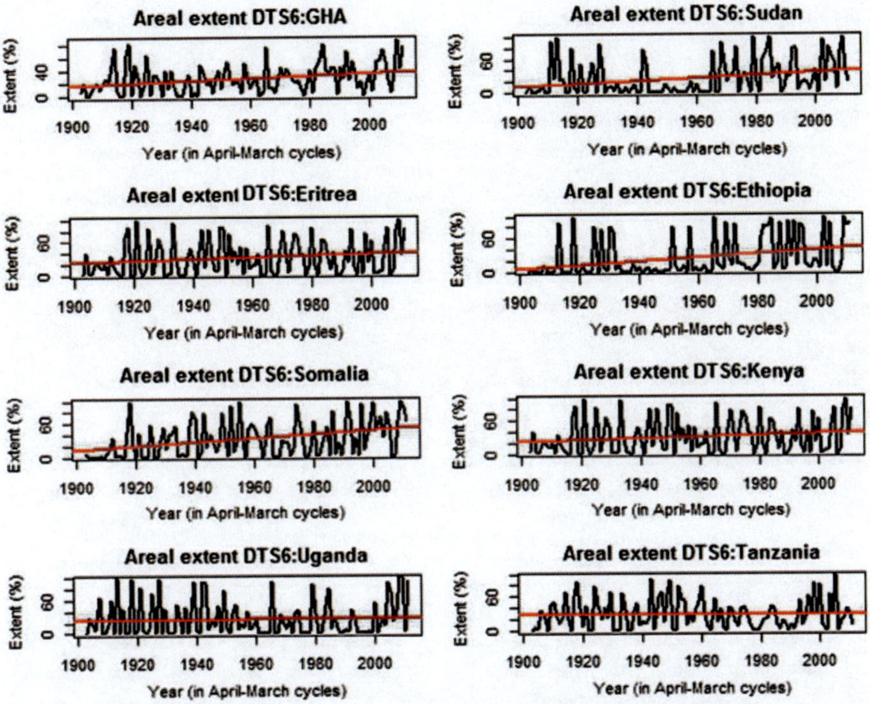

Fig. 10.7 Areal-extent (% of total land area) (black) and a trend line (red) of +0.20, +0.28, +0.10, +0.36, +0.35, +0.15, +0.04 and −0.01 (percent/annum) over the GHA region Members States: Sudan, Eritrea, Ethiopia, Somalia, Kenya, Uganda and Tanzania, respectively, for the 6-month timescale droughts of 1903–2012 annual (April-March) cycles. *Source* Mpelesoka et al. [31]

for South Sudan, Ethiopia, and some parts of Kenya. For example, Ethiopia appealed for US$ 948 million to address emergency needs for 5.6 million people in 2017. This was almost a 50-percent reduction in the number of food insecure people, which reached 10.2 million in 2016, as reported by the UNISDR [47].

10.7 Trends in Rainfall

The spatial distribution of mean annual rainfall and its trend in Fig. 10.9, panels (a) and (b), respectively, show that relatively wet regions such as the Lake Victoria basin are getting wetter and dry regions like Somalia and northeastern Ethiopia are getting drier. This process is often referred to as "amplification of the water cycle", associated with global warming in previous studies. Earlier studies indicated that the amplification of the water cycle was happening at 7% per 1 °C of global warming [16]. However, a new study [42] suggests that the amplification is happening at about

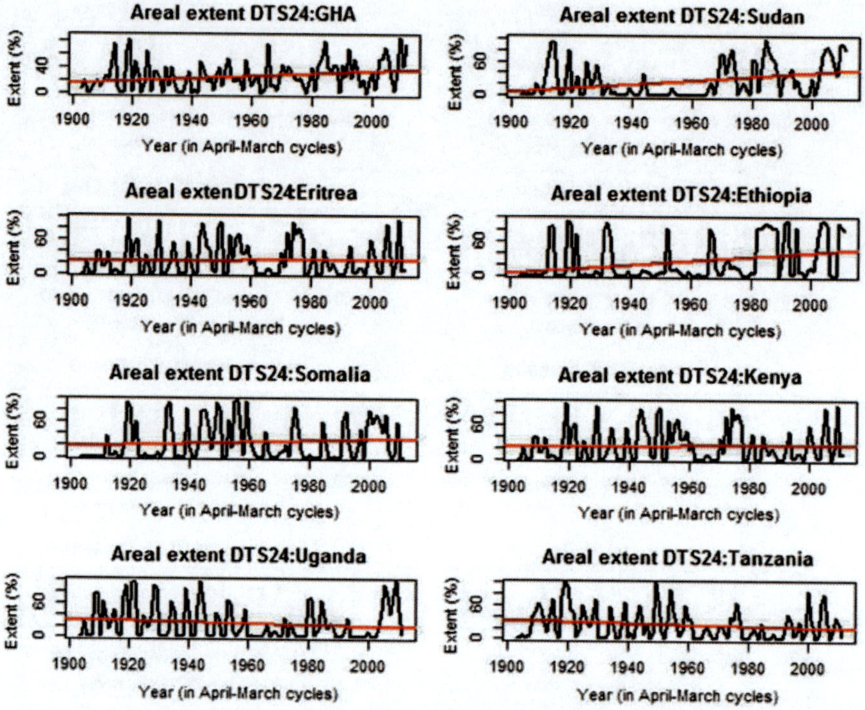

Fig. 10.8 Areal-extent (% of total land area) (black) and trend line (red) of +0.20, +0.34, +0.10, +0.35, +0.13, +0.04, −0.10 and −0.13 (percent/annum) over the GHA region Member States, Sudan, Eritrea, Ethiopia, Somalia, Kenya, Uganda and Tanzania, respectively, for the 24-month timescale droughts of 1903–2012 annual (April-March) cycles. *Source* Mpelesoka et al. [31]

3–4% per 1 °C. This is probably due to a weakening of the atmospheric circulation, which transports freshwater from the dry to wet regions of the globe.

Nevertheless, the low level of statistically significant rainfall trends demonstrated over the GHA, particularly the drying trend, also suggests a slower amplification process than first thought. The agreement between recent results from climate models [42] and observations (analyzed here) over the recent past is another important finding of this study because it adds confidence to climate model projections of water cycle amplification under greenhouse gas emission scenarios.

10.8 Summary of GHA's Drought Characteristics

The results show remarkable spatial variations in the association of the probability of drought occurrences with ENSO, IOD, IPO and NAO. These climate variability drivers tend to either enhance or suppress or displace the time/region dependent effect

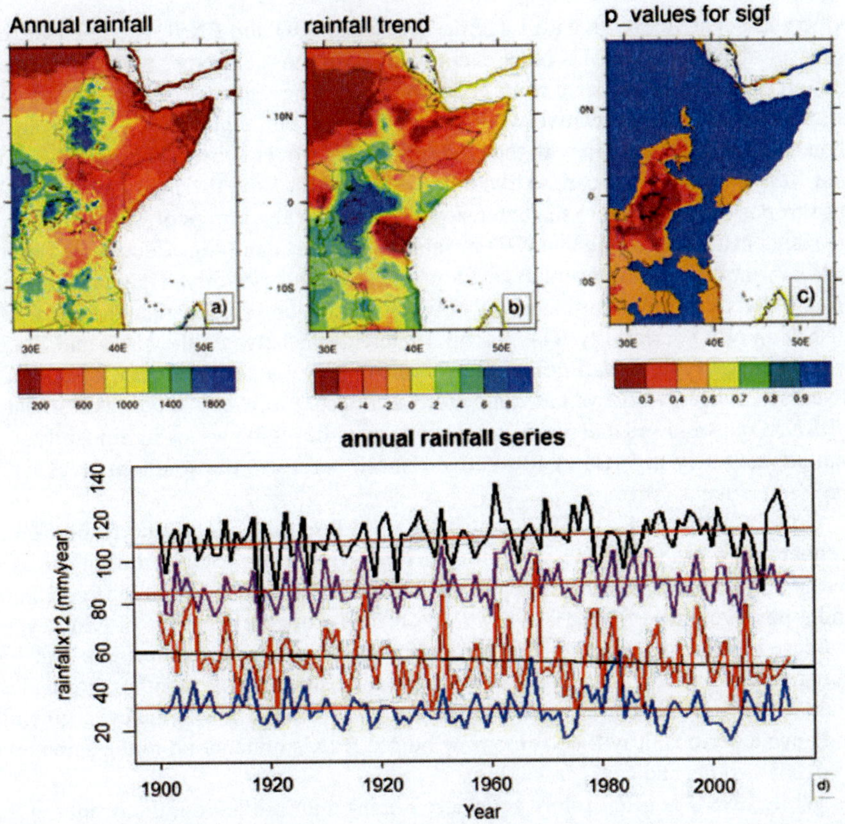

Fig. 10.9 Spatial distribution of annual **a** mean rainfall (mm/year); **b** rainfall trend (mm/decade); **c** p-values for trend statistical significance; and **d** rainfall time series (1900–2014) for selected grid-cells: Uganda (2.4°N, 33.2°E) black; Tanzania (4.3°N, 30.7°E) purple; Kenya (-2.7°N, 38.3°E) red; and Ethiopia (7.6°N, 42.5°E) blue. Panels **a** and **b** portray a tendency of relatively dry areas becoming drier and relatively wet areas getting wetter. However, panel **c** shows only a low level of statistical significance in these trends. *Source* Mpelesoka et al. [31]

of weather systems [39]. The abnormalities imposed on the effects of weather systems that include (1) the inter-tropical convergence zone (ITCZ), (2) monsoon/trade winds, (3) ridges, and (4), jet systems, are dependent on the status of modes of the prevailing drivers at a time. Naturally, complex interactions manifest between modes of these four major climate variability drivers. For example, ENSO extends its influence on modes of IOD and NAO, which in turn feed back onto ENSO [22]. The interactions between pairs of modes (Fig. 10.3) can alter their strength, periodicity, seasonality, and ultimately their predictability. It is thought that the IOD has a link with ENSO events through an extension of the Walker Circulation to the west and associated Indonesian flow of warm tropical ocean water from the Pacific into the Indian Ocean [22]. Hence, positive IOD events are often associated with El

Niño and negative events with La Niña. When the IOD and ENSO are in phase, the impacts of El Niño and La Niña events are often most extreme, while when they are out of phase the impacts of El Niño and La Niña events can be diminished. For example, the observed relatively bigger (areal-extent) influence of IOD on droughts than anomalies associated with the ONI, particularly over Eritrea, Ethiopia, Somalia and Kenya (Table 10.1) can be explained as a result of the thermodynamics in the mid-troposphere related to the extension of the tropical warm pool westwards [52].

Kajtar et al., [22] found that IOD variability has a net damping effect on ENSO and NAO variability, and conversely, each promote IOD variability. They also demonstrated a weak damping influence by NAO on ENSO, and an enhancing influence by ENSO on NAO variability. The potential interactions between the NAO and ENSO in influencing regional latitudinal inter-tropical convergence zone (ITCZ) shifts are discussed in [29]. They evaluated moisture and circulation field variations associated with NAO and showed that precipitable water over the region varies significantly such that anomalously high (low) convective rainfall occurs to the southern areas when the NAO is weak (strong).

On the other front, the IPO fluctuations are less understood due to being less frequent than other climate variability drivers. However, notably during the 20th century, there were negative IPO phases in 1925–1946 and 1978–1998 (Fig. 10.3b); and a positive phase from 1947 to 1976. Of interest is the fact that El Niño events increased between 1978 and 1998, therefore it is reasonable to speculate that IPO may be linked to ENSO events and to SST changes in general. In other words, SST changes may cause the IPO and not the other way around [41]. However, given its long cycle periods, it will be some time before IPO is understood to the same level of detail as other indices.

While ENSO is often solely considered in drought prediction, the results in this chapter demonstrated that not all droughts are ENSO driven. For the GHA, ENSO-driven droughts were found dominant only over Sudan, Eritrea, most of Ethiopia except the south-eastern areas, Somalia, western and south-western Tanzania. However, the highest proportion of ENSO-driven drought-year occurrences is about 70% for the 3- and 6-month timescales, otherwise generally much lower proportions (<30%) for the 12- and 24-month timescale droughts.

The spatial distribution of drought-risk parameters and the patterns of drought occurrence association with climate variability drivers revealed complex zonality. For example, over the northern and north-eastern areas drought occurrences are associated with El Niño, positive mode IOD, negative modes of IPO and NAO (with well-defined patterns at short timescales). The converse holds for the southern and south-western areas (e.g., occurrences of droughts are associated with La Niña). While the range of long-term probabilities of drought-year occurrences across the GHA region is understandably small (20–30%), the northeastern areas experience more persistent droughts than elsewhere. Similarly, the tendency of increases in drought areal-extent is more pronounced to the northern and north-eastern sectors of the region.

In addition, the trends in annual rainfall show that relatively wet regions (e.g. the Lake Victoria basin) are getting wetter while the converse is true for dry regions (e.g. Somalia and northeastern Ethiopia). However, these trends are of low statistical significance over the study period. This is consistent with findings of [42], suggesting a slower water cycle amplification process associated with climate change than first thought. The agreement between climate models and observations over the recent past is another important finding as it adds confidence to climate model projections. However, in the context of drought occurrences, this systematic rainfall reduction does not necessarily translate into drought events. Arguably, it could simply mean a change in rainfall regime towards new aridity levels. Thus, the management strategies could be different from those relevant to drought events.

10.9 Concluding Remarks

This chapter sought to quantify the variation of influence among climate variability drivers across the region; establish drought characteristics that underpin drought impacts management; and examine consistency of increase in drought-like crises with trends in drought characteristics over the GHA region. The chapter concludes that:

1. El Niño driven droughts affected 19% of the area for Eritrea and Ethiopia at 3- and 6-month timescales, respectively; 14–40% for Somalia; and 7–43% for Kenya across all drought timescales while the La Niña driven droughts affected 7–15% of Sudan, across the 4 drought time-scales; 4% and 8% for Kenya, at 3-, and 12-month drought timescales, respectively. The La Niña driven droughts affect Uganda most, affecting 30–60% of the area across all drought time-scales; and about 10–38% for Tanzania. Similarly, there is substantial variations in area proportions affected by IOD, IPO and NAO phenomena.
2. The long-term probability of drought-year occurrences across the GHA region range from 20 to 30%, except for isolated areas, which exhibit probabilities as low as 10% and up to 40% on the lower and higher cases, respectively. There is no statistically significant trend in the probability of drought-year occurrences.
3. The duration of more than half of the drought events is up to 4, 7, 14 and 24 months for the 3-, 6-, 12- and 24-month timescales, respectively. On the other hand, the 1 out of 10 drought events, can persist for up to 18 months (the 3- and 6-month timescales). Droughts at the 12- and 24-month timescales tend to persist much longer ($\geq$36 months). However, there is no significant trend in drought duration. By contrast, the drought areal-extent has shown generally increasing trends. On the average (across drought timescales), trends per decade of 3%, 1%, 3.7%, 2.4%, 0.7%, −0.3% and −0.6% were found for Sudan, Eritrea, Ethiopia, Somalia, Kenya, Uganda and Tanzania, respectively.
4. The results also show weak trends for the relatively wet areas (e.g. the Lake Victoria basin) getting wetter while the dry areas (e.g., Somalia and northeastern Ethiopia) becoming drier. These weak trends are consistent with Skliris et al., [42] findings and indicate a slower water cycle amplification process than first

thought. The agreement between climate models and observations over the recent past is an important finding, as it adds confidence to climate model projections.

5. In addition, since there is no evidence of substantially significant changes in drought characteristics, the peculiarity of drought-like crises in the GHA can be attributed to unaccounted for systematic rainfall reduction (among other factors reported elsewhere). This highlights the importance of distinguishing drought impacts from those associated with new levels of aridity. In principle drought is a temporary phenomenon while aridity is permanent, require deferent management strategies.

References

1. Awange JL, Mpelasoka F, Goncalves RM (2016) When every drop counts: analysis of droughts in Brazil for the 1901–2013 period. Sci Total Environ 566–567:1472–1488
2. Baroody J (1995) Anatomy of an overthrow: Why a revered African leader was toppled. http://adst.org/2016/10/anatomy-overthrow-revered-african-leader-toppled/. Association for Diplomatic Studies and Training (Accessed 6th February 2017)
3. Behera SK, Luo J, Masson S, Rao SA, Sakuma H, Yamagata T (2006) A CGCM study on the interaction between IOD and ENSO. J Clim 19:1688–1705
4. Beltrando G, Camberlin P (1993) Interannual variability of rainfall in eastern horn of Africa and indicators of atmospheric circulation. Int J Climatol 13:533–546
5. BOM (2017) The Three Phases of the El Niño–Southern Oscillation (ENSO). http://www.bom.gov.au/climate/enso/history/ln-2010-12/three-phases-of-ENSO.shtml. Beurau of Meteorology, Australia (Accessed 12 April 2017)
6. Cai W (2015) What is El Niño and why does it have so much influence over our weather? http://www.abc.net.au/news/science/2015-12-09/el-nino-what-is-it-and-whydoes-it-matter/6908126. ABC Science (Accessed 8 March 2017)
7. Cook ER, J. E., Cole RD, D'arrigo DWS, Villalba R (eds) (1999) Tree ring records of past ENSO variability and forcing. Cambridge University Press, Cambridge, U.K
8. Cordery I, Mccall M (2000) A model for forecasting drought from teleconnections. Water Res Res 36:763–768
9. Evans M, Hastings N, Peacock B (2000) Statistical distributions. Beta distribution. 3rd ed. Wiley, New York
10. FAO (2007) Evaluation of the FAO emergency and rehabilitation assistance in the Greater Horn of Africa (2004–2007). Final Report, FAO
11. FAO (2017) Millions of people face food shortages in the Horn of Africa. http://www.fao.org/news/story/en/item/468941/icode/. [Accessed 25 April 2017]
12. Funk C, Nicholson SE, Landsfeld M, Klotter D, Peterson P, Harrison L (2015) The Centennial Trends Greater Horn of Africa Precipitation Dataset SCIENTIFIC DATA 2:150050. https://doi.org/10.1038/sdata.2015.50
13. Gupta AK, Nadarajah S (eds) (2004). Handbook of beta distribution and its applications (Statistics: A Series of textbooks and monographs), 1st edn. CRC Press
14. Hafez Y (2016) Study on the relationship between the oceanic Niño Index and surface air temperature and precipitation rate over the Kingdom of Saudi Arabia. J Geosci Environ Protect 4:146–162
15. Henley BJ, Gergis J, Karoly DJ, Power SB, Kennedy J, Folland CK (2015) A tripole index for the interdecadal pacific oscillation. Clim Dyn 45:3077–3090
16. Huntington TG (2006) Evidence for intensification of the global water cycle: review and synthesis. J Hydrol 319:83–95

17. IGAD (2017) http://reliefweb.int/report/somalia/statement-igad-executive-secretary-current-drought-greater-horn-africa. [Accessed 06 March 2017]
18. IPCC (2012) In: II, I.W.G. (Ed.), Special report on managing the risks of extreme events and disasters to advance climate change adaptation (SREX). Stanford, CA 94305, USA
19. Jansson K, Harris M, Penrose A (1987) The Ethiopian famine. London Zed Press
20. Johnson NL, Kotz S, Balakrishnan N (1995) Continuous univariate distributions. 2nd edn. Wiley
21. Jones PD, Jonsson T, Wheeler D (1997) Extension to the North Atlantic oscillation using instrumental pressure observations from Gibraltar and south-west Iceland. Int J Climatol 17:1433–1450
22. Kajtar JB, Santoso A, England MH, Cai W (2017) Tropical climate variability: interactions across the Pacific, Indian, and Atlantic oceans. Clim Dyn 48:2173–2190
23. Kalnay E, Kanamitsu M, Kistler R, Collins W, Deaven D, Gondin L, Iredell M, Saha S, Whilte G, Woollen J, Zhu Y, Chelliah M, Ebisuzaki W, Higgins W, Janowiak J, Mo KC, Ropelewski C, Wang J, Leetmaa A, Reynolds R, Jenne R, Joseph D (1996) The NCEP/NCAR 40-year reanalysis project. Bull Am Meteor Soc 77:437–470
24. Kane RP (1997) Prediction of droughts in north-east Brazil: role of ENSO and use of periodicities. Atmos Sci Lett 17:655–665
25. LAT (1991) Chance for Peace in Ethiopia. http://articles.latimes.com/keyword/president-mengistu-haile-mariam. Los Angels Times [Accessed 5th February 2017]
26. Liebmann B, Hoerling MP, Funk C, Bladé I, Dole RM, Allured D, Quan X, Pegion P, Eischeid JK (2014) Understanding recent Eastern Horn of Africa rainfall variability and change. J Clim 27:8630–8645
27. Lyon B, Dewitt D (2012) A recent and abrupt decline in the East African long rains. Geophys Res Lett 39:L02702
28. Mccabe GJ, Wolock DM (2015) Variability and trends in global drought. Earth Space Sci 2:223–228
29. McHugh MJ, Rogers JC (2001) North Atlantic Oscillation influence on precipitation variability around the Southeast African convergence zone. J Clim 14:3631–3642
30. Mckee TB, Doesken NJ, Kleist J (1993) The relationship of drought frequency and duration to time scales. In: Proceedings of the eighth conference on applied climatology. American Meteorological Society, Boston, pp 179–184
31. Mpelasoka F, Awange JL, Zerihun A (2018) Influence of coupled ocean-atmosphere phenomena on the Greater Horn of Africa droughts and their implications. Sci Total Environ 610–611:691–702. https://doi.org/10.1016/j.scitotenv.2017.08.109
32. Mpelasoka F, Hennessy K, Jones R, Bates B (2008) Comparison of suitable drought indices for climate change impacts assessment over Australia towards resource management. Int J Climatol 28:1283–1292
33. Murphy BF, Timbal B (2008) A reviewof recent climate variability and climate change in southeastern Australia. Int J Climatol 28:859–879
34. NCEP (2017) Equatorial pacific sea surface temperatures. https://www.ncdc.noaa.gov/teleconnections/enso/indicators/sst.php. [Accessed 10 July 2017]
35. Nicholson ES (2014) A detailed look at the recent drought situation in the Greater Horn of Africa. J Arid Environ 103:71–79
36. Osborn TJ, Briffa KR, Tett SFB, Jones PD, Trigo RM (1999) Evaluation of the North Atlantic Oscillation as simulated by a climate model. Clim Dyn 15:685–702
37. Otake M, Prentice RL (1984) The analysis of chromosomally aberrant cells based on beta-binomial distribution. Radiat Res 98:456–470
38. Pozo-Vázquez D, Esteban-Parra MJ, Rodrigo FS, Castro-Díez Y (2000) An analysis of the variability of the North Atlantic oscillation in the time and the frequency domains. Int J Climatol 20:1675–1692
39. Riddle EE, Cook KH (2008) Abrupt rainfall transitions over the Greater Horn of Africa: observations and regional model simulations. J Geophys Res: Atmos 113:D15109

40. Saji NH, Ambrizzi T, Ferraz SET (2005) Indian Ocean Dipole Mode events and austral surface air temperature anomalies. Dyn Atmos Oceans 39:87–10
41. Salinger MJ, Renwick JA, Mullan AB (2001) Interdecadal Pacific Oscillation and South Pacific climate. Int J Climatol 21:1705–1721
42. Skliris N, Zika JD, Nurser G, Josey SA, Marsh R (2016) Global water cycle amplifying at less than the Clausius-Clapeyron rate. Sci Rep 6:38752
43. Swetnam TW, Betancourt JL (1998) Mesoscale disturbance and ecological response to decadal climatic variability in the American southwest. J Clim 11:3128–3147
44. Tierney JE, Ummenhofer CC, Demenocal PB (2015) Past and future rainfall in the Horn of Africa. Sci Adv 1:e1500682
45. UN (2014) http://www.un.org/waterforlifedecade/water_and_energy_2014/
46. UN (2017) Drought in Botswana is learning opportunity to achieve water security - UN rights expert. http://www.un.org/apps/news/story.asp?NewsID=52578#.WOSH2qFs2po:, Accessed date: 5 March 2017
47. UNISDR (2017) Prolonged drought threatens Greater Horn of Africa. United Nations Office for disaster risk reduction. https://www.unisdr.org/archive/52011. Accessed date: 26 March 2017
48. Venton P (2012) Drought risk management: practitioners's perspectives from Africa and Asia. UNDP
49. Weil CS (1970) Selection of valid number of sampling units and a consideration of their combination in toxicological studies involving reproduction, teratogenesis or carcinogenesis reproduction, teratogenesis. Food Cosmet Toxicol 8:177–182
50. Wilhite DA, Hayes MJ, Knutson C, Smith KH (2000) Planning for drought: moving from crisis to risk management. J Am Water Res Assoc 36:697–710
51. Williams DA (1975) The analysis of binary responses from toxicological experiments involving reproduction and teratogenicity. Biometrics 31:949–952
52. Williams A, Funk C (2011) A westward extension of the warm pool leads to a westward extension of theWalker circulation, drying eastern Africa. Clim Dyn 37:2417–2435
53. WMO (2012) Standardized precipitation index user guide Geneva
54. Yang W, Seager R, Cane MA, L, B, (2014) The East African long rains in observations and models. J Clim 27:7185–7202

Chapter 11
Extreme Climate: Food Security in GHA

The climate within the region is highly variable with a higher natural susceptibility to frequent climate extremes than neighbouring regions. In the past, the majority of interventions in the region have been in flood management. However, drought, which is a slow and progressive extreme climatic event, has until recently been the least anticipated extreme in this region filled with wetlands. As a result, its impacts are more adverse than those of floods because of the inadequacy of existing coping mechanisms. Although drought is a natural, recurring phenomenon which cannot be controlled, prediction of its frequency, severity and probability is essential for farmers in that it helps in planning and management. In the absence of such information, food security is threatened.—Awange et al., [9].

11.1 Summary

Given the significance of extreme climate (droughts and floods) on GHA's waters, using monthly and seasonal precipitation data for the period 1961–1999, this Chapter presents the work of [9] who established drought frequency and severity in relation to food security in the Lake Victoria part of GHA. Use is made of percentiles together with time series analysis for a 40 yr period, with the lower 25% of the quartile designated as the threshold values that indicate drought years. The percentiles and Drought Severity Index (DSI) are then used to ascertain the physical conditions and severity of the drought years. The 1980s and 1990s are found to be drier decades (the 1980s had more drought events than any other decade considered herein), with lower negative anomalies in comparison to the 1960s and 1970s, and the number of drought events and their severity have been increasing in recent years (especially during the 1990s). In general, drought affects the crop planting seasons of March to May (MAM) and September to November (SON), and leads to increased food imports in the years following the drought. Incidentally, MAM and OND are the main rainy seasons in the lake Victoria region (with MAM being the main rainy

J. Awange, *Food Insecurity & Hydroclimate in Greater Horn of Africa*,
https://doi.org/10.1007/978-3-030-91002-0_11

season) that replenishes the lake [24]. If the trends persist, the Lake Victoria region will face not only more severe drought events, but also significant reductions in its food security [9].

11.2 Droughts and Floods: Threat to Food Security

As already discussed in Chap. 5, droughts have contributed to declines noticed in Lake Victoria water levels, which has had a number of social and economic effects, e.g., affecting power production, with resulting economic losses [25]. Combined with other existing factors, these droughts contribute to a worsening food situation in the region where a majority of the population are poor and live below poverty line, see e.g., [2]. Food insecurity (i.e. a lack of basic food to provide enough energy for day to day activities) in the region could lead to a catastrophe [29], since many of the local communities living around Lake Victoria are among the poorest [7, 12] and most food insecure in the region, due to high population densities [2, 16, 30]. Moreover, widespread poverty, recurrent droughts, crop failures, high mortality rates (from tuberculosis, malaria, HIV/AIDS and more recently COVID-19), and environmental degradation (contamination, pollution, land/forest degradation, biodiversity degradation, introduction of exotic species are common in the region; [2]. Shipton[28] reviewed the famine and food shortage situation in Africa and focused specifically on those who suffer and the action they take, information which may be vital to the lake region part of GHA.

Food security is dependent on social, economic as well as climatic factors. Climatic extremes, particularly droughts and floods, affect the state of food security in Africa [19, 27]. Climatic extremes are normal climatic occurrences and their impacts on society vary with respect to vulnerability and the success or failure of coping mechanisms at both the household and national level. Absence of adaptation measures to climate variability results in negative impacts on food security, health, and ultimately, life. An example is the 1999–2001 drought that affected most parts of Kenya, including some areas that normally receive high rainfall. At its peak, 4.5 million people in Kenya lost their livelihoods and ability to cope, and were subsequently entirely dependent on relief food provided jointly by the government and donors [14]. Adger[1] discussed adaptation to climate change related phenomena in developing countries while [13] described food assistance programs. In addressing the Brazilian drought, Awange et al. [8] proposed the concept of living with droughts. The impacts of such extremes (in the context of climate change) on Lake Victoria Basin (LVB) has been assessed in [10, 23, 26], while Agutu et al. [3] proposed remotely sensed approach to monitoring agricultural droughts in the region to improve food security (see e.g., Chap. 14).

In the LVB, floods are more frequent than drought and have had adverse impacts. The climate within the region is highly variable with a higher natural susceptibility to frequent climate extremes than neighbouring regions. In the past, the majority of interventions in the region have been in flood management. However, drought, which is a slow and progressive extreme climatic event, has until recently been the least anticipated extreme in this region filled with wetlands. As a result, its impacts are more adverse than those of floods because of the inadequacy of existing coping mechanisms.

Although drought is a natural, recurring phenomenon which cannot be controlled [4, 15, 17, 18, 32, 33], prediction of its frequency, severity and probability, e.g., [8, 10, 22, 23] is essential for farmers in that it helps in planning and management. In the absence of such information, food security is threatened. Lack of such information in Uganda, for instance, exacerbated the impact of drought on maize yield and led Tumwesigye [31] to characterize drought patterns to help in the development of drought resistant maize cultivars. This Chapter focuses on the severity and frequency of drought events in the Kenyan part of the LVB, with the hope of providing mitigation and/or adaptation measures necessary for key stakeholders to improve drought management.

11.3 Drought Resistant Crops and the Planting Seasons

The study of [9] was centred on the Kenyan part of LVB, with data taken from Kisumu station, 0.1° S, 34.75° E (Fig. 11.1), which is approximately 1500 m above mean sea level. Lake Victoria Basin has 4 seasons: the hot dry season (December to February, DJF); the major (long) rainy season (March to May, MAM); the short dry season (June to August, JJA); and the short rainy season (October to December, OND) [9, 12]. In this chapter, however, September to November (SON) is used to coincide with the planting season. Kisumu has an annual average rainfall of 1340 mm, and experiences relatively high rainfall in the region. For instance, [5] found that both annual rainfall and rainfall during the long rainy season were greater at Kisumu than at any of the 8 stations in Kenya that they analysed. The highest rainfall in the LVB occurs in the north-western parts and gradually decreases towards the southeast, see [11]. The average annual rainfall for the LVB is 1424 mm, and varies between 891 mm in parts of Mara catchments to a maximum of 2168 mm in the middle of Yala Basin [6]. The principal occupation in the LVB is subsistence agriculture (e.g. maize, sorghum, beans, cassava), and planting takes place largely during the long and short rainy seasons (February to March and September to October, respectively). This contrasts with the rest of Kenya, where planting generally occurs in March to April and October to November [5]. Recently, however, this pattern has changed. During the 2005 long dry spell, for example, the planting season started as late as April–May in most parts of LVB (Kenya), and in 2006 it started in late March in most places.

As a response to these changing patterns, farmers have intensified the planting of drought resistant crops (e.g. sweet potatoes and cassavas). Cassava crops are, however, in decline, as consumer preferences shift towards maize and rice; poor cassava yields are also attributed to destruction by moles [21]. This further exacerbates food insecurity. Maize is the preferred crop, and (together with legumes) is often planted

Fig. 11.1 Lake Victoria basin (*Source* Kayombo and Jorgensen [20]) and weather stations (black triangles) in the Kenyan part of the region

in both rainy seasons, due to the fact that one yield is often insufficient to sustain farmers throughout the year. In addition to the problem of drought, poor soil fertility resulting from a lack of crop rotation has also contributed to serious food insecurity in the region. To address these emerging threats of drought and soil infertility, research is being done on drought-resistant varieties of maize.

In the study of Awange et al., [9], the data used consisted of:

1. Rainfall data (1961–1999). Kisumu station is selected because of its position in relation to other stations in the region (see Fig. 11.1). There are a number of other meteorological stations in the LVB, but all are found in the same (Lake Victoria) climatological zone. Awange et al. [9] therefore used one station to represent the region in order to avoid duplication of results. Kisumu is very close to Lake Victoria, and thus can sufficiently account for both the catchment and over-lake rainfall. Rainfall anomaly data are presented in Table 11.1.
2. Food import data (1961–2002). Secondary data on food imports was obtained from abstracts provided by the Central Bureau of Statistics (CBS) (available at the Kenyan Ministry of Agriculture's Library) and are presented in Table 11.2. Maize being the staple food in the region, the quantities of maize imports from 1961–2002 were used to represent overall food imports. In Kenya, food is not normally imported for specific regions affected by droughts; rather, the affected region triggers imports of food to the national reserve. When drought is severe in

Table 11.1 Seasonal rainfall anomalies for the Kisumu meteorological station. *Source* [9]

YEAR	DJF	MAM	JJA	SON
1961	–	−1.8	2.6	6.2
1962	4.3	3.2	−0.2	0.7
1963	1.7	1.4	−1.0	−2.1
1964	1.4	0.5	1.1	−1.1
1965	−0.5	1.7	−3.0	2.1
1966	1.2	0.2	0.8	−2.3
1967	−3.1	−1.1	−3.7	4.0
1968	0.1	1.1	1.6	1.2
1969	3.3	−1.8	−1.8	−0.9
1970	1.5	−0.3	−2.1	−2.1
1971	−2.0	2.2	−0.8	−0.9
1972	−0.3	−0.6	2.5	1.9
1973	0.7	−1.0	−2.6	−2.1
1974	−3.1	−0.3	1.8	−0.8
1975	−2.1	0.9	0.0	−1.4
1976	−0.2	−1.5	0.2	1.0
1977	−0.1	2.5	1.4	−1.3
1978	2.3	1.4	1.7	0.8
1979	1.9	−0.3	−0.7	−0.8
1980	−0.5	−0.3	−1.0	−1.4
1981	−2.0	0.4	0.8	−2.9
1982	−1.2	−3.7	2.7	5.0
1983	−1.9	−1.7	−0.4	−0.1
1984	−1.5	−3.1	3.2	0.9
1985	−0.1	2.0	0.5	−2.6
1986	−0.6	1.4	−0.6	−0.5
1987	0.1	−1.7	−0.2	1.5
1988	0.0	1.1	−0.5	0.1
1989	0.2	0.2	−1.9	0.2
1990	0.9	−1.0	−0.3	−2.5
1991	−0.2	0.8	−2.2	0.4
1992	−2.3	−1.8	4.0	0.4
1993	1.6	−0.4	0.3	−2.9
1994	−2.2	3.5	−0.4	0.1
1995	−0.7	0.9	−0.9	1.2
1996	0.0	1.5	−0.2	1.6
1997	−1.9	−0.3	−1.1	−1.4
1998	6.7	−1.4	−1.0	−2.1
1999	−1.5	1.7	2.3	−0.2

Table 11.2 Food imports (t) in Kenya (Central Bureau of statistics, statistical abstract, 2003. Extract on net imports, principle articles, Table 52b. *Source* [9])

YEAR	Maize (tons)	Wheat (tons)	Rice(tons)
1961	102130	13004	4037
1962	25590	52470	4128
1963	36	8523	4594
1964	264	6	672
1965	81452	5999	2096
1966	143444	24965	2021
1967	73	4512	207
1968	39	–	2995
1969	268	45	245
1970	14334	–	1149
1971	29077	13000	10203
1972	104	64821	2150
1973	82	77083	23
1974	728 1	3744	1502
1975	357 8	2917	4
1976	32	50	10001
1977	32	33	24
1978	80	90888	11
1979	18	21152	241
1980	323873	48462	1239
1981	77394	49239	4573
1982	89056	139326	11880
1983	–	81946	44768
1984	405443	149906	507
1985	125454	143793	562
1986	700	115281	61745
1987	–	217857	39129
1988	–	75578	10000
1989	2	123535	30006
1990	–	322632	27983
1991	–	242612	61163
1992	414929	100808	58920
1993	12874	314410	37150
1994	650387	353076	93519
1995	11965	249134	28177
1996	6759	486917	39159
1997	1101105	388138	62435
1998	368761	478865	62893
1999	75520	578543	53358
2000	409416	636045	105803
2001	308606	617542	172246
2002	229288	515179	148705

the LVB, for example, the Government checks the national reserve for surplus. If the food stored in the national reserve begins to dwindle, and the other regions are not in a position to supply the national reserve, then an order is made to import food.

11.4 Drought Analysis

For easier analysis and comparison purposes, both rainfall and food data are standardized in [9]. The daily records data are averaged to get the monthly mean

$$X_m = \frac{\sum d}{n}, \tag{11.1}$$

where d is the daily measurements and n the number of days in a given month. The monthly means are averaged for the 3 months in a season to obtain the seasonal mean. To get the annual means, the monthly means for all the 12 months of the year are averaged.

11.4.1 *Determination of Drought Years*

Once the rainfall mean is computed in (11.1), the "percent of normal" (quartile) is calculated by dividing actual precipitation by normal precipitation—typically considered to be a 30 yr mean -and multiplying by 100% (Eq. 11.3). Usually time scales range from a single month, or a group of months representing a particular season, to an annual or water year. Normal precipitation for a specific location is considered to be 100%. The normal precipitation N is calculated by

$$N = \frac{\sum X_A}{30}, \tag{11.2}$$

i.e. 30 yr mean of a season, where X_A is the annual mean summed over 30 years. Percent of normal P_n is then given by

$$P_n = \frac{\sum A_p}{N} \times 100\%, \tag{11.3}$$

where A_p is the actual precipitation. The Drought Severity Index (DSI) is then determined by considering as "dry" all those observations <25% (first quartile) of the ranked historical records, while those which were >75% (fourth quartile) are considered as "wet".

Table 11.3 Percentage range for dry and wet spells. *Source* [9]

Percentage rainfall relative to mean	Physical conditions
>175	Extremely wet
>150	Very wet
125	Near normal
100	Normal
75	Near normal
<50	Severe drought
<25	Extreme drought

11.4.2 Standardization of Data

In order to compare the data, the rainfall and food data are standardized using

$$Z = \frac{X_t - X_{30}}{\sigma}, \tag{11.4}$$

where Z is the standardized value, X_t is the observed value at a particular time, X_{30} the 30-year mean for the parameter and σ is the standard deviation. This gave the anomalies that are used for plotting the time series charts (see Table 11.1).

11.4.3 Data Analysis Methods

Anomalies of the parameters are used in the plotting for comparison between rainfall and food variables from the standardized values of Z in Eq. 11.4. Time series and relational analysis are then performed. The data are sorted in descending order; 25% of the normal and below represented the "lower threshold". The threshold value is then plotted in the time series, identifying severe and extreme drought as the 2 main categories (Table 11.3). This method is used to determine the frequency, severity and duration of droughts in LVB.

11.5 Drought Years and Food Security

11.5.1 Drought Years

Fig. 11.2 indicates that the period from 1960 to 1970 is wetter with higher positive anomalies. The period 1970–1980 shows a mixed spell of wet and dry years, while 1983 to 1997 shows a largely dry spell with lower negative anomalies. Table 11.4

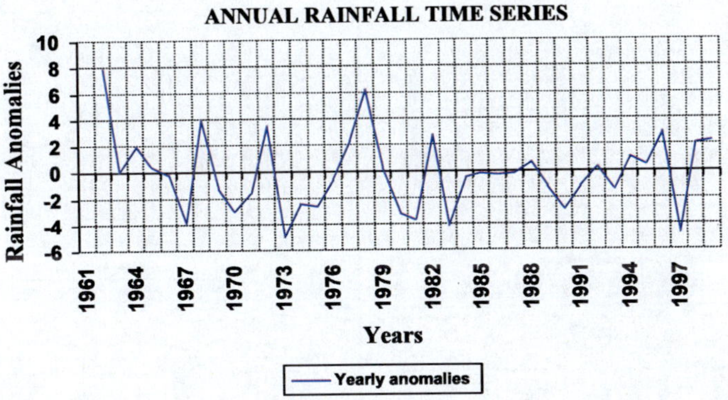

Fig. 11.2 Time series of annual rainfall anomalies at Kisumu during 1961–1999. *Source* Awange et al. [9]

Table 11.4 Seasonal drought years. *Source* [9]

Seasons	Drought years
DJF	1967, 1971, 1974, 1975, 1981, 1982, 1983, 1984,1992, 1994, 1997, 1999
MAM	1961, 1967, 1969, 1976, 1982, 1983, 1984, 1987,1992, 1998
JJA	1963, 1965, 1967, 1969, 1970, 1973, 1980, 1989,1991, 1995, 1997, 1998
SON	1963, 1966, 1970, 1973, 1975, 1977, 1980, 1981,1985, 1990, 1993, 1997, 1998

gives drought years obtained from observing the lower threshold values in Fig. 11.3, which are calculated as 1.2, 1.05, 0.98 and 1.38 for DJF, MAM, JJA and SON, respectively. Figure 11.3 and Table 11.5 show the severity and duration of drought years. The worst drought in terms of severity is 1980–1985, and drought events have been increasing in recent years. Because of the El Ñino rains of 1998, the 1990s are not particularly dry, but significant droughts did occur in 1990 and in 1997.

A general pattern of drought emerges:

- During the hot dry season (DJF), severe drought can be expected approximately every 7–8 years.
- During the long rainy season (MAM), severe. drought can be expected every 5–8 years.
- During the dry season (JJA) there is no clear cycle of drought events.
- During the short rainy season (SON), severe. drought can be expected every 3–4 years.

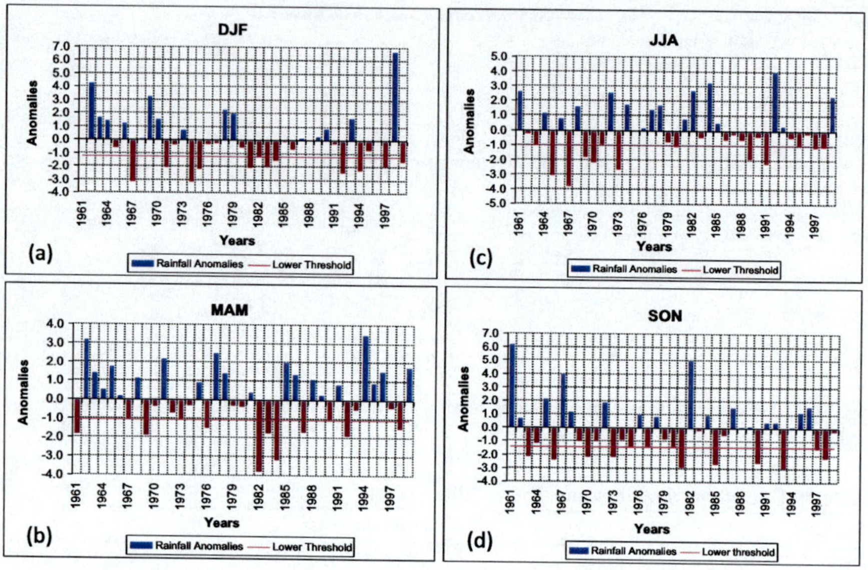

Fig. 11.3 Seasonal droughts from 1961–1999 (see Sect. 11.3 for season abbreviations). Blue bars indicate positive rainfall anomalies (i.e., increase in rainfall) while pink bars indicate negative anomalies (i.e., decrease in rainfall). *Source* Awange et al. [9]

Table 11.5 Periods of drought and their intensity. Extreme: rainfall <25% of normal; severe: <50% to 25% of normal. *Source* [9]

Drought	Year
Extreme	1973–1975, 1981–1985, 1990–1993, 1996–1997
Severe	1963, 1967, 1970–1971, 1980

11.5.2 Drought in Relation to Food Security

In the LVB the long and short rainy seasons (i.e. MAM and SON, respectively) are critical to food security. Figure 11.4 shows the relationship between the various seasonal droughts and food imports. For all the seasons considered, there is a significant rise in food imports in the years 1966, 1970, 1980, 1984, 1992, 1994 and 1997. These years followed the drought years identified in Tables 11.4 and 11.5, with a lag of 1 yr. From the drought profile in Table 11.6, a relationship between drought severity and populations affected by food relief is established (Fig. 11.5), which indicates that as drought severity increases, population receiving food relief also increases, and it also shows that after a significant drop of drought severity from 1971 to 1975, the population receiving food relief during this time also decreased.

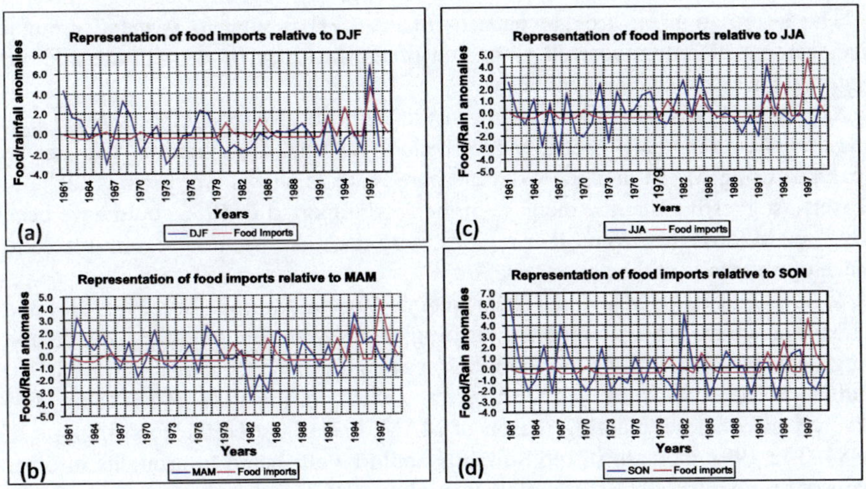

Fig. 11.4 Time series of food imports relative to the 4 seasons during 1961–1999. *Source* Awange et al. [9]

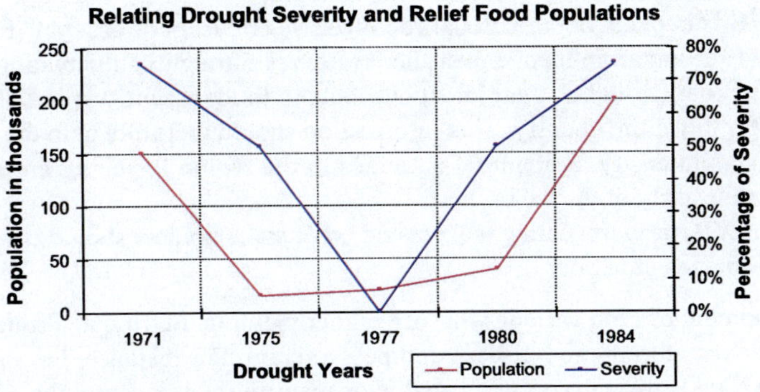

Fig. 11.5 Relationship between drought severity and food-insecure population in Kenya. *Source* Awange et al. [9]

Table 11.6 Drought profile in relation to numbers of people receiving food relief in the Lake Victoria region of Kenya (all droughts are geographically widespread). *Source* Kenya Government Disaster Preparedness Paper. *Source* [9]

Year	Drought	No. of people
1983–1984	Extreme	200 000
1980	Severe	40 000
1975	Severe	16 000
1971	Extreme	150 000

The information on drought patterns in the LVB is vital to farmers' drought management planning, as well as to that of the Ministry of Agriculture and the National Disaster Management body.

The variability of rain influences not only drought but also food imports (Fig. 11.5), and droughts in the LVB probably contribute to national food imports. This can be analysed using MAM and SON data. Figure 11.4b, d shows wet periods for both seasons, suggesting that the increase of the food imported in 1966 could have been triggered by droughts from other regions, with the Lake Victoria region unable to make up shortfalls in the national basket.

In 1969 and 1979, MAM and SON were dry (negative anomalies, Figs. 11.4 and 11.5). This could have affected the planting of maize during February–March and August–September, leading to a food shortage that triggered the food imports of 1970 and 1980. The extreme drought for Kenya as a whole during 1981–1985 (Table 11.5) severely affected the planting season of MAM, and triggered the food imports of 1984. The 1992 increase in food imports could be attributed to droughts in other parts of the country, since the rainfall anomalies of 1991 indicate a wet season in the LVB for MAM and SON. Similar to the El'Ñino year of 1997, rainfall in 1994 was above normal, which could have led to crop failure due to floods. The food imports of 1994 therefore could not be attributed to drought.

The LVB is dependent on national food reserves in years of drought (Fig. 11.5), while also exerting an influence over these reserves through its fluctuating level of food production. Droughts, therefore, contribute to food insecurity, not only for the LVB, but for the entire country. This information should therefore help the planners and managers intensify agricultural activities in the region for better crop yield to ensure food security in the basin.

For the LVB, farmers, policy makers and other stake holders should consider the following courses of action.

1. Development of crop varieties that are higher yielding, nutritious, drought resistant, early-maturing, and disease and pest tolerant. For instance, the PIONEER and PANNAR maize types are capable of maturing within 3 months, even with low rainfall. The setback is that the crop is favoured by weevils, and significant losses could occur through lack of proper treatment (this is a particular danger where yields are large, since farmers may fail to treat the entire crop in an attempt to reduce treatment costs). Information on such types of seed needs to be disseminated to farmers and the seeds made accessible and affordable to them.
2. The need to educate the people on the drought patterns ascertained in [9, 23], and the need to embrace drought resistant crops. For example, cassava—a favourite crop from the 1960s to the 1980s—is rapidly losing favour among farmers, because of its relatively poor performance in drought conditions and because of shifts in consumer preference.
3. Research and development of improved technologies in agriculture; for example, irrigation, such as has been achieved by the Yala Swamp project (despite doubts that were initially expressed concerning its environmental consequences)—mostly on wetlands—see [12].

4. Studies to establish the stored water potential of LVB (see Chap. 5). This will unravel the potential of irrigated agriculture within the region.

11.6 Concluding Remarks

Increased precipitation variability in the LVB has led to an increase in drought events in the recent past. Drought in the region is inversely related to food security (though the relationship is an indirect one), and if the drought trend persists, it is likely to have further negative effects. In the past, negative effects of drought may have stemmed from an inability to anticipate and cope with drought events. However, severe climate extremes were not as frequent as in recent decades, and since the trend of rainfall variability is expected to increase, drought severity is likely to be of greater magnitude, and with much greater extremes. The effect of future droughts on food security are thus likely to be severe, and the task of recovering from them is costly, if communities in this region do not anticipate trends of increased drought severity, and if they fail to adopt coping strategies, e.g., [22]. Drought analysis is therefore vital for proper food management.

As already stated in Sect. 1.2, food insecurity in GHA can also be linked to lack of education on precise farming methods, lack of technology and a general lack of information, i.e., of appropriate crops that could perform well in a given region. In most GHA countries, information can be accessed from Agricultural offices but rarely do farmers utilize such services. For example, Holo in Kisumu Kenya is known to perennially rely on vegetables supplied from more than 200 km away as inhabitants claim poor crop yield. A simple experiment was undertaken by the author where information on the correct vegetable type to be planted was obtained from the local agricultural office (Fig. 11.6). From a nearly stream, water was pumped and used to irrigate the planted vegetable (Fig. 11.6a). Figures (Fig. 11.6b–d) show the outcome, which dispels the notion that vegetables cannot perform well in the region on the one hand, while on the other hand, showcase that if the locals could be educated and utilize the available information, a great step would be advanced towards tackling food insecurity. Regional governments therefore need to close the gap between information availability from, e.g., agricultural offices and the same reaching and being implemented by farmers. In most cases, employees within agricultural offices wait for those who reach out to them as did the author in the experiment above. Perhaps the reverse should be more enhanced where they reach out to farmers or through use of modern social media. Information on its own without appropriate technological support, e.g., of water pumping tools as shown in Fig. 11.6a would not yield much. In this regard, the governments need to subsidize the cost of such tools, seeds and fertilizers in order to empower the locals.

Furthermore, since droughts are expected to become severe due to increase in rainfall variability, this will negatively affect the GHA's economy. Measures proposed in Sect. 1.3 such as broadening and maximising other opportunities (e.g., fish farming within Lake Victoria) need to be in place. Even with drought, Lake Victoria's waters

Fig. 11.6 Vegetables planted at Holo in Kisumu Kenya **a** the farmer irrigating the farm from a nearby river, **b** outcome of the proper use of technology, i.e., correct type of vegetable recommended by a local agricultural office accompanied by the proper irrigation technique, **c** zoomed version of **b**, and **d** alternative type of vegetable planted while the seedlings of **a** are maturing as recommended by the agricultural office

will never dry and its proper and sustainable utilization to provide fish can cushion the riparian communities around it and the broader GHA during famine. Appropriate technologies that support fish farming within the Lake therefore need to be developed as a measure to provide both food and employment to the locals, which is independent of rain-fed and irrigated agriculture.

References

1. Adger WN, Huq S, Brown K, Conway D, Hulme M (2003) Adaptation to climate change in the developing world. Prog Dev Stud 3:179–195
2. Agola NO and Awange JL (2014) Globalized poverty and environment. 21st century challenges and innovative solutions. Springer, Berlin
3. Agutu NO, Awange JL, Zerihun A, Ndehedehe CE, Kuhn M, Fukuda Y (2017) Assessing multi-satellite remote sensing, reanalysis, and land surface models' products in characterizing agricultural drought in East Africa. Remote Sens Environ 194:287–302. https://doi.org/10.1016/j.rse.2017.03.041
4. Akong JJ, Downing TE (1988) Small holder vulnerability and response to drought. In: Parry ML, Carter TR, Konjin NT (eds) The impact of climate variations on agriculture, vol 2. Assessments in semi-arid regions. Kluwer, Dordrecht, pp 221–247
5. Amissah-Arthur A, Jagtap S, Rosenzweig C (2002) Spatiotemporal effects of El Niño events on rainfall and maize yield in Kenya. Int J Climatol 22:1849–1860
6. APFM (Associate Programme on Flood Management) (2004) Strategy for flood management for Lake Victoria Basin, Kenya. www.apfm.info/pdf/strategy_kenya_e.pdf
7. Aseto O, Ong'ang'a O, Awange JL (2003) Poverty reduction: a challange of Lake Victoria Basin (Kenya). African Herald Publishing House, Kendu Bay
8. Awange JL, Mpelasoka F, Goncalves R (2016) When every drop counts: analysis of droughts in Brazil for the 1901–2013 period. Sci Total Environ 566–567:1472–88. https://doi.org/10.1016/j.scitotenv.2016.06.031
9. Awange JL, Aluoch J, Ogallo L, Omulo M, Omondi P (2007) An assessment of frequency and severity of drought in the Lake Victoria region (Kenya) and its impact on food security. Climate Res 33:135–142. https://doi.org/10.3354/cr033135
10. Awange JL, Khandu Forootan E, Schumacher M, Heck B (2016) Exploring hydro-meteorological drought patterns over the Greater Horn of Africa (1979–2014) using remote sensing and reanalysis products. Adv Water Res 94:45–59. https://doi.org/10.1016/j.advwatres.2016.04.005
11. Awange JL (2021) Lake Victoria monitored from space. Springer Nature International
12. Awange JL, Ong'ang'a O (2006) Lake Victoria: ecology, resources and environment. Springer, Heidelberg
13. Barrett CB (2002) Food security and food assistance programs. In: Gardner B, Rausser G (eds) Handbook of agricultural economics, vol 2. Elsevier Science, Amsterdam, pp 2103–2190
14. DMCN (Drought Monitoring Centre Nairobi) (2002) Factoring of weather and climate information and products into disaster management policy. A contribution to strategies for disaster reduction in Kenya. UNDP, Government of Kenya, and WMO, Nairobi
15. Downing TE, Kangethe W, Karnaw C (1989) Coping with drought in Kenya: national and local strategies. Lynne Rienner, Boulder CO
16. Ehrlich PR, Ehrlich AH, Daily GC (1993) Food security, population and environment. Pop Dev Rev 19:1–32
17. Garcia R, Escudero J (1982) The constant catastrophe: Malnutrition, famines and drought. Drought and man series, Vol. 2. IFIAS Project. Pergamon Press, New York

18. Glantz HM (1987) Drought and hunger in Africa: denying famine a future. Cambridge University Press, Cambridge
19. Kadomura H (1994) Climate changes, drought, desertification and land degradation in the Sudano-Sahelian region: a historic geographical perspective. In: Kadomura H (ed) Savannization process in tropical Africa. Tokyo Metropolitan University, II. Country briefs, pp 203–228
20. Kayombo S, Jorgensen SE (2006) Lake Victoria: experience and lessons learned brief. International Lake Environment Committee, Lake Basin Management Initiative, Kusatsu, Japan. www.ilec.or.jp/eg/lbmi/reports/27_Lake_Victoria_27February2006.pdf
21. Kiptot E, Franzel S, Hebinck P, Richards P (2006) Sharing seed and knowledge: Farmer to farmer dissemination of agroforestry technologies in western Kenya. Agrofor Syst 68:167–179
22. Mpelasoka F, Awange JL, Goncalves RM (2018) Accounting for dynamics of mean precipitation in drought projections: A case study of Brazil for the 2050 and 2070 periods. Sci Total Environ. https://doi.org/10.1016/j.scitotenv.2017.10.032
23. Mpelasoka F, Awange JL, Zerihun A (2018) Influence of coupled ocean-atmosphere phenomena on the Greater Horn of Africa droughts and their implications. Sci Total Environ 610–611:691–702. https://doi.org/10.1016/j.scitotenv.2017.08.109
24. Nicholson SE (1998) Historical fluctuations of Lake Victoria and other lakes in the Northern Rift Valley of East Africa. In: Lehman JT (ed) Environmental change and response in East African lakes. Kluwer, Dordrecht, pp 7–35
25. Otieno HO, Awange JL (2006) Energy resources in East Africa. Springer, Heidelberg
26. Phoon SY, Shamseldin AY, Vairavamoorthy K (2004) Assessing impacts of climate change on Lake Victoria Basin, Africa: people-centred approaches to water and environmental sanitation. 30th Water Engineering and Development Centre (WEDC) Int Conf, Vientiane, Lao PDR, pp 392–397
27. Scoones L, Chibudu C, Chikura S, Jeranyama P, 8 others, (1996) Hazards and opportunities: farming livelihoods in dryland Africa. Zed Books, London, Lessons from Zimbabwe
28. Shipton P (1990) African famines and food insecurity: anthropological perspectives. Ann Rev Anthropol 19:353–394
29. Skees JR (2000) A role for capital markets in natural disasters: a piece of the food security puzzle. Food Pol 25:365–378
30. Swaminathan MS (2001) Food security and sustainable development. Curr Sci 81:948–954
31. Tumwesigye EK, Musiitwa F (2004) Characterizing drought patterns for appropriate development and transfer of drought resistant maize cultivars in Uganda. In: Fiesen DK, Palmer AFE (eds) Integrated approaches to higher maize productivity in the new millennium: Proc 7th Eastern and Southern Africa Regional Maize Conf, 5–11 February 2002, Nairobi, Kenya. CIMMYT (International Maize and Wheat Improvement Center) and KARI (Kenya Agricultural Research Institute), Nairobi, pp 260–262
32. Awange JL (2012) Environmental monitoring using GNSS, global navigation satellite system. Springer, Berlin, New York
33. Awange JL (2018) GNSS environmental sensing, revolutionizing environmental monitoring. Springer, Berlin, New York

Chapter 12
Hydrometeorological Droughts over GHA

Since there is no evidence of significant changes in drought characteristics, the peculiarity of drought-like crises in the GHA can be attributed (at least in part) to unaccounted for systematic rainfall reduction. This highlights the importance of distinguishing drought impacts from those associated with new levels of aridity. In principle drought is a temporary phenomenon while aridity is permanent, a difference that managers and decision-makers should be more aware.—Mpelasoka et al., [6]. *Of interest in this Chapter is the question "is there any possible irreversible impact of this hydrometeorological drought on aquatic species considering the fact that LVB is home to Tilapia (Oreochromis niloticus) and Nile Perch (Lates niloticus) Fish? What can be done to avoid or minimize these impact?"*

12.1 Summary

Spatio-temporal patterns of hydrological droughts over the Greater Horn of Africa (GHA) are explored based on total water storage (TWS) changes derived from time-variable gravity field solutions of Gravity Recovery And Climate Experiment (GRACE, 2002–2014), together with those simulated by Modern Retrospective Analysis for Research Application (MERRA, 1980–2014). These hydrological extremes are then related to meteorological drought events estimated from observed monthly precipitation products of Global Precipitation Climatology Center (GPCC, 1979–2010) and Tropical Rainfall Measuring Mission (TRMM, 1998–2014). The major focus of this chapter, which is based on the work of [3] lies on the application of spatial Independent Component Analysis (sICA) to extract distinguished regions with similar rainfall and total water storage (TWS; surface, groundwater, soil moisture and vegetation water) with similar overall trend and seasonality. Rainfall and TWS are used to estimate Standard Precipitation Indices (SPIs) and Total Storage Deficit Indices (TSDIs), respectively, that are employed to characterize frequency and intensity of hydro-meteorological droughts over GHA. Significant positive (negative) changes in monthly rainfall over Ethiopia (Sudan) between 2002 and 2010 leading to

J. Awange, *Food Insecurity & Hydroclimate in Greater Horn of Africa*,
https://doi.org/10.1007/978-3-030-91002-0_12

a significant increase in TWS over the central GHA region are noted in both MERRA and GRACE TWS (2002–2014). However, these trends are completely reversed in the long-term (1980–2010) records of rainfall (GPCC) and TWS (MERRA). The four independent hydrological sub-regions extracted based on sICA (i.e., Lake Victoria Basin, Ethiopia-Sudanese border, South Sudan, and Tanzania) indicate fairly distinct temporal patterns that matched reasonably well between precipitation and TWS changes. While meteorological droughts are found to be consistent with most previous studies in all sub-regions, their impacts are clearly observed in the TWS changes resulting in multiple years of extreme hydrological droughts. Correlations between SPI and TSDI are found to be significant over Lake Victoria Basin, South Sudan, and Tanzania. The low correlations between SPI and TSDI over Ethiopia are likely related to inconsistency between TWS and precipitation signals. Further, hydrological droughts in these regions are found to be significantly associated with Indian Ocean Dipole (IOD) events while El'Niño Southern Oscillation (ENSO) played a secondary role.

12.2 Greater Horn of Africa's Drought

As stated in Chap. 1, the people of the semi-arid region of Greater Horn of Africa (GHA) largely depend on rain-fed agriculture and livestock that is increasingly coming under threat from the intensified frequency and severity of drought events over the past decades, e.g., [36]. Since most crops in GHA are planted during rainy seasons, i.e., March-May (MAM) and October–December (OND; see Chap. 11), food security is increasingly coming under jeopardy given the fact that these seasons too are now reported to experience drought episodes [44]. This is further exacerbated by the continued warming of the Indian-Pacific Oceans possibly linked to anthropogenic influences or multi-decadal climate variability (see Chaps. 7 and 8), which has been shown to contribute to more frequent droughts in GHA over the past 30 years during the spring and summer seasons, see e.g., [32, 42, 69].

Even with the reality of the threat posed by droughts in the GHA, see e.g., [42], effective drought monitoring in this region is, however, challenged by multiple factors including: inadequate spatial coverage and hydro-meteorological observations (e.g., precipitation, temperature, soil moisture, and groundwater storage) that are not readily available beyond the local meteorological services, poor communication frameworks, and national red tapes in accessing data.

Furthermore, the large spatial extent of GHA makes the collection of hydrological data quite challenging and hence the reason why most studies e.g., [67] have focused largely on meteorological droughts. This view is supported, e.g., by [49] who pointed out that the lack of reliable hydro-meteorological data hold up development of effective real-time drought monitoring and early warning systems in the region. To account for various drought conditions over GHA, therefore, an integrated drought index that involves various meteorological and hydrological variables is desirable, e.g., the African Drought Monitor (ADM[1]) project [61]. This, therefore, highlights the significant role played by satellite remote sensing in data deficient regions (see e.g., Chap. 3).

To this extent, the Gravity Recovery and Climate Experiment (GRACE) [63] satellite remote sensing mission (see Chap. 3), launched in 2002, offers the possibility to characterize droughts at a wider spatial coverage albeit for shorter time span (from 2002 onwards). The value of GRACE-derived total water storage (TWS) changes for drought studies and estimating, e.g., a Total Storage Deficient Index (TSDI), has been demonstrated e.g., in [7, 11, 19, 20, 33, 39–41, 64, 72].

Although GRACE has the capability to offer a wider spatial coverage, spatio-temporal analysis of drought in GHA using a single areal-average as a representation of the entire GHA region is challenging. This is mainly due to the meteorological observations that are influenced by spatial inhomogeneity of climate over the GHA caused by uneven topography, north-south migration of Inter-tropical Convergence Zone (ITCZ), and spatially varying influences of large-scale climate events among others, see e.g., [67].

In this chapter, as opposed to [12, 14, 16, 17] that focused mainly on TWS changes in Lake Victoria Basin (LVB), Nile River Basin (NRB) and Ethiopia, GRACE satellites and sICA method [29–31] are employed for the first time to characterize meteorological and hydrological droughts in the entire GHA. The novelty is that for the first time, the long-term (1980–2014) meteorological and hydrological drought in GHA is extracted using the Standardized Precipitation Index (SPI) and TSDI, respectively. The spatio-temporal variability of SPI and TSDI are also compared and a comprehensive interpretation of various drought events across the GHA is provided.

Specifically, the chapter (i) applies sICA to partition the GHA into various sub-regions based on precipitation and TWS changes, (ii) employs GRACE (2002–2014) and Modern Retrospective Analysis for Research Application (MERRA, 1980–2014) TWS changes to derive hydrological drought indices for the four independent sub-regions (as computed in i above), (iii) uses observed precipitation products, Global Precipitation Climatology Center (GPCC, 1979–2010) and Tropical Rainfall Measuring Mission (TRMM, 1998–2014) to compute remotely sensed SPI, and (iv), examines the relative influence of ENSO and IOD events on hydro-meteorological droughts (reflected in TSDI and SPI) within the GHA sub-regions.

The remainder of the chapter is organized as follows. In Sect. 12.3, an overview of the data as well as the analysis methods used is presented. In Sect. 12.4, the drought indicators used are described, while in Sect. 12.5, the results are analysed and interpreted. The chapter is concludes by looking at the possible impacts of this hydrometeorological drought on aquatic species considering the fact that LVB is home to Tilapia (*Oreochromis niloticus*) and Nile Perch (*Lates niloticus*) fish, which is a source of livelihood and food security in GHA in Sect. 12.6.

[1] http://stream.princeton.edu/AWCM/WEBPAGE/interface.php?locale=en.

12.3 Hydrometeorological Products

The hydrometeorological products used in this chapter include monthly precipitation obtained from global gridded rain gauge and near-global satellite-based estimates, as well as total water storage (TWS) changes from a reanalysis model and Gravity Recovery and Climate Experiment (GRACE) time-variable gravity field solutions.

12.3.1 Precipitation Products

1. *GPCC v6*: The Global Precipitation Climatology Center (GPCC) provides gridded precipitation products at various temporal and spatial resolutions derived from up to 67,000 quality controlled station data around the world between 1901 and 2010 [60]. This chapter use monthly precipitation data at $0.50° \times 0.50°$ (latitude $\times$ longitude) spatial resolution covering the period 1979–2010. GPCC products have been found to be consistent with other rainfall products in Africa, see e.g., [2, 13]. The distribution of rain gauges over the GHA region, however, is rather sparse over certain areas such as Somalia, Sudan, and Tanzania, and as such, may not provide accurate representation of the precipitation variability over these regions.
2. *TRMM 3B43*: To complement the GPCC v6 data, monthly precipitation estimates from the Tropical Rainfall Measuring Mission (TRMM) Multi-satellite Precipitation Analysis (TMPA) [35] version 7 for the period 1998 to 2014 are used. These products at $0.25° \times 0.25°$ spatial resolution are interpolated to $0.50° \times 0.50°$ to make them consistent with those of GPCC. Awange et al., [2, 13] assessed various precipitation products over Africa covering 2003–2010, from which TRMM 3B43 (hence forth referred to as TRMM) indicated similar skills compared to GPCC v6 (Fig. 12.1).

12.3.2 Water Storage Change Products

The reprocessed GRACE Level-2 RL05a time-series for the period 2002 to 2014 from the Gravity Recovery and Climate Experiment GRACE, e.g., [63] is used. The GRACE data is provided by GeoForschungsZentrum (GFZ, Potsdam) [23] as sets of (approximately) monthly fully normalized geopotential spherical harmonic coefficients up to degree and order 90 (see, ftp://podaac.jpl.nasa.gov/allData/grace/L2/).

GRACE degree one (C_{10}, C_{11}, S_{10}) coefficients are replaced by those from [21], and the more accurate degree two coefficients derived from satellite laser ranging (SLR) [22] replaced C_{20}. A commonly used non-isotropic de-correlation filter DDK3 [28, 37] is applied to reduce the high degree/order correlated errors, which are mani-

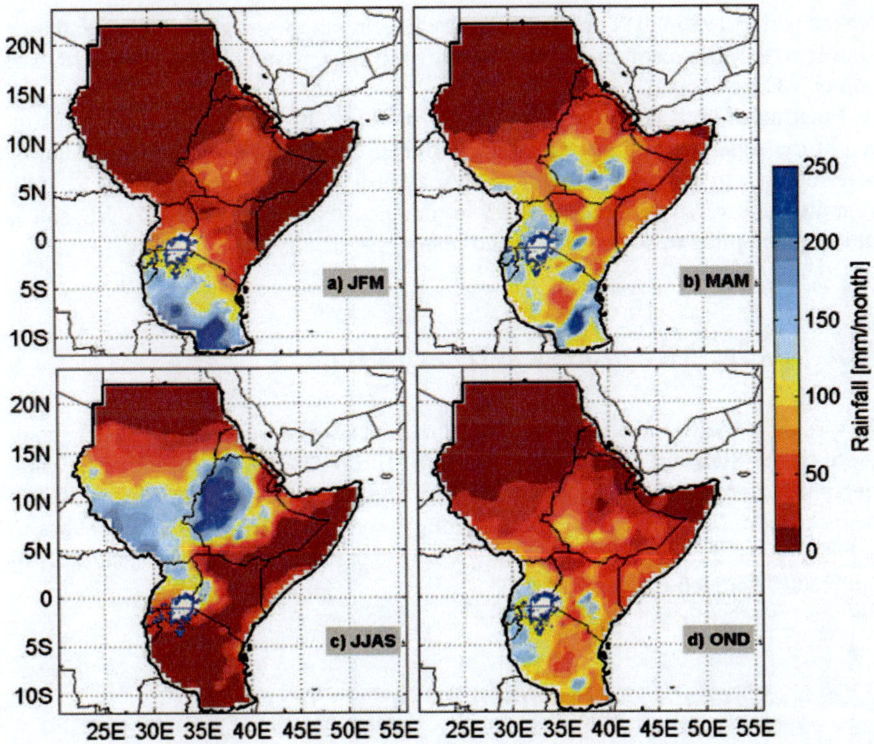

Fig. 12.1 Spatial variability of seasonal rainfall over GHA from 1979 to 2010 based on GPCC v6 data: **a** January–March (JFM), **b** March–May (MAM), **c** June–September (JJAS), and **d** October–December (OND). Note that the signals over Lake Victoria has been masked out in the analysis to avoid spurious trends due to varying lake levels. *Source* [3]

fested as stripes in the spatial domain. This filter removes most of north-south stripes and seems to be suitable to process GFZ RL05a products over the GHA. To be consistent with GPCC, the time-variable spherical harmonic coefficients are converted to $0.50° \times 0.50°$ (latitude × longitude) TWS grids using the approach of [68]. The selected spatial grid is optimistic for GRACE products whose spatial resolution is $\sim 1°$.

12.3.3 Reanalysis Products

GRACE TWS estimates are compared with those simulated by the Modern Retrospective Analysis for Research Application (MERRA) [59]. MERRA reanalysis are run globally at a relatively high spatial resolution ($0.67° \times 0.50°$) and are available as monthly products from 1979-present. Monthly MERRA-LAND TWS estimates for the period 1980–2014 are processed by filtering in the spectral domain and then converted to a grid resolution of 0.50° × 0.50° to be consistent with those of GRACE and GPCC products.

Filtering of both GRACE and MERRA products, however, causes some damping of the signal amplitude and might introduce spatial leakages, which should be restored by introducing a multiplicative scaler (or a gridded) gain factor e.g., [38]. A multiplicative scale factor of 1.05 is obtained from the MERRA TWS and is uniformly applied to both GRACE and reanalysis data used.

12.4 Hydro-Meteorological Drought Indices

To assess the hydro-meteorological drought events, two drought indicators are analysed: (i) the Standard Precipitation Index (SPI) estimated from monthly precipitation products of GPCC and TRMM, and (ii), the Total Storage Deficit Index (TSDI) deduced from MERRA and GRACE. Drought conditions are assessed over various sub-regions of the GHA extracted using the spatial independent component analysis approach discussed in Sect. 12.4.3.

12.4.1 *Standardized Precipitation Index (SPI)*

The Standardized Precipitation Index (SPI) is a widely-used meteorological drought indicator based on probability distribution of long-term rainfall time-series developed by [45, 46] to provide a spatially and temporally invariant measure of rainfall deficit (or excess) over a variety of accumulation timescales (e.g., 3, 6, 12, 24 months). SPI is derived by fitting a parametric cumulative probability distribution (CDF) of e.g., γ-distribution to the precipitation time-series, which are then transformed to obtain the SPI using the inverse normal (Gaussian) distribution function, e.g., [67]. SPI provides the value of standard anomalies from the median indicating negative for drought and positive for wet conditions (see details in Table 12.1). SPI is computed for the 12-month accumulation time periods to capture the long-term trend of meteorological droughts over the GHA. Detailed discussion on SPI are presented in Chap. 10.

12.4.2 *Total Storage Deficit Index (TSDI)*

The Total Storage Deficit Index (TSDI) is a renamed version of Soil Moisture Deficient Index (SMDI) [48] by [72]. TWS changes, which comprise information of the changes in all water components (surface, groundwater, soil moisture, biomass, ice and snow) are used to compute TSDI. The procedure to estimate TSDI (based on GRACE data) involves an estimation of the Total Storage Deficit as (TSD%) [72]:

Table 12.1 Various categories of droughts and wet conditions based on the SPI Index [45, 67]. *Source* [3]

SPI	Category
+2.0 and above	Extreme wet
+1.5 to +1.99	Very wet
+1.0 to +1.49	Moderately wet
+0.99 to −0.99	Normal
−1.0 to −1.49	Moderate drought
−1.5 to −1.99	Severe drought
−2 and below	Extreme drought

$$TSD(k, j - 2001) = \frac{TWS(k, j) - Mean(TWS(k, :))}{Max(TWS(k, :)) - Min(TWS(k, :))} \times 100, \qquad (12.1)$$
$$k = 1, 2, \ldots, 12,\ j = 2002, 2003, \ldots, 2014$$

where $TWS(k, j)$ is the kth month of total water storage (anomaly) in each year ($j = 2002, 2003, \ldots, 2014$) obtained from GRACE. For MERRA, the left hand side of Eq. 12.1 becomes $TSD(k, j - 1979)$ and $j = 1980, 1981, \ldots, 2014$. $Mean(TWS(k, :))$, $Max(TWS(k, :))$, and $Min(TWS(k, :))$ are respectively the mean, maximum, and minimum of the k'th month of TWS over the study period (for GRACE covering 2002–2014 and for MERRA 1980–2014). The TSD values in Eq. 12.1 can be stored in a vector TSD_i ($i = 1, 2, \ldots,$ length of TWS time series). Consequently, TSDI can be computed as

$$TSDI_i = p \times TSDI_{i-1} + q \times TSD_i, \qquad (12.2)$$

where the drought severity and duration factors p and q respectively are defined from the cumulative TSD plot (e.g., Fig. 12.2) based on the following relation [72]:

$$p = 1 - \frac{m}{m + b}, \qquad q = -\frac{C}{m + b}. \qquad (12.3)$$

In Eq. 12.3, C represents drought intensity, which is obtained from the best fit line of the cumulative TSD (from the drought monograph) during the dryness period (see e.g., line 3 in Fig. 12.2). In general, C can take on any of the four drought classifications values (−4.0 for extreme drought, −3.0 for severe drought, −2.0 for moderate drought, and −1.0 for mild drought) defined according to [56]. The other elements of the regression line are the slope m and the intercept b of the best fit of the drought monograph for the dryness period of the cumulative TSD curve.

Using an example, it is illustrated how the TSDI values of this chapter are estimated. Figure 12.2a shows a TSD (%) plot over Ethiopia derived from GRACE TWS

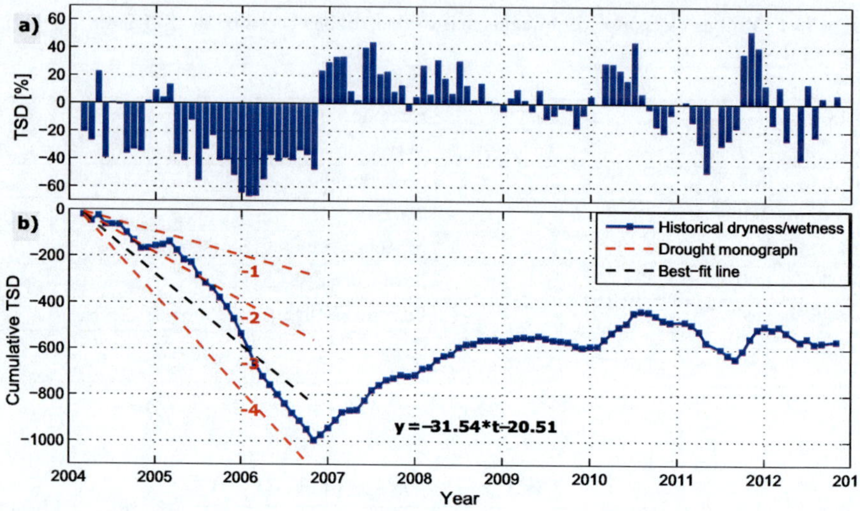

Fig. 12.2 Total storage deficit over eastern GHA region from 2004 to 2013 **a** and cumulative total storage deficit (%) over the same period (**b**). Considering the horizontal line of zero in the figure **b** to represent near normal condition, following [56] classification, the interval from near normal to severe dryness is divided into four equal intervals, with the part above the best fit line having two lines (1 and 2) and the best fit is shown by line 3. To complete the [56] classification, a fourth line 4 is plotted. All the four lines are drawn at equal intervals to define the extreme (Line 4), severe (Line 3), moderate (Line 2), and normal (Line 1) drought conditions. *Source* [3]

changes based on Eq. 12.1 between 2004 and 2013. A prolonged drought pattern from March 2004 to November 2006 (33 months) has been detected with the TSD values reaching as low as 66%. Assuming this scenario, the cumulative TSDs in Fig. 12.2b are plotted to derive the parameters in Eqs. 12.2–12.3. Using the linear regression parameters during the declining TWS period (i.e., 2004–2007), TSDI is computed for the period of decline. The slope ($m = -31.54$) and y-intercept ($b = -20.51$) parameters derived from the cumulative TSD based on 33 months from March 2004 to November 2006 are subsequently used to calculate the critical parameters p and q. Therefore, $C = -3$ is the category of drought event as can be seen in Fig. 12.2. Correspondingly p and q are derived as 0.394 and 0.058, respectively. $TSDI_{1,1}$ is assumed to be 2% of TSD_1 in line with [72]. The TWS droughts, categorized based on the "C" values of Fig. 12.2, are reported in Table 12.2. This procedure is used to compute TSDI from GRACE and MERRA TWS changes in order to explore hydrological droughts over the GHA for the entire study period (1980–2014).

Table 12.2 Various categories of droughts and wet conditions based on the TSDI Index [56, 72]. *Source* [3]

TSDI	Category
+4.0 and above	Extreme wet
+3.0 to +3.99	Very wet
+2.0 to +2.99	Moderate wet
+1.99 to −1.99	Normal
−2.0 to −2.99	Moderate drought
−3.0 to −3.99	Severe drought
−4 and below	Extreme drought

12.4.3 Spatial Independent Component Analysis

The spatial independent component analysis (sICA) [30] approach is applied to extract statistically independent modes of precipitation and TWS changes. Note that sICA patterns are essentially regions with similar seasonality (and not typical spatial patterns of variability) and as such, there could be significant bias in the analysis from the wettest stations. As the estimated modes are statistically independent, they can be separately analysed without considering other modes, see an application of sICA over the Nile River Basin (NRB) in [14]. Suppose $\mathbf{X}(t, s)$ is a gridded time series of precipitation or TWS changes after removing their dominant annual and semi-annual cycles. sICA decomposes the time-series of $\mathbf{X}$ into j spatial $\mathbf{S}_j$ and temporal $\mathbf{A}_j$ modes as:

$$\mathbf{X}(t, s) = \mathbf{A}_j(t)\ \mathbf{S}_j(s), \tag{12.4}$$

where t is the time and s represents the grid points. By applying sICA the rows of $\mathbf{S}$ are statistically as independent as possible, and the columns of $\mathbf{A}$ are their corresponding temporal evolutions. The use of sICA as opposed to principal component analysis (PCA) or its rotated version (RPCA) makes sense since it helps to identify (or localize) specific hydrological areas impacted by droughts or extreme wet conditions over GHA, which is not possible when using PCA [4] or RPCA [8]. The details of sICA estimation are reported in [28]. Here, only the first four dominant modes of both precipitation and TWS changes are found to be statistically significant and consequently are considered separately to reconstruct precipitation and TWS changes over various sub-regions of GHA. Both SPI and TSDI (as discussed in Sects. 12.4.1 and 12.4.2) are estimated using the four dominant sICA modes.

12.5 Hydrometeorological Drought Characterization

12.5.1 Changes in Precipitation and TWS

Changes in TWS depend on the variability of precipitation and climate extremes. Rainfall over GHA region has been reported to show considerable inter-seasonal and inter-annual variability over the past few decades beside the long-term changes, e.g., [2, 6, 13, 42, 54, 70]. Figure 12.3 shows the variability of rainfall and TWS changes. Trends are estimated from monthly rainfall/TWS anomalies (y) over various common time periods (T) using multilinear regression technique

$$y = \hat{\beta}_0 + \hat{\beta}_1 T + \hat{\beta}_2.\sin(2\pi T) + \hat{\beta}_3.\cos(2\pi T) + \hat{\beta}_4.\sin(4\pi T) + \hat{\beta}_5.\cos(4\pi T) + \epsilon(T), \tag{12.5}$$

where $\hat{\beta}_0$ is a constant term, $\hat{\beta}_1$ represents the linear trend, and the remaining terms represent the annual and semi-annual signals, ϵ represents the error term. Only changes that are significant at 95% confidence level based on student's t-test are shown.

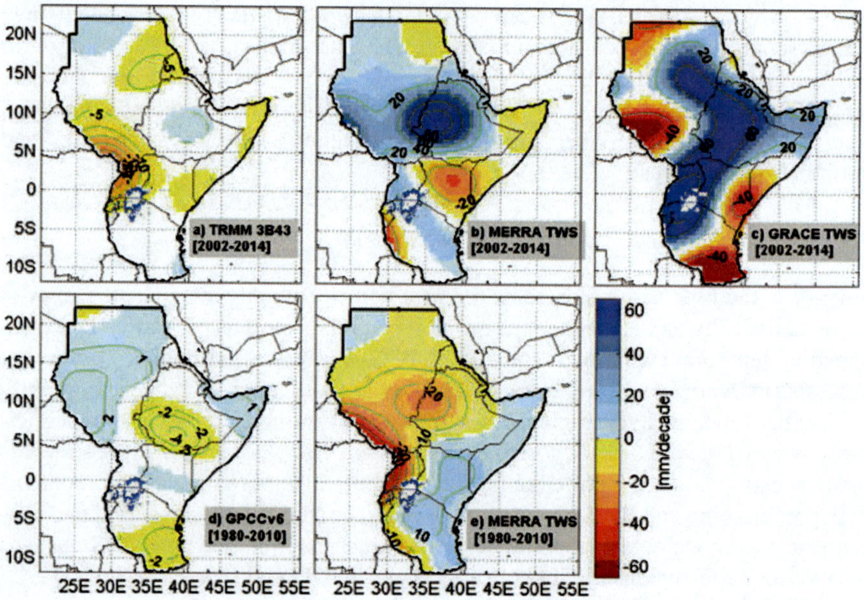

Fig. 12.3 Spatial variability of linear trend over GHA based on a full complement of monthly data at 95% confidence level from 2002 to 2014 for **a** TRMM rainfall, **b** MERRA TWS, and **c** GRACE TWS. Long-term variability are also assessed from 1980 to 2010 for **d** GPCC rainfall, and **e** MERRA TWS. Note that signals over Lake Victoria were not included (i.e., they are masked). *Source* [3]

Figure 12.3a shows significant decline (up to 20 mm/decade) in rainfall over southern and northern Sudan as well as increase in rainfall over central Ethiopia between 2002 and 2014 based on monthly TRMM precipitation estimates. Equivalent increase (decrease) in TWS over central Ethiopia (eastern Kenya, Somalia, and northern Sudan) are noted in MERRA (Fig. 12.3b). GRACE also indicated decreasing trends over southern Sudan and eastern Kenya (similar to rainfall) as well as anomalously large positive trends (up to 20 mm/year) over the Lake Victoria region (Fig. 12.3c). However, a completely opposite sign is observed in the long-term records of GPCC precipitation products between 1980 and 2010 (Fig. 12.3d; i.e., multidecadal oscillations may be influencing the sign of the trends), which indicate significant negative (positive) changes over Ethiopia and Tanzania (Sudan and Somalia). This might have led to a overall negative trend in TWS over the northern parts of GHA during the same period as shown by MERRA in Fig. 12.3e.

Figure 12.4 shows the area-averaged seasonal anomalies of TWS based on MERRA (1980–2014) and GRACE (2002–2014) data. Substantial inter-seasonal variations in TWS over the past 30–40 years are observed. Besides decadal trends, prolonged periods of negative TWS anomalies in the late 1980s to early 1990s and late 1990s and early 2000s are observed. Although, GRACE data covers a very short time period, the inter-seasonal variations are visible in all the seasons. While the

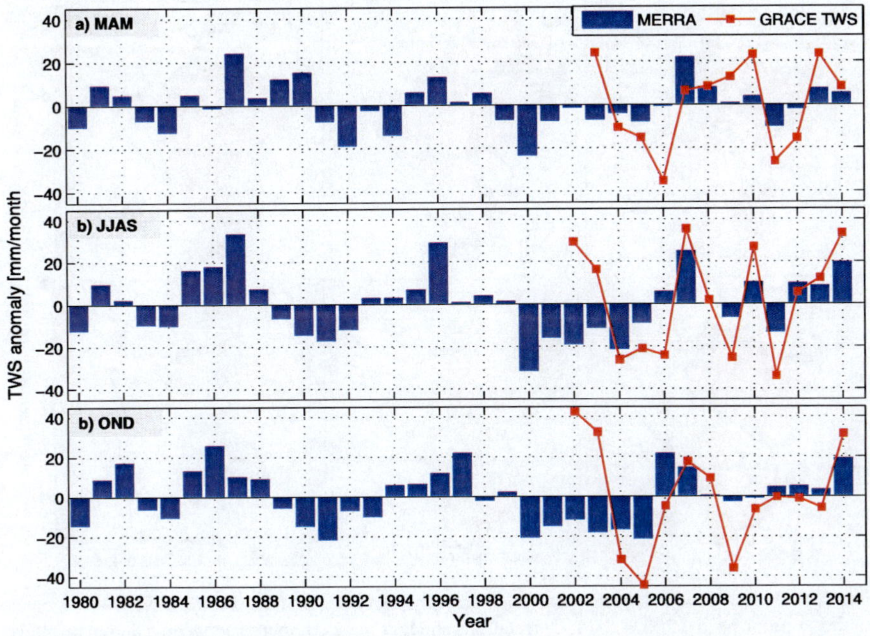

Fig. 12.4 Regional-averaged seasonal anomalies of TWS over the GHA region based on MERRA (1980–2014) and GRACE (2002–2014). The seasonal averages are shown for **a** MAM, **b** JJAS, and **c** OND corresponding to Fig. 12.1b–d. *Source* [3]

TWS changes in 2009 may have occurred as a result of the historical 2009–2011 drought, TWS changes between 2002 and 2006 at a rate of 6.20 mm/month seen in GRACE data has been linked to other processes such as anthropogenic, evaporations, and rainfall deficiency [14, 17].

To assess the overall relationship between rainfall and GRACE TWS changes over the region, the temporal correlations are computed and presented in Fig. 12.5, together with their lags. High correlation between rainfall and TWS (with a 2-month time lag) observed in Fig. 12.5a indicate that TWS changes are primarily driven by precipitation. Areas of low correlation (over central GHA region) show a time-lag of about one month (see, Fig. 12.5d). This low correlations could be related to seasonality, where the rainfall is usually bimodal over these regions.

MERRA TWS changes on its part show considerably better correlations with rainfall indicating that TWS changes simulated by MERRA are mainly driven by precipitation (Fig. 12.5b). It should be noted that MERRA does not simulate changes in surface water and groundwater, and therefore, does not represent the integrated changes in TWS as observed by GRACE satellites [59]. The large-scale high correlation patterns observed in Fig. 12.5b are further illustrated by the time lags, where the majority of the region indicated a lag of only one month. Further, the correlations

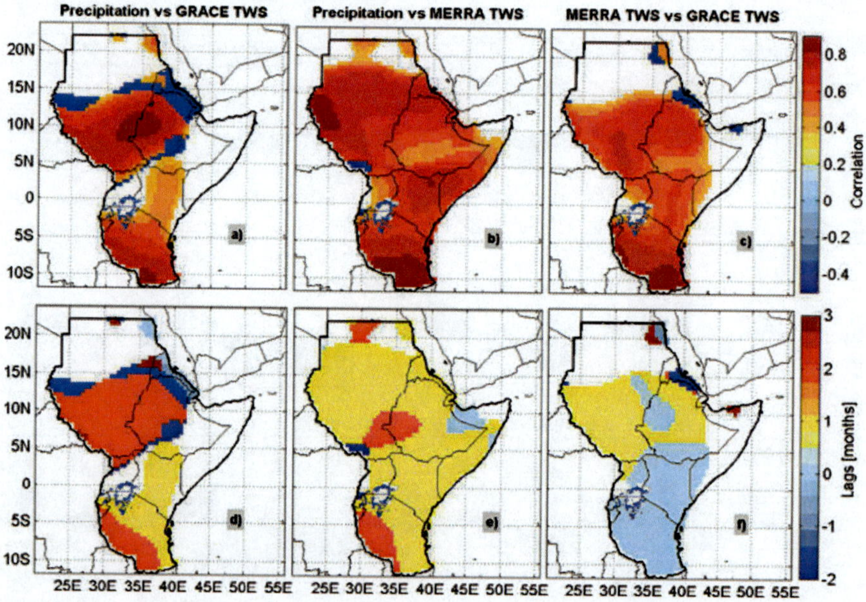

Fig. 12.5 Temporal correlations (with lags in months) between monthly rainfall and TWS changes for the common period 2002–2014. Correlations and lags were computed between **a** and **d**; monthly precipitation estimates of TRMM and TWS changes derived from GRACE Level 2 data, **b** and **e**; monthly precipitation estimates of TRMM 3B43 and TWS changes simulated MERRA, and **c** and **f**; GRACE and MERRA TWS changes over the entire GHA region. Note that the values which are not significant at 95% were not shown. *Source* [3]

between GRACE and MERRA (Fig. 12.5c) closely resemble those of rainfall and GRACE, although over a different time lag (see, Fig. 12.5f).

12.5.2 Spatio-Temporal Drought Patterns over GHA

To study the temporal drought patterns over the GHA, sICA was applied on the non-seasonal rainfall/TWS anomalies to derive region-specific patterns of temporal variability over the period 1979 to 2014. Figure 12.6 shows the standard deviations derived from the first four leading modes of rainfall variability, together with their corresponding temporal patterns (ICs). The first independent mode of rainfall variability based on GPCC indicates the largest anomalies over central GHA (Kenya, parts of Ethiopia and Somalia), accounting for 23.8% of the variability (Fig. 12.6a). The second mode (20.2%) is localized over central Ethiopia indicating a maximum standard deviation of about 30 mm (Fig. 12.6b) while the third mode (20.0%) is concentrated over Tanzania (Fig. 12.6c). The fourth mode (13.3%) is localized over northern Ethiopia and Sudan (Fig. 12.6d). TRMM showed similar spatial patterns to GPCC (Fig. 12.6e–h) except that the second and third modes were inter-changed (cf. that of GPCC in Fig. 12.6b, c). The maximum variability in the third mode was found to be localized completely over Ethiopia (Fig. 12.6g–h).

The corresponding temporal patterns in Fig. 12.6i–l, indicate considerable inter-annual variability and their temporal variations appeared to be considerably different from each other signifying the complexity of the region. Note that the ICs are only shown for the period 1998–2014 in order to highlight the differences between the two precipitation products.

The four dominant modes of TWS variability are shown in Fig. 12.7 with both MERRA (1980–2014) and GRACE (2002–2014) indicating distinct patterns of maximum variability over various parts of the GHA region. In order to highlight the differences in temporal evolutions between MERRA and GRACE, the ICs are plotted from 2002 to 2014 only. MERRA TWS show the largest anomalies over western Ethiopia (explaining about 30% of the variability; Fig. 12.7a) while the second mode (18.2%) is localized over western Tanzania (Fig. 12.7b). The third independent mode show maximum variations over the northern Lake Victoria region accounting for 17.3% of the variability (Fig. 12.7c). The fourth mode (14.3%) is mainly concentrated over South Sudan (Fig. 12.7d).

GRACE showed maximum variability (~27%) over the Lake Victoria region (Fig. 12.7e). Its second mode (16%) indicates maximum variations between Sudan and Ethiopia (Fig. 12.7f) while its third mode (~14%) is localized over South Sudan (Fig. 12.7g). The fourth mode (13.8%), although not very distinctive in its spatial pattern, shows a similar temporal pattern to that of MERRA over Tanzania. It should be noted here that the spatial patterns of maximum variability shown by MERRA do not exactly match with those from GRACE data. This could be possibly due to limitations in MERRA (errors in TWS modelling) and GRACE data (short time period). One obvious limitation in MERRA lies in its inability to capture the maximum

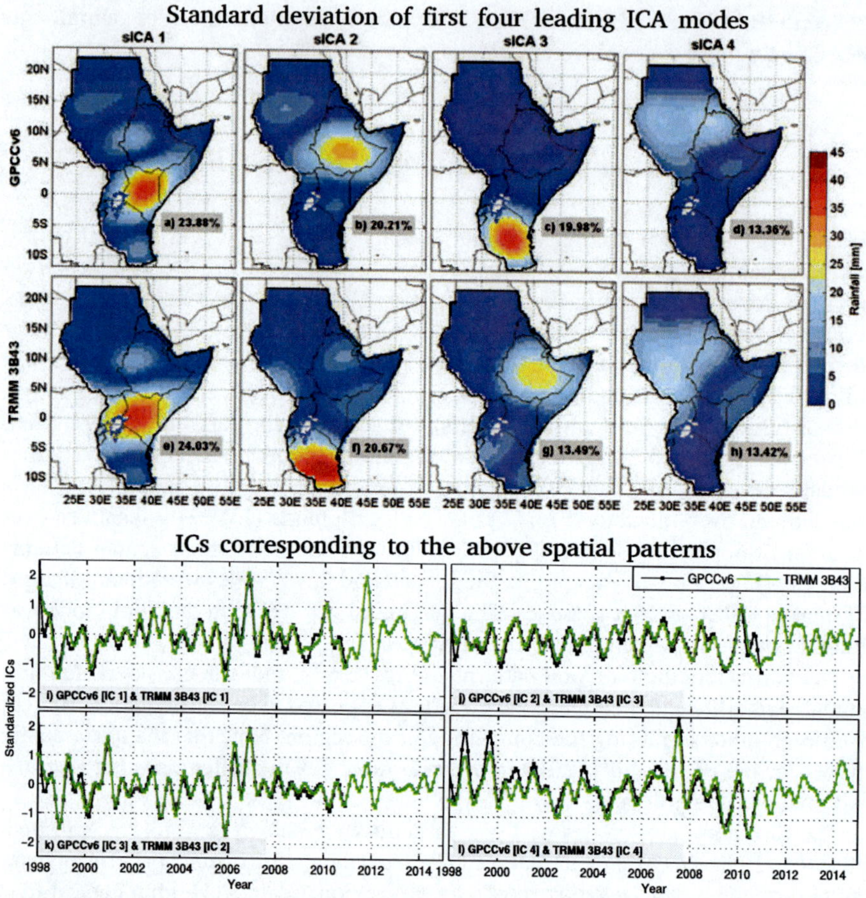

Fig. 12.6 The first four dominant modes of rainfall variability over the GHA region based on **a–d** GPCC (1979–2010), and **e–h** TRMM (1998–2014). The corresponding temporal patterns (or ICs) are shown in the bottom panels **i–l**), where a low-pass filter is applied to enhance their interpretation. (i) Temporal patterns corresponding to a & e, (j) b & g, (k) c & f, and (l), d & h. *Source* [3]

variations resulting from the Lake Victoria Basin (see, e.g., sICA 1 in Fig. 12.7e). Note though that for GHA, the regions of maximum variations often overlap among the four leading modes. For example, the maximum variations over South Sudan shown by MERRA (Fig. 12.7d) match quite well with sICA 3 of GRACE data (see, Fig. 12.7g).

In this way, the spatial patterns of sICA from GRACE and MERRA are matched to plot their ICs in Fig. 12.7i–l. It can be seen that the matched temporal patterns between GRACE and MERRA are very close. The corresponding ICs plotted between 2002 and 2014 show considerably inter-annual variability, indicating the hydrological extremes over parts of the GHA region. For example, the temporal patterns over the

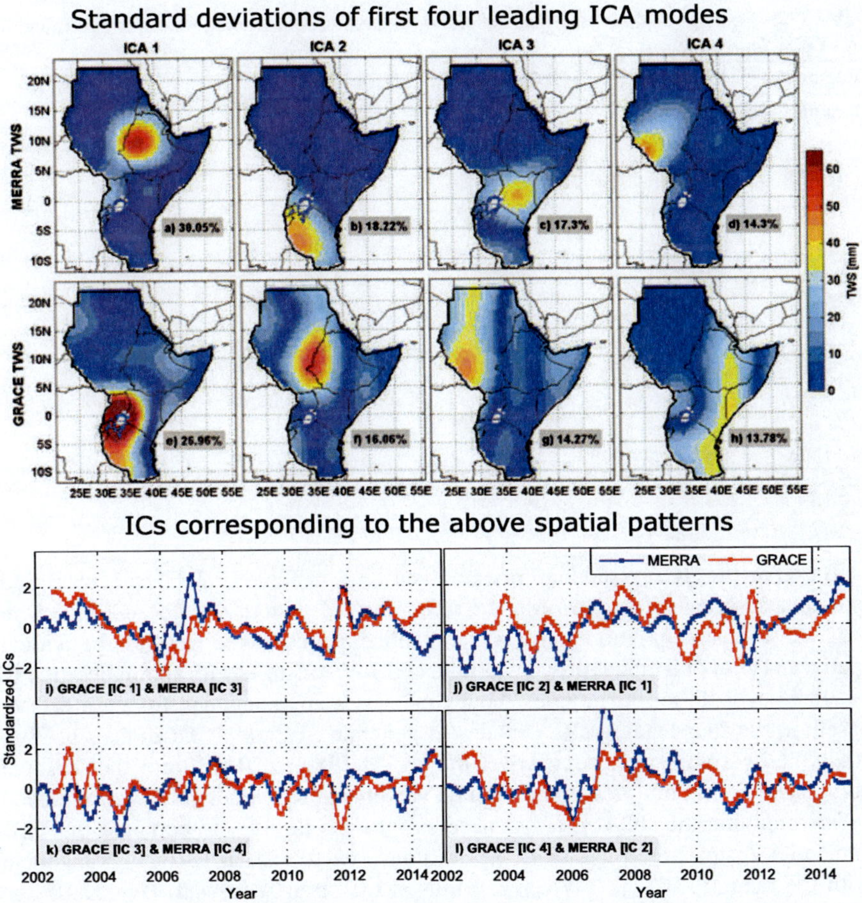

Fig. 12.7 The first four dominant modes of TWS variability over the GHA region based on **a–d** MERRA (1980–2014) and **e–h** GRACE data (2002–2014). The corresponding temporal patterns (or ICs) are shown in the bottom panels **(e–h)**, where a 6-month low-pass filter is applied to enhance their interpretation. (i) Temporal patterns corresponding to c & e, (j) a & f, (k) d & g, and (l), b & h. *Source* [3]

Lake Victoria region (Fig. 12.7e) are reproduced reasonably well by GRACE and MERRA (Fig. 12.7i), indicating a rapid decline in TWS from 2002 to 2006 and steadily increasing thereafter, see e.g., [17, 62]. GRACE (IC2, Fig. 12.7j, Ethiopia-Sudanese border) indicates the typical hydrological extremes, i.e., droughts of 2004, 2009, and 2011 while GRACE (IC3) fairly representes the dry periods over South Sudan (Fig. 12.7k). GRACE (IC4), which is mainly localized over the eastern GHA region (Fig. 12.7l) indicates a rapid decline of TWS from 2002 to 2006 and 2007 to 2010.

Table 12.3 Four sub-regions classified according to the first four leading modes of precipitation and TWS. *Source* [3]

Region	Area	GPCC v6	TRMM 3B43	MERRA	GRACE
Region 1	Kenya	ICA1	ICA1	ICA3	ICA1
	Uganda				
	Rwanda				
	Burundi				
	Northern Tanzania				
Region 2	Ethiopia	ICA2	ICA3	ICA1	ICA2
	Sudan				
Region 3	South Sudan	ICA4	ICA4	ICA4	ICA3
Region 4	Tanzania	ICA3	ICA2	ICA2	ICA4
	Eastern Ethiopia				

Based on the sICA results in Figs. 12.6 and 12.7, in Table 12.3, GHA is classified into four sub-regions in the order of sICA-derived patterns of GRACE TWS in Figs. 12.7e–h. Precipitation and TWS anomalies reconstructed from the individual spatial and temporal modes provide rainfall and TWS changes with anomalies mainly localized over these four regions. Four time-series are calculated for each dataset based on the four leading sICA modes to generate SPI and TSDI indices used to study the hydrological extremes over the period 1980 to 2014. Since the seasonal component from both precipitation and TWS changes have already been removed before applying the sICA, it is not necessary to use Eq. 12.1 to compute the TSD anomalies. Instead, the TWS anomalies are converted to TSD (in %) by dividing them with the range (TWS_{max}-TWS_{min}). While SPI is straight forward (Fig. 12.8), the calculation of TSDI index is rather complicated depending on the period of drought events, which is important for deriving the slope and y-intercept parameters from the cumulative TSDs. By analyzing the individual time-series, the TSDIs are estimated for all the four regions and plotted in Fig. 12.9. For example, TSDI (of GRACE TWS) over Lake Victoria region (Region 1) was calculated from cumulative TSDs between June 2006 and November 2006 while the TSDI over South Sudan (Region 3) was calculated for August 2009 to May 2010 (see Table 12.3).

Figure 12.8 shows the 12-month SPI indices calculated from GPCC (1979–2010) and TRMM (1998–2014) over the four regions as indicated in Table 12.3 and Fig. 12.7e–h. These are (a) Region 1 covering the Lake Victoria Basin, (b) Region 2, including the Ethiopia-Sudan border, (c) Region 3 covering South Sudan, and (d), Region 4, including Tanzania and parts of Ethiopia. The first 12 months of GPCC (i.e., 1979) and TRMM (i.e., 1998) are removed during the SPI computation. Table 12.4 summarizes the duration and intensity of severe to extreme meteorological drought episodes for the four sub-regions between 1980 and 2014. The SPI indices over Region 1 in Fig. 12.8a (corresponding to Fig. 12.6a, e) shows extreme drought events

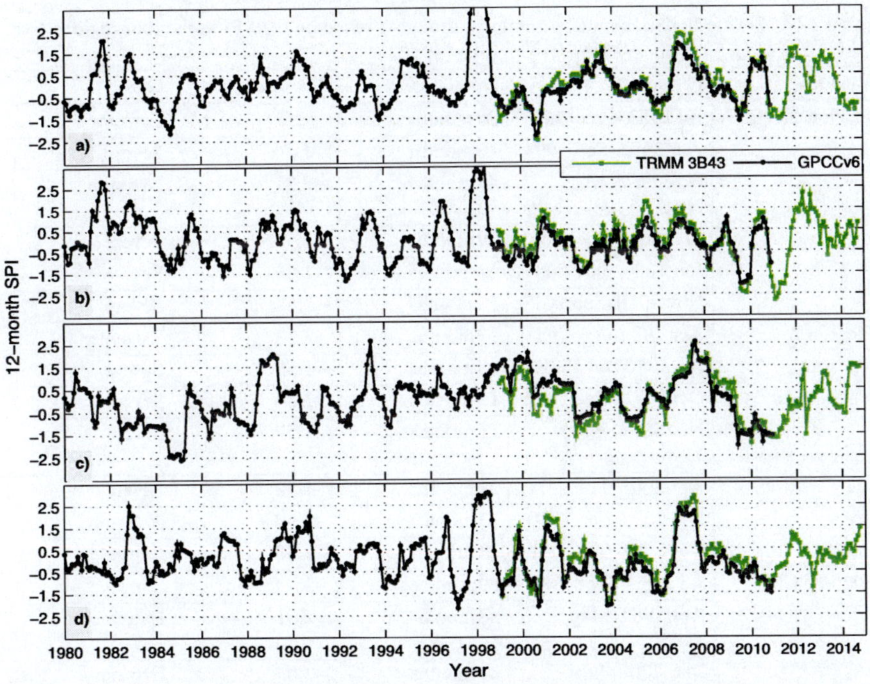

Fig. 12.8 12-month SPI indices estimated from monthly precipitation products of GPCC (1979–2010) and TRMM (1998–2014) over the four homogenous regions classified based on sICA (see also, Table 12.3). **a** Region 1, **b** Region 2, **c** Region 3, and **d** Region 4. Note that SPI is not calculated for the first 12 months in both datasets. *Source* [3]

in 1984, 1999–2000, 2009, and 2011, consistent with the findings of [42]. The most notable droughts over Ethiopia and eastern Sudan (Region 2, Table 12.3) include the extreme drought events of 2009 and some moderate droughts in 1984, 1986, 1992, 1994, and 2002 (Fig. 12.8b). SPI patterns over South Sudan (Fig. 12.8c, see also, Table 12.3) indicate prolonged droughts from 1982 to early 1985 and between 2009 and 2010. The SPI indices over Tanzania (Region 4, Fig. 12.8d) indicate considerable variability with severe/extreme droughts lasting from one to two years (e.g., 1988–1989, 1997–1998, 2003–2004, and 2005–2006).

Although there were indications of decreasing rainfall from 1980–1982, it did not lead to extreme droughts over Tanzania. The region exhibited only a moderate drought from 2010–2011, which caused devastating effects in other parts of the GHA [65, 71]. The two precipitation products, GPCC and TRMM are found to be fairly consistent across all the four regions between 1998 and 2010, differing only slightly over Region 3 (over South Sudan, Fig. 12.8c), where rainfall variability of GPCC is highest over central Sudan and northern Ethiopia (see, Fig. 12.6d) in contrast to South Sudan as indicated by TRMM (Fig. 12.6g).

Table 12.4 Severe to extreme meteorological droughts based on the SPI in the four regions (see, Table 12.3) of GHA from 1979 to 2014. *Source* [3]

Region	Periods	Duration	SPI [Max]	Class
Region 1	Jul 1983–Feb 1985	20 months	−2.13	Extreme
	Apr 1993–Sep 1994	18 months	−1.54	Severe
	Nov 1998–Mar 2002	43 months	−2.31	Extreme
	Mar 2004–Jul 2006	29 months	−1.53	Severe
	Jun 2008–Nov 2009	19 months	−1.66	Severe
	Nov 2010–Sep 2011	12 months	−1.62	Severe
Region 2	Apr 1984–Apr 1985	13 months	−2.13	Extreme
	Jan 1986–Jul 1988	28 months	−1.60	Severe
	Oct 1991–Dec 1992	15 months	−1.84	Severe
	Nov 1993–Nov 1994	13 months	−1.59	Severe
	Apr 2002–Jan 2004	22 months	−1.52	Severe
	Feb 2009–Feb 2010	13 months	−2.50	Extreme
	Oct 2010–Oct 2011	13 months	−2.80	Extreme
Region 3	Jul 1982–Sept 1985	39 months	−2.57	Extreme
	Apr 1982–May 1985	38 months	−2.61	Extreme
	Mar 2009–July 2012	41 months	−2.03	Extreme
Region 4	Dec 1987–Mar 1989	16 months	−1.54	Severe
	Nov 1996–Sep 1997	9 months	−2.18	Extreme
	Nov 1998–Oct 2000	24 months	−2.14	Extreme
	Mar 2003–Jul 2006	41 months	−2.10	Extreme
	Dec 2008–Jul 2011	32 months	−1.52	Severe

Figure 12.9 shows the TSDIs of the four regions (see Table 12.3). Major hydrological drought events between 1980 and 2014 based on TWS changes derived from MERRA (1980–2014) and GRACE (2002–2014) are summarized in Table 12.5. For MERRA, only severe to extreme droughts are shown for brevity. Clearly, three or more severe to extreme hydrological droughts occur over various regions of GHA with some of them lasting for 75 months (Table 12.5). Over the Lake Victoria Basin, especially over Kenya and northern Tanzania, four major hydrological droughts are detected, which in total lasted for 20 or more months (Table 12.5, Fig. 12.9a). All the four droughts occur as a result of prolonged meteorological dry episodes (see, Table 12.4) over the same region. While MERRA indicate only moderate droughts from 2004–2006, GRACE show severe droughts from September 2004 to November 2006, which lasted for 27 months, the period that coincide with the fall of Lake Victoria water level between 2002 and 2006 due to the expansion of the Owen Falls Dam/Nalubaale, see e.g.,[17]. From Table 12.4 (see also, Fig. 12.8a), it can be seen

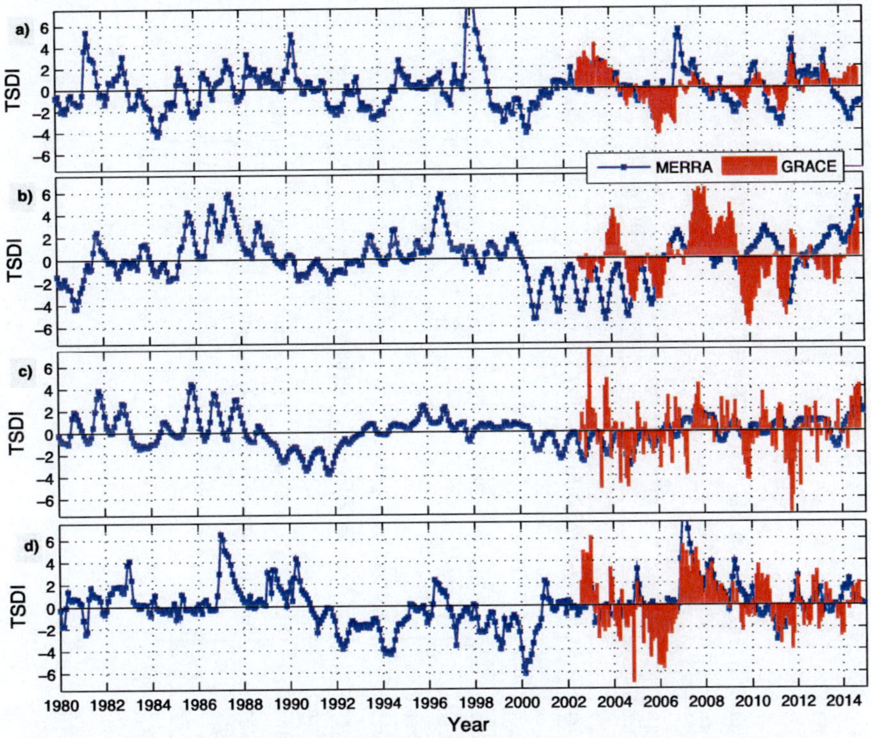

Fig. 12.9 TSDI indices computed based on sICA-derived modes over **a** Region 1, **b** Region 2, **c** Region 3, and **d**, Region 4 (see Table 12.3). *Source* [3]

that a severe meteorological drought also occurred from October further exacerbating the TWS decline. However, the severe meteorological droughts of 2008–2009 and 2010–2011 (see, Table 12.4) only caused moderate hydrological droughts over the region (Table 12.5).

Over Ethiopia and Sudan (Region 2), very different scenarios of TWS variability between GRACE and MERRA are observed especially over 2002 to 2014, where GRACE data show large positive TSDIs between 2007 and 2009 (2007 and 2011), which were not represented well by MERRA (negative; Fig. 12.9b). MERRA also indicates prolonged droughts from April 2000 to June 2006 lasting for about 75 months (Fig. 12.9b and Table 12.5), which could be spurious. However, it is also observed that two meteorological droughts occurred (1999–2000, and 2002–2004), which might have affected the TWS variability in MERRA. Extreme hydrological droughts have been found from July 2011-July 2012 (MERRA) and September 2009-October 2011 (GRACE) in response to the prolonged meteorological drought conditions over the region (see, Fig. 12.9b). MERRA also indicate extreme droughts between 1990 and 1993, which was again, in response to the severe meteorolog-

Table 12.5 Severe to extreme hydrological droughts based on the TSDI in the four regions (see Table 12.3) of GHA derived from MERRA (1980–2014) and GRACE (2002–2014) products. *Source* [3]

	MERRA (1980–2014)				GRACE (2002–2014)			
Region	Periods	Duration	TSDI [Max]	Category	Periods	Duration	TSDI [Max]	Category
Region 1	Nov 1983–May 1987	43 months	−4.33	Severe	Sep 2004–Nov 2006	27 months	−4.31	Severe
	Mar 1992–Mar 1996	49 months	−4.05	Extreme	Jun 2008–Feb 2010	21 months	−2.08	Moderate
	Feb 2000–Nov 2002	34 months	−3.34	Severe	Sept 2010–Sep 2011	13 months	−2.58	Moderate
	Oct 2010–May 2012	20 months	−3.68	Severe				
Region 2	Jan 1980–Jul 1981	19 months	−5.02	Extreme	Mar 2004–Nov 2006	33 months	−3.36	Severe
	Apr 2000–Jun 2006	75 months	−6.25	Extreme	Sep 2009–Oct 2011	26 months	4.60	Extreme
	Jul 2011–Jul 2012	13 months	−4.72	Extreme				
Region 3	Dec 1998–Mar 1993	52 months	−5.93	Extreme	May 2004–Nov 2006	31 months	−3.26	Extreme
	Jul 2000–Mar 2002	21 months	−3.9	Severe	Jun 2009–Sep 2011	28 months	−4.40	Extreme
	Jun 2000–Feb 2006	45 months	−4.81	Extreme				
Region 4	Feb 1991–Mar 1996	62 months	−5.84	Extreme	Mar 2004–Oct 2006	32 months	−5.70	Extreme
	Jan 1997–Jan 2001	49 months	−6.93	Extreme	Nov 2011–Jan 2014	27 months	−3.98	Severe
	Nov 2005–Nov 2006	13 months	−3.47	Severe				
	May 2010–Nov 2011	19 months	−3.17	Severe				

ical droughts that occurred from 1991–1992 (Table 12.4). The region also experienced the largest TWS decline (extreme hydrological drought) in 2009–2011 (Table 12.5), corresponding to the historical meteorological drought events from 2009–2011 (Table 12.4). These events are well-captured by the two precipitation products (Fig. 12.8b) and GRACE-derived TWS changes (Fig. 12.9b). During these two periods, SPI dropped below -2.5 while the TSDI was round −6.0 during the peak period (i.e., December to February 2009–2010). It is also interesting to observe that both the meteorological droughts eventuated from rainfall deficits during the OND season.

TSDI patterns over Region 3 representing the South Sudan region are quite complicated as both GRACE and MERRA indicate high intra-seasonal variability (especially observed in GRACE) but the extreme drought of 2011–2012 is well captured by GRACE (Fig. 12.9c). This could have occurred as a result of prolonged meteorological droughts that lasted for 39 months from May 2009 to July 2012. Note that MERRA also indicate prolonged hydrological droughts between July 1988 and December 1992. Region 4 covering Tanzania and parts of Ethiopia experienced several instances of droughts with varying intensities between 1980 and 2014 (Fig. 12.9d). The most notable drought events shown by MERRA are the prolonged droughts that occurred in the early 1990, the late 1990s, and in 2011. Similar to the Lake Victoria Basin (Region 1), this region also exhibited significant decline in TWS from 2003 to mid-2006 leading to a prolonged (and extreme) hydrological drought over the region. It should be noted here that the region also suffered two meteorological droughts (2003–2004 and 2005–2006) during the period (see, Fig. 12.8d). The 2010–2011 extreme drought is also captured very well by GRACE data.

Figure 12.10 illustrates the spatial evolution of the 2009–2010 drought computed as a sum of rainfall/TWS over three important seasons. It is observed that the entire GHA experienced anomalously low rainfall during OND 2010 (Fig. 12.10a), which remained relatively dry during MAM 2011 over Kenya, Ethiopia, and South Sudan (Fig. 12.10b) before recovering to positive changes. TWS anomalies are mostly negative in the southern and western parts of GHA during OND 2010 (Fig. 12.10d) but their magnitudes further increased during MAM 2011 (Fig. 12.10e). High negative anomalies shifted to Region 3 during JJAS 2011 but the majority of the regions still indicate negative anomalies (Fig. 12.10f) as a result of deficit rainfall in the two preceding months.

It is observed that almost all the hydrological droughts over the GHA region resulted from extreme (and/or prolonged) meteorological droughts with both SPI and TSDI indicating consistent temporal variability during the period 1980–2014. To illustrate the relationship between SPI and TSDI, the SPI and TSDI values are plotted based on TRMM precipitation estimates (1998–2014) and GRACE-derived TWS changes (2002–2014) for the period 1999–2014 in Fig. 12.11, with TSDI values scaled to the SPI values. The temporal patterns of SPI and TSDI are closely matched for Regions 1, 3, and 4 while the major drought events of 2009–2010 and 2010–2011 over Region 2 are also represented well by SPI and TSDI, where hydrological droughts have occurred almost 5–6 months after the meteorological drought.

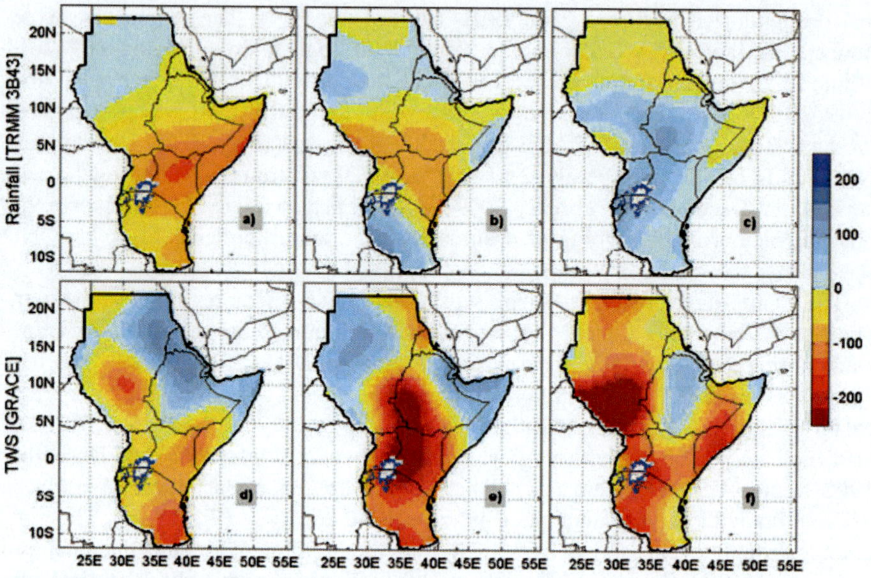

Fig. 12.10 Seasonal rainfall anomaly and TWS changes (mm/season) over the GHA region during **a–d** October–December 2010, **b–e** March–May 2011, and **c–f** June–September 2011. *Source* [3]

The correlations between SPI and TSDI are computed for the period 2002–2014 (Table 12.6). Consistent with Fig. 12.11, significant correlations (at 95% confidence interval) are found between SPI and TSDI during the period with a correlation of 0.44 (at one month lag), 0.52 (at 3 month lag), and 0.59 (at one month lag) for Region 1, Region 3, and Region 4, respectively. The correlation between SPI and TSDI for Region 2 is not significant. There exists a high seasonal correlation over the region between rainfall and TWS changes (see, Fig. 12.5a). The correlations between MERRA-derived TSDI and GRACE-derived TSDI are found to be significant over all the four regions with a lag of up to 1 month (see, Table 12.6). In quantifying changes in TWS and the resulting hydrological drought events (as indicated by TSDI) over GHA, contributions due to evapotranspiration and increased use of freshwater (groundwater and surface water) in the region should be taken into account.

The chapter further explore global teleconnection [50] relationships to SPI and TSDI indices. Table 12.7 shows the correlation coefficients and time lags (in months) between SPI (and TSDI) indices and Niño3.4 and the Indian Ocean Dipole (IOD) over various time periods between 1979 and 2014. Both SPI and TSDI (GRACE) of Region 1 are significantly correlated with ENSO and IOD, and are more related to IOD than ENSO (with a maximum correlation of 0.53 between TRMM and IOD). SPI (GPCC) and TSDI (GRACE) also show significant correlations with ENSO (IOD) while both SPI and TSDI over Region 4 show significant correlations with IOD with the highest correlation of 0.41 for the GRACE-derived TSDI. SPI and TSDI over Region 3 (representing South Sudan) show the least correlation against ENSO (IOD)

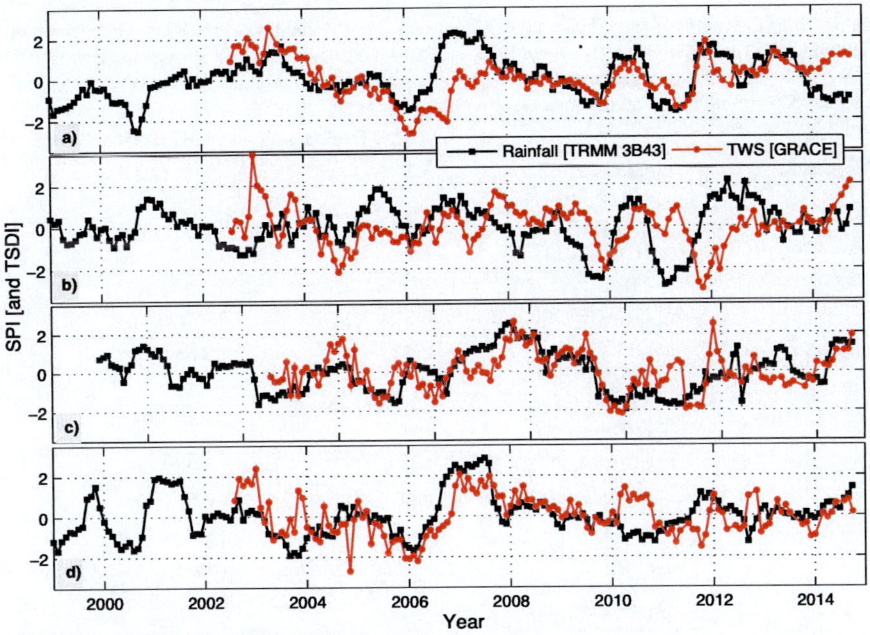

Fig. 12.11 SPI and TSDI indices of TRMM precipitation product and GRACE TWS changes for the period 1999 to 2014, i.e., **a** region 1, **b** region 2, **c** region 3, and **d** region 4. The TSDI indices were normalized to the SPI to highlight their temporal pattern over the four regions. *Source* [3]

Table 12.6 Correlations between TSDI derived from GRACE TWS changes and SPI derived from TRMM precipitation estimates between 2002 and 2014. Correlations between GRACE and MERRA TSDIs are also shown to indicate the skills of MERRA with respect to TWS changes derived from GRACE observations. Correlation coefficients that are significant at 95% confidence level are indicated in bold. *Source* [3]

Region	Region 1	Region 2	Region 3	Region 4
SPI and TSDI (GRACE)	**0.44** (1 month)	0.24 (2 month)	**0.52** (3 month)	**0.59** (1 month)
MERRA verses GRACE (TSDI)	**0.47** (1 month)	**0.51** (0 month)	**0.46** (1 month)	**0.57** (0 month)

indicating that rainfall and TWS variability over South Sudan is not associated with large-scale climate events as the region is located further inland compared to the other three regions.

It is observed that MERRA product shows low but opposite correlations with ENSO (IOD) compared to GRACE, which could be another limitation of the reanalysis product. Furthermore, IOD is seen to be the leading driver of rainfall and TWS variability especially over the eastern GHA region with the highest impact over Tanzania and Lake Victoria [1, 57] (see, Table 12.7). This is consistent with previous

Table 12.7 Correlations (significant at 95% confidence level) and lags (in brackets) between rainfall (and TWS) and ENSO/IOD between 1979 and 2014. Negative months' lags indicate ENSO/IOD occurs before rainfall or variations of TWS and vice versa. Correlation coefficients that are significant at 95% confidence level are indicated in bold. *Source* [3]

Region	Products	ENSO (Niño3.4)	IOD (DMI)
Region 1	GPCC v6 (1979–2010)	**0.33** (−1)	**0.48** (−1)
	TRMM 3B43 (1998–2014)	**0.44** (−1)	**0.53** (−1)
	MERRA (1980–2014)	−0.23 (5)	−0.18 (6)
	GRACE (2002–2014)	**0.38** (−6)	**-0.42** (−6)
Region 2	GPCC v6 (1979–2010)	**0.44** (−1)	**0.53** (−1)
	TRMM 3B43 (1998–2014)	−0.26 (−6)	−0.07 (−2)
	MERRA (1980–2014)	−0.26 (1)	−0.24 (1)
	GRACE (2002–2014)	**0.48** (−1)	**0.57** (−5)
Region 3	GPCC v6 (1979–2010)	−0.23 (1)	−0.18 (−6)
	TRMM 3B43 (1998–2014)	−0.26 (−1)	−0.20 (−1)
	MERRA (1980–2014)	−**0.37** (3)	−**0.44** (3)
	GRACE (2002–2014)	−0.21 (2)	−**0.34** (−6)
Region 4	GPCC v6 (1979–2010)	−0.23 (−1)	**0.39** (−2)
	TRMM 3B43 (1998–2014)	−0.20 (−6)	**0.34** (−1)
	MERRA (1980–2014)	−0.22 (7)	−**0.34** (6)
	GRACE (2002–2014)	**0.34** (−6)	**0.41** (−6)

studies, which reported that the East African climate is highly governed by the Indian Ocean SST variability e.g., [6, 42, 65, 69].

12.6 Hydrometeorological Impacts on Aquatic Species

The chapter found significant changes in long-term and decadal rainfall based on monthly estimates from GPCC and TRMM over 1979–2014. Long-term rainfall changes between 1980 and 2010 indicated significant positive (negative) changes over Sudan (Ethiopia and Tanzania) while changes from 2002 to 2014 showed completely opposite trends over these two regions. GRACE-derived total water storage TWS changes showed high positive (negative) trends over central GHA (South Sudan and Tanzania) with a maximum increase of about 20 mm/yr over the Lake Victoria

Basin and central Ethiopia. High correlations (up to 0.9) were noted between monthly rainfall and GRACE TWS changes over South Sudan, western Ethiopia, and Tanzania, with a lag of 1–2 months while rainfall over arid and semi-arid regions such as northern Sudan and Somalia were least correlated, indicating that large-scale variations of TWS are mainly influenced by seasonal rainfall variations.

SPI and TSDI indices estimated over the four sICA-derived spatial regions (Lake Victoria Basin, Ethiopia-Sudanese border, South Sudan, and Tanzania) indicated several instances of severe to extreme meteorological droughts (SPI < 1.5) resulting in extreme hydrological droughts that were observed in MERRA and GRACE TWS changes (indicated by TSDIs). Correlations between meteorological droughts (SPI) and hydrological droughts (GRACE-derived TSDI) were found to be significant especially over Lake Victoria region (0.44), Tanzania (0.52), and South Sudan (0.59), signifying that precipitation plays an important role in the hydrological budget of the regions on the one hand, and the effectiveness and reliability of the TSDI in indicating hydrological droughts over the LVB. An alternative method that uses standardised anomaly is presented, e.g., in [9, 10].

While the Lake Victoria Basin experienced six major meteorological droughts between 1979 and 2014, South Sudan and Ethiopia suffered seven meteorological droughts lasting between one year to 3 and half years. The impact of prolonged meteorological droughts is clearly evident in the TWS changes over all the four regions, which resulted in multiple years of hydrological droughts. Consistent with the previous studies e.g., [65], the Indian Ocean SST variations were found to play a huge influence on the regional TWS changes compared to the ENSO mode. It should be noted here that correlations between MERRA and ENSO (IOD) indices were considerably lower (and often opposite). Further verifications are required to assess the quality of reanalysis products to study the long-term changes and hydrological extremes over the GHA region.

Threats posed by these changes on freshwater over Lake Victoria Basin (LVB) include increased river siltation resulting from high soil erosion in the basin, recurrent destructive floods in the low-lying areas, riparian land encroachment, degradation of river banks, eutrophication, and proliferation of the invasive water hyacinth, all which pose additional threats to the future ecological and community well-being in the Lake Victoria Basin (LVB) [58]. To minimize these impacts, [58] suggest among others multiscalar present and future climate change studies in order to advance scientific understanding and providing information for decision making in adaptation and mitigation strategies to deal with widening variability in river flows.

Economic losses associated with drought impacts have been reported over West Africa, Australia, California, and Southern Africa, e.g., in [24–27, 47] while over East Africa, losses associated with rainfall deficit is captured by [15, 54]. Such losses emphasize the economic connection of climate and hydrology to water and sanitation, agriculture, fisheries, and energy sectors [58]. Furthermore, [58] point out that high frequency of droughts in response to changes in seasonal and annual rainfall will not only directly impact freshwater availability, base flow, storm flow, and groundwater recharge in all the river catchments within the LVB but also on the hydrology of the lake, which is likely to follow the rainfall patterns because recharge to the lake is predominantly (80%) by direct rainfall whereas river inflow contributes only 20% [1]. Hydrometeorological droughts would impact on LVB's water bodies (both incoming rivers and the lake itself). For the flowing rivers, a decline in water levels occur and with less riparian shading and high air temperatures, high water temperatures results in fish kills and the loss of macroinvertebrates, see e.g., [18]. On the lake itself, drought impacts on the vegetation and wetlands, which provide breeding sites for the fish [5]. This is supported by [53] who notes that decreasing water levels, either in shallow lakes or shallow areas of deep lakes, have been linked to disappearance of vegetation and interference with fish reproduction through loss of breeding areas [34, 43, 51].

Indeed, although most studies have not shown a conclusive cause-effect relationship between climate change and fishery productivity, past events in some lakes do show a link between climate extremes (droughts and floods) and fish yield [53]. For deep lakes, the stability of the water column, which is associated with climate variability and change would result into less nutrient transport from the permanently anoxic hypolminion into the upper seasonally mixed phototrophic zone leading to alteration of primary production [53], where the composition of phytoplankton is expected to shift to species that can readily persist or adapt to reduced nutrient conditions [53]. This was evidenced in Lake Tanganyika within GHA where 20% decrease in algal productivity due to climate warming contributed to a 30% reduction in fish production [52, 53, 66]. Considering that LVB is home to Tilapia (*Oreochromis niloticus*) and Nile Perch (*Lates niloticus*), hydrometeorological drought could therefore have irreversible impact on aquatic species as climate change has been singled out to shift species distribution, composition, and phenology [58]. Olaka et al., [58] point out that the viability of populations of the remaining species and overall biodiversity will most strongly depend upon the conservation of their habitats and the prevention of pollution of the water bodies, which intensifies under flooding conditions. Further adaptation and mitigation strategies to minimize hydroclimatic impacts on aquatic species are discussed, e.g., by [53] who state:

> Although mitigation of effects of climate change by fishers may be inconsequential on the global scale, a shift from smoking fish using wood to other low cost and environmentally innocuous preservation methods, such as sun drying and salting, as well as planting trees and preserving the riparian wetlands can help in modulation of local climate change. These, also, can reduce the cumulative impacts from other stressors such as sedimentation. Limited effort, however, has been registered by African fishers or managers to mitigate the effects from climate change, at least on a local basis. The application of land-based mitigation measures, including mulching and planting trees, on shores of Lake Wamala is still trivial and would need to be scaled up to have a local impact

In general, drought is a temporary climate feature and cannot be prevented. Its impacts, i.e., drought related disasters can be prevented. Opere and Ogallo [55] opine that coping with drought hazards can be developed through adequately addressing the answers to the following three questions:

- How frequently or extensively does a certain type of drought occur in a given region?

- What are the vulnerabilities and expected losses, which are associated with that type of drought?
- What are the costs for implementing the plausible strategies or options for mitigating the disaster that can be caused by that type of drought?

In summary, Opere and Ogallo [55] write:

> Strategies and drought mitigation options that are set to address the above questions involve integration of a wide spectrum of drought issues such as enhanced drought monitoring, drought research and forecasting, drought risk and vulnerability assessments, capacity building, drought awareness and education and dissemination of drought information. The use of such strategies aims to reduce vulnerability of the drought prone communities by either altering their land use practices or by modifying the drought severity through provision of the required water supplies for food production or through the implementation of relief programmes that enhance water and food securities on both short and long-term basis. The slow onset of drought combined with drought forecasting capabilities also enables the implementation of contingency plans and measures in advance of the occurrence of the drought. The improvement in recent years in the seasonal and long-term climate predictions, such as those issued by many national and regional institutes in Africa, including the Drought Monitoring Centres have assisted in the implementation of drought disaster mitigation and effective contingency plans. There are many other types of drought coping mechanisms. Some of these that have been found to be important in drought-coping include:
>
> - Encouraging the use of drought resilient crops (see Chap. 11);
> - Strengthening land tenure policies to discourage settlements in marginal lands;
> - Revival of traditional customs, traditions and land-use practices that are important in the mitigation of the severity of droughts; and,
> - Relocation of communities to other areas, which are less drought-prone.
>
> For drought, prediction is still heavily reliant on monitoring of observed patterns of monthly and seasonal rainfall, stream flow, reservoir and ground water levels, and soil moisture. Widespread drought evolves relatively slowly when compared with floods and may persist for months or years and it can be difficult, if not impossible, to predict its end. During drought situations, it is difficult to implement many remedial actions and those measures, which can be taken, are often site and situation specific. Statistical approaches are more relevant for the development of predictive tools for large geographic regions on monthly and seasonal time scales. Statistical techniques offer promise for increasingly useful forecasts of the onset, severity and duration of drought. First-order Markov process has been used to estimate critical wet and dry spells in Kano plains while the possible application of climate variability to agricultural planning and operations including avoidance of drought and rainfall (flood) hazards has been presented. The potential benefits of accurate forecasts on these time scales are evidently enormous.

References

1. Awange JL (2021) Lake Victoria monitored from space. Springer Nature International
2. Awange JL, Hu KX, Khaki M (2019) The newly merged satellite remotely sensed, gauge and reanalysis-based Multi-Source Weighted-Ensemble Precipitation: Evaluation over Australia and Africa (1981–2016). Sci Total Environ 670:448–465. https://doi.org/10.1016/j.scitotenv.2019.03.148
3. Awange JL, Khandu Forootan E, Schumacher M, Heck B (2016) Exploring hydrometeorological drought patterns over the Greater Horn of Africa (1979–2014) using remote sensing and reanalysis products. Adv Water Res 94:45–59. https://doi.org/10.1016/j.advwatres.2016.04.005
4. Awange JL, Palancz B, Völgyesi L (2020) Hybrid imaging and visualization. employing machine learning with mathematica - python. Springer Nature International, Berlin. 978-3-030-26152-8, https://doi.org/10.1007/978-3-030-26153-5
5. Aseto O, Ong'ang'a O (2003) Lake Victoria (Kenya) and its environs: resource, opportunites and challenges. Africa Herald Publishing House, Kendu Bay, Kenya
6. Mpelasoka F, Awange JL, Zerihun A (2018) Influence of coupled ocean-atmosphere phenomena on the Greater Horn of Africa droughts and their implications. Sci Total Environ 610–611:691–702. https://doi.org/10.1016/j.scitotenv.2017.08.109
7. Agboma CO, Yirdaw SZ, Snelgrove KR (2009) Intercomparison of the total storage deficit index (TSDI) over two canadian prairie catchments. J Hydrol 374(3–4):351–359. https://doi.org/10.1016/j.jhydrol.2009.06.034
8. Agutu N, Awange JL, Zerihun A, Ndehedehe C, Kuhn M, Fukuda Y (2017) Assessing multi-satellite remote sensing, reanalysis, and land surface models' products in characterizing agricultural drought in East Africa. Remote Sens Environ 194:287–302. https://doi.org/10.1016/j.rse.2017.03.041
9. Agutu N, Awange J, Zerihun A, Ndehedehe C, Kuhn M, Fukuda Y (2017) Assessing multi-satellite remote sensing, reanalysis, and land surface models' products in characterizing agricultural drought in East Africa. Remote Sens Environ 194:287–302. https://doi.org/10.1016/j.rse.2017.03.041
10. Agutu NO, Awange JL, Ndehedehe C, Mwaniki MW (2020) Consistency of agricultural drought characterization over Upper Greater Horn of Africa (1982–2013): Topographical, gauge density, and model forcing influence. Sci Total Environ 709. https://doi.org/10.1016/j.scitotenv.2019.135149
11. Andersen OB, Seneviratne SI, Hinderer J, Viterbo P (2005) GRACE derived terrestrial water storage depletion associated with the 2003 european heat wave. Geophy Res Lett 32(18):n/a–n/a. https://doi.org/10.1029/2005GL023574
12. Awange JL, Anyah R, Agola N, Forootan E, Omondi P (2013) Potential impacts of climate and environmental change on the stored water of lake victoria basin and economic implications. Water Res Res 49(12):8160–8173. https://doi.org/10.1002/2013WR014350
13. Awange JL, Ferreira VG, Khandu Andam-Akorful SA, Forootan E, Agutu NO, He XF (2015) Uncertainties in remotely-sensed precipitation data over Africa. Int J Climatol. https://doi.org/10.1002/joc.4346
14. Awange JL, Forootan E, Kuhn M, Kusche J, Heck B (2014) Water storage changes and climate variability within the Nile basin between 2002 and 2011. Adv Water Resour 73:1–15. https://doi.org/10.1016/j.advwatres.2014.06.010
15. Awange JL, Anyah R, Agola N, Forootan E, Omondi P (2013) Potential impacts of climate and environmental change on the stored water of Lake Victoria Basin and economic implications. Water Res Res 49:8160–8173. https://doi.org/10.1002/2013WR014350
16. Awange JL, Gebremichael M, Forootan E, Wakbulcho G, Anyah R, Ferreira VG, Alemayehu T (2014) Characterization of Ethiopian mega hydrogeological regimes using GRACE, TRMM and GLDAS datasets. Adv Water Res 74:64–78. https://doi.org/10.1016/j.advwatres.2014.07.012

17. Awange JL, Sharifi MA, Ogonda G, Wickert J, Grafarend E, Omullo M (2008) The falling lake victoria water level: GRACE, TRIMM and CHAMP satellite analysis of the lake basin. Water Res Manag 22:775–796. https://doi.org/10.1007/s11269-007-9191-y
18. Bond NR, Lake PS, Arthington AH (2008) The impacts of drought on freshwater ecosystems: an Australian perspective. Hydrobiologia 600:3–16. https://doi.org/10.1007/s10750-008-9326-z
19. Chen JL, Wilson CR, Tapley BD, Longuevergne L, Yang ZL, Scanlon BR (2010) Recent la plata basin drought conditions observed by satellite gravimetry. J Geophys Res Atmos 115:D22108. https://doi.org/10.1029/2010JD014689
20. Chen JL, Wilson CR, Tapley BD, Yang ZL, Niu GY (2009) 2005 drought event in the Amazon River basin as measured by GRACE and estimated by climate models. J Geophys Res: Solid Earth 114:B05404. https://doi.org/10.1029/2008JB006056
21. Cheng MK, Ries JC, Tapley BD (2013) Geocenter variations from analysis of SLR data. In: Reference Frames for Applications in Geosciences. In: International association of geodesy symposia, vol 138. Springer, Berlin, Heidelberg, pp 19–25
22. Cheng MK, Tapley BD (2004) Variations in the earth's oblateness during the past 28 years. Geophys Res Lett 109(B9). https://doi.org/10.1029/2004JB003028
23. Dahle C, Flechtner F, Gruber C, König D, König R, Michalak G, Neumayer K-H (2013) GFZ GRACE Level–2 processing standards document for Level–2 Product Release 0 0 05: revised edition, January 2013. Scientific Technical Report STR12/02 – Data, rev. ed.. Deutsches GeoForschungsZentrum GFZ. Potsdam https://doi.org/10.2312/GFZ.b103-1202-25
24. Ding Y, Widhalm M, Hayes MJ (2011) Measuring economic impacts of drought: a review and discussion. Disaster Prev Manag Int J 20:434–446
25. van Dijk AI, Beck HE, Crosbie RS, de Jeu RA, Liu YY, Podger GM, Viney NR (2013) The Millennium Drought in southeast Australia (2001–2009): Natural and human causes and implications for water resources, ecosystems, economy, and society. Water Res Res 49:1040–1057. [CrossRef]
26. Howitt R, Medellín-Azuara J, MacEwan D, Lund JR (2014) Sumner, D. Economic Analysis of the (2014) Drought for California agriculture. Center for Watershed Sciences University of California, Davis CA, USA
27. Roudier P, Ducharne A, Feyen L (2014) Climate change impacts on runoff in West Africa: a review. Hydrol Earth Syst Sci 18:2789–2801
28. Forootan E (2014) Statistical signal decomposition techniques for analyzing time-variable satellite gravimetry data. University of Bonn, Germany PhD thesis. http://hss.ulb.uni-bonn.de/2014/3766/3766.htm
29. Forootan E, Awange J, Kusche J, Heck B, Eicker A (2012) Independent patterns of water mass anomalies over Australia from satellite data and models. Remote Sens Environ. 124(1):427–443. https://doi.org/10.1016/j.rse.2012.05.023
30. Forootan E, Kusche J (2012) Separation of global time-variable gravity signals into maximally independent components. J Geodesy 86(7):477–497. https://doi.org/10.1007/s00190-011-0532-5
31. Forootan E, Kusche J (2013) Separation of deterministic signals using independent component analysis (ICA). Stud Geophys Geod 57(1):17–26. https://doi.org/10.1007/s11200-012-0718-1
32. Funk C, Hoell A, Shukla S, Bladé I, Liebmann B, Roberts JB, Robertson FR, Husak G (2014) Predicting east African spring droughts using Pacific and Indian ocean sea surface temperature indices. Hydrol Earth Syst Sci Discuss 11:3111–3136. https://doi.org/10.5194/hessd-11-3111-2014
33. Houborg R, Rodell M, Li B, Reichle R, Zaitchik BF (2012) Drought indicators based on model-assimilated Gravity Recovery and Climate Experiment (GRACE) terrestrial water storage observations. Water Res Res 48(7). https://doi.org/10.1029/2011WR011291
34. Hickley P, Muchiri M, Boar R, Britton R, Adams C, Gichuru N, Harper D (2004) Habitat degradation and subsequent fishery collapse in Lakes Naivasha and Baringo. Kenya Ecohydrol Hydrobiol 4:503–517

35. Huffman GJ, Bolvin DT, Nelkin EJ, Wolff DB, Adler RF, Gu G, Hong Y, Bowman KP, Stocker EF (2007) The TRMM multisatellite precipitation analysis (TMPA): Quasi-global, multiyear, combined-sensor precipitation estimates at fine scales. J Hydrometeorol 8(1):38–55. https://doi.org/10.1175/JHM560.1
36. Kurnik B, Barbosa P, Vogt J (2011) Testing two different precipitation datasets to compute the standardized precipitation index over the Horn of Africa. Int J Remote Sensing 32(21):5947–5964. https://doi.org/10.1080/01431161.2010.499380
37. Kusche J, Schmidt R, Petrovic S, Rietbroek R (2009) Decorrelated GRACE time-variable gravity solutions by GFZ, and their validation using a hydrological model. J Geod 83(10):903–913. https://doi.org/10.1007/s00190-009-0308-3
38. Landerer FW, Swenson SC (2012) Accuracy of scaled GRACE terrestrial water storage estimates. Water Res Res 48(W04531). https://doi.org/10.1029/2011WR011453
39. Leblanc MJ, Tregoning P, Ramillien G, Tweed SO, Fakes A (2009) Basin-scale, integrated observations of the early 21st century multiyear drought in southeast Australia. Water Res Res 45(4):1–10. https://doi.org/10.1029/2008WR007333
40. Li B, Rodell M (2014) Evaluation of a model-based groundwater drought indicator in the conterminous U.S. J Hydrol 526:78–88. https://doi.org/10.1016/j.jhydrol.2014.09.027
41. Long D, Scanlon BR, Longuevergne L, Sun AY, Fernando DN, Save H (2013) GRACE satellite monitoring of large depletion in water storage in response to the 2011 drought in Texas. Geophys Res Lett 40(13):3395–3401. https://doi.org/10.1002/grl.50655
42. Lyon B (2014) Seasonal drought in the Greater Horn of Africa and its recent increase during the March-May long rains. J Climate 27(21):7953–7975. https://doi.org/10.1175/JCLI-D-13-00459.1
43. Luxereau A, Genthon P, Karimou JMA (2012) Fluctuations in the size of Lake Chad: consequences on the livelihoods of the riverine peoples in Eastern Niger. Reg Environ Chang 12:507–521
44. Marthews TR, Otto FEL, Mitchell D, Dadson SJ, Jones RG (2015) The 2014 drought in the horn of Africa: attribution of meteorological drivers. Bull Am Meteor Soc 96:83–88. https://doi.org/10.1175/BAMS-EEE_2014_ch17.1
45. McKee TB , Doesken NJ , Kliest J (1993) The relationship of drought frequency and duration to time scales. In: Proceedings of the 8th conference of applied climatology, 17–22 January, Anaheim, CA. American Meterological Society, Boston, MA, pp. 184–194
46. McKee TB, Doesken NJ, Kliest J (1995) Drought monitoring with multiple time scales. In: 9th AMS conference on applied climatology. American Meterological Society, Boston, MA, pp 233–236
47. Mukwada G, Manatsa D (2018) Is climate change the nemesis of rural development? An analysis of patterns and trends of Zimbabwean droughts. climate change, extreme events and disaster risk reduction. Springer, Berlin/Heidelberg, Germany, pp 173–182
48. Narasimhan B, Srinivasan R (2005) Development and evaluation of Soil Moisture Deficit Index (SMDI) and Evapotranspiration Deficit Index (ETDI) for agricultural drought monitoring. Agri Forest Meteorol 133(1–4):69–88. https://doi.org/10.1016/j.agrformet.2005.07.012
49. Naumann G, Dutra E, Barbosa P, Pappenberger F, Wetterhall F, Vogt VJ (2014) Comparison of drought indicators derived from multiple data sets over Africa. Hydrol Earth Syst Sci 18:1625–1640. https://doi.org/10.5194/hess-18-1625-2014
50. Nicholson SE (1997) An analysis of the enso signal in the tropical Atlantic and west- ern INDIAN oceans. Int J Clim 17(4):345–375. https://doi.org/10.1002/(SICI)1097-0088(19970330)17:4(345::AID-JOC127)3.0.CO;2-3
51. Njaya F, Snyder KA, Jamu D, Wilson J, Howard-Williams C, Allison EH, Neil L, Andrew NL (2011) The natural history and fisheries ecology of Lake Chilwa, southern Malawi. J Great Lakes Res 37:15–25
52. O'Reilly CM, Alin SR, Pilsnier PD, Cohen AS, McKee BA (2003) Climate change decreases aquatic ecosystem productivity of Lake Tanganyika, Africa. Nature 424:766–768
53. Ogutu-Ohwayo R, Natugonza V, Musinguzi L, Olokotum M, Naigaga S (2016) Implications of climate variability and change for African lake ecosystems, fisheries productivity, and livelihoods. J Great Lakes Res 3:498–510. https://doi.org/10.1016/j.jglr.2016.03.004

54. Omondi PA, Awange JL, Forootan E, Ogallo LA, Barakiza R, Girmaw GB, Fesseha I, Kululetera V, Kilembe C, Mbati MM, Kilavi M, King'uyu SM, Omeny PA, Njogu A, Badr EM, Musa TA, Muchiri P, Bamanya D, Komu- tunga, E. (2014) Changes in temperature and precipitation extremes over the greater horn of Africa region from 1961 to 2010. Int J Clim 34(4):1262–1277. https://doi.org/10.1002/joc.3763
55. Opere A, Ogallo LA (2006) Natural disasters in Lake Victoria Basin (Kenya): Causes and impacts on environment and livelihoods. In: Odada EO, Olago DO, Ochola WO (eds) Environment for development: an ecosystems assessment of Lake Victoria Basin environmental and socio-economic status, trends and human vulnerabilities, pp 135–148
56. Palmer WC (1965) Meteorological drought. Technical Report Research Paper 45. US Department of Commerce, Weather Bureau. Washington DC. https://www.ncdc.noaa.gov/temp-and-precip/drought/docs/palmer.pdf
57. Khaki M, Awange J (2021) The 2019–2020 rise in Lake Victoria monitored from space: exploiting the state-of-the-art grace-fo and the newly released ERA-5 reanalysis products. Sensors 21(13):4304. https://doi.org/10.3390/s21134304
58. Olaka LA, Ogutu JO, Said MY, Oludhe C (2019) Projected climatic and hydrologic changes to Lake Victoria basin rivers under three RCP emission scenarios for 2015–2100 and impacts on the water sector. Water 11:1449. https://doi.org/10.3390/w11071449
59. Rienecker MM, Suarez MJ, Gelaro R, Todling R, Bacmeister J, Liu E, Bosilovich MG, Schubert SD, Takacs L, Kim GK, Bloom S, Chen J, Collins D, Conaty A, da Silva A, Gu W, Joiner J, Koster RD, Lucchesi R, Molod A, Owens T, Pawson S, Pegion P, Redder CR, Reichle R, Robertson FR, Ruddick AG, Sienkiewicz M, Woollen J (2011) Merra: NASA's modern-era retrospective analysis for research and applications. J Clim 24:3624–3648. https://doi.org/10.1175/jcli-d-11-00015.1
60. Schneider U, Becker A, Finger P, Meyer-Christoffer A, Ziese M, Rudolf B (2014) GPCC's new land surface precipitation climatology based on quality-controlled in-situ data and its role in quantifying the global water cycle. Theor Appl Clim 115(1–2):15–40. https://doi.org/10.1007/s00704-013-0860-x
61. Sheffield J, Wood EF, Chaney N, Guan K, Sadri S, Yuan X, Olang L, Amani A, Ali A, Demuth S, Ogallo L (2014) A drought monitoring and forecasting system for Sub-Sahara African water resources and food security. Bull Am Meteor Soc 95:861–882. https://doi.org/10.1175/BAMS-D-12-00124.1
62. Swenson S, Wahr J (2009) Monitoring the water balance of lake victoria, east Africa, from space. J Hydrol 370(1–4):163–176. https://doi.org/10.1016/j.jhydrol.2009.03.008
63. Tapley BD, Bettadpur S, Watkins M, Reigber C (2004) The gravity recovery and climate experiment: Mission overview and early results. Geophys Res Lett 33. https://doi.org/10.1029/2004GL019779
64. Thomas AC, Reager JT, Famiglietti JS, Rodell M (2014) A GRACE-based water storage deficit approach for hydrological drought characterization. Geophys Res Lett 41(5):1537–1545. https://doi.org/10.1002/2014GL059323
65. Tierney JE, Smerdon JE, Anchukaitis KJ, Seager R (2013) Multidecadal variability in East African hydroclimate controlled by the Indian Ocean. Nature 493:389–392. https://doi.org/10.1038/nature11785
66. Verburg P, Hecky RE, Kling H (2003) Ecological consequences of warming in Lake Tanganyika. Science 301:505–507
67. Viste E, Korecha D, Sorteberg A (2013) Recent drought and precipitation tendencies in Ethiopia. Theor Appl Clim 112(3):535–551. https://doi.org/10.1007/s00704-012-0746-3
68. Wahr J, Molenaar M, Bryan F (1998) Time variability of the Earth's gravity field: hydrological and oceanic effects and their possible detection using GRACE. J Geophys Res 103(B12):30205–30229. https://doi.org/10.1029/98JB02844
69. Williams AP, Funk C (2011) A westward extension of the warm pool leads to a westward extension of the walker circulation, drying eastern Africa. Clim Dyn 37:2417–2435. https://doi.org/10.1007/s00382-010-0984-y

70. Williams AP, Funk C, Michaelsen J, Rauscher SA, Robertson I, Wils THG, Es- hetu Z, Loader NJ (2012) Recent summer precipitation trends in the Greater Horn of Africa and the emerging role of Indian Oceansea surface temperature. Clim Dyn 39:2307–2328. https://doi.org/10.1007/s00382-011-1222-y
71. WMO, 2012. WMO statement on the status of the global climate in 2011. Technical Report WMO-No. 1085. World Meteorological Organization. Geneva, Switzer- land. Yirdaw, S.Z., Snelgrove, K.R., Agboma, C.O (2008) GRACE satellite observations of terrestrial moisture changes for drought characterization in the Canadian prairie. J Hydrol 356(1–2):84–92. https://doi.org/10.1016/j.jhydrol.2008.04.004
72. Yirdaw SZ, Snelgrove KR, Agboma CO (2008) GRACE satellite observations of terrestrial moisture changes for drought characterization in the Canadian prairie. J Hydrol 356(1–2):84–92. https://doi.org/10.1016/j.jhydrol.2008.04.004

Part IV
Potential of Irrigated Agriculture in GHA

Chapter 13
Potential for Irrigated Agriculture: Groundwater

The limited use of groundwater for irrigated agriculture in the wider GHA region could be attributed to various factors such as; lack of detailed analysis of country-wide groundwater irrigation potential, existence of large swathes of land with high potential for surface water irrigated agriculture, limited information on aquifer potentials and yields of individual boreholes, inadequate groundwater resource management plans, and inadequate groundwater related knowledge by majority of the population leading to inappropriate groundwater development. For GHA, a region experiencing groundwater data deficiency, use of GRACE products to provide a broader picture on changes in groundwater at a regional scale could be useful in; better understanding of the long-term sustainability of groundwater (aquifers) as sources of water supply, making appropriate policy formulation and/or choices, and assessing the impacts of climate change (and/or variability) on groundwater by decision makers and/or regional bodies, e.g., Intergovernmental Authority on Development (IGAD) and the East African Community (EAC) secretariat. —[1].

13.1 Summary

Greater Horn of Africa (GHA) is projected to face negative impacts on per capita food production due to dwindling nature of water resources forced by climate change and rising population growth (see Chap. 1). The region has limited groundwater irrigated agriculture and also lacks groundwater monitoring infrastructure. This Chapter presents the work of [1] who (i) employs Independent Component Analysis (ICA) to localize Gravity Recovery and Climate Experiment (GRACE)-derived groundwater changes and analyses the corresponding temporal variabilities and their link to climate variability indices (Indian Ocean Dipole (IOD) and El Niño-Southern Oscillation (ENSO)), and (ii), explores the irrigation potentials of the localized groundwater (e.g., Fig. 13.1). Monthly GRACE-derived groundwater changes show similar temporal variability to WaterGap Hydrological Model (WGHM), i.e., a correlation of 0.7 (significant at 95% confidence level), highlighting GRACE's potential to pro-

J. Awange, *Food Insecurity & Hydroclimate in Greater Horn of Africa*,
https://doi.org/10.1007/978-3-030-91002-0_13

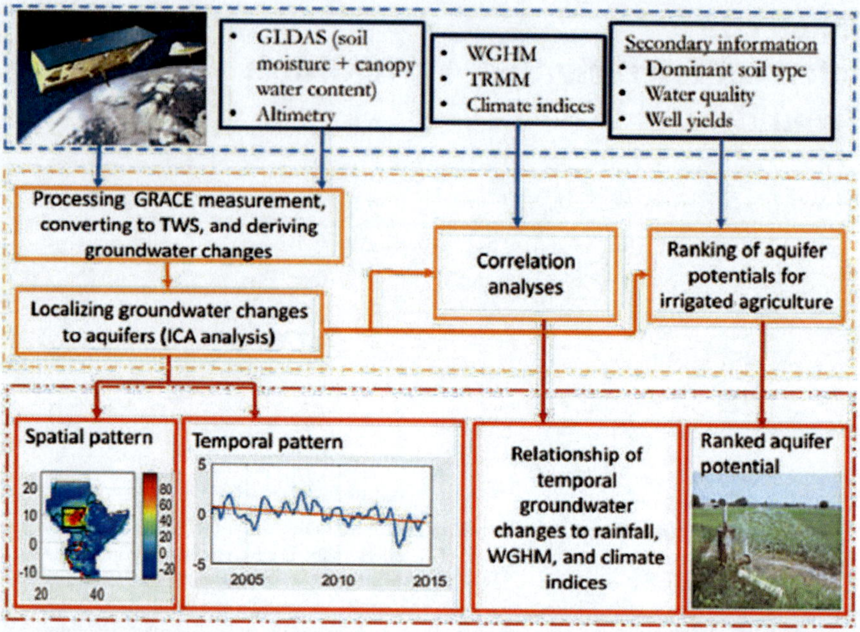

Fig. 13.1 Graphic abstract summarizing the chapter's work. *Source* [1]

vide GHA-wide changes in groundwater. Based on GHA aquifer location maps, [1] associate the localized groundwater changes to nine major aquifers namely; Nubian sandstone, Karoo Carbonate, Upper Nile, Ethiopian highlands, Lake Tana region, Kenya-Somalia, Central Tanzania, Karoo sandstone, and Ruvuma. All temporal groundwater changes, except Nubian sandstone and Kenya-Somalia, show an annual (cyclic) pattern indicating an annual (yearly) recharge cycle. Weak relationships with rainfall and both climate variability indices are noted. Maximum correlation occur when rainfall precede the temporal groundwater changes by several months. Based on water availability (from GRACE), water quality (indicated by the total dissolved substance), and dominant soil types, potential for groundwater irrigated agriculture results show: low potentials for Nubian Sandstone and Kenya-Somalia areas; low to moderate potentials for Karoo Carbonate, Lake Tana region, central Tanzania, and Ruvuma; *moderate to high potentials* for Upper Nile and Karoo Sandstone; and *high potential* for Ethiopian highland. Even though [1] considers relatively short time period (10 years), these results are critical to the sustainable management of the region's groundwater resources and appropriate/informed policy formulation.

13.2 GHA's Groundwater: Potential and Challenges

Groundwater is a critical source of freshwater for human related uses [89]. Globally, it contributes 42%, 36%, and 27% of water used for irrigation, households, and manufacturing, respectively [31], and is the third largest water storage after oceans (salty water) and cryosphere (water in its solid state of snow and ice), but by far the largest available reservoir of freshwater [54]. In Africa, groundwater is estimated to be about 0.66 million km^3 [61], compared to annual average rainfall of about 0.02 million km^3 [70], while freshwater stored in lakes is estimated to be 0.03 million km^3 [83]. Groundwater holds numerous advantages over surface water sources, e.g., reliable availability during drought periods due to slow response to climate changes [19, 20], less vulnerability to pollution [54], and a reliable source of safe and clean drinking water in arid and semi-arid regions [30]. In addition, groundwater offers better accessibility as surface water sources are normally far and widely spaced, seasonal, or are not available in arid and semi-arid areas [31]. Despite the significant role played by groundwater in irrigation globally, and its numerous advantages listed above, see also [94], its use in irrigated agriculture in Africa has remained rather low.

The level of irrigation in Africa is generally low and agriculture is mainly rain-fed. On average, 5.5% of the cultivated land over the continent is equipped for irrigated agriculture while for East Africa within the Greater Horn of Africa (GHA), this is only 1.5%. Of this 1.5%, only 3.4% is equipped for irrigation using groundwater [84]. The limited use of groundwater for irrigated agriculture in the wider GHA region could be attributed to various factors such as; lack of detailed analysis of country-wide groundwater irrigation potential, existence of large swathes of land with high potential for surface water irrigated agriculture, limited information on aquifer potentials and yields of individual boreholes, inadequate groundwater resource management plans, and inadequate groundwater related knowledge by majority of the population leading to inappropriate groundwater development, see, e.g., [46, 67]. However, with a high population growth rate, see, e.g., [8, 13, 14], and impact of climate change on rain-fed agriculture and surface water sources, further development and utilization of groundwater may become essential for increased food production [40, 54, 61, 65]. Indeed, [5] have pointed to the fact that for Sub-Sahara Africa, increased groundwater use could significantly improve the well being of its inhabitants. Further, [99] have shown the potential for profitable small holder irrigation expansion in sub-Sahara Africa.

Furthermore, the GHA region has limited long-term groundwater monitoring infrastructure, resulting in little or no long-term groundwater monitoring data. In cases where they are available, often inconsistencies exist due to different methodologies used at different times making comparisons rather difficult, and the resulting trends unclear [27]. This infrastructural limitation has led to sparse *in-situ* measurements contributing to lack of or limited knowledge on groundwater level changes. The lack of long-term groundwater level changes has contributed to decision makers having inadequate understanding of the long-term sustainability of the aquifers as sources of water supply. In addition, without information on the long-term groundwater level changes, the impacts of climate change (and/or variability) on groundwater levels over the region remains largely unknown, see, e.g., [4, 16]. This situation calls for additional and/or further studies on groundwater variability (level changes) and its relation to climate change (and/or variability).

Groundwater level changes have been determined using various methods, e.g., groundwater well level observations in combination with estimated storage capacities, using models, e.g., WaterGap Hydrological Model (WGHM; [32] and satellite based approaches, e.g., Gravity Recovery and Climate Change Experiment (GRACE) [87]. GRACE satellite mission and its follow on GRACE Follow-ON (GRACE-FO) have been providing global monthly temporal gravity variations in the form of spherical harmonic coefficients since 2002 (see Sect. 3.3.3), which when synthesised provide monthly global mass changes. The potentials and limitations of using GRACE satellites to assess groundwater depletion and dynamics have been presented in, e.g., [24]. Furthermore, its ability to provide a wider picture that could benefit the setting of broader regional and national policies has earned it consideration as one of the hydrologists' toolbox, see, e.g., [5]. Over Africa, its application to investigate trends and seasonality cycles of groundwater is reported, e.g., in [65]. Other studies in the region that present groundwater related information include, e.g., [2, 6, 27, 41, 49, 61, 88].

For GHA, a region experiencing groundwater data deficiency [27], use of GRACE products to provide a broader picture on changes in groundwater at a regional scale could be useful in; better understanding of the long-term sustainability of groundwater (aquifers) as sources of water supply, making appropriate policy formulation and/or choices, and assessing the impacts of climate change (and/or variability) on groundwater by decision makers and/or regional bodies, e.g., Intergovernmental Authority on Development (IGAD) and the East African Community (EAC) secretariat. To this end, this chapter exploits the advantages of GRACE to study groundwater changes over the entire GHA region and the associated impacts of climate variabilities. The chapter exploits the advantage of Independent Component Analysis (ICA) [22, 26, 38], a fourth-order statistical method, to localize the GRACE-derived groundwater changes into smaller locations associated with aquifers. This enables a more localized analysis of groundwater temporal variability and their association to climate variability, a feat currently not possible due to scarce groundwater change monitoring information in the region. Specifically, this chapter aims at (i) employing ICA method to localize GRACE-derived groundwater changes into their respective groundwater locations and analyse the corresponding temporal variabilities and their links to climate variability, and (ii), exploring the irrigation potentials of the localized groundwater regions in (i) above.

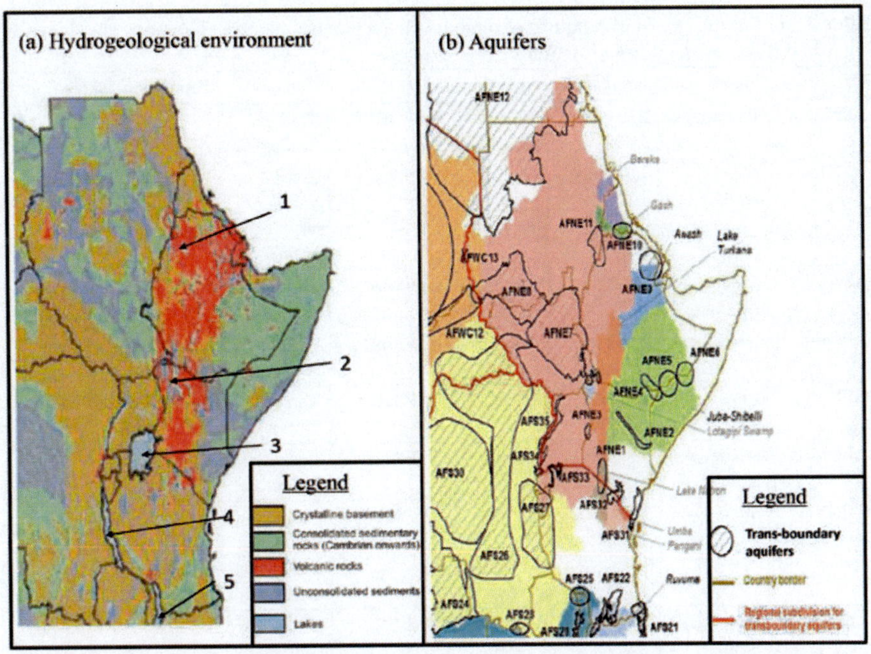

Fig. 13.2 Greater Horn of Africa **a** hydrogeological characteristics (adapted from [62]), and **b** selected aquifers (adapted from [6]). Numbers in **a** depict dominant Lakes (e.g., 1 = Lake Tana, 2 = Lake Turkana, 3 = Lake Victoria, 4 = Lake Tanganyika, and 5 = Lake Malawi). The aquifers characteristics are presented in Table 13.1. Colored regions in **b** depict river and/or lake basins. *Source* [1]

13.3 GHA's Hydrogeology and Groundwater Data

13.3.1 Hydrogeology

Greater Horn of Africa (Fig. 13.2) is prone to impacts of climate change extremes (droughts and floods), which are projected to impact negatively on per capita food production, see e.g., [18, 40, 71] and Sect. 1.2.2. Groundwater, if sustainably exploited could help increase food production through groundwater-based irrigated agriculture.

Groundwater occurrence depends on geology, history of weathering, and the recharge, i.e., dependent on rainfall, rock types, and land use and/or land cover over the recharge region; [62]. Over Africa, most (high quantities) of groundwater occur in large sedimentary aquifers in Northern Africa and were recharged more than 5000 years ago when the climate over the Sahel was wet [81]. Over GHA and the wider African continent, groundwater can be categorised to occur in four major rock types namely; crystalline basement rocks, consolidated sedimentary rocks, volcanic rocks, and unconsolidated sediments (Fig. 13.2a). These blocks have the following

Table 13.1 Summary of GHA aquifers' regional characteristics. Aquifer locations are shown in Fig. 13.2b. Table adapted from [6]. *Source* [1]

ID	Name	Population	Area (km^2)	ID	Name	Population	Area (km^2)
AFNE1	Rift aquifer	279 000	21 150	AFNE12	Nubian Sandstone	67 320 500	2 608 000
AFNE2	Merti aquifer	129 000	13 500	AFS21	Coastal sedimentary basin3	794 000	23 000
AFNE3	Mount Elgon	806 550	5400	AFS22	Karoo Sandstone	214 500	40 000
AFNE4	Dawa	223 150	24 000	AFS25	Weathered basement	852 000	25 842
AFNE5	Juba aquifer	197 600	34 600	AFS26	Karoo Carbonate	9 400 000	941 100
AFNE6	Shebelle aquifer	334 400	31 000	AFS27	Tanganyika aquifer	11 940 000	222 300
AFNE7	Sudd basin	2 926 500	331 600	AFS31	Coastal sedimentary basin1	2 150 000	16 800
AFNE8	Baggara basin	2 433 500	239 300	AFS32	Kilimanjaro aquifer	1 396 000	14 600
AFNE9	Awash valley basin	627 400	50 700	AFS33	Kagere aquifer	493 500	5800
AFNE10	Maerb aquifer	1 827 900	22 800	AFS34	Mgahinga	1 151 000	4400
AFNE11	Gedaref	732 000	38 700	AFS35	Western Rift Valley sediment	1 151 000	29 500

average groundwater potentials; crystalline (low to moderate), consolidated (moderate to high), volcanic (low to high), and unconsolidated (low to high), with the potentials being dependent on level of fracturing and/or weathering, and individual composition/rock elements, e.g., porosity and permeability [62]. The quality of groundwater is inherently dependent on the chemical composition of the aquifer rocks. Some of the major aquifers over GHA are shown in Fig. 13.2b, with their regional characteristics summarised in Table 13.1 (see [6], for additional information).

Table 13.2 A summary of the dataset employed in this chapter. *Source* [1]

Data	Period	Spatial resolution
GRACE	2003–2014	1° × 1°
GLDAS (TWS, soil moisture, and canopy water content)	2003–2014	1° × 1°
Satellite altimetry	2003–2014	N/A
WGHM groundwater changes	2003–2009	0.5° × 0.5°
Climate indices (IOD and ENSO)	2003–2014	N/A
TRMM	2003–2014	0.25° × 0.25°

13.3.2 Groundwater Monitoring Products

The following datasets are used (see Table 13.2 for a summary); GRACE-derived total water storage (TWS; surface water, groundwater, soil moisture, vegetation water), Global Land Data Assimilation System (GLDAS; moisture, canopy water content, and ice and snow water content), satellite altimetry, WaterGap Hydrological Model derived groundwater changes, and monthly Tropical Rainfall Measuring Mission (TRMM) precipitation estimates.

13.3.2.1 GRACE-Derived Terrestrial Water Storage (TWS)

The GRACE satellite mission launched in 2002 provided global monthly temporal gravity variations in the form of spherical harmonic coefficients, see e.g., [87, 98] up to 2017 when it was deactivated. Currently, GRACE Follow-On (GRACE-FO) launched in 2018 is spearheading a similar mission. Centre for Space Research's (CSR) release five (RL05) monthly spherical harmonic coefficients for the duration 2003–2014, downloaded from International Centre for Global Earth Models portal[1] are used in this chapter. Since GRACE does not sense the geo-center motion, degree one coefficients are replaced by those provided by [86] while degree two coefficients, which are affected by tidal aliasing errors are replaced by those provided by [25]. The gravitational potential is then converted into equivalent water heights following [92]. In order to remove stripes and spurious patterns in the spherical harmonic coefficients, a decorrelation and non-isotropic filter (DDK2), see, e.g., [55, 56] is applied. The impacts of the filter on the resulting GRACE synthesised TWS is alleviated through the application of a scale factor developed from GLDAS TWS following [58]. The synthesised GRACE-derived TWS over GHA comprise changes in accumulated soil moisture, groundwater, surface water, and biomass/canopy water content. In order to retain groundwater, surface water changes from the major lakes in the region are removed through the integration of lake level changes from satellite altimetry and

[1] http://icgem.gfz-potsdam.de/series/01_GRACE_monthly/CSR%20Release%2005.

lake kernel (coefficient) function to generate a gridded time series of surface water changes. In addition, correction is applied to the biomass/canopy and soil moisture changes using GLDAS model estimates.

13.3.2.2 Global Land Data Assimilation System (GLDAS)

GLDAS is a joint product of NASA's Goddard Space Flight Center (GSFC) and NOAA's National Centers for Environmental Prediction (NCEP). It is a global, high resolution offline (uncoupled to the atmosphere) terrestrial modeling system that incorporates satellite and ground-based observations in order to produce optimal fields of land surface states and fluxes in real time [77]. It drives the following land surface models; Mosaic, Noah, Community Land Model (CLM), and Variable Infiltration Capacity Model. For a detailed description see, [77]. GLDAS version 1, NOAH, monthly 1° by 1° soil moisture (0–2 depth), TWS, and canopy water content products from 2003 to 2014[2] are used as outlined for GRACE above; GLDAS TWS is used to derive the scaling factor (for mitigating spatial leakage, see, e.g., [58]) while GLDAS (soil moisture, canopy and snow water equivalent) are used to estimate these components from the GRACE TWS in order to retain GRACE-derived groundwater.

13.3.2.3 Satellite Altimetry

Monthly lake level variations from Jason-1, Jason-2/OSTM, and TOPEX/POSEIDON altimetry, products of United States Department of Agriculture (USDA)[3] are used to remove surface water variations from the GRACE TWS as outlined in Sect. 13.4.1. The monthly lake variations, from 2003 to 2014 for Lake Victoria, Lake Turkana, Lake Tana, Lake Tanganyika, and Lake Malawi are used as they are the dominant surface water bodies in the region and are the only ones with up-to-date altimetry level variations.

13.3.2.4 WaterGap Hydrological Model (WGHM)

WGHM simulates monthly continental water flows and storages as well as human water use for all land areas of the globe except Antarctica at a spatial resolution of 0.5° by 0.5° [32]. Its input include GPCC v6 monthly rainfall, monthly data on temperature, cloud cover, and the number of wet days. The model applies a simple water balance approach (see, [32], for more detailed description). WGHM groundwater model variants IRR_70_S (deficit irrigation at only 70% of optimal irrigation with groundwater recharge from surface bodies) and NOUSE_S (no water use at all assumed with groundwater recharge from surface bodies) are used in this chapter to evaluate

[2] Downloaded from https://disc.gsfc.nasa.gov/datasets?keywords=GLDAS.

[3] Downloaded from http://www.pecad.fas.usda.gov/cropexplorer/global_reservoir/.

(for comparison purposes) GRACE-derived groundwater changes. IRR_70_S has been found to best represent groundwater changes over the globe by [32]. WGHM groundwater changes from 2003 to 2009 were downloaded from http://www.uni-frankfurt.de/49903932/7_GWdepletion?.

13.3.2.5 Climate Indices

Due to the link between the sea surface temperatures and the rainfall in the Greater Horn of Africa [64], climate variability indices Indian Ocean Dipole (IOD), [7, 78] and El'Niño Southern Oscillation (ENSO, [97] from 2003 to 2014[4] are used to assess climate variability impact on groundwater. These two global climate indices are selected due to their dominance in the region, see e.g., [9, 57] and Chaps. 7, 8. IOD, defined by Dipole Mode Index (DMI), indicates the east-west temperature gradient across the tropical Indian Ocean, while ENSO indicates sea surface temperature anomalies in the Pacific. In this chapter, ENSO is defined by Multivariate ENSO Index (MEI), which is based on the following six parameters; sea-level pressure, zonal and meridional components of the surface wind, sea surface temperature, surface air temperature, and total cloudiness fraction of the sky, see e.g. [97].

13.3.2.6 Tropical Rainfall Measuring Mission (TRMM)

TRMM [44, 53] precipitation estimate for 2003–2014 downloaded from National Aeronautics and Space Administration's Goddard Earth Sciences Data and Information Services Center[5] is used to aid the understanding of groundwater trends and the impact of climate variability. TRMM 3B43 (V7) given as monthly precipitation rate (mm/h) at 0.25° spatial resolution is converted to monthly totals before subsequent analysis. Dinku et al. [29] have shown that it is suitable to be used in the region, see also [10, 12].

13.4 Groundwater Changes and Agricultural Potential

13.4.1 GRACE-Derived Groundwater Changes

GRACE provides monthly global mass changes, which over the continents have been equated to changes in terrestrial water storage (TWS). These changes are quantified as combinations of different components, i.e., changes in biomass/canopy water content, accumulated soil moisture, changes in surface water storage, and changes

[4] Downloaded from https://climatedataguide.ucar.edu/climate-data.

[5] https://pmm.nasa.gov/dataaccess/downloads/trmm.

in groundwater, see e.g., [60, 65, and the references therein]. Over GHA, changes in groundwater (after removal of the undesired components of TWS) can be quantified, e.g., [24] as

$$\Delta GW = \Delta TWS - (\Delta SW + \Delta CW + \Delta SM), \tag{13.1}$$

with ΔGW being changes in groundwater, ΔTWS changes in terrestrial water storage, ΔSW changes in surface water storage, ΔCW changes in canopy water content, and ΔSM changes in soil moisture. Changes in surface water (ΔSW) is derived through defining lake Kernel function using lakes' geodetic boundary coordinates with 1 (one) unit of water over the respective lake and 0 (zero) elsewhere over the region as

$$\partial_{lake}(\phi, \lambda) = \begin{Bmatrix} 1, \texttt{over lake} \\ 0, \texttt{elsewhere} \end{Bmatrix}, \tag{13.2}$$

where θ and λ are the geodetic latitude and longitude, respectively. Since GRACE fields are represented in terms of spherical harmonics, the lake kernel function (Eq. 13.2) is then expressed as [65]

$$\partial_{lake}(\phi, \lambda) = \frac{1}{4\pi} \sum_{l=0}^{l_{max}} \sum_{m=0}^{l} (C_{l,m}^{lake} \cos m\lambda + S_{l,m}^{lake} \sin m\lambda) P_{l,m}(cos\theta), \tag{13.3}$$

where $P_{l,m}$ is the normalised associated Legendre function of degree l and order m, and $C_{l,m}^{lake}$ and $S_{l,m}^{lake}$ are the dimensionless lakes' spherical harmonic coefficients. The summation in Eq. (13.3) is restricted to $l_{max} = 60$ in line with CSR RL05 monthly spherical harmonic coefficients used herein. Figure 13.3 shows the kernel functions of the considered lakes as per Eq. (13.3), whose coefficients are scaled with respective altimetry level changes to give changes in surface water, ΔSW.

Following the accounting of ΔSW using satellite altimetry scaled lake kernel functions, and ΔCW and ΔSM using GLDAS hydrological model, the derived groundwater changes (ΔGW) still contain some uncertainties (δGW). These uncertainties arise from GLDAS-derived components, surface water variability corrections, and other unaccounted for smaller surface water bodies, e.g., Lakes Edward and Kyoga, the Sudd-wetlands, etc. Assuming that ΔGW is sufficiently larger than δGW (i.e., $\Delta GW >>> \delta GW$, see more details in Sect. 13.5), the estimated ΔGW can sufficiently be assumed to represent groundwater changes over the GHA region, which following [32] can be expressed as

$$\Delta GW = R - Q - B + L, \tag{13.4}$$

where R is diffuse recharge (largely soil moisture percolating into groundwater), Q is net abstraction (groundwater withdrawals minus return flow from irrigation with both surface water and groundwater), B is base-flow from groundwater to surface water bodies, and L recharge from underground streams/runoff.

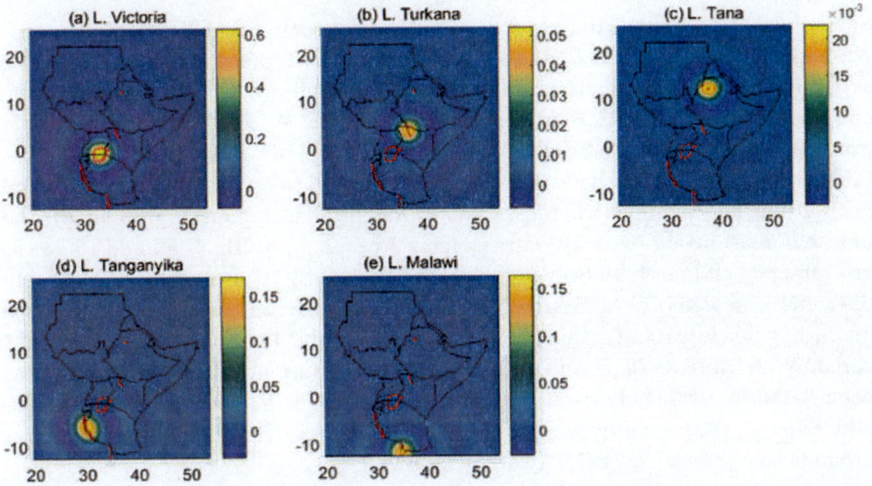

Fig. 13.3 Greater Horn of Africa Lakes' kernel (coefficient) functions. The areas and minimum/maximum values for the kernel functions are **a** lake Victoria (68,800 km^2, min/max—0.0674/0.6235), **b** lake Turkana (6405 km^2, min/max—0.0074/0.0529), **c** lake Tana (2156 km^2, min/max—0.003/0.0220), **d** lake Tanganyika (39,200 km^2, min/max—0.0298/0.1722), **e** lake Malawi (29,600 km^2, min/max—0.0264/0.1647). Lake coefficient function depends on the area and shape of the lake. Circular/spherical shapes results in higher lake coefficients compared to narrow elongated ones. *Source* [1]

13.5 Groundwater Changes: Hydrological Model Evaluation

Due to lack of in-situ groundwater data over GHA, the GRACE-derived groundwater changes (ΔGW) are compared to those obtained from WGHM model as the model has been found to adequately capture groundwater changes, see [32]). In [32], WGHM groundwater changes are compared with independent estimates (observations, local scale modeling, and GRACE-derived estimates) across various locations globally that included North Western Sahara Aquifer and Nubian Sandstone Aquifer, which falls within the GHA region. Although WGHM model has been found to adequately capture groundwater changes over various locations globally, being a model, it suffers from various uncertainties thus does not fully represent in-situ groundwater changes. Nevertheless, such comparison is standard and has been applied, e.g., in [32, 42, 76, 100]. The results of GRACE-derived groundwater changes capture similar temporal variability over the region as those derived from WGHM variants (WGHM IRR_70_S (limited irrigation) and WGHM_NOUSE), albeit with stronger amplitudes (Fig. 13.4a). The correlation between GRACE-derived groundwater changes and those of WGHM IRR_70_S and WGHM_NOUSE are 0.7147 and 0.7173, respectively (significant at 95% confidence level). Such strong correlation validates the assumption of relatively small δGW (i.e., $\Delta GW >>> \delta GW$, considered in

Sect. 13.4.1. In addition, the close relationship between GRACE-derived groundwater changes and WGHM IRR_70_S/WGHM_NOUSE points to low groundwater extraction in the region as already reported in other studies, e.g., [84, 90]. The spatial correlation between GRACE-derived groundwater and WGHM variants are similar except for areas north of latitude 15° where the WGHM NOUSE model variant has (almost) constant values hence no correlation (Fig. 13.4b, c). The correlation are fairly strong and significant in most areas within the GHA region (Fig. 13.4d, e). The low and/or strong anti-correlation in isolated areas (e.g., North of latitude 15°, eastern Ethiopia, and northern Somalia) could be caused by (phase) shifts due to depth differences between GRACE and WGHM since GRACE sense greater depths. Based on the largely close relationship between GRACE-derived groundwater and WGHM variant WGHM IRR_70_S, and the fact that WGHM variant WGHM IRR_70_S has been found to adequately capture groundwater changes in various locations globally, see e.g., [32], it suffices to imply that GRACE-derived groundwater captures groundwater changes over GHA sufficiently well.

13.5.1 Groundwater Sustainability

Due to lack of recharge data (both in-situ and from WGHM), this section compares GRACE TWS and model derived water flux in a bid to understand the sustainability of groundwater for irrigation;

$$\frac{d(TWS)}{dt} = P - R - E, \tag{13.5}$$

where time (dt) is per month, P is precipitation, R runoff, and E-evapotranspiration.[6] On average, GRACE TWS shows similar temporal variability with model derived water flux albeit with larger amplitudes (Fig. 13.5a). The overall average correlation stands at 0.5508 at zero lag while with water flux proceeding GRACE TWS with one month, the correlation improves to 0.8390. Figure 13.5b shows the spatial variation in the correlation at zero lag while Fig. 13.6a shows maximum correlation at various lags shown in Fig. 13.6b. The high correlation over Ethiopian highlands, South Sudan, Lower Sudan, Tanzania and Western Uganda points to GRACE TWS accurately representing water flux and could by extension imply the dependency of groundwater (since TWS has groundwater as one of the components) on rainfall. The areas with low and/ or negative correlations, e.g., Upper Sudan (North of latitude 15°), Somalia, Eastern Ethiopia and Eastern Kenya have relatively high evapotranspiration coupled with groundwater recharge from outside the local areas hence GRACE TWS and model-derived flux have weak relationships. With climate studies (see, e.g., [71] and the references therein) pointing towards possible future increase in extreme events, the close relationship between GRACE TWS and model derived water flux over

[6] P from TRMM while R and E are from GLDAS.

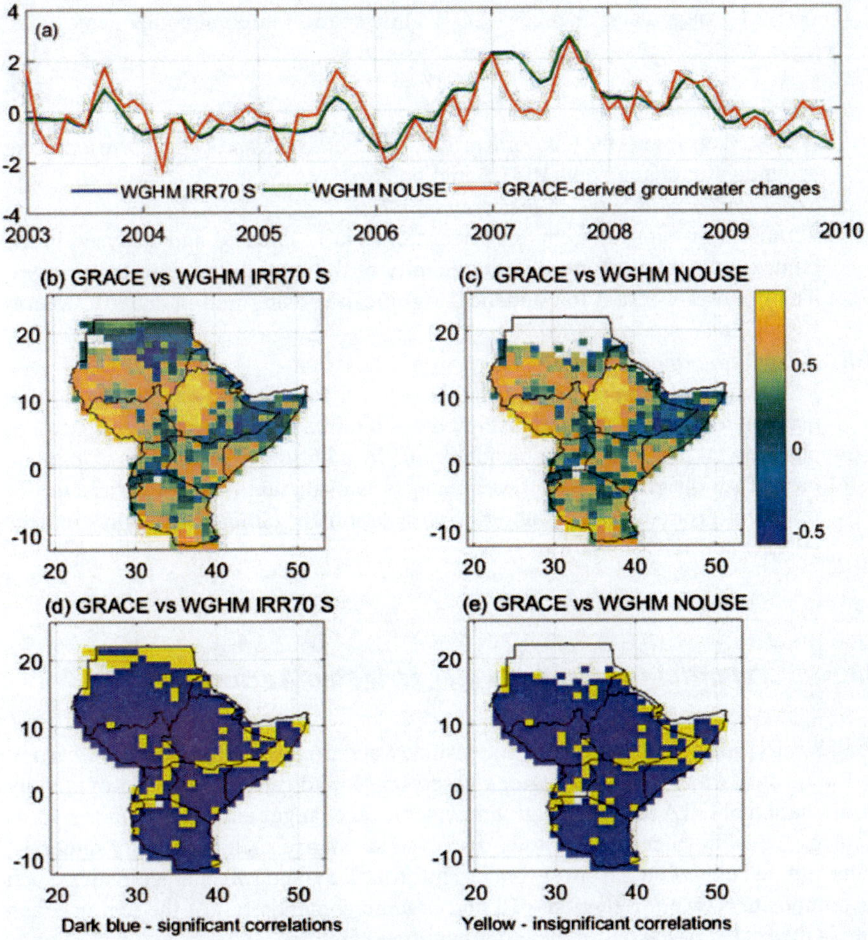

Fig. 13.4 A comparison between GRACE-derived groundwater changes and those of WaterGap Hydrological Model (WGHM). **a** Temporal variability of standardized (spatially averaged) groundwater changes from 2003 to 2009 (y-axis indicates anomalies), **b** and **c** are correlation coefficients between GRACE derived groundwater changes and those of WGHM variants, **d** and **e** are significance of the correlations. GRACE-derived groundwater changes sufficiently captures groundwater changes over GHA. The similarity in variability of the two WGHM variants indicates limited groundwater exploitation in the region. The correlations are significant at $p < 0.05$. In **b–e**, the y-axis indicates the latitudes while the x-axis indicates the longitudes. *Source* [1]

several areas points to sustainability of groundwater based on rainfall. Further, the GRACE-derived groundwater changes are analysed as follows;

(i) *Localization of GRACE-derived groundwater:*
Independent Component Analysis (ICA; see, e.g., [11, 17, 21, 23, 38, 39, 69] for formulation and description) is used to decompose and localize GRACE-

derived groundwater changes (ΔGW) into spatio-temporal components as

$$\Delta GW = PQ, \tag{13.6}$$

where P represents the temporal groundwater changes and Q the corresponding groundwater change patterns (spatial patterns).

(ii) *Temporal groundwater change trend analysis:*
Temporal groundwater change linear trends are evaluated and analysed in the context of rainfall changes in the vicinity of the spatial groundwater changes. The trends are tested for statistical significance using non-parametric Mann-Kendall test, see, e.g., [48].

(iii) *Correlation analysis:*
Pearson correlation analysis, (see [37, 48] for formulation and description) is used to determine the relationship between rainfall and temporal groundwater changes taking lags into consideration. In addition, the impacts of climate variability on groundwater level changes is evaluated through correlation of temporal groundwater variabilities with dominant climate variability indices (ENSO and IOD) at various lags.

13.5.2 Potential for Groundwater Irrigated Agriculture

Preliminary analysis of potential for groundwater irrigated agriculture is carried out on groundwater localized regions based on groundwater quantity (provided by combination of GRACE-derived groundwater level changes and groundwater hydrogeology), quality (measured by total dissolved substance such as primary minerals, salts, organic matter etc.; from secondary information), and soil characteristics (such as composition, water retention, pH, and organic content, etc.) of the region. Each site is then rated based on all the aforementioned factors.

13.6 Spatio-Temporal Variability of Groundwater Changes

The decomposition of GRACE-derived groundwater changes using ICA results in spatio-temporal groundwater changes. Cross comparison of the locations of spatial groundwater change patterns with aquifer (and/or groundwater) location maps of the region (see, e.g., Fig. 13.2b; [2, 6, 47] reveals several spatial groundwater change patterns in regions belonging to various aquifers. These spatial groundwater change patterns are thereafter referred to by names of the aquifer (and/or basins) in whose regions they occur (Fig. 13.7).

Majority of the identified spatial groundwater change patterns (aquifers) are transboundary, i.e., their extent transverse two or more countries. Only two in Ethiopia

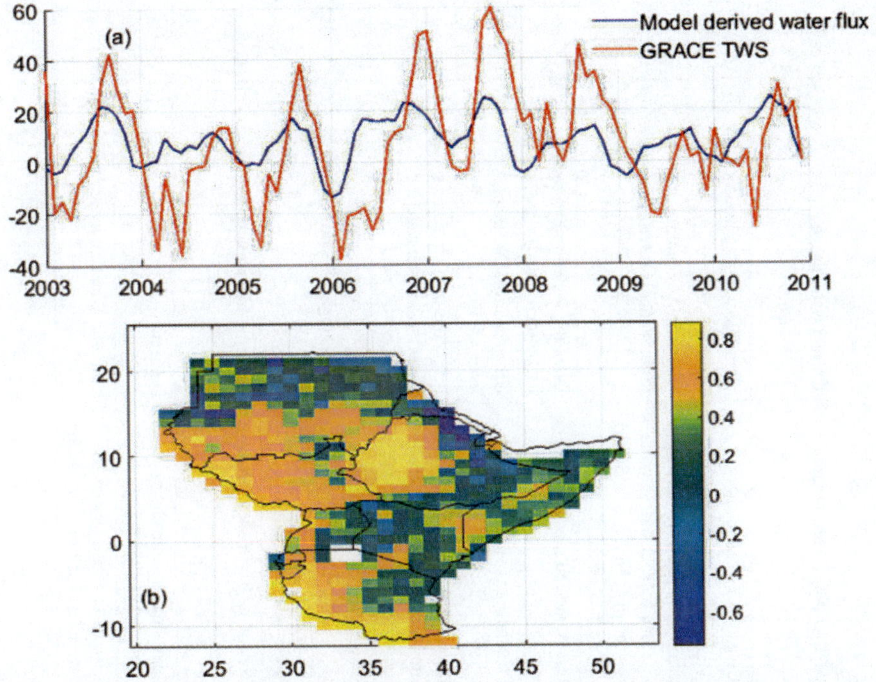

Fig. 13.5 Relationship between GRACE-TWS and model-derived water flux; **a** time series plot [mm/month], **b** correlation. The water flux is derived from rainfall (CHIRPS), runoff (GLDAS) and evapotranspiration (GLDAS). The y-axis indicates the latitudes while the x-axis indicates the longitudes. *Source* [1]

(Fig. 13.7d, e) and one in Tanzania (Fig. 13.7g) are within countries. Due to the limited resources, political will, and the cross-country boundary location of most of these aquifers, very limited resources are allocated if any for their monitoring, e.g., [6]. GRACE thus provides an efficient and affordable way for large-scale monitoring (i.e., several aquifers at the same time) of these aquifers. The general characteristics of these aquifers (spatial groundwater change patterns) are summarised in Table 13.3.

The temporal evolutions of the groundwater changes are presented in Fig. 13.8. Other than Nubian Sandstone and Kenya-Somalia (Fig. 13.8a, f), the temporal evolutions of groundwater changes for the rest of the aquifers show an annual pattern, which could be due to the fact that recharge takes place once every year during main rainfall seasons. Also, the groundwater temporal evolutions seem to peak approximately 6 months, on average, after rainfall (Fig. 13.8b–e and g–i). The temporal evolution of the groundwater changes in upper Greater Horn of Africa (UGHA; Nubian Sandstone, Upper Nile, Karoo Carbonate, Ethiopian highland, and Lake Tana Basin) have low maximum (peak) values in 2005, 2010, 2012, and 2014 (Fig. 13.8a–e), which could be attributed to low rainfall (see, Fig. 13.8a–e for rainfall), hence reduced recharge and/ or relatively more usage of groundwater on those years. These

Table 13.3 Groundwater characteristics and usage. *Source* [1]

Groundwater changes	Recharge rate mm/annum	Recharge source or mode	Aquifer type (characteristic)	Current water usage	Source
Nubian Sandstone (Sahara Nubian Basin)	0–2	Occasional rainfall	Consolidated sedimentary	Domestic watering livestock	[6, 73, 79]
Upper Nile (Sudd basin, Baggara basin, and other small aquifers)	2–100	Rivers Wadies Direct rainfall	Aquifers in precambrian and volcanic rocks. Umm Rawaba formation confined, unconfined, and semi confined	Domestic Agricultural (Subsistence) Watering livestock	[2, 6, 82]
Karoo Carbonate	100->300	Rainfall River Congo	Limestone and sandstone	Domestic Watering livestock Irrigation (limited)	[6]
Ethiopia highlands	150–250	Direct deffuse recharge through soil (rainfall)	Precambrian metamorphic crystalline basement	Urban and rural Watering livestock Irrigation	[47]
Lake Tana region	20–300	Summer rains Losing streams	Deep and shallow volcanic aquifers. Shallow basement aquifers.	Irrigation (both large and small scales). Domestic watering livestock	[47]
Kenya-Somalia	2–100	Underground run-off. River Ewaso Ngiro. Dera drainage (river and swamp). Limited direct recharge from rainfall	Semi-consolidated sedimentary	Refugee camps. Domestic public water supply watering livestock	[6, 51, 67, 68, 74, 85]
Central Tanzania	2–20 (periodic)	Direct rainfall infiltration, preferential flow, and through fractures	Crystalline basement aquifers	Domestic Irrigation watering livestock	[46, 75]
Karoo Sandstone	20->300 (periodic)	Rivers Rainfall	Consolidated sedimentary	Domestic	[6, 93]
Ruvuma	20–300	River Ruvuma rainfall	Un/semi-confined quaternary consolidated sedimentary rocks	Domestic Watering livestock	[6, 93]

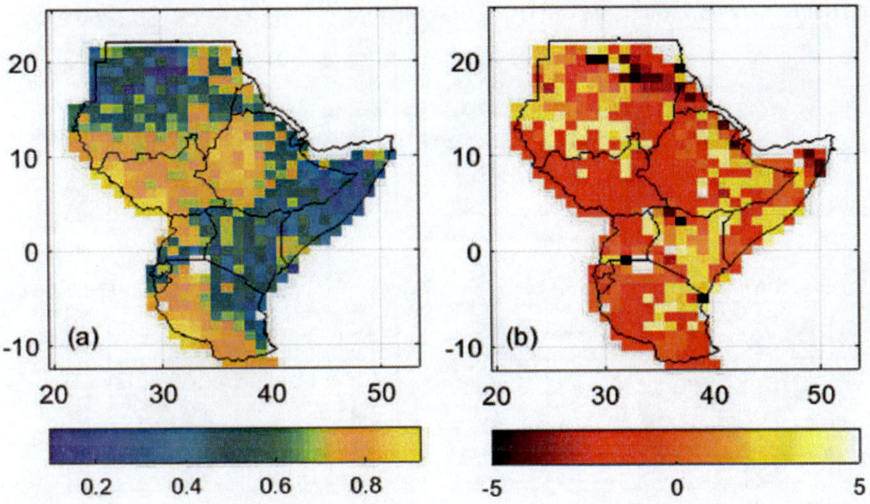

Fig. 13.6 Lag cross correlation; **a** maximum correlation, **b** lag at maximum correlation [month]. Negative lag values indicate water flux leading while positive indicate water flux lagging GRACE TWS. The y-axis indicates the latitudes while the x-axis indicates the longitudes. *Source* [1]

years are preceded by low maximum (peak) rainfall in 2004, 2009, and 2010 (see, Fig. 13.8a–e for rainfall), hence drought as has been reported in other studies, e.g., [34, 63, 91], and the references therein.

The temporal evolutions over East Africa (EA; Kenya-Somalia, Central Tanzania, Karoo sandstone, and Ruvuma basin) record relatively higher maximum (peak) values due to high recharge from the impact of El'Niño induced rain that occurred in the region in 2006/2007 (Fig. 13.8f–i; [16, 50, 88]. In addition, the temporal evolutions have low maximum (peak) values in 2004, 2006, 2011, and 2013 (Fig. 13.8f–i), years that are preceded by relatively lower rainfall (Fig. 13.8f–i) and are reported as having been drought years, see e.g., [45, 59, 63, 72, and the references therein].

The close relationship between groundwater change temporal evolution and rainfall as observed above could be attributed to the fact that due to limited groundwater exploitation in the region (see, e.g., [84, 90], the temporal evolution is to a large extent influenced by recharge (see, Eq. 13.4). Recharge in itself is dependent on rainfall (intensity, duration, and volume), hydrogeological characteristics of the groundwater region (e.g., geomorphology, geology, and pedology) [43], and vegetation cover and/or land use in the recharge regions [27, 84]. Land use land cover changes e.g., deforestation (afforestation) leads to increased (reduced) run-off and/or evapotranspiration and has a site specific influence on recharge while continued urbanization leads to lower long-term average groundwater recharge [27, 54, 66].

Further, the temporal evolutions of GRACE-derived groundwater changes are found to be largely similar to spatially averaged WGHM IRR_70_S (averaged over the boxed areas in Fig. 13.7) though with relatively larger amplitudes (Fig. 13.9). Over

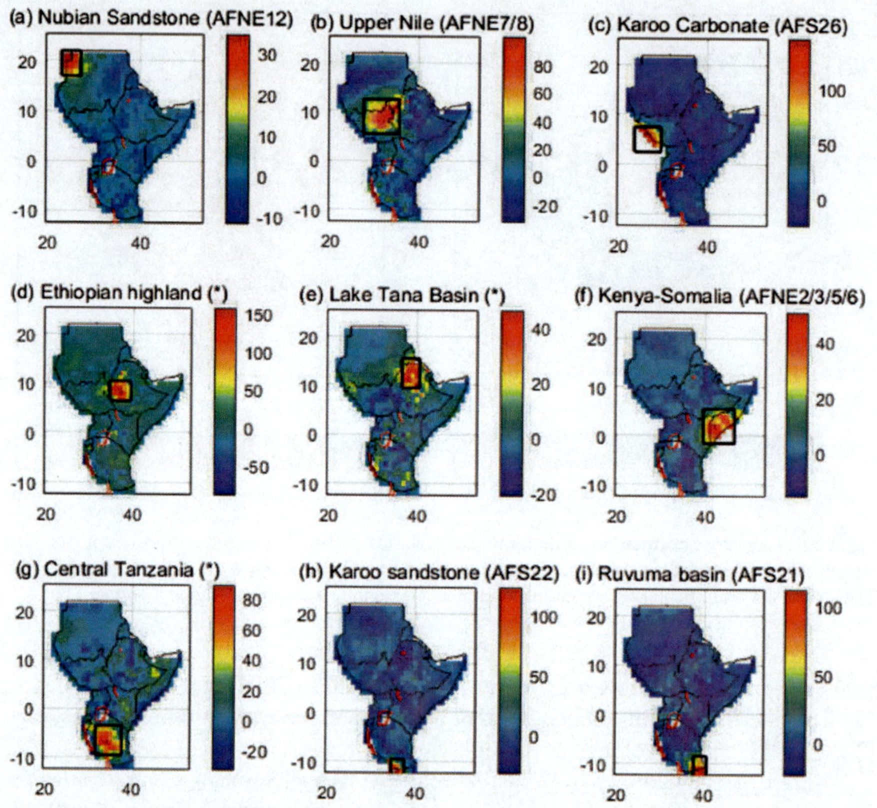

Fig. 13.7 Selected spatial groundwater change patterns over GHA [mm/month]. The spatial patterns result from regionalization/localization of GRACE-derived groundwater changes (2003–2014) using Independent Component Analysis (ICA). The names in brackets correspond to aquifers' names as referred in Fig. 13.2, while * denotes those aquifers not in Fig. 13.2 and are only found entirely within a country. The rectangles (black) within each figure indicate areas over which precipitation and WGHM-derived groundwater have been spatially averaged. *Source* [1]

East Africa, both GRACE-derived groundwater changes and WGHM IRR_70_S temporal evolutions clearly depict the high 2007 and low 2006 maximum (peak) values corresponding to the El'Niño induced rains of 2007 and the droughts of 2005–2006, respectively. Over UGHA, distinctive feature like the low maximum (peak) value in 2005 is clearly visible in all the figures.

Over the entire 2003–2014 period, the relationships between the temporal groundwater changes and spatially average rainfall (averaged over extents shown in Fig. 13.7) show weak relationships with varied lags (Table 13.4). The weak relationships could be attributed to the fact that most of the aquifers considered are recharged through additional means, e.g., losing streams, preferential flows and through fractures, and underground rivers, other than direct rainfall infiltration (see, Table 13.3

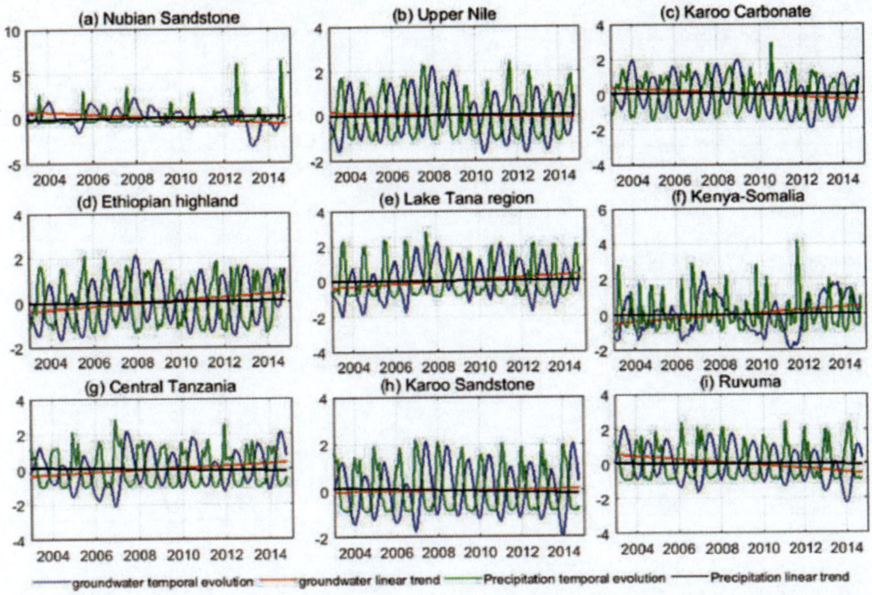

Fig. 13.8 Temporal variability of spatial groundwater change patterns (in Fig. 13.7) and precipitation spatial averages (2003–2014; [mm/month]), fitted with linear trends. The groundwater temporal evolutions are the independent components from ICA localization of GRACE-derived groundwater changes while precipitation spatial averages (standardized) are done over the rectangles in Fig. 13.7. The magnitude of the trends are tabulated in Table 13.4. The groundwater temporal evolutions largely reflects the average rainfall with approximately 6 months shift between the signals. The y-axis indicate anomalies while the x-axis are years. *Source* [1]

and the references therein). Probably, a stronger relationship would be between rainfall and recharge as opposed to the relationship between rainfall and groundwater level changes. This follows from the evidence of episodic recharge link to extreme rainfall events in an aquifer in Central Tanzania, see e.g., [88]. Other than the Nubian Sandstone and Kenya-Somalia with lags of 80 and 105 months, respectively, the rest of the aquifers have lags in the range of 2–27 months (Table 13.4). These lags are largely dependent on among other things; the depth and characteristics of the associated aquifers (e.g., confined vs unconfined, volcanic vs Precambrian, etc.), the source and mode of the recharge (Table 13.3), and the land use land cover in the recharge region. The reflection of rainfall in groundwater level changes after durations of time (lag) forms one of the major advantages of groundwater as a source of water due to the fact that it can be used without worry even during longer drought events, see e.g., [19, 20].

Least squares trend analysis results of GRACE-derived temporal groundwater changes, and spatially averaged rainfall and WGHM IRR_70_S are shown in Table 13.4 (also see, Figs. 13.8 and 13.9). GRACE derived temporal groundwater changes have: significant positive trends over Ethiopian highlands, Lake Tana region,

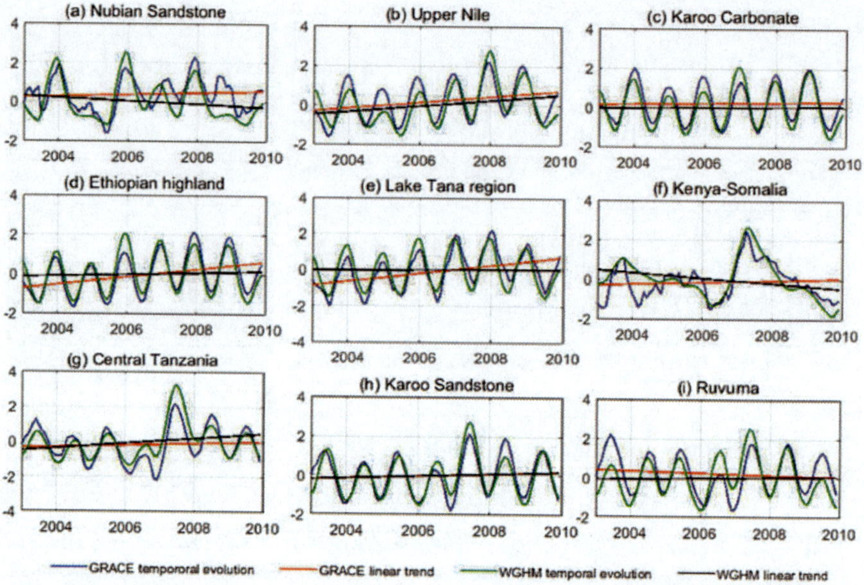

Fig. 13.9 Temporal variability of spatial groundwater change patterns (in Fig. 13.7) and WGHMIRR_70_S groundwater change spatial averages (2003–2010; [mm/month]), fitted with linear trends. The groundwater temporal evolutions are the independent components from ICA localization of GRACE-derived groundwater changes while WGHMIRR_70_S groundwater change spatial averages (standardized) are done over the rectangles in Fig. 13.7. The magnitude of the trends are tabulated in Table 13.4. The WGHM temporal evolutions are consistent with those of GRACE-derived groundwater changes. The y-axis indicate anomalies while the x-axis are years. *Source* Fig. [1]

and Kenya-Somalia; significant negative trends over Nubian Sandstone and Ruvuma; and stable temporal groundwater changes over the remaining aquifers, i.e., recharge balances base flow and usage over the considered duration of time (2003–2014).

The trends in averaged WGHM IRR_70_S are all insignificant except for Upper Nile and Central Tanzania, while average rainfall has insignificant trends over all the groundwater localized areas. The lack of significance in these trends especially for WGHM IRR_70_S could be attributed to the short duration of time under consideration (7 years), while the rainfall appears to be stable over the considered duration (2003–2014).

Considering Eq. (13.4), the insignificant rainfall trends, and the average annual rainfall (Table 13.4), the positive trends in GRACE-derived temporal groundwater changes over Ethiopian highlands and Lake Tana region could be attributed to a combination of steady recharge (evidenced by stable average annual rainfall), limited abstraction, see, e.g., [47, 84], and very low base flow in comparison to recharge amounts. The case for Kenya-Somalia is special due to low annual rainfall coupled with high evapotranspiration over the region (e.g., 432 mm/annum of average annual rainfall (Table 13.4) vis-a-vis $\simeq$ 2330 mm/annum of evapotranspiration; [68].

Table 13.4 Trend magnitudes and lag relationships. Ethiopia highlands, Lake Tana region, and Kenya-Somalia have increasing groundwater trends, Nubian Sandstone and Ruvuma have reducing groundwater trends, while the rest have stable groundwater levels over the considered duration (2003–2014). The relationships between temporal groundwater changes and average rainfall are weak. Maximum correlations are found when rainfall preceded groundwater changes by the tabulated months. Significant trends are in bold ($p < 0.05$). *Source* [1]

Groundwater	Groundwater trend mm/month	WGHMIRR_70_S trend mm/month	Rainfall trend mm/month	Maximum correlation with groundwater	Lag at maximum correlation (months)	Average annual rainfall (mm)
Nubian sandstone	**−1.328**	−0.001	0.016	0.1540	80	35
Upper Nile	0.963	**0.069**	0.034	0.4542	2	907
Karoo Carbonate	−2.362	−0.013	−0.044	0.3299	3	1480
Ethiopia highlands	**3.134**	0.022	0.071	0.3974	27	1385
Lake Tana region	**1.228**	0.002	0.041	0.4560	27	887
Kenya-Somalia	**1.582**	−0.025	0.015	0.2047	105	432
Central Tanzania	2.163	**0.125**	−0.095	0.3734	3	916
Karoo sandstone	0.283	0.073	−0.191	0.4778	2	1230
Ruvuma	**−2.753**	0.026	−0.040	0.4181	9	1050

The recharge for groundwater in this region is from River Ewaso Ng'iro (originating from Mount Kenya region) and underground runoff from other regions e.g., Ethiopian highlands, e.g., [2, 67, 74]. The Nubian Sandstone is characterized by very limited active recharge as indicated by low correlation, long lags, and very low annual average rainfall (Table 13.4) hence its significant negative trend (depletion; as there is no recharge) could be driven by base flow and abstraction. This follows from Eq. (13.4) in which the groundwater level variation is governed by a balance of recharge, abstraction and base flow and since in this region recharge (derived rainfall) is negligible, the temporal groundwater variation (trend) is therefore governed by abstraction (water usage) and base flow. This is in line with other studies, see e.g., [32, 41, 81], which have documented the negative trend in the aquifer as being driven by abstraction without a recharge. The negative trend for Ruvuma is driven by higher base flow as water flows back into the river tributaries in several locations in the region [93], coupled with low net abstraction.

Finally, in exploring the impact of climate variability on temporal groundwater changes, the correlation between temporal groundwater changes and the dominant climate variability indices (ENSO and IOD) shows weak relationships (Table 13.5). Since climate variability impact on groundwater through its influence on rainfall, this

Table 13.5 Relationship between temporal groundwater changes and climate indices. All the correlations are statistically significant ($p < 0.05$) but largely weak. Both Nubian Sandstone and Kenya-Somalia do not show any meaningful relationship with ENSO while Nubian Sandstone, Kenya-Somalia, Ethiopia highlands, and Lake Tana regions do not show any meaningful relationship with IOD. *Source* [1]

Groundwater	Maximum correlation with IOD	Maximum correlation with ENSO	Lag at maximum correlation with IOD	Lag at maximum correlation with ENSO
Nubian Sandstone	0.5652	0.5404	19	29
Upper Nile	0.3176	0.3052	19	16
Karoo Carbonate	0.2822	0.2427	61	16
Ethiopia highlands	0.2373	0.1658	19	51
Lake Tana region	0.3026	0.1928	19	52
Kenya-Somalia	0.4387	0.3974	95	5
Central Tanzania	0.3525	0.3132	37	17
Karoo Sandstone	0.2537	0.2234	37	15
Ruvuma	0.2655	0.3503	37	23

chapter considered only the correlations when groundwater was lagging climate variability indices. Due to the short duration considered, approximately 10 years, these results should be analysed with caution as longer duration datasets may be necessary to give more reliable results. The association of groundwater temporal evolution with IOD is relatively higher than that with ENSO, which could be attributed to the relatively stronger influence of IOD on the region's rainfall, see e.g., [7, 78, 95]. Both Nubian Sandstone and Kenya-Somalia did not show any meaningful relationship with ENSO, while the following temporal groundwater changes did not show any meaningful relationship with IOD; Nubian Sandstone, Ethiopian highlands, Lake Tana region, and Kenya-Somalia. The lack of meaningful relationships follows from the fact that climate variability impacts on groundwater change through rainfall, thus the lag at maximum correlation are expected to be higher in the correlation of temporal groundwater changes with climate variability indices than with rainfall (Table 13.5 vis-a-vis Table 13.4). For Nubian Sandstone and Kenya-Somalia, their lack of meaningful relationships could be attributed to low or no recharge from rainfall while for Ethiopian highlands and Lake Tana region, their lack of meaningful relationships could be attributed to the fact that IOD does not have much influence on rainfall in those regions. The rainfall over these regions (Ethiopian highland and northern Ethiopia) results from moisture migration from rain forest regions over Congo Basin as a result of westerly winds and thermal low pressure of ITCZ [96]. Longer duration of temporal groundwater changes (groundwater level changes) would be necessary in order to understand with certainty the impact of climate variability (and/or change) on groundwater changes hence long-term resource sustainability.

13.7 Potential of Groundwater Irrigated Agriculture

The GRACE-derived groundwater changes characteristics from Sect. 13.6 are combined with existing secondary information (literature) e.g., groundwater potential, groundwater quality, and soil types, with a view to examining the potential for groundwater irrigated agriculture, which is explored based on water availability (quantity and quality) and dominant soil types (Table 13.6). An indication of water availability for sustainable exploitation is provided by temporal groundwater changes, trends (Sect. 13.6) and yield information (Table 13.7), while quality is based on either physicochemical properties of total dissolved solids (TDS) or electrical conductivity (EC); both measures of dissolved minerals content in water. The chapter assumes that the TDS/EC values cover the entire spatial groundwater change areas. For a comprehensive suitability analysis, a more detailed investigation will have to be carried out at a larger scale considering specific nutrients, well-specific yields, and finer soil classification.

TDS of <450 mg/l, 450–2000 mg/l, and >2000 mg/l are considered suitable, slight to moderately suitable, and severe, respectively, in degrees of restriction of use for irrigated agriculture while EC of <700 mS/cm, 700–3000 mS/cm, and >3000 mS/cm are considered excellent, good, and fair (would greatly affect yield), respectively, see e.g., [15, 52, 80]. Based on water availability (Fig. 13.8 and Table 13.4), water quality (TDS) and dominant soil types (Tables 13.6 and 13.7), the potential for groundwater irrigated agriculture for various groundwater localized areas is presented in Table 13.7.

Kenya-Somalia has good water availability (quantity) considering the temporal groundwater changes (Fig. 13.8d), trends (Table 13.4), and average yield (Table 13.7) that can easily support subsistence agriculture. The freshwater within the center of Merti aquifer has TDS of less than 1000 mg/l with a pH value of between 7.2 to 7.8 with the rest of the area overlain by Precambrian crystalline or consolidated sedimentary having TDS of 2000–4000 mg/l (Fig. 13.2a; [85]. The dominant soil types over the region are Solonetz and Calciols (Table 13.7), poor soils that requires expensive agricultural management, e.g., leaching with freshwater, liming, fertilization, and construction of engineering drainage systems in order to be suitable for crop production, hence low potential for groundwater irrigated agriculture (Table 13.7). However, the area could be used for sodium and calcium tolerant crops. The limited freshwater at the center is under pressure from the community and the refugee centers, and could run the risk of saline intrusion if subjected to an additional pressure of supporting irrigated agriculture in areas of suitable soils.

Although Nubian Sandstone (Nubian Sahara) has falling (groundwater) trends (Fig. 13.8b and Table 13.4) due to limited recharge, it has a good yield (Table 13.7) with one of the largest deposits of fossil water in Africa, see e.g., [62]. The fall is due to water use by other states sharing the aquifer as the Sudanese side comprises only of pastoralist communities [3]. It has a fair quality of water with TDS values of between 500 and 800 mg/l [79] and a dominant soil type of Alisols (Table 13.7), which like Solonetz and Calciols found in Kenya-Somalia has poor agricultural potential.

Table 13.6 Soil characteristics. *Source* [1] adopted from [33, 35]

Soil type	Characteristics (summary)
Acrisols	Suitable for production of rain-fed and irrigated crops only after liming and application of fertilizer. Its limitations include; poor chemical properties, low levels of plant nutrients, and aluminium toxicity and P-sorption
Solonetz	Characterized by high concentration of sodium which impacts negatively on plants. Its use for agriculture is dependent on its depth and property of surface soil. Reclamation is possible but expensive
Vertisols	They have high natural fertility and positive response to management, however, their moisture control problem imposes critical constraints to low input agriculture. Under rain-fed conditions and depending on temperature, they have been used to produce wheat, maize, sorghum, soya-beans, cassava e.t.c.
Ferrisols	They have great soil depth, good permeability, and stable micro-structure. They are well drained but have low water holding capacity. They have poor chemical properties hence careful selection of fertilizer and liming would be necessary for better returns
Plinthosols	Their potential for agriculture depends on how deep they are. Requires management intervention due to poor natural soil fertility and water logging issues
Nitisols	Are among the most productive soils of the humid tropics. They are of good work-ability, good internal drainage, fair water holding properties, and of good fertility. They contain high concentration of weathered materials and are suitable for a wide range of crops
Cambisols	They have natural to weakly acid retention, satisfactory chemical fertility, and active soil fauna. They make good agricultural lands and are intensively used
Leptosols	Are common in mountainous regions. Their shallowness and/or stoniness and implicit low water holding capacity are some of their limitations. Erosion is the greatest threat to steep slope soil areas. Hill slopes are normally fertile
Alisols	Strongly acidic soils with accumulated high activity of clay at sub-soils. They have toxic levels of aluminium at shallow depths and poor natural soil fertility. Use restricted to acid tolerant plants/crops and low volume grazing. Productivity for subsistence agriculture is low as these soils do not recover easily from chemical exhaustion. Crop production is possible if fully limed and fertilized
Lixisols	Agricultural use is subject to use of fertilizer and/or lime addition due to low level of plant nutrient and low cation retention
Calciols	Have substantial secondary lime accumulation. Suitable for lime tolerant crops. They reach their full potential when carefully irrigated; furrow irrigation performs better
Luvisols	Most Luvisols are fertile soils and are suitable for a wide range of agricultural activities

Alisols requires liming and addition of fertilizer for productive use though they do not easily recover from chemical exhaustion (Table 13.6), hence low potential for groundwater irrigated agriculture (Table 13.7).

Upper Nile, comprising of several aquifers, has good water availability (Fig. 13.8a and Tables 13.4 and 13.7) with varying TDS in the ranges of 200–500 mg/l and 100–400 mg/l over Sudd and Baggara basins, respectively [79]. The dominant soil type is Vertisols (Table 13.7), which despite containing a high level of plant nutrients, may require additional management practice due to its high clay content. The clay content is normally montmorillinitic with high cation exchange capacity (CEC) but introduces problems such as lowering the hydraulic conductivity of soil when wet due to expansion of montmorillonite. If the dissolved salt includes Na, montmorillonite will disperse and significantly lower the hydraulic conductivity though liming and increased organic matter should help. The high clay content, therefore, presents a moisture control challenge that may impose a critical constraint on low input agriculture, hence moderate to high potential for groundwater irrigated agriculture (Table 13.7).

Ethiopian highland is characterized by good water availability (Fig. 13.8a and Table 13.4) of suitable quality with high agricultural potential soil hence high potential for groundwater irrigated agriculture (Table 13.7).

Lake Tana has both good water availability and soils of good agricultural potential. Considering Fig. 13.8c and Tables 13.4 and 13.7), coupled with TDS of less than 500 mg/l [47] the region has low to high agricultural potential. On areas of the crystalline basement with low groundwater potential, the agricultural potential is likewise low while over areas of volcanic hydrogeology, the agricultural potential is moderate (Table 13.7).

Karoo Carbonate is characterized by large volumes of good quality water (Fig. 13.8c and Tables 13.4 and 13.7) with poor quality soils (Plinthosols and Lixisols) requiring additional management practices, e.g., the addition of fertilizer and/or lime, water logging issues, etc. Its irrigated agricultural potential is thus low to moderate (Table 13.7). Central Tanzania has a stable quantity of groundwater (Fig. 13.8c and Table 13.4) though with relatively high TDS of 1000–3000 mg/l [46] and varying groundwater potential and yield depending on the hydrogeology (Table 13.7). Due to varied dominant soils e.g., Cambisols, Solonetz, Acrisols, and Vertisols its groundwater irrigated potential ranges from low to moderate as additional intervention measures could be necessary for good agricultural returns (Table 13.7). Karoo Sandstone is dominated by good agricultural soil, which coupled with groundwater availability (Tables 13.4 and 13.7) with TDS of 400–800 mg/l [93] make the region to have moderate to high potential for groundwater irrigated agriculture (Table 13.7).

Finally, Ruvuma though characterized by high potential of groundwater with good quality (Table 13.7), the falling trend (Table 13.4) coupled with average soils (Cambisols and Acrisols) qualifies it for low to moderate potential for groundwater irrigated agriculture. The regions dominated with Acrisols, soils with low levels of plant nutrients and poor chemical properties, have low potentials for agriculture (Table 13.7).

Based on the yield, groundwater potential, and recharge rates (Table 13.7), large-scale groundwater irrigated agriculture may not be possible but subsistence, medium to small-scale groundwater irrigated agriculture would be sufficiently supported in these areas. Groundwater with high TDS or EC levels results in plants being unable to absorb sufficient water from the salty solution despite the presence of sufficient moisture hence reduced yield/productivity [36]. On the same note, soils with high concentration of salts result in the salts leaching into the ground during rainfall events. This leads to increase in concentration of salts in the soil (i.e., salinity), thus impeding plants from absorbing water and nutrients. In addition, in the case of salty water, as it is used up (some evaporates) the salt remains deposited in the soil and is washed back and the cycle continues making the soil unsuitable for agricultural activities. Also, salt resistant crops could be tried in regions of salty water and/or soils with excess salts. Finally, soil quality in such regions could be improved with the installment of properly designed drainage systems though this would be beyond small scale farmers.

13.8 Concluding Remarks

The chapter derived groundwater changes from GRACE TWS and subsequently used ICA method to localize them into aquifer regions and corresponding temporal variabilities. The temporal groundwater changes were then analysed and correlated with dominant climate variability indices (ENSO and IOD) in order to explore the impact of climate variability. Finally, the potential for groundwater irrigated agriculture in the aquifer regions were explored. The major findings of the chapter are:

(i) GRACE-derived groundwater changes showed similar temporal variability to WGHM variants IRR_70_S and WGHM_NOUSE with Pearson correlation coefficients of 0.7147 and 0.7173 (significant at 95% confidence level), respectively. This indicates the potential of using GRACE satellites to study groundwater changes in this data deficient region.

(ii) Based on the aquifer (and/or groundwater) location maps of the region, the chapter identified the following aquifers (and/or groundwater areas); Nubian sandstone, Karoo Carbonate, Upper Nile, Ethiopian highlands, Lake Tana region, Kenya-Somalia, Central Tanzania, Karoo sandstone, and Ruvuma.

(iii) The temporal evolution of the identified GRACE-derived groundwater changes showed largely similar variability to those of spatially averaged WGHM variants IRR_70_S and reflected changes in annual average rainfall. All the temporal groundwater changes, except Nubian sandstone and Kenya-Somalia, showed an annual (cyclic) pattern indicating an annual (yearly) recharge cycle.

(iv) Least squares trend analysis of the temporal groundwater changes had; increasing groundwater levels for Ethiopian highlands, Lake Tana region, and Kenya-Somalia; stable groundwater levels for Upper Nile, Karoo Carbonate, and Cen-

Table 13.7 Potential for groundwater irrigated agriculture based on groundwater characteristics and dominant soil types. Potential for groundwater irrigated agriculture varies from low to high across the groundwater availability regions. The hydrogeology information has been sourced from [62], the soil type from [28] and the water quality from the references in Table 13.3. *Source* [1]

Groundwater region	Materials/ properties	Groundwater potential	Average yield (l/s)	Water quality	Dominant soil type	Agricultural potential
Nubian	Sandstone	Moderate to high	1–20 ranges of 500–800 ppm patches of areas with high sodium and bicarbonate concentration	Dissolved particles in the	Alisols	Low
Upper Nile	Precambrian basement. Volcanic rocks. Unconsolidated and semiconsolidated sediments (gravel, sane, clay, etc.)	Moderate to high	1–40	Varied with dissolved particles reaching 500–800 ppm over the Umm Rawaba formation	Vertisols	Moderate to High
Karoo Carbonate	Limestone Sandstone	Moderate to high	1–100	Good quality	Plinthosols Lixisols	Low to Moderate
Ethiopia highlands	Precambrian-metaomorphic Tertiary volacanic	Moderate	1–40	Dissolved particles approximately 500–800 ppm dominated by calcium bicarbonate ions	Nitisols	High

(continued)

Table 13.7 (continued)

Groundwater region	Materials/ properties	Groundwater potential	Average yield (l/s)	Water quality	Dominant soil type	Agricultural potential
Lake Tana region	Deep and shallow volcanic rocks. Crystalline basement	Low to high	0.1–100	Low level of dissolved solids suitable for drinking and irrigation	Leptosols Luvisols Vertisols	Low to Moderate
Kenya-Somalia	Semi-consolidated sedimentary	Low to moderate	1–10	Freshwater along the aquifer center (limited area) and brackish water in the remaining sections	Solonetz Calciols	Low
Central Tanzania	Crystalline basement precambrian formation (Granite and/or Gneiss)	Moderate	0.1–40	TDS in the range 1000–3000 p.p.m. Varied with high alkalinity some areas in	Cambisols Solonetz Acrisols Vertisols	Low to Moderate
Karoo Sandstone	Sandstone (consolidated)	Moderate to high	1–20	Generally good. Dissolved particles in the range of 400–800 ppm	Loamy sandy soils	Moderate to High
Ruvuma	Aluvium Sedimentary	High	1–40	Water with good quality though some locations are brackish or saline	Cambisols Acrisols	Low to Moderate

tral Tanzania; and decreasing groundwater levels for Nubian Sandstone and Ruvuma.

(v) Correlation analysis showed weak associations/relationships between the temporal groundwater changes and the dominant climate indices (IOD and ENSO), an indication of the limited impact of climate variability and/or change on the groundwater level changes. Both Nubian Sandstone and Kenya-Somalia did not show any meaningful relationship with ENSO, while Nubian Sandstone, Ethiopian highlands, Lake Tana region, and Kenya-Somalia did not show any meaningful relationship with IOD. Due to the short duration of the groundwater changes considered, longer duration of groundwater changes would be required for certainty of the observed (and/or lack of) association.

(vi) Based on water availability (from GRACE), water quality (indicated by the total dissolved substance) and dominant soil types, potential for groundwater irrigated agriculture results showed: low potentials for Nubian Sandstone and Kenya-Somalia areas; low to moderate potentials for Karoo Carbonate, Lake Tana region, central Tanzania, and Ruvuma; moderate to high potentials for Upper Nile and Karoo Sandstone; and high potential for Ethiopian highland. Due to the short duration of the dataset considered, approximately 10 years, the results of this chapter should be interpreted with caution as longer duration datasets may be necessary to give more reliable results.

On the spatio-temporal scales considered, the results show the potential of monitoring groundwater changes using GRACE and the capability of groundwater irrigated agriculture to reduce the negative impact on per capita food production if implemented. Future studies should consider analyses at larger scales, e.g., crop areas and/or sub country, well specific yields, and finer soil classifications in order to translate the results on to practical platforms and inform decision making at local levels.

References

1. Agutu NO, Awange JL, Ndehedehe C, Kirimi F, Kuhn M (2019) GRACE-derived groundwater changes over Greater Horn of Africa: temporal variability and the potential for irrigated agriculture. Sci Total Environ. https://doi.org/10.1016/j.scitotenv.2019.07.273
2. Abiye TA (2010) An overview of the transboundary aquifers in East Africa. J Afr Earth Sci 58(4):684–691. https://doi.org/10.1016/j.jafrearsci.2009.10.003
3. Alker M (2007) The nubian sandstone aquifer system: a case study for the research project - transboundary groundwater management in Africa. German Development Institute, pp 233–273 Accessed from http://mercury.ethz.ch/serviceengine/Files/ISN/103367/ichaptersection_singledocument/86a7e72a-f661-4151-a978-11727d7e45ac/en/8.pdf, On August 29, 2016
4. Alley WM (2007) Sustainable management of groundwater in Mexico: proceedings of a workshop (series: strengthening science-based decision making in developing countries). chap. The importance of monitoring to groundwater management. The National Academies Press, Washington, DC, pp 76–85. https://doi.org/10.17226/11875
5. Alley WM, Konikow LF (2015) Bringing GRACE down to earth. Groundwater 53(6):826–829. https://doi.org/10.1111/gwat.12379

6. Altchenko Y, Villholth KG (2013) Transboundary aquifer mapping and management in Africa: a harmonised approach. Hydrgeol J 21(7):1497–1517. https://doi.org/10.1007/s10040-013-1002-3
7. Ashok K, Guan Z, Yamagata T (2003) A look at the relationship between the ENSO and the Indian Ocean Dipole. J Meteorol Soc Jpn Ser II 81(1):41–56. https://doi.org/10.2151/jmsj.81.41
8. Ashton P, Turton A (2009) Facing global environmental change: environmental, human, energy, food, health and water security concepts. In: Water and security in Sub-Saharan Africa: emerging concepts and their implications for effective water resource management in the Southern African Region. Springer, Berlin, Heidelberg, pp 661–674. https://doi.org/10.1007/978-3-540-68488-6_50
9. Anyah RO, Forootan E, Awange JL, Khaki M (2018) Understanding linkages between global climate indices and terrestrial water storage changes over Africa using GRACE products. Sci Total Environ 635:1405–1416. https://doi.org/10.1016/j.scitotenv.2018.04.159
10. Awange J, Ferreira V, Forootan E, Khandu Andam-Akorful S, Agutu N, He X (2016) Uncertainties in remotely sensed precipitation data over Africa. Int J Climatol 36(1):303–323. https://doi.org/10.1002/joc.4346
11. Awange JL, Forootan E, Kuhn M, Kusche J, Heck B (2014) Water storage changes and climate variability within the Nile Basin between 2002 and 2011. Adv Water Res 73:1–15. https://doi.org/10.1016/j.advwatres.2014.06.010
12. Awange JL, Hu KX, Khaki M (2019) The newly merged satellite remotely sensed, gauge and reanalysis-based Multi-Source Weighted-Ensemble Precipitation: Evaluation over Australia and Africa (1981–2016). Sci Total Environ 670:448–465. https://doi.org/10.1016/j.scitotenv.2019.03.148
13. Awange JL (2021) Lake Victoria monitored from space. Springer Nature International
14. Awange JL (2021) Nile waters. Weighed from space. Springer Nature International, Berlin
15. Ayers RS, Westcot DW, Food and Agriculture Organization of the United Nations (1976) Water quality for agriculture. Food and agriculture organization of the United Nations, Rome
16. Becker M, LLovel W, Cazenave A, Güntner A, Cráetaux J (2010) Recent hydrological behaviour of the East African great lakes region inferred from GRACE, satellite altimetry and rainfall observations. C R Geosci 342:223–233. https://doi.org/10.1016/j.crte.2009.12.010
17. Boergens E, Rangelova E, Sideris MG, Kusche J (2014) Assessment of the capabilities of the temporal and spatiotemporal ICA method for geophysical signal separation in GRACE data. J Geophys Res Solid Earth 119(5):4429–4447. https://doi.org/10.1002/2013JB010452
18. Boko M, Niang I, Nyong A, Vogel C, Githeko A, Medany M, Osman-Elasha B, Tabo R, Yanda P (2007) Climate Change 2007: Impacts, adaptation and vulnerability. contribution of working group II to the fourth assessment report of the intergovernmental panel on climate change. Africa Cambridge University Press, Cambridge, United Kingdom, pp 433–467
19. Calow RC, Robins NS, Macdonald AM, Macdonald DMJ, Gibbs BR, Orpen WRG, Mtembezeka P, Andrews AJ, Appiah SO (1997) Groundwater management in drought-prone areas of Africa. Int J Water Res Dev 13(2):241–262. https://doi.org/10.1080/07900629749863
20. Calow RC, MacDonald AM, Nicol AL, Robins NS (2010) Groundwater security and drought in Africa: linking availability, access, and demand. Ground Water 48(2):246–256. https://doi.org/10.1111/j.1745-6584.2009.00558.x
21. Cardoso JF (1998) Blind signal separation: statistical principles. Proc IEEE 86(10):2009–2025
22. Cardoso JF (1999) High-order contrast for independent component analysis. Neural Comput 11:157–192
23. Cardoso JF, Souloumian A (1993) Blind beamforming for non-Gaussian signals. IEEE Proc 140(6):362–370
24. Castellazzi P, Martel R, Galloway DL, Longuevergne L, Rivera A (2016) Assessing groundwater depletion and dynamics using GRACE and InSAR: potential and limitations. Groundwater. https://doi.org/10.1111/gwat.12453

25. Cheng M, Tapley BD, Ries JC (2013) Deceleration in the Earth's oblateness. J Geophys Res Solid Earth 118(2):740–747. https://doi.org/10.1002/jgrb.50058
26. Comon P (1994) Higher order statistics independent component analysis, a new concept? Signal Process 36(3):287–314. https://doi.org/10.1016/0165-1684(94)90029-9
27. Comte JC, Cassidy R, Obando J, Robins N, Ibrahim K, Melchioly S, Mjemah I, Shauri H, Bourhane A, Mohamed I, Noe C, Mwega B, Makokha M, Join JL, Banton O, Davies J (2016) Challenges in groundwater resource management in coastal aquifers of East Africa: investigations and lessons learnt in the Comoros Islands, Kenya and Tanzania. J Hydrol Reg Stud 5:179–199. https://doi.org/10.1016/j.ejrh.2015.12.065
28. Dewitte O, Jones A, Spaargaren O, Breuning-Madsen H, Brossard M, Dampha A, Deckers J, Gallali T, Hallett S, Jones R, Kilasara M, Roux PL, Micheli E, Montanarella L, Thiombiano L, Ranst EV, Yemefack M, Zougmore R (2013) Harmonisation of the soil map of Africa at the continental scale. Geoderma 211–212:138–153. https://doi.org/10.1016/j.geoderma.2013.07.007
29. Dinku T, Ceccato P, Grover-Kopec E, Lemma M, Connor SJ, Ropelewski CF (2007) Validation of satellite rainfall products over East Africa's complex topography. Int J Remote Sens 28(7):1503–1526. https://doi.org/10.1080/01431160600954688
30. Döll P (2009) Vulnerability to the impact of climate change on renewable groundwater resources: a global-scale assessment. Environ Res Lett 4(3):035006. https://doi.org/10.1088/1748-9326/4/3/035006
31. Döll P, Hoffmann-Dobrev H, Portmann F, Siebert S, Eicker A, Rodell M, Strassberg G, Scanlon B (2012) Impact of water withdrawals from groundwater and surface water on continental water storage variations. J Geodyn 59–60:143–156. https://doi.org/10.1016/j.jog.2011.05.001
32. Döll P, Müller Schmied H, Schuh C, Portmann FT, Eicker A (2014) Global-scale assessment of groundwater depletion and related groundwater abstractions: combining hydrological modeling with information from well observations and GRACE satellites. Water Res Res 50(7):5698–5720. https://doi.org/10.1002/2014WR015595
33. Driessen P, Deckers O (2001) Lecture notes on the major soils of the world. FAO corporate document repository Accessed from http://www.fao.org/docrep/003/y1899e/y1899e00.htm#toc,_On24/02/2017
34. Elagib NA (2013) Meteorological drought and crop yield in sub-Sahara Sudan. Int J Water Res Arid Environ 2(3):164–171
35. Eswaran H, Cook T (1987) Classification and management-related properties of Vertisols. In: Jutzi SC, Haque I, McIntire J, Stares JES (Eds), Management of vertisols in Sub-Saharan Africa, Proceedings of a conference held at ILCA, Addis Ababa, Ethiopia, 31 August–4 September 1987. ILCA, Addis Ababa
36. Fipps G (2003) Irrigation water quality standards and salinity management strategies. Texas agricultural extension servicesB 1667(4-03), Texas A&M University System, College Station, TX (USA), pp. 1–19. Accessed from http://oaktrust.library.tamu.edu/bitstream/handle/1969.1/87829/pdf_94.pdf?sequence=1 on August 29, 2016
37. Fisher R (1944) Statistical methods for research workers. Oliver and Boyd
38. Forootan E, Kusche J (2013) Separation of deterministic signals using independent component analysis (ICA). Stud Geophys Geod 57(1):17–26. https://doi.org/10.1007/s11200-012-0718-1
39. Frappart F, Ramillien G, Leblanc M, Tweed SO, Bonnet M-P, Maisongrande P (2011) An independent component analysis filtering approach for estimating continental hydrology in the GRACE gravity data. Remote Sens Environ 115(1):187–204. https://doi.org/10.1016/j.rse.2010.08.017
40. Funk CC, Brown ME (2009) Declining global per capita agricultural production and warming oceans threaten food security. Food Sec 1(3):271–289. https://doi.org/10.1007/s12571-009-0026-y
41. Gossel W, Ebraheem AM, Wycisk P (2004) A very large scale GIS-based groundwater flow model for the Nubian sandstone aquifer in Eastern Sahara (Egypt, northern Sudan and eastern Libya). Hydrobiol J 12(6):698–713. https://doi.org/10.1007/s10040-004-0379-4

42. Hu K, Awange JL, Khandu Forootan E, Goncalves RM, Fleming K (2017) Hydrogeological characterisation of groundwater over Brazil using remotely sensed and model products. Sci Total Environ 599:372–386. https://doi.org/10.1016/j.scitotenv.2017.04.188
43. Hu K, Awange JL, Kuhn M (2021) Inference of the spatio-temporal variability and storage potential of groundwater in data-deficient regions through groundwater models and inversion of impact factors on groundwater, as exemplified by the Lake Victoria Basin. Science of the Total Environment 800. https://doi.org/10.1016/j.scitotenv.2021.149355
44. Huffman GJ, Adler RF, Bolvin DT, Gu G, Nelkin EJ, Bowman KP, Hong Y, Stocker EF, Wolff DB (2007) The TRMM multi-satellite precipitation analysis (TMPA): quasi-global, multi year, combined-sensor precipitation estimates at fine scales. J Hydrometeorol 8:38–55. https://doi.org/10.1175/JHM560.1
45. IFRC (2011) Drought in the horn of Africa: preventing the next disaster. International Federation of Red Cross and Red Crescent Societies, Geneva Accessed from http://www.ifrc.org on March 15, 2015
46. Kashaigili JJ (2010) Assessment of groundwater availability and its current and potential use and impact in tanzania. Final report International Water Management Institute (IWMI) Accessed from http://gw745africa.iwmi.org/Data/Sites/24/media/pdf/CountryReport-Tanzania.pdf on May 15, 2016
47. Kebede S (2013) Groundwater in Ethiopia, 1st edn. Springer hydrogeology. Springer, Berlin, Heidelberg
48. Kendall M, Stuart A (1963) The advanced theory of statistics. In: The advanced theory of statistics, 2nd edn. Griffin
49. Khaki M, Awange J, Forootan E, Kuhn M (2018) Understanding the association between climate variability and the Nile's water level fluctuations and water storage changes during 1992–2016. Sci Total Environ 645:1509–1521. https://doi.org/10.1016/j.scitotenv.2018.07.212
50. Kijazi AL, Reason CJC (2009) Analysis of the 2006 floods over northern Tanzania. Int J Climatol 29(7):955–970. https://doi.org/10.1002/joc.1846
51. Krhoda GO (1989) Groundwater assessment in sedimentary basins of Eastern Kenya, Africa. In: Proceedings of the baltimore symposium, May 1989182, IAHS. Accessed from http://www.oceandocs.org/bitstream/handle/1834/7780/ktf0465.pdf?sequence=2 on May 15, 2016
52. Kumar M, Kumari K, Ramanathan A, Saxena R (2007) A comparative evaluation of groundwater suitability for irrigation and drinking purposes in two intensively cultivated districts of Punjab. India Environ Geol 53(3):553–574. https://doi.org/10.1007/s00254-007-0672-3
53. Kummerow C, Simpson J, Thiele O, Bernes W, Chang ATC, Stocker E, Adler RF, Hou A, Kakar R, Wentz F, Ashcroft P, Kozu T, Hong Y, Okamoto K, Iguchi T, Kuroiwa F, Im E, Haddad Z, Huffman G, Ferrier B, Olson WS, Zipser E, Smith EA, Wilheit TT, North G, Krishnamurti T, Nakamura K (2000) The status of the tropical rainfall measuring mission (TRMM) after two years in orbit. J Appl Meteorol 39:1965–1982. https://doi.org/10.1175/1520-0450 (2001)040<1965:TSOTTR>2.0.CO;2
54. Kundzewicz ZW, Döll P (2009) Will groundwater ease freshwater stress under climate change? Hydrol Sci J 54(4):665–675. https://doi.org/10.1623/hysj.54.4.665
55. Kusche J (2007) Approximate decorrelation and non-isotropic smoothing of time-variable GRACE-type gravity field models. J Geod 81:733–749. https://doi.org/10.1007/s00190-007-0143-3
56. Kusche J, Schmidt R, Petrovic S, Rietbroek R (2009) Decorrelated GRACE time-variable gravity solution by GFZ, and their validation using a hydrological model. J Geod 83:903–913. https://doi.org/10.1007/s00190-009-0308-3
57. Khaki M, Awange J (2021) The 2019–2020 rise in Lake Victoria monitored from space: exploiting the state-of-the-art grace-fo and the newly released ERA-5 reanalysis products. Sensors 21(13):4304. https://doi.org/10.3390/s21134304
58. Landerer FW, Swenson SC (2012) Accuracy of scaled GRACE terrestrial water storage estimates. Water Res Res 48(4). https://doi.org/10.1029/2011WR011453W04531

59. Loewenberg S (2011) Humanitarian response inadequate in Horn of Africa crisis. The Lancet 378(9791):555–558. https://doi.org/10.1016/S0140-6736(11)61276-2
60. Longuevergne L, Scanlon BR, Wilson CR (2010) GRACE hydrological estimates for small basins: evaluating processing approaches on the High Plains Aquifer, USA. n/a-n/a Water Res Res 46(11). https://doi.org/10.1029/2009WR008564W11517
61. MacDonald AM, Bonsor HC, Dochartaigh BEO, Taylor RG (2012) Quantitative maps of groundwater resources in Africa. Water Res Res 7(2):024009. https://doi.org/10.1088/1748-9326/7/2/024009
62. MacDonald AM, Davies J, Calow RC (2008) Applied groundwater studies in Africa: IAH selected papers on hydrogeology. African hydrogeology and rural water supply IAH - Selected Papers on Hydrogeology. 13. CRC Press
63. Masih I, Maskey S, Muss'a FEF, Trambauer P (2014) A review of droughts on the African continent: a geospatial and long-term perspective. Hydrol Earth Syst Sci 18(9):3635–3649. https://doi.org/10.5194/hess-18-3635-2014
64. Mpelasoka F, Awange JL, Zerihun A (2018) Influence of coupled ocean-atmosphere phenomena on the Greater Horn of Africa droughts and their implications. Sci Total Environ 610–611:691–702. https://doi.org/10.1016/j.scitotenv.2017.08.109
65. Moore P, Williams SDP (2014) Integration of altimetric lake levels and GRACE gravimetry over Africa: inferences for terrestrial water storage change 2003–2011. Water Res Res 50:9696–9720. https://doi.org/10.1002/2014WR015506
66. Morgan B, Awange JL, Saleem A, Hu K (2020) Understanding vegetation variability and their "hotspots" within Lake Victoria Basin (LVB: 2003–2018), 122. https://doi.org/10.1016/j.apgeog.2020.102238.
67. Mumma A, Lane M, Kairu E, Tuinhof A, Hirji R (2011) Kenya groundwater governance Case Study. Water papers: World Bank Accessed from www.worldbank.org/water on May 15, 2016
68. Mwango FK, Muhang'u BC, Juma CO, Githae IT (2004) Groundwater resources in Kenya. In: Appelgren G (ed) Managing shared aquifer resources in Africa. United Nations Educational, Scientific and Cultural Organization, p 216
69. Ndehedehe CE, Awange JL, Corner RJ, Kuhn M, Okwuashi O (2016) On the potentials of multiple climate variables in assessing the spatio-temporal characteristics of hydrological droughts over the Volta Basin. Sci Total Environ 557–558:819–837. https://doi.org/10.1016/j.scitotenv.2016.03.004
70. New M, Hulme M, Jones P (2000) Representing twentieth-century space-time climate variability. Part II: development of 1901-96 monthly grids of terrestrial surface climate. J Clim 13(13), 2217–2238. https://doi.org/10.1175/1520-0442(2000)013<2217:RTCSTC>2.0.CO;2
71. Niang I, Ruppel OC, Abdrabo MA, Essel A, Lennard C, Padgham J, Urquhart P (2014) Climate change 2014: Impacts, adaptation, and vulnerability. Part B: regional aspects. Contribution of working group ii to the fifth assessment report of the intergovernmental panel on climate change. Africa. Cambridge University Press, Cambridge, United Kingdom and New York, NY, USA, pp 1199–1265
72. Nicholson SE (2014) A detailed look at the recent drought situation in the Greater Horn of Africa. J Arid Environ 103:71–79. https://doi.org/10.1016/j.jaridenv.2013.12.003
73. Omer A (2002) Focus on groundwater in Sudan. Environ Geol 41(8):972–976. https://doi.org/10.1007/s00254-001-0476-9
74. Oord A, Collenteur R, Tolk L (2014) Hydrological assessment of the Merti Aquifer. Kenya. Draft Report, NERC-UPGRO
75. Pavelic P, Keraita M, Ramesh V, Rao T (Eds) (2012) Groundwater availability and use in Sub-Sahara Africa: a review of 15 countries. International Water Management Institute (IWMI), Colombo, Sri Lanka
76. Rajner M, Liwosz T (2017) Analysis of seasonal position variation for selected GNSS sites in Poland using loading modelling and GRACE data. Geod Geodyn 8(4):253–259. https://doi.org/10.1016/j.geog.2017.04.001

77. Rodell M, Houser PR, Jambor U, Gottschalck J, Mitchell K, Meng CJ, Arsenault K, Cosgrove B, Radakovich J, Bosilovich M, Entin JK, Walker JP, Lohmann D, Toll D (2004) The global land data assimilation system. Bull Am Meteorol Soc 85(3):381–394. https://doi.org/10.1175/BAMS-85-3-381
78. Saji NH, Goswami BN, Vinayachandran PN, Yamagata T (1999) A dipole mode in the tropical Indian Ocean. Lett Nat 401:360–363
79. Salama BR (1988) Groundwater resources of Sudan. Rural water corporation 824SD76-474, Accessed from http://www.ircwash.org/sites/default/files/824SD76-474.pdf on May 15, 2016
80. Salifu M, Aidoo F, Hayford MS, Adomako D, Asare E (2015) Evaluating the suitability of groundwater for irrigational purposes in some selected districts of the Upper West region of Ghana. Appl Water Sci 1–10. https://doi.org/10.1007/s13201-015-0277-z
81. Scanlon BR, Keese KE, Flint AL, Flint LE, Gaye CB, Edmunds WM, Simmers I (2006) Global synthesis of groundwater recharge in semiarid and arid regions. Hydrol Process 20(15):3335–3370. https://doi.org/10.1002/hyp.6335
82. Selley R (1997) African basins. Sedimentary basins of the world. Elsevier Science
83. Shiklomanov I, Rodda J (2004) World water resources at the beginning of the twenty-first century. International hydrology series. Cambridge University Press, Cambridge
84. Siebert S, Burke J, Faures JM, Frenken K, Hoogeveen J, Döll P, Portmann FT (2010) Groundwater use for irrigation - a global inventory. Hydrol Earth Syst Sci 14(10):1863–1880. https://doi.org/10.5194/hess-14-1863-2010
85. Swarzenski WV, Mundorff MJ (1973) Geohydrology of North Eastern Province, Kenya. Accessed from http://pubs.usgs.gov/wsp/1757n/report.pdf on May 15, 2016 USGS, Water Supply Paper 1757-N
86. Swenson S, Chambers D, Wahr J (2008) Estimating geocenter variations from a combination of GRACE and ocean model output. n/a-n/a J Geophys Res Solid Earth 113(B8). https://doi.org/10.1029/2007JB005338B08410
87. Tapley B, Belabour S, Watkins M, Reigber C (2004) The gravity recovery and climate experiment: mission overview and early results. Geophys Res Lett 31:1–4. https://doi.org/10.1029/2004GL019920
88. Taylor GR, Todd MC, Kongola L, Mourice L, Nahozya E, Sanga H, MacDonald AM (2012) Evidence of the dependence of groundwater resources on extreme rainfall in East Africa. Nat Clim Chang 3:374–378. https://doi.org/10.1038/nclimate1731
89. Taylor RG, Scanlon B, Döll P, Rodell M, van Beek R, Wada Y, Longuevergne L, Leblanc M, Famiglietti JS, Edmunds M, Konikow L, Green TR, Chen JY, Taniguchi M, Bierkens MFP, MacDonald A, Fan Y, Maxwell RM, Yechieli Y, Gurdak JJ, Allen DM, Shamsudduha M, Hiscock K, Yeh PJF, Holman I, Treidel H (2013) Ground water and climate change. Nat Clim Chang 3(4):322–329. https://doi.org/10.1038/nclimate1744
90. Villholth KG (2013) Groundwater irrigation for smallholders in sub-Saharan Africa - a synthesis of current knowledge to guide sustainable outcomes. Water Int 38(4):369–391. https://doi.org/10.1080/02508060.2013.821644
91. Viste E, Korecha D, Sorteberg A (2013) Recent drought and precipitation tendencies in Ethiopia. Theor Appl Climatol 112:535–551. https://doi.org/10.1007/s00704-012-0746-3
92. Wahr J, Molenaar M, Bryan F (1998) Time variability of the Earth's gravity field: hydrological and oceanic effects and their possible detection using GRACE. J Geophys Res Solid Earth 103(B12):30205–30229. https://doi.org/10.1029/98JB02844
93. Wellfield C, British GS (2011) Groundwater and drought management project: regional groundwater monitoring network. Transboundary Aquifer Report, Southern Africa Development Community
94. Wichelns D (2014) Investing in small, private irrigation to increase production and enhance livelihoods. Agric Water Manag 131:163–166. https://doi.org/10.1016/j.agwat.2013.09.003
95. Williams AP, Funk C (2011) A westward extension of the warm pool leads to a westward extension of the Walker circulation, drying Eastern Africa. Clim Dyn 37(11):2417–2435. https://doi.org/10.1007/s00382-010-0984-y

96. Williams AP, Funk C, Michaelsen J, Rauscher SA, Robertson I, Wils THG, Koprowski M, Eshetu Z, Loader NJ (2012) Recent summer precipitation trends in the Greater Horn of Africa and the emerging role of Indian Ocean Sea surface temperature. J Clim Dyn 39:2307–2328. https://doi.org/10.1007/s00382-011-1222-y
97. Wolter K, Timlin MS (2011) El Ni no/Southern Oscillation behaviour since 1871 as diagnosed in an extended multivariate ENSO index (MEI.ext). Int J Climatol 31(7):1074–1087. https://doi.org/10.1002/joc.2336
98. Wouters B, Bonin JA, Chambers DP, Riva REM, Sasgen I, Wahr J (2014) GRACE time-varying gravity, earth system dynamics and climate change. Rep Prog Phys 77:41. https://doi.org/10.1088/0034-4885/77/11/116801
99. Xie H, You L, Wielgosz B, Ringler C (2014) Estimating the potential for expanding smallholder irrigation in sub-Saharan Africa. Agric Water Manag 131:183–193. https://doi.org/10.1016/j.agwat.2013.08.011
100. Zhang L, Dobslaw H, Stacke T, Güntner A, Dill R, Thomas M (2017) Validation of terrestrial water storage variations as simulated by different global numerical models with GRACE satellite observations. Hydrol Earth Syst Sci 21(2):821–837. https://doi.org/10.5194/hess-21-821-2017

Chapter 14
Agricultural Drought's Indicators: Assessment

It is increasingly alarming that being in dire need of food assistance in the GHA is becoming a permanent feature of the region. Almost every year, including 2014, 2015, 2016 and 2017 famine headlines appear in the news as drought related crisis—[39].

14.1 Summary

Heavy reliance of East Africa (EA) on rain-fed agriculture makes it vulnerable to drought-induced famine. Yet, most research on East Africa drought focuses on meteorological aspects with little attention paid on agricultural drought impacts. The inadequacy of in-situ rainfall data across East Africa has also hampered detailed agricultural drought impact analysis. Recently, however, there has been increased data availability from remote sensing (rainfall, vegetation condition index—VCI, terrestrial water storage—TWS), reanalysis (soil moisture and TWS), and land surface models (soil moisture). In the work of [5], these products are employed to characterise East Africa droughts between 1982 and 2013 in terms of severity, duration, and spatial extent. Furthermore, the capability of these products to capture agricultural drought impacts is assessed using maize and wheat production data. This chapter presents the work of [5] whose results show that while all products were similar in drought characterisation in dry areas, the similarity of CHIRPS and GPCC rainfall product extended over the whole East Africa. CHIRPS and GPCC also identified the highest proportion of areas under drought followed closely by soil moisture products whereas VCI had the least coverage. Drought onset was marked first by a decline/lack of rainfall, followed by VCI/soil moisture, and then TWS. VCI indicate drought lag at 0–4 months following rainfall while soil moisture and TWS products have variable lags vis-á-vis rainfall. GLDAS mischaracterize the 2005–2006 drought vis-á-vis other soil moisture products. Based on the annual crop production variabilities explained, CHIRPS, GPCC, FLDAS, and VCI are identified as suitable

J. Awange, *Food Insecurity & Hydroclimate in Greater Horn of Africa*,
https://doi.org/10.1007/978-3-030-91002-0_14

for agricultural drought monitoring/characterization in the region for the 1983–2013 period. Finally, GLDAS explains the lowest percentages of the Kenyan and Ugandan annual crop production variances. These findings are important for the gauge data deficient East Africa region as they provide alternatives for monitoring agricultural drought.

14.2 East African Drought

East Africa (EA, defined as Kenya, Uganda, Tanzania, Rwanda, and Burundi) relies heavily on rain-fed subsistence agriculture, which is increasingly becoming vulnerable to frequent drought events, see e.g., [63, 94, 106]. Furthermore, the impacts of drought are compounded by high levels of poverty, conflicts, population migration, and lack of social infrastructure across the region (see Chap. 1), triggering famine cycles every time an episode occurs [51, 59, 63, 79–81]. As drought is in part a naturally recurrent feature in East Africa, there is a need for comprehensive and reliable monitoring in order to aid planning and mitigation of its impacts. Since frequency and severity of droughts are likely to intensify with climate change, e.g., [121], the need to characterize droughts in terms of duration, severity, frequency and spatial extent is critical.

Comprehensive characterization of drought in East Africa, like in many other places around the world, faces a number of challenges with respect to use of in-situ precipitation data. For instance, often spatial variability in precipitation cannot be adequately captured due to sparse and uneven spatial distribution of rain gauges. Furthermore, gaps in individual rainfall records, and at times lack of consistency due to poor handling complicate the use of precipitation data [78, 79, 94]. In many studies, this led to the replacement or augmentation of in-situ rainfall data with remotely sensed precipitation, reanalysis, and model outputs, providing consistent and homogeneous data with global coverage at various spatial scales that are suitable for drought monitoring [27]. However, these products can have considerable discrepancies and limitations in representing rainfall at local and regional scales [2, 27, 50, 78, 94].

In addition to satellite and model-based precipitation products, normalised difference vegetation index NDVI [82, 96, 113] and Gravity Recovery and Climate Experiment (GRACE) total water storage TWS [10, 110] have been used to monitor drought. NDVI has been used directly or in its derivative form to monitor impacts of drought on vegetation health, e.g., [18, 58, 90]. In East Africa, it has been used by [8, 9, 74, 79, 82], while the use of GRACE satellite temporal gravity measurements, see e.g., [110, 124] in East Africa has been limited to monitoring changes in TWS, e.g., [11, 12, 14, 15, 19, 109], and drought analysis [10].

Drought studies carried out in the East Africa region range from purely precipitation based, e.g., [26, 59, 78], a combination of precipitation and climate models, e.g., [34, 126], to precipitation in combination with soil moisture and/or NDVI, e.g., [1, 8, 79]. Some of the aforementioned studies and few others, see e.g., [8, 76, 94, 103] have examined agricultural drought using standardised precipitation index (SPI), NDVI, and/or soil moisture. However, for a region like East African part of GHA, where the majority of the population depends on subsistence rain-fed agriculture, additional studies focusing on agricultural drought impacts, e.g., related to crop production, would be more relevant and beneficial to the population. Therefore, this chapter focuses on both the characterization of drought behavior in general and agricultural drought in particular using various indicators (precipitation, soil moisture, and total water storage) derived from multi-satellite remote sensing, reanalysis, and model products. Further, this chapter evaluates the utility of these products using annual crop production, which has so far not been done by the aforementioned studies.

To support agricultural drought monitoring from diverse indicators, it is imperative to identify and provide information on the most effective agricultural drought indicator or a combination of indicators for the East Africa region. Therefore, the objectives of this chapter are: (i) to characterise agricultural drought in terms of severity, duration, and spatial (areal) extent using satellite remote sensing, reanalysis, and modelled soil moisture data, and (ii), evaluate how well these products capture agricultural drought in the region as reflected by national crop production data (wheat and maize) during the 1982–2013 period.

This chapter, therefore, provides the first comprehensive assessment of the potential of these remotely sensed products, reanalysis data, and land surface model outputs to monitor agricultural drought in the East Africa region of GHA. Moreover, this contribution proposes for the first time the possibility of using GRACE satellite products for agricultural drought monitoring in East Africa thus providing a link between TWS and crop production.

14.3 East Africa: Background and Drought Products

14.3.1 GHA: The East African Part

The East Africa region (Fig. 14.1) has a bimodal rainfall regime, the March-April-May (MAM; long rains) and the October-November-December (OND; short rains) with the MAM contributing over 70% of the annual rainfall while the OND contributing less than 20% [72]. The rainfall regime is controlled by the inter-tropical convergence zone ITCZ, effects of climate variability such as El Niño Southern Oscillation (ENSO) and the Indian Ocean Dipole (IOD), and sea surface temperature variations in the Indian and Pacific oceans [26, 36, 57, 68, 86, 111, 122].

The amount of the MAM rainfall has been declining in the region since 1999, with the 1990s to 2000s mean being below the 1980s mean [68, 122], while the frequency and duration of drought episodes have increased since 1998 [67, 75, 79]. Drought events have been observed in 2000–2001, 2005–2006, 2008–2009, and 2010–2011, with the latter being the worst in 60 years due to failure of short rains in 2010 and

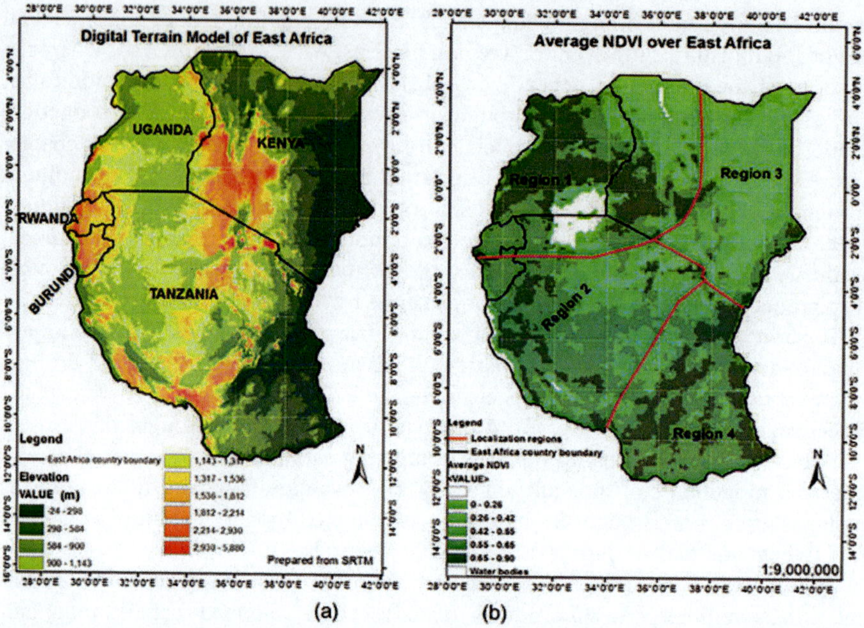

Fig. 14.1 East Africa (EA) region; **a** Elevation variation from Shuttle Radar Topographical Mission (SRTM, *source* http://www.cgiar-csi.org/data/srtm-90m-digital-elevation-database), and **b**, Temporal NDVI average (1983–2014) with standardised indices localization regions (see Fig. 14.2, Table for region details). Note that the white portions in **b** indicate the water bodies (Lakes Victoria and Tana, respectively). *Source* [5]

long rains in 2011. This particular drought affected over 12 million people bringing untold sufferings to the region [51, 63].

14.3.2 Agricultural Drought Characterization Products

The following data sets are used (see Table 14.1 for a summary): precipitation products from the Global Precipitation Climatology Centre (GPCC) and Climate Hazard Group (Climate Hazard Group InfraRed Precipitation with Stations (CHIRPS)); soil moisture products from the Global Land Data Assimilation System (GLDAS), Climate Prediction Center (CPC), the European Centre for Medium-Range Weather Forecasts Interim Re-Analysis (ERA-Interim), the second Modern-Era Retrospective analysis for Research and Applications (MERRA-2), and Famine Early Warning System Network (FEWS NET) Land Data Assimilation System (FLDAS); Global Inventory Monitoring and Modelling Studies (GIMMS) NDVI; and terrestrial water storage (TWS) from MERRA-2 and GRACE. Although ERA-Interim is herein used, it should be pointed out that it has been superseded by ERA5, see e.g., [57, 129].

14.3.3 Precipitation Products

1. **CHIRPS** is a quasi-global (50° S—50° N) high resolution, 0.05°, daily, pentad, and monthly precipitation data set produced from a combination of in-situ station observations and satellite precipitation estimates based on Cold Cloud Duration (CCD) observations to represent sparsely gauged regions. It has been developed to primarily support agricultural drought monitoring, see [44] for a detailed description). Monthly precipitation data, version 2.0, from 1982 to 2013 is used.[1] CHIRPS precipitation was found to have correlation of greater than 0.75 with GPCC over East Africa region, see e.g., [44] and has subsequently been used in a number of drought and hydrology related studies in the region, see, e.g., [10, 71, 86, 103].
2. **GPCC** [99] full data reanalysis version 7, 0.5° spatial resolution, monthly land surface precipitation from 1982 to 2013[2] is used in addition to CHIRPS for drought analysis. It is a purely gauge gridded product based on 75,000 rain gauge stations worldwide, that feature record durations of 10 years or longer, see [99]. It has been used in several drought related studies both globally and in East Africa region, see e.g., [10, 35, 43, 59, 130].

14.3.4 Soil Moisture Products

Soil moisture nominal depths considered in this chapter are root zone for MERRA-2; aggregation of 0–1 m depth layers for ERA-Interim, GLDAS, and FLDAS; and whole column depth ($\cong$ 0.76 m) for CPC since its a single bucket layer product.

1. **MERRA-2** is a NASA atmospheric reanalysis from 1980 that replaced the original MERRA reanalysis [28, 92] using upgraded version of the Goddard Earth Observing System Model, version 5.12.4 (GEOS 5.12.4) data assimilation system [21]. Monthly 0.625° by 0.5° root zone soil moisture from 1982 to 2013 is used.[3] Because of the improved assimilation system (updates to the treatment of canopy interception) and better forcing data (use of observation-corrected precipitation), MERRA-2 has improved soil moisture estimates over MERRA [22].
2. **ERA-Interim** [28, 29] monthly means of daily averages of soil moisture from 1982 to 2013, at 0.25° spatial resolution[4] is used. The three layers of soil moisture from 0 to 1 m are aggregated into one value before application to drought analysis. ERA-Interim has been found to have good skills in capturing surface soil moisture variability, though it tends to overestimate, especially over dry lands [7]. In addition, it has been used in a number of studies globally and in East Africa

[1] Downloaded from http://ftp.chg.ucsb.edu/pub/org/chg/products/CHIRPS-2.0/.

[2] Downloaded from http://ftp.dwd.de/pub/data/gpcc/html/fulldata_v7_doi_download.html.

[3] Downloaded from https://gmao.gsfc.nasa.gov/reanalysis/MERRA-2/data_access/.

[4] Downloaded from http://apps.ecmwf.int/datasets/data/interim-full-moda/levtype=sfc/.

region, e.g., [16, 28, 29, 34, 76, 117]. As mentioned earlier, this product has now been superseded by ERA5.

3. **GLDAS** [93] version 2, Noah, monthly 1° spatial resolution soil moisture product from 1982 to 2010[5] is employed. Like ERA-Interim, the three layers of soil moisture from 0 to 1 m depth are aggregated into one product before further processing.
4. **FLDAS** is a custom instance of NASA Land Information System (LIS), adapted to work with domains, data streams and monitoring, and forecast systems associated with food security assessment in data sparse, developing country settings [97]. FLDAS is driven by Noah and VIC land surface models. FLDAS Noah 0.1° spatial resolution, monthly soil moisture from 1982 to 2013[6] is used. This soil moisture resulted from simulation run forced by a combination of MERRA-2 and CHIRPS dataset. Noah model (GLDAS and FLDAS) is chosen due to its wide use by atmospheric and land modelling communities hence model parameters are well tested [71]. In addition, various studies have used it (Noah) over East Africa region e.g., [8, 71, 128].
5. **CPC** [40, 114] global monthly mean 0.5° spatial resolution soil moisture, version 2, for the duration 1982–2013 downloaded from the National Oceanic & Atmospheric Administration's (NOAA) Earth System Research Laboratory database[7] is used in this chapter because it incorporates in-situ rainfall as one of its inputs, hence likely to be closer to real soil moisture. CPC soil moisture simulates the seasonal and inter-seasonal annual variability reasonably well over East Africa region, see [31].

14.3.5 Total Water Storage (TWS)

1. **GRACE** satellite mission has been in operation from 2002 providing global monthly temporal gravity variations, see e.g., [110, 124]. These gravity variations are provided in terms of spherical harmonic coefficients (see Sect. 3.3.3). The Centre for Space Research's (CSR) release five (RL05) monthly spherical harmonic coefficients for the duration 2003–2013 downloaded from International Centre for Global Earth Models ICGEM[8] are processed following the approach of [119] and used in this chapter. During the processing, the coefficients are filtered using a decorrelation and non-isotropic filter, see e.g., [60, 61] in order to remove stripes and spurious patterns. This is followed by the application of a scaling factor, derived using GLDAS TWS following the approach of [62], onto the synthesised GRACE TWS to remove the leakage effect due to filter-

[5] Downloaded from http://disc.sci.gsfc.nasa.gov/services/grads-gds/gldas.

[6] Downloaded from http://hydro1.sci.gsfc.nasa.gov/data/s4pa/FLDAS/FLDAS_NOAH01_C_EA_M.001/.

[7] http://www.esrl.noaa.gov/psd/data/gridded/data.cpcsoil.html.

[8] http://icgem.gfz-potsdam.de/ICGEM/shms/monthly/csr-rl05/.

ing. The synthesised GRACE-derived TWS over East Africa comprises changes from accumulated soil moisture, groundwater, surface water, and biomass/canopy water content. It is referred to as GTWS in the remainder of the chapter. GRACE measurements agree with Earth rotation-derived changes and geophysical model estimates [23], and have a global root mean square error of 2 cm to degree and order 70, uniformly over land and ocean [110]. It has been used in a number of drought related studies both globally and in East Africa region, see e.g., [10, 24, 64].

2. **MERRA-2** total land water storage from 1982 to 2013, at 0.5° latitude by 0.625° longitude[9] is used in addition to GTWS. It does not include canopy water content and groundwater. It is referred to as MTWS in the remainder of the chapter.

14.3.6 Vegetation Condition Index (VCI)

Long-term series of NOAA Advanced Very High Resolution Radiometer (AVHRR) NDVI dataset 1982–2013, from NASA's Global Inventory and Modelling Systems (GIMMS)[10] is used to compute VCI [58]. The data is comprised of 15 days maximum composites at 5-arc-minute spatial resolution (for a detailed description see [84, 112]). VCI is advantageous as it is able to isolate weather related vegetation stress [58, 88, 94], which within East Africa, would correspond to water availability. It is computed as [58]

$$\mathrm{VCI}_i = 100 \times \frac{\mathrm{NDVI}_i - \mathrm{NDVI}_{min}}{\mathrm{NDVI}_{max} - \mathrm{NDVI}_{min}}, \tag{14.1}$$

where $NDVI_i$ is the monthly NDVI, $NDVI_{max}$ and $NDVI_{min}$ are the multi-year maximum and minimum NDVI, respectively. AVHRR NDVI has been used extensively globally and over Africa for drought and other related studies, see e.g., [10, 25, 32, 46, 94, 116].

14.3.7 National Annual Crop Production

National annual maize and wheat production data for Kenya, Uganda, and Tanzania downloaded from Food and Agriculture Organization (FAO) data portal[11] is used to evaluate the effectiveness of various satellite/model drought indices in capturing agricultural droughts. Even though this data set undergoes several quality checks along

---

[9] Downloaded from https://gmao.gsfc.nasa.gov/reanalysis/MERRA-2/data_access/.

[10] Downloaded from http://ecocast.arc.nasa.gov/data/pub/gimms/3g.v0/.

[11] http://www.fao.org/faostat/en/#data/QC.

Table 14.1 Overview of data-set employed in the chapter. *Source* [5]

	Data	Temporal resolution	Spatial resolution	Period used	References/ Studies used
Precipitation	GPCC	Monthly	0.5° × 0.5°	1982–2013	[35, 43, 59, 99, 130]
	CHIRPS	Monthly	0.05° × 0.05°	1982–2013	[44, 71, 86, 103]
Soil moisture	MERRA-2	Monthly	0.625° × 0.5°	1982–2013	[21, 22]
	ERA-Interim	Monthly	0.25° × 0.25°	1982–2013	[7, 16, 28, 29, 34, 117]
	GLDAS	Monthly	1° × 1°	1982–2013	[8, 93, 128]
	FLDAS Noah	Monthly	0.1° × 0.1°	1982–2013	[8, 71, 97, 128]
	CPC	Monthly	0.5° × 0.5°	1982–2013	[31, 40, 114]
TWS	GRACE	Monthly	1° × 1°	2003–2013	[10, 23, 24, 64, 110, 124]
	MERRA-2	Monthly	0.625° × 0.5°	1982–2013	[21, 22].
VCI	NDVI	15 d	5-arc-minute	1982–2013	[25, 32, 46, 84, 94, 112, 116]
Crop production	Crop production	Annual	National	1982–2013	http://www.fao.org/faostat/en/#data/QC

the processing chain, see e.g., [55], lack of direct production/yield reporting from farmers to government agencies in developing countries (e.g., East Africa region) means there is some level of uncertainty in the production data used. Even with the uncertainties, this data is still the most credible, readily available production data.

14.4 Agricultural Drought Characterization

Due to the existence of a link between agricultural drought and 1–6 months precipitation anomalies, e.g., [37, 59, 95, 108], Standardized Indices (SI) (e.g., SPI, [70]) are derived to characterize agricultural drought using precipitation, VCI, TWS, and soil moisture products. Similarly, Standardized Anomalies (SA)/Z-scores [125] are computed to characterize drought from GTWS due to its short duration. The resulting SI and SA indices are then subjected to rotated principal component analysis to obtain their most dominant spatial and temporal drought variabilities. Finally, the temporal variabilities are subjected to partial least-squares regression analysis to determine how well they captured drought variability. Other than GRACE and GLDAS, all the other data sets are spatially aggregated to 1° by 1° before standardization for

Table 14.2 Drought categories according to SPI valuess [4]. *Source* [5]

SPI	Drought Category
>1.65	Extremely wet
>1.28	Severely wet
>0.84	Moderately wet
>− 0.84 and < 0.84	Normal
>−0.84	Moderate drought
>−1.28	Severe drought
>−1.65	Extreme drought

consistency. For all products, the unstandardized data are tested using W-statistics [101] and found to be normally distributed.

Given the differences in the variables used in this chapter, comparison of drought information is primarily carried out between various products of the same variables, e.g., between precipitation products, or soil moisture products, or TWS. Notwithstanding the differences between the variables, links/relations in drought information across the various products are explored since drought progresses from deficiencies in rainfall followed by moisture through to TWS.

14.4.1 Standardized Precipitation Index (SPI)

SPI [70], one of the most commonly used drought indices due to its numerous advantages [108], expresses precipitation anomalies with respect to its long-term average. Its computation involves fitting a gamma probability distribution function to precipitation time series followed by the transformation of the accumulated gamma probability distribution to the cumulative distribution function of the standard normal distribution, see e.g., [41, 77]. Due to the sensitivity of the computed SPI values to the fitted parametric distributions, especially at the tail ends of the distribution, see [87], a non-parametric SPI fitting method was adopted in this chapter, see e.g., [41, and the references therein for the formulation]. This approach is implemented using the Standardized Drought Analysis Toolbox (SDAT, [41]) and the SPI drought limit categories (intensities) proposed by [4] (Table 14.2) are used. For this chapter, a drought episode begins any time SPI is continuously less than −0.84 for a period of at least three months, and ends when SPI value exceeds −0.84. The various drought intensities (moderate, severe, and extreme) are then said to occur when the values in Table 14.2 are attained (see [75] for alternative formulation). The resulting standardized indices in this chapter are SPI, standardized soil moisture index (SSI), standardized vegetation condition index (SVCI), and standardized terrestrial water storage index (STWSI).

14.4.2 Standardized Anomalies (SA)

As already pointed out, due to the short time frame of the GRACE products, SA instead of SI is computed for characterizing agricultural drought. Here, the 3 and 6 month GTWS time series cumulations are obtained in a manner similar to those of the Standardized Precipitation Index [70]. Due to seasonality in precipitation, soil moisture, TWS, and NDVI dataset [127], GTWS anomalies are calculated by removing the monthly mean from the 1, 3, and 6 month time series. The anomalies are then divided by the standard deviation for the duration of the data, e.g., [83], i.e.,

$$XS_{ijk} = \frac{X_{ijk} - \frac{1}{n}\sum_{k=1}^{n} X_{ijk}}{\sigma_{ij}}, \tag{14.2}$$

where XS_{ijk} is the monthly standardized GTWS anomaly for location i, month j, and year k; X_{ijk} is the monthly GTWS for location i, month j, and year k; n is the length of GTWS in years; and σ_{ij} is the multi-year standard deviation for location i, month j. The resulting standardized anomalies (z-scores), express the deviation of the GTWS above or below the mean value and has been used to monitor drought in various studies, e.g., [3, 56, 66, 125]. Positive values indicate wet conditions, 0 indicate normal (average) conditions while negative values indicate drought conditions [125].

In order to demonstrate the consistency between SPI and SA in characterizing drought over the region, this chapter compares the spatio-temporal decompositions of CHIRPS-derived SPI and CHIRPS-derived SA over East Africa. This comparison showed similar spatio-temporal drought patterns (See Figs. 14.2 and 14.4 for CHIRP-derived SPI spatio-temporal drought patterns). Further, Pearson correlations between the SI and SA temporal patterns are greater than 0.95 over the region. Due to the close association between SA and SI, see [125], the SPI drought limit categories (Table 14.2) are used to differentiate the various SA drought intensities.

14.4.3 Principal Component Analysis (PCA)

Principal component analysis PCA, [13, 48, 53, 65, 85, 120] is one of the most widely used methods in atmospheric sciences for pattern extraction and dimensionality reduction. It has been used in drought studies, e.g., [89, 98, 104] to decompose spatial-temporal fields such as SPI, SSI, SVCI, and STWSI, into spatial patterns and their corresponding temporal evolutions.

In this chapter, PCA is applied to the 1, 3, 6 month time scales of SI and SA. Log-eigenvalue (LEV) diagrams [53] are used to determine and retain the significant components that are then rotated through Varimax rotation [42, 52, 54] for better localization (for more information on rotated PCA, see e.g., [48, 91, 100, 118]. The resulting spatial patterns, normalised by multiplying with the standard devia-

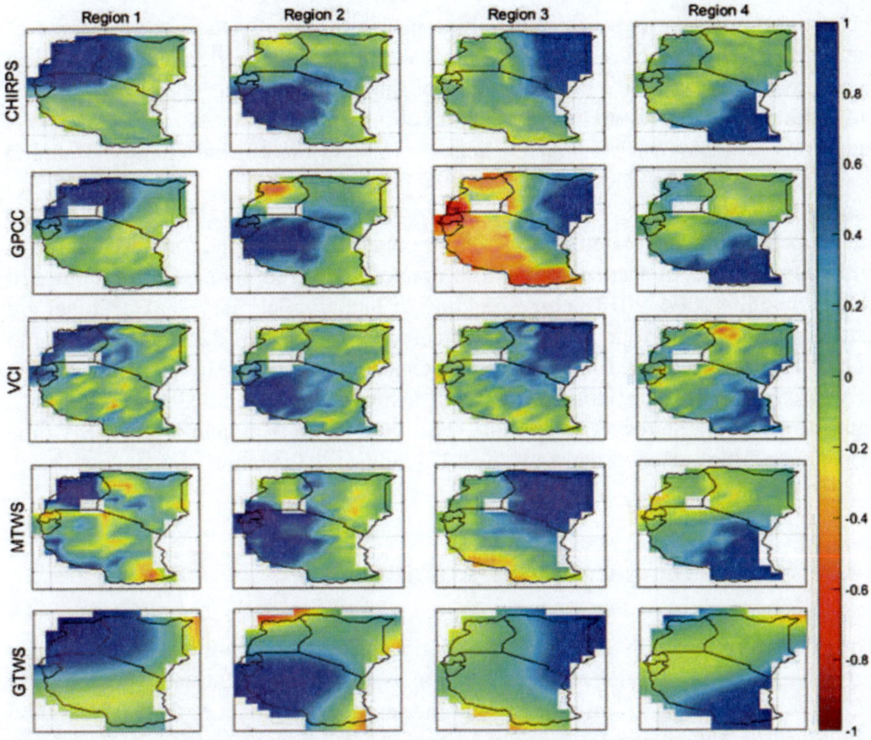

Fig. 14.2 Rotated principal component spatial patterns of standardized index/anomalies (SI/SA). Rows denote products while columns denote regions (also see Table 14.3). The spatial patterns have been scaled to ± 1, thus the temporal evolutions shown in Fig. 14.4 indicate the actual magnitude of SA/SA for regions where the spatial patterns have values close to ± 1. The spatial patterns are interpreted in conjunction with temporal evolutions in Fig. 14.4 and represent drought spatial patterns any time the temporal evolutions falls below -0.84, as in Table 14.2. The white rectangular area in all the images except CHIRPS and GTWS is Lake Victoria.*Source* [5]

tion of their corresponding rotated principal components (RPC) series, represent the correlation between the original data (in our case 1, 3, 6 month SI or SA at single grid point) and the corresponding RPC. Normalised RPC (divided by its standard deviation) represent SI/SA in each case, see [20].

14.4.4 Partial Least Squares Regression (PLSR)

PLSR is a regression technique in which the response variables are regressed on the predictor scores. The scores (few new variables) are linear combinations of the original predictor variables [45, 123]. The generation of the scores takes into account the variability in the dependent variable ensuring that only those components of the independent variables that are related to the dependent variables are used in the regression [45]. It is a generalization of the multiple linear regression (MLR), but unlike MLR, it can analyze data with collinearity (correlated), noisy, and with numerous predictor variables [123] hence its use in the current chapter. Detailed description and formulation can be found in [49, 115].

For each country, SI/SA values for each month of the year over the entire duration are extracted from the rotated principal components. For example, considering Kenya with four GRACE SA rotated principal components, each component comprising 120 values/months (2004–2013), corresponding to January, February, ..., December are extracted resulting in four 10 by 12 matrices, i.e., 10 years of data for every month of the year. The resulting four matrices are concatenated to a 10 by 48 matrix, which serve as the predictor variable in the PLSR against national annual production data (maize/wheat) as the response variable. This is done for 1, 3, and 6 month SA time scales for all the variables across Kenya, Uganda, and Tanzania.

14.5 Spatio-Temporal Drought Patterns

The PCA decomposition of SI/SA show spatial and temporal patterns, which became very distinct upon applying varimax rotation as compared to unrotated components (data not shown). Integrating the spatial and temporal patterns of the SIs/SAs, and using the drought category definitions in Table 14.2, percentages of areas under various drought intensities are evaluated. The results presented and discussed here are for the 3-month time scale only, as this is representative of the results for the 1 and 6 month time scales. Note that this time scale is vital for crop growth in the region.

14.5.1 Spatial Variability

The four most significant components in terms of explaining the total variability from RPCA of SI/SA revealed four distinct spatial patterns across all products (Figs. 14.2 and 14.3). The geographical coverage of these spatial patterns are summarized in Table 14.3. The spatial patterns of ERA-Interim and to some extent of GLDAS in region 2 are different from those of other products.

The four RPCs explained between 38% (SVCI) and 96% (GTWS SA) of the total variance of the respective original SI/SA variables (Table 14.4). Most of the products have the highest and the lowest variabilities explained in regions 3 and 4, respectively (Table 14.4). This could be attributed to the fact that region 3 covering almost the entire region of Kenya (in those indicators showing it highest) is wet and dry on the western and eastern parts of the country, respectively, hence has high variability due to the presence of wet and dry extremes. On the other hand, region 4 is relatively wet and receives consistent rainfall resulting in a smaller variation in SI/SA.

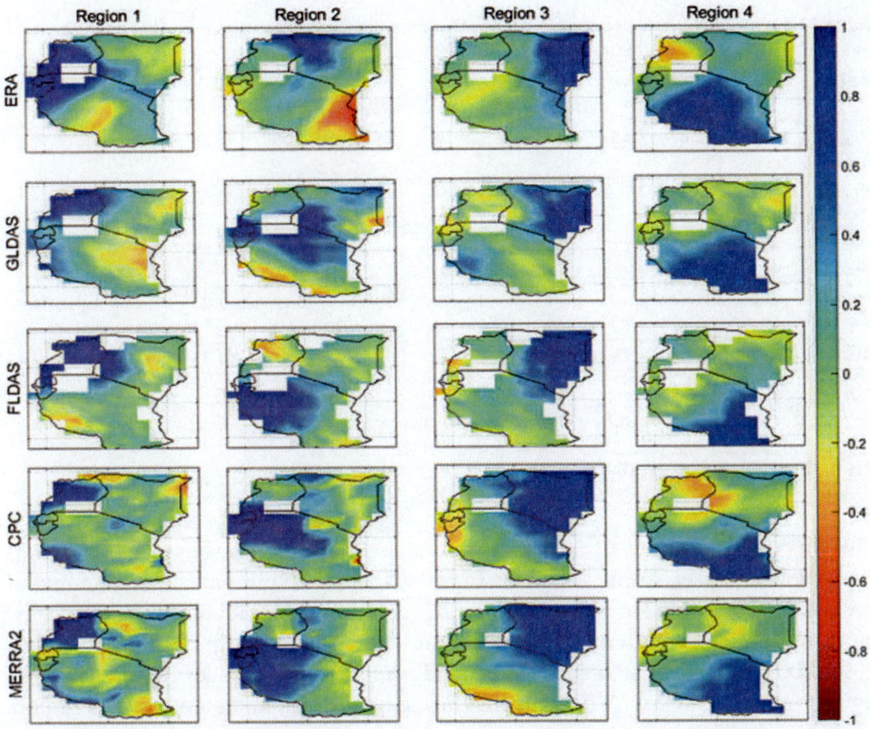

Fig. 14.3 Rotated principal component spatial patterns of standardized soil moisture indices (SSI). Rows denote products while columns denote regions (also see Table 14.3). The spatial patterns have been scaled to $\pm$ 1, thus the temporal evolutions shown in Fig. 14.5 indicate the actual magnitude of SSI for regions where the spatial patterns have values close to ± 1. The spatial patterns are interpreted in conjunction with temporal evolutions in Fig. 14.5 and represent drought spatial patterns any time the temporal evolutions falls below -0.84, as in Table 14.2. Patterns are consistent with those in Fig. 14.2 except for ERA-interim and to some extent GLDAS in region 2. The white rectangular area in all the images is Lake Victoria. *Source* [5]

Table 14.3 Geographical coverage of SI/SA spatial patterns. *Source* [5]

Region	Countries/Areas
1	Lake Victoria, Uganda, and western Kenya
2	Western Tanzania, Rwanda, and Burundi
3	Eastern Kenya
4	Eastern and southern Tanzania

Table 14.4 Variances explained in each region by each product. Regions are as shown in Figs. 14.2 and 14.3. *Source* [5]

Region	CHIRPS	GPCC	VCI	MTWS	GTWS	ERA-interim	GLDAS	CPC	MERRA-2	FLDAS
Region 1	15.26	12.07	8.24	13.86	29.88	18.91	11.79	10.29	13.61	14.86
Region 2	14.62	13.15	10.01	18.13	24.58	10.78	15.75	16.46	17.72	15.77
Region 3	15.26	13.67	12.57	21.44	21.58	14.32	14.22	21.80	21.98	18.21
Region 4	10.60	11.64	6.94	13.67	19.77	20.05	16.72	13.27	14.64	12.64
Total	55.53	50.53	37.77	67.10	95.81	64.06	58.48	61.83	67.95	61.48

14.5.2 Temporal Patterns

The temporal evolutions of the spatial patterns in regions 1–4 (Figs. 14.2 and 14.3) from the rotated PCA are shown in Figs. 14.4 and 14.5. In general, the temporal evolutions (interpreted in conjunction with Figs. 14.2 and 14.3, and Table 14.2) show most of the regions suffering from severe to extreme drought in 1984/1985, 1999, 2000, 2005/2006, and 2010/2011. These and other drought episodes captured in these figures are consistent with documented drought episodes in the East Africa region, e.g., [51, 69, 79].

All products have similar performance in region 3 (Figs. 14.4c and 14.5c), which may be attributed to the relatively flat terrain (Fig. 14.1a) coupled with relatively less rainfall hence good performance by the models and rainfall products. The performance of the rainfall products (CHIRPS and GPCC) are almost identical over the entire East African region as a result of both containing in-situ rainfall (CHIRPS has satellite-derived precipitation estimates in addition to in-situ data while GPCC is purely gridded in-situ product, see e.g., [44, 99]). In relation to the rainfall products, the remaining products (VCI, soil moisture, and TWS) show delayed (lagged) response in the temporal evolution. This is clearly visible in Fig. 14.4a in which MTWS appears like a low pass filtered version of the CHIRPS/GPCC signals. This behavior could be due to a delayed response of terrestrial water storage changes to rainfall and soil moisture changes. Finally, the soil moisture products seem to be from largely two classes/categories of models with ERA-Interim, FLDAS, and GLDAS in one category and CPC and MERRA-2 on the other, especially considering region 1 (Fig. 14.5).

Further, correlation analysis between the drought indices revealed close relationships between various products e.g., CHIRPS and GPCC, MERRA-2, MTWS, and CPC, across the regions (Table 14.5). The close relationship between MTWS and MERRA-2 is similar to that between CHIRPS and GPCC, since MTWS include aspects of soil moisture captured by MERRA-2 in addition to greater depth of soil water content. Furthermore, the significant and high correlations between the drought

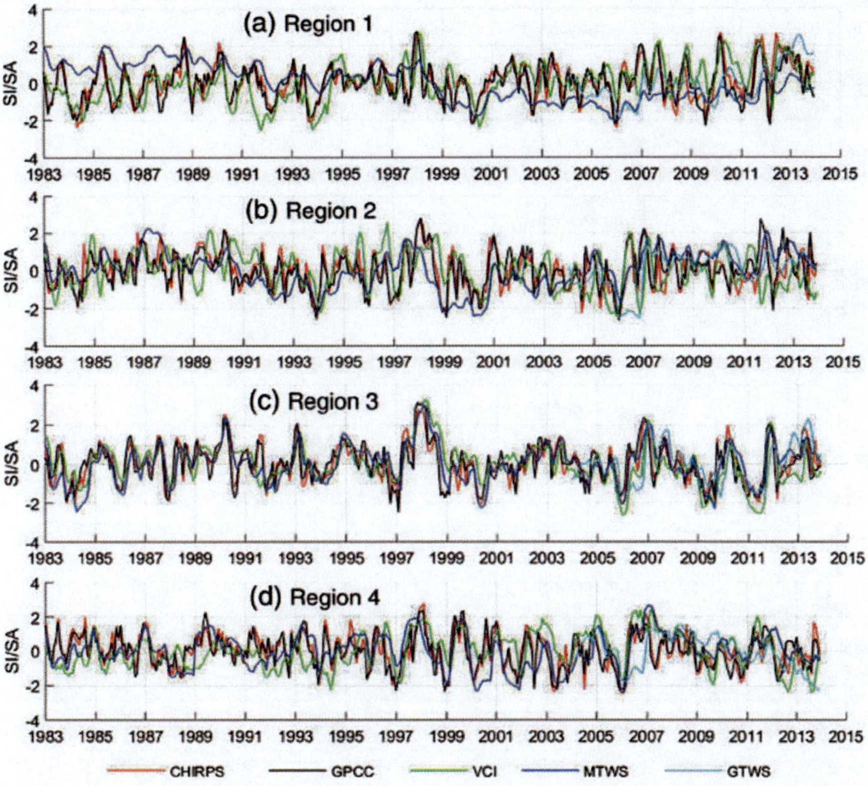

Fig. 14.4 Temporal evolutions of SA/SI spatial patterns in Fig. 14.2. The temporal evolutions are interpreted in conjunction with Table 14.2, to classify drought and/or wet conditions. Rainfall products (CHIRPS and GPCC) exhibit similar consistent performance across the region. Also, all the products exhibit consistent performance in region 3, while VCI and GTWS show some lag in relation to rainfall. *Source* [5]

indices in region 3 support the similar performance observed in Figs. 14.4c and 14.5c. VCI has weak negative correlation trends with the following products: MTWS, MERRA-2 and CPC in region 1 due to these products showing a pre-dominantly wet pre-1993 and dry post-1999 that is opposite to the general VCI trend.

14.5.3 Drought Intensity Area Analyses

In order to gain further insight into the spatial extent of the drought events and their intensities, the spatial and temporal patterns (Figs. 14.2, 14.3, 14.4, and 14.5) are integrated and using drought limit (intensity) categories in Table 14.2, percentages of areas under drought (by intensity) are evaluated and results presented in Fig. 14.6.

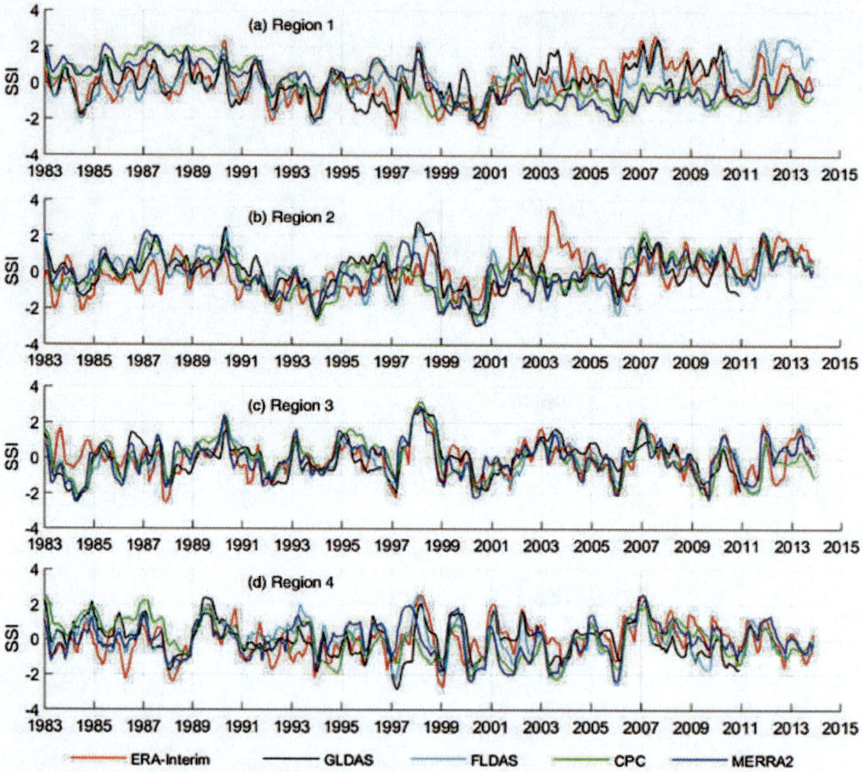

Fig. 14.5 Temporal evolutions of SSI spatial patterns in Fig. 14.3. The temporal evolutions are interpreted in conjunction with Table 14.2, to classify drought and/or wet conditions. All the moisture products have consistent performance in region 3 while in the rest of the regions, CPC is similar to MERRA-2 and similarly, ERA-Interim is closer to GLDAS. *Source* [5]

Rainfall, soil moisture and VIC products produce significantly different estimates of percentage of area under drought ($F_{2,2212} = 19.7$, $p < 0.0001$). In particular, the estimates from the rainfall (CHIRPS and GPCC at 13.09%) and soil moisture (11.90%) products are more than twice those of VCI (5.5%). One reason for the rainfall products showing more areas as being under drought may be that meteorological drought is a binary event (present or absent), which is not affected by modulating factors unlike the other drought indicators, e.g., VCI. VCI-based drought, unlike meteorological drought, is modulated by soil characteristics (water holding capacity) and/or plant (vegetation) type. Thus, for example, there could be meteorological drought over an area but due to soil water retention capacity and/or vegetation with deep roots capable of drawing water from deep soils (underground), VCI indicates no drought condition, hence the smaller area under drought. Since the soil moisture products and MTWS are modelled from rainfall and other additional inputs, their estimates of percentage areas under drought are likely to follow closely those of rainfall. However,

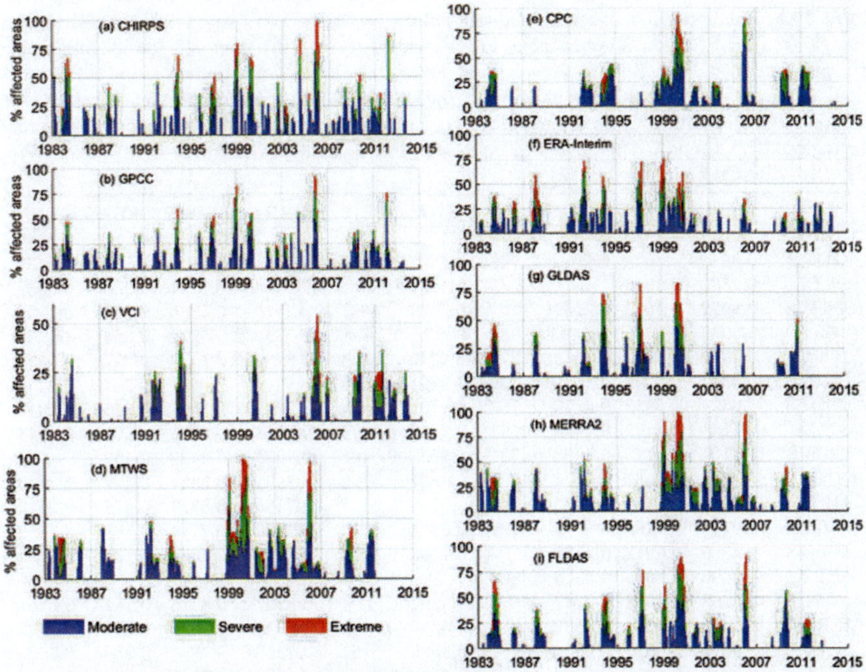

Fig. 14.6 Percentage of areas affected by various drought intensities during the period 1983–2013. Percentage areas are computed by integrating the regional spatial and temporal patterns (Figs. 14.2, 14.3, 14.4, and 14.5) then determining percentage of pixels under each drought category as per Table 14.2. The rainfall products have the highest percentage areas under drought followed closely by soil moisture products and finally the lowest percentage areas are by VCI. In addition to the soil moisture products having different percentage areas under drought, CPC is consistent with MERRA-2, GLDAS is consistent with ERA-interim while FLDAS is in between. *Source* [5]

the soil moisture products had statistically significantly different percent of areas under drought among themselves as is determined by one way ANOVA ($F_{4,1250} = 3.5410$, p = 0.0070).

The observed differences in percentage of drought areas between the various soil moisture products arise from differences in; (i) forcing precipitation, (ii) the ways in which the individual hydrological models partition precipitation into run-off and evapotranspiration, and (iii), water holding capacities, the last two of which impact on the modelled soil moisture sensitivities to precipitation variability [103]. The contribution of forcing precipitation on the differences in percentage of areas is highlighted by the differences in areas presented by GLDAS and FLDAS, products of the same model (Noah) but different forcing precipitation, hence different drought spatial extents and cycles. Of all the model forcing parameters, precipitation is the key factor determining the characteristics of the resulting soil moisture, see e.g., [30, 31, 38, 73], hence the areal extents under drought.

Table 14.5 Relationship between the drought indices by regions: (i) Region-1 upper table, upper triangle (red), (ii) Region-2 upper table, lower triangle (blue), (iii) Region-3 lower table, upper triangle (green), and (iv), Region-4 lower table, lower triangle (brown). Regions are as in Fig. 14.2. Non-significant correlations are in italics ($p < 0.05$). Region 3 has the strongest relationships with all values being significant. Also, note the high correlations between the following products across the regions: GPCC and CHIRPS; and MTWS, MERRA-2, and CPC. (MTWS—MERRA-2 TWS, GTWS—GRACE TWS). *Source* [5]

	CHIRPS	GPCC	VCI	MTWS	ERA	CPC	MERRA2	FLDAS	GLDAS	GTWS
CHIRPS	1	0.8991	0.4907	0.1203	0.4255	0.1073	0.2052	0.7300	0.5106	0.3196
GPCC	0.8619	1	0.4250	0.1288	0.4505	0.1266	0.2161	0.6120	0.5454	0.2433
VCI	0.2832	0.1918	1	−0.1082	0.4366	−0.1091	−0.0690	0.6737	0.527	0.3658
MTWS	0.3922	0.4683	0.2152	1	0.0526	0.8437	0.9910	0.1888	0.0359	0.7939
ERA	0.0947	0.2525	−0.1041	0.3628	1	0.1788	0.1138	0.5229	0.6664	−0.1708
CPC	0.2970	0.3962	0.2410	0.8714	0.4240	1	0.8482	0.1370	0.0976	0.5036
MERRA2	0.4546	0.5310	0.1889	0.9864	0.3791	0.8513	1	0.2431	0.0810	0.7352
FLDAS	0.6324	0.6352	0.4079	0.7372	0.2704	0.6844	0.7212	1	0.6365	0.7047
GLDAS	0.4141	0.4499	0.4494	0.5911	0.3352	0.6274	0.5583	0.6940	1	0.0643
GTWS	0.1101	0.0776	0.1863	0.7310	0.5553	0.6465	0.6651	0.4429	0.4505	1
CHIRPS	1	0.9267	0.5771	0.7388	0.6827	0.5622	0.7921	0.5914	0.4727	0.3796
GPCC	0.8959	1	0.5733	0.7023	0.6797	0.5583	0.7589	0.5377	0.4782	0.2643
VCI	0.2740	0.2788	1	0.7895	0.6614	0.8072	0.7636	0.7144	0.6837	0.6688
MTWS	0.4808	0.4919	0.4735	1	0.7195	0.8676	0.9912	0.8604	0.7316	0.6209
ERA	0.4095	0.3615	0.5310	0.5082	1	0.6686	0.7307	0.6937	0.6083	0.5950
CPC	0.4003	0.3978	0.2787	0.6799	0.3201	1	0.8267	0.8335	0.7513	0.6861
MERRA2	0.5482	0.5534	0.4569	0.9837	0.5250	0.6692	1	0.8256	0.6903	0.5743
FLDAS	0.6516	0.6312	0.4592	0.7627	0.5595	0.6161	0.7465	1	0.8302	0.7693
GLDAS	0.5047	0.5016	0.4130	0.4736	0.5432	0.5496	0.4578	0.7438	1	0.5598
GTWS	0.0196	0.0609	0.4195	0.4637	0.4334	0.4860	0.4200	0.2699	−0.0189	1

The MERRA-2 products show similar patterns and are closer to CPC (Fig. 14.6d, e, and h) while GLDAS is closer to ERA-interim as had been observed from the correlations (Table 14.5) and in the temporal evolutions (Figs. 14.4, 14.5). FLDAS appear to be in between the two groups. Also, the lag in drought detection (already noted in Figs. 14.4, 14.5) becomes more evident with the rainfall products detecting drought onset and duration first, followed by VCI/soil moisture products, and finally the TWS products. This would be attributed to time delayed response in moisture accumulation from rainfall through soil moisture, vegetation, and finally to changes in TWS during both the start and cessation of rainfall. Generally, the results also indicate the post-1999 period as having more drought events with higher intensity than the pre-1999 period except for ERA-Interim and GLDAS indicators. This is in line with other drought and climate studies that observed a decline in rainfall since 1999 and increased drought frequencies, see [67, 68, 126]. Also, GLDAS seems to have underestimated the 2005–2006 drought in terms of both duration and intensity as compared to the rest of the soil moisture products.

Further, GTWS returns higher percentage of areas under drought on average than MTWS as confirmed by one way ANOVA (25.307 vs. 9.8147 at F_1, 138 = 16.1064,

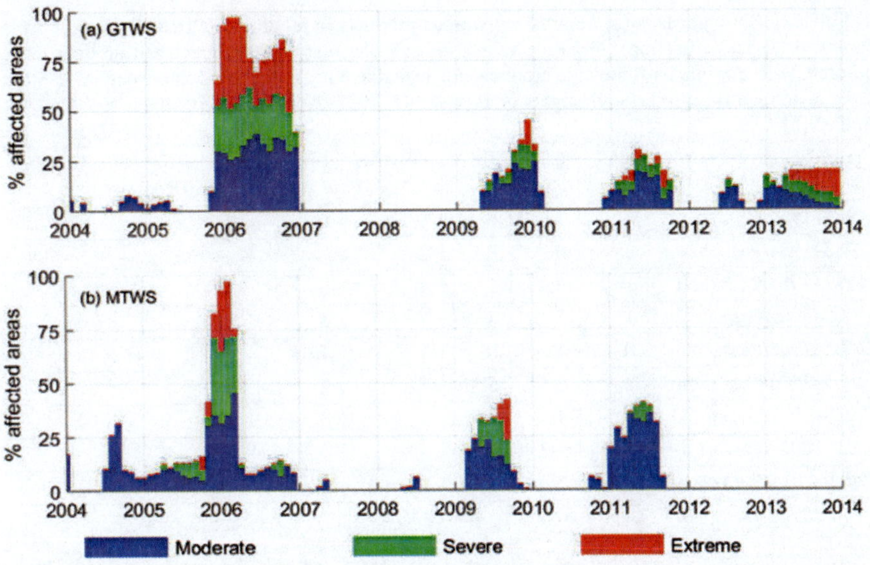

Fig. 14.7 Comparison of performance between GTWS and MTWS in terms of percentage of areas affected by various drought categories. Percentage areas are computed as in Fig. 14.6. They have consistent performance, with GTWS having a lag in drought detection probably due to groundwater that is lacking in MTWS. *Source* [5]

$p = 0.0001$) though with almost equal percentage of areas at drought peaks, at which GTWS lags MTWS by 0–3 months in the detection of drought onset and cessation (Fig. 14.7a, b). Since MTWS is modelled on precipitation and other input without groundwater while GTWS is observed, the lack of groundwater in MTWS probably explains why it does not properly account for the buffer effect, hence possible lag by GRACE in detecting the onset and cessation of drought. In addition, GTWS shows drought episodes in the post 2012 period while MTWS does not (see Fig. 14.7).

The drought severity is well captured by all the products as evidenced by the majority of the areas being under moderate drought followed by severe drought and then extreme according to the definition of SPI (see, e.g., Figs. 14.6 and 14.7, [70]). All the products capture different severity levels except MTWS and MERRA-2, which has similar severity levels as a result of overlapping formulation. The differences in severity levels among the other products could be attributed to the different formulation of the products and to the fact that they represent droughts in different environments with different impacting factors, e.g., soil properties influence the severity of drought as captured by the soil moisture products while rainfall characteristics (amount, intensity and duration) influence the drought severity as captured by rainfall products.

Finally, from the knowledge gained in the analyses above, the droughts of 1983–1984, 2005–2006, and 2010–2011 are examined closely using selected indicators in order to quantify the above-observed lags in drought cycles (Fig. 14.8,

Table 14.6 Drought lags (in months) by various products in relation to CHIRPS drought cycle (onset, peak, and cessation). Negative values indicate the respective product had drought cycle before CHIRPS while dash indicate products not available during that particular drought. The lags were quantified from selected droughts of 1983–1984, 2005–2006, and 2010–2011, see Fig. 14.8. *Source* [5]

Year/drought cycle	VCI	CPC	ERA	GLDAS	FLDAS	MTWS/ MERRA2	GTWS
1983–1984/Onset	1	1	6	−1	0	0	–
1983–1984/Peak	3	0	3	2	1	−6	–
1983–1984/Cessation	2	4	5	4	5	3	–
2005–2006/Onset	1	5	4	7	5	−1	6
2005–2006/Peak	0	1	0	1	1	1	1
2005–2006/Cessation	2	>5	5	1	2	>5	>5
2010–2011/Onset	3	1	2	–	3	2	4
2010–2011/Peak	3	−3	−7	–	0	0	−1
2010–2011/Cessation	4	3	−4	–	3	2	3

Table 14.6). These drought years have been selected for further analysis because they had more severe impacts in the region, see e.g., [69, 103, 105]. From this analysis, VCI has a lag of 0–4 months in relation to CHIRPS in picking drought stages (onset, peak, and cessation) while the soil moisture products (CPC, ERA-Interim, GLDAS, FLDAS, and MERRA-2) has inconsistent lags among themselves, and in relation to CHIRPS for the considered drought episodes (e.g., Fig. 14.8, Table 14.6). Soil moisture, being an integration of rainfall anomalies over time [33, 102], is expected to have a lag in response to rainfall behavior throughout the hydrological cycle hence the soil moisture products and VCI (an indicator of moisture availability to vegetation) lag rainfall in the analysis. The inconsistency in the lags by the soil moisture products, similar to observed inconsistency in the percentage of areas under drought (Fig. 14.6), could be due to the different model forcing parameters used in generating various products in addition to different model thresholds as discussed above. Finally, the TWS products has different lags with GTWS having longer lag (Fig. 14.7). This longest lag from GTWS could be due to the fact that it is the last in the transition from rain event to moisture accumulation, and eventually groundwater change over time. Also, the under-characterization of the 2005–2006 drought by GLDAS already observed in Fig. 14.6g is clearer in Fig. 14.8b.

14.6 Effectiveness of Drought Indicators: Crop Production Assessment

In order to assess the effectiveness of the indicators in capturing agricultural drought, partial least square regression (PLSR) models are fitted with indices as the predictors

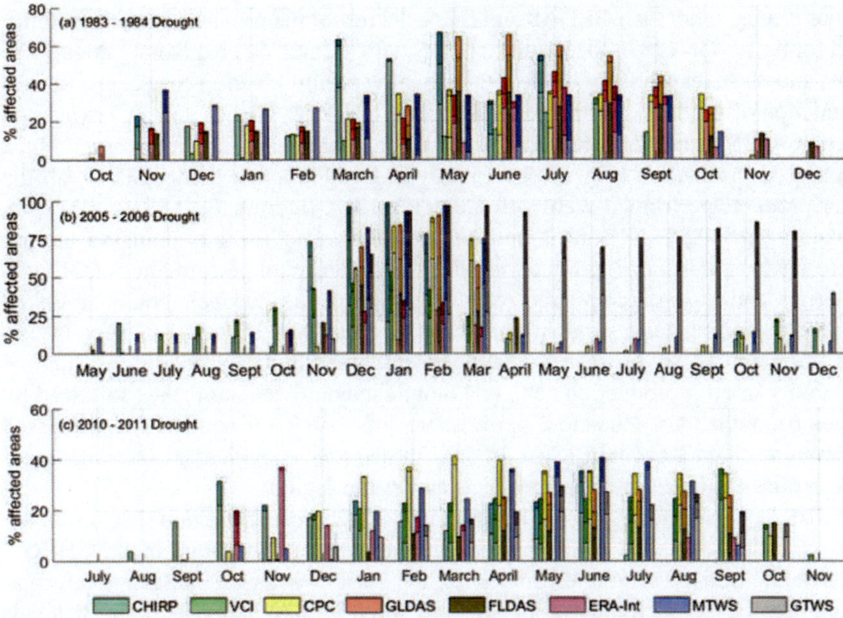

Fig. 14.8 Percentage of areas affected by various drought intensites during the 1983–1984, 2005–2006, and 2010–2011 drought episodes. Each bar has up to 3 colour grades (gradients) representing from bottom moderate, severe, and extreme droughts at the top. Percentage areas are computed as in Fig. 14.6 but only for the duration of drought. VCI has a lag of about 2–3 months in identifying the drought cycle in relation to CHIRPS. The rest of the products have inconsistent lags in relation to rainfall across the three drought episodes. *Source* [5]

and annual crop production data as the responses. The model with the lowest estimated mean squared prediction error is adopted in each case and the proportion of variability explained (R^2) used for comparison. As production is known to be related to water availability at various stages of crops' growth [47, 107], and water being a major growth determinant in the East Africa region [17], a good relationship is expected between drought indices that capture (characterize) drought well and crop production over the considered duration of time.

Because crop production data is reported at country level (national) while the generalization in Sect. 14.5 (Figs. 14.2 and 14.3) has signals across countries, SIs/SAs are re-computed for each country and the resulting rotated principal components reconstructed and used in PLSR with country level crop production data. The SIs are computed for the long-term duration (1983–2013) and SAs for the short-term (2004–2013). The latter duration though shorter, is necessitated by the need to compare the performance of GRACE SA against the other products. The proportions of variabilities explained (R^2) from the regression using the short duration (SAs) should be interpreted with care due to the short length of the data used.

For Kenya, other than GLDAS and ERA, the rest of the products performed fairly well for the period 1983–2013 with CHIRPS, GPCC, and VCI explaining up to 94%, 73%, and 89%, respectively of the total annual variability in crop (wheat and maize) production (Fig. 14.9a). Similarly for Tanzania, CHIRPS, GPCC, and VCI explained up to 96%, 85%, and 89%, respectively of the total annual variability in crop (wheat and maize) production (Fig. 14.9b). Finally in Uganda, other than GLDAS, all the other products performed well with CHIRPS, ERA-Interim, FLDAS, and MTWS explaining up to 88%, 92%, 84%, and 77%, respectively of the total annual variability in crop (wheat and maize) production (Fig. 14.9c). The poor performance of GLDAS in Kenya and Uganda compared to other soil moisture products could be linked to poor performance in drought characterization as was observed in Sect. 14.5.3, Fig. 14.6g and 14.8b. Most of the products explained higher proportions of annual variability in crop production (R^2) in 1 month standardized anomalies followed by 3 then 6 months. Also, the close performance of MERRA-2 and MTWS witnessed in drought characterization (Sect. 14.5.2, Table 14.5) is evident in the amount of variabilities explained by these products across the region.

CHIRPS performed generally better than GPCC across the region (Fig. 14.9a–c). This could be attributed to the fact that in addition to rain gauge input, CHIRPS has satellite-derived rainfall estimates for areas with less or no rain gauge information unlike GPCC with only rain gauge measured rainfall hence its performance is dependent on gauge density and terrain changes, see e.g., [6, 44, 99]. In relation to the rainfall products (CHIRPS and GPCC), the soil moisture products explained less variability in annual crop production over East Africa region except in Uganda where the performance of ERA-Interim was almost as good as rainfall-derived indicators. Since soil moisture products represent the rainfall that remains after run-off and evaporation, the effective water available to the plants (crops), they are expected to explain higher variabilities in the annual crop production than rainfall thus their poor performance could be linked to how well they fit the region. In addition, the inconsistent performance of the soil moisture products (CPC, ERA-Interim, GLDAS, FLDAS, and MERRA-2) and MTWS across the East Africa region in explaining the annual variability in crop production could be linked to the inconsistencies observed in the drought characterization as discussed in Sect. 14.5.2.

Overall, the good performance of FLDAS over GLDAS across the East African region despite both being products of the same model (Noah) is due to the fact that for FLDAS, the Noah model was forced by CHIRPS, a precipitation product designed for the region. The magnitudes of the annual variabilities in crop production explained by FLDAS could be a pointer to difficulties faced by Noah in correctly partitioning precipitation into moisture, run-off, and evapotranspiration as per natural occurrence in the East Africa region.

Though based on a short duration data set (10 years), GRACE SA has mixed performance between wheat and maize across the countries but does better than or equals to soil moisture products across the region (Fig. 14.9d–f). The performance could be attributed to the fact that over a shorter duration of time such as the one considered (i.e., 1-, 3-, and 6 month anomalies), the bulk of the variation in the GRACE TWS occurs in the soil moisture compartment, which is more sensitive to

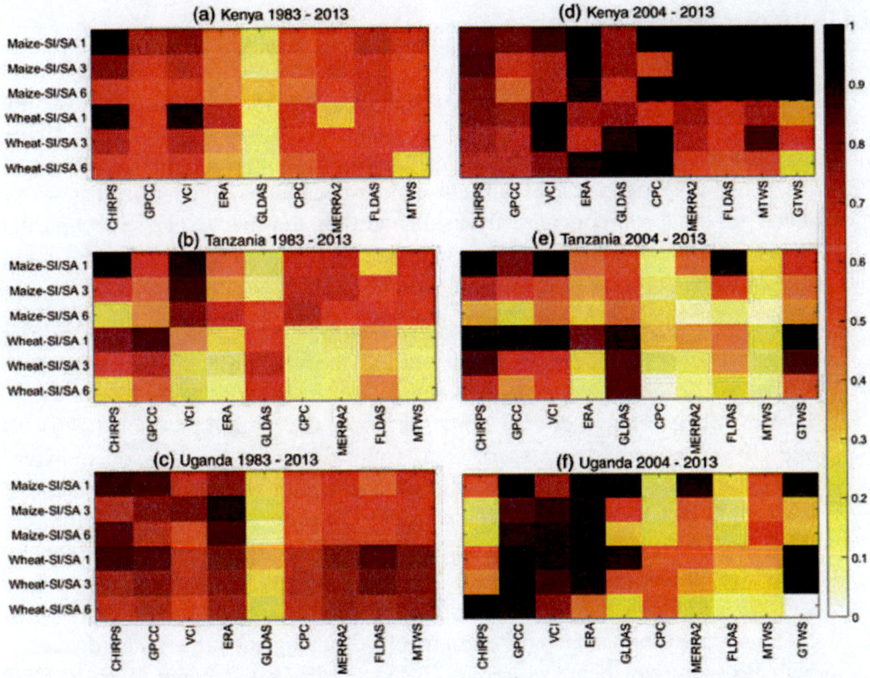

Fig. 14.9 Proportions of variability in national annual crop (maize and wheat) production (R^2) explained by various drought indices for Kenya, Tanzania, and Uganda. CHIRPS, GPCC, and VCI consistently explains relatively higher variability in crop production, while the soil moisture products have inconsistent performance (**a**, **b**, and **c**). The figures **d**, **e**, and **f** should be interpreted with care as the datasets used to fit the models are only 10 years long. Both SI and SA are computed at annual scales. (The y axis indicates the crop (maize/wheat), SI (for **a**, **b**, and **c**), and SA (for **d**, **e** and **f**) while 1,3 and 6 indicate the standardization time scales for the indicators on the x axis). *Source* [5]

climate variability than groundwater change, e.g., [127]. This shows the potential of GRACE product to monitor agricultural drought although longer duration of dataset is essential.

Results from regression analysis should be interpreted with caution though, as the relationship between production and climate conditions (water availability) only hold if other factors in the production chain are held constant, e.g., areas under cultivation over the period considered and technical factors of production (e.g., fertilizers, crop cultivars, pesticides). In addition, production response to water at any stage of growth can be modified by various factors, e.g., diseases, weeds, insects, crop variety [47, 107], hence, results should not be generalized to other areas.

14.7 Concluding Remarks

This chapter characterized agricultural drought over East Africa part of GHA using precipitation products (CHIRPS and GPCC), soil moisture products (CPC, ERA-Interim, MERRA-2, FLDAS, and GLDAS), and TWS products (MERRA-2 and GRACE). This was accomplished through standardized index/standardized anomaly and rotated principal component analyses. In addition, the chapter carried out partial least squares regression (PLSR) analysis over Kenya, Uganda, and Tanzania to assess the utility of these products in capturing agricultural drought in these countries.

(i) Drought characterization results showed CHIRPS and GPCC as being similar and consistent over the entire region, while all the other products were consistent for region 3 (dry lowland eastern Kenya).

(ii) In terms of percentage of areas under drought, the rainfall products (CHIRPS and GPCC) covered the highest areas followed closely by the soil moisture products, while VCI covered the least percentage areas under drought. Results further indicated drought cycle detection in the order; rainfall, VCI/soil moisture, and TWS. VCI had 0–4 months of lag in detecting drought cycle (onset, peak, and cessation) in relation to rainfall products while the soil moisture and TWS products had inconsistent lag varying from one drought to the next. Soil moisture products had different results (both lag and areas under drought), with ERA-Interim being closer to GLDAS, MERRA-2 being close to CPC while FLDAS was in between. GLDAS under-characterized the 2005–2006 drought to under 2 months in comparison to over 7 months of ERA and CPC. Finally, the TWS products were consistent with GTWS having few months' lag probably due to groundwater that is missing in MTWS.

(iii) From the PLSR analysis, consistent performances by CHIRPS, GPCC, and VCI in explaining relatively high proportions of variabilities in annual crop production in Kenya, Tanzania, and Uganda over the period 1983–2013 was noted. In addition, the lack of consistency observed from the soil moisture products in drought characterization also was evident in the amount of annual crop production variability explained by them (soil moisture products) across the region.

(iv) The chapter identified the following indicators as suitable for agricultural drought monitoring/characterization for the region during the 1983–2013 period; (a) for Kenya: CHIRPS, GPCC, VCI, MERRA-2, FLDAS and MTWS; (b) for Uganda: CHIRPS, GPCC, VCI, FLDAS, ERA, MERRA-2, and MTWS; and (c), for Tanzania: CHIRPS, GPCC, VCI, FLDAS, GLDAS and ERA.

(v) Also, GTWS showed potential in explaining the annual variability in crop production, albeit a longer period of dataset is required to evaluate its potential.

Further studies need to be undertaken to determine how well the model soil moisture products (CPC, ERA-Interim, MERRA-2, FLDAS, and GLDAS) and MTWS fit the region. Also, care should be taken in generalizing these results as production response to water at any different stages of crop growth can be modified by several factors.

References

1. AghaKouchak A (2015) A multivariate approach for persistence-based drought prediction: application to the 2010–2011 East African Drought. J Hydrol 526:127–135. http://dx.doi.org/10.1016/j.jhydrol.2014.09.063
2. AghaKouchak A, Nasrollahi N, Habib E (2009) Accounting for uncertainties of the TRMM satellite estimates. Remote Sens 1:606–619. http://dx.doi.org/10.3390/rs1030606
3. Agnew C, Chappell A (1999) Drought in the Sahel. GeoJournal 48(4):299–311. http://dx.doi.org/10.1023/A:1007059403077
4. Agnew CT (2000) Using the SPI to identify drought. Drought Netw News 12:6–12
5. Agutu N, Awange J, Zerihun A, Ndehedehe C, Kuhn M, Fukuda Y (2017) Assessing multi-satellite remote sensing, reanalysis, and land surface models' products in characterizing agricultural drought in East Africa. Remote Sens Environ 194:287–302. https://doi.org/10.1016/j.rse.2017.03.041
6. Agutu NO, Awange JL, Ndehedehe C, Mwaniki MW (2020) Consistency of agricultural drought characterization over Upper Greater Horn of Africa (1982–2013): Topographical, gauge density, and model forcing influence. Sci Total Environ 709. https://doi.org/10.1016/j.scitotenv.2019.135149
7. Albergel C, de Rosnay P, Balsamo G, Isaksen L, Muñoz-Sabater J (2012) Soil moisture analyses at ECMWF: evaluation using global ground-based *in-situ* observations. J Hydrometeorol 13(5):1442–1460. http://dx.doi.org/10.1175/JHM-D-11-0107.1
8. Anderson WB, Zaitchik BF, hain CR, Anderson MC, Yilmaz MT, Mecikalski J, Schultz L (2012) Towards an integrated soil moisture drought monitor for East Africa. Hydrol Earth Syst Sci 16:2893–2913. http://dx.doi.org/10.5194/hess-16-2893-2012
9. Anyamba A, Tucker CJ (2005) Analysis of Sahelian vegetation dynamics using NOAAAVHRR NDVI data from 1981–2003. J Arid Environ 63(3):596–614. http://dx.doi.org/10.1016/j.jaridenv.2005.03.007
10. Awange J, Khandu Schumacher M, Forootan E, Heck B (2016) Exploring hydrometeorological drought patterns over the Greater Horn of Africa (1979–2014) using remote sensing and reanalysis products. Adv Water Res 94:45–59. http://dx.doi.org/10.1016/j.advwatres.2016.04.005
11. Awange JL, Sharifi MA, Ogonda G, Wickert J, Grafarend EW, Omulo MA (2008) The falling lake Victoria water level: GRACE, TRIMM and CHAMP satellite analysis of the lake basin Water Res Manag 22(7):775–796. http://dx.doi.org/10.1007/s11269-007-9191-y
12. Awange JL, Anyah R, Agola N, Forootan E, Omondi P (2013) Potential impacts of climate change and environmental change on the stored water of Lake Victoria Basin and economic implications. Water Res Res 49:8160–8173. http://dx.doi.org/10.1002/2013WR014350
13. Awange JL, Palancz B, Völgyesi L, (2020) Hybrid Imaging and Visualization. Employing Machine Learning with Mathematica - Python, Springer Nature International, Berlin. 978-3-030-26152-8, https://doi.org/10.1007/978-3-030-26153-5
14. Awange JL (2021) Lake Victoria monitored from space. Springer Nature International
15. Awange JL (2021) Nile Waters. Weighed from space, Springer Nature International
16. Balsamo G, Beljaars A, Scipal K, Viterbo P, van den Hurk B, Hirschi M, Betts AK (2009). A revised hydrology for the ECMWF model: verification from field site to terrestrial water storage and impact in the integrated forecast system. J Hydrometeorol 10(3):623–643. http://dx.doi.org/10.1175/2008JHM1068.1
17. Barron J, Rockström J, Gichuki F, Hatibu N (2003) Dry spell analysis and maize yields for two semi-arid locations in east Africa. Agric. For Meteorol 117:23–37. http://dx.doi.org/10.1016/S0168-1923(03)00037-6
18. Bayarjargal YAK, Bayasgalan M, Khudulmur S, Gandush C, Tucker CJ (2006) Comparative study of NOAA AVHRR derived vegetation indices using change vector analysis. Remote Sens Environ 105: 9–22. http://dx.doi.org/10.1016/j.rse.2006.06.003

19. Becker M, LLovel W, Cazenave A, Güntner A, Crétaux J (2010) Recent hydrological behaviour of the East African great lakes region inferred from GRACE, satellite altimetry and rainfall observations. C. R. Geoscience 342:223–233. http://dx.doi.org/10.1016/j.crte.2009.12.010
20. Bordi I, Fraedrich K, Petitta M, Sutera A (2006) Large-scale assessment of drought variability based on NCEP/NCAR and ERA-40 re-analyses. Water Res Manag 20(6):899–915. http://dx.doi.org/10.1007/s11269-005-9013-z
21. Bosilovich M, Lucchesi G, Suarez M (2016) MERRA-2: file specification. GMAO Office Note No. 9 (Version 1.1). 73 pp. http://gmao.gsfc.nasa.gov/pubs/office_notes
22. Bosilovich MG, Akella S, Coy L, Cullather R, Draper C, Gelaro R, Kovach R, Liu Q, Molod A, Norris P, Chao W, Reichle R, Takacs L, Todling R, Vikhliaev Y, Bloom S, Collow A, Partyka G, Firth S, Labow G, Pawson S, Reale O, Schubert S, Suarez M (2015) Merra-2: Initial evaluation of the climate. Technical report series on global modeling and data assimilation NASA/TM2015-104606/Vol. 43. NASA:GSFCG., https://gmao.gsfc.nasa.gov/pubs/docs/Mahanama804.pdf
23. Chen JL, Wilson CR, Tapley BD, Ries JC (2004) Low degree gravitational changes from GRACE: validation and interpretation. Geophys Res Lett 31(22). http://dx.doi.org/10.1029/2004GL021670. (n/a–n/a, l22607)
24. Chen JL, Wilson CR, Tapley BD, Yang ZL, Niu GY (2009) 2005 drought event in the Amazon River basin as measured by GRACE and estimated by climate models. J Geophys Res Solid Earth 114:(B5). http://dx.doi.org/10.1029/2008JB006056. (n/a–n/a, B05404)
25. Chen T, de Jeu R, Liu Y, van der Werf G, Dolman A (2014) Using satellite based soil moisture to quantify the water driven variability in NDVI: a case study over mainland Australia. Remote Sens Environ 140:330–338. http://dx.doi.org/10.1016/j.rse.2013.08.022
26. Clark CO, Webster PJ, Cole JE (2003.)Interdecadal variability of the relationship between the Indian Ocean Zonal Mode and East African Coastal Rainfall Anomalies. J Clim 16:548–554. http://dx.doi.org/10.1175/1520-0442(2003)016¡0548:IVOTRB¿2.0.CO;2
27. Damberg L, AghaKouchak A (2014) Global trends and patterns of drought from space. Theor Appl Climatol 117:441–448. http://dx.doi.org/10.1007/s00704-013-1019-5
28. Decker M, Brunke MA, Wang Z, Sakaguchi K, Zeng X, Bosilovich MG (2012) Evaluation of the reanalysis products from GSFC, NCEP, and ECMWF using flux tower observations. J Clim 25:1916–1944. http://dx.doi.org/10.1175/JCLI-D-11-00004.1
29. Dee DP, Uppala SM, Simmons AJ, Berrisford P, Poli P, Kobayashi S, Andrae U, Balmaseda MA, Balsamo G, Bauer P, Bechtold P, Beljaars ACM, van de Berg L, Bidlot J, Bormann N, Delsol C, Dragani R, Fuentes M, Geer AJ Haimberger L, Healy SB Hersbach, H, Hólm EV, Isaksen L Kållberg, P, Köhler M, Matricardi M, McNally AP, Monge-Sanz BM, Morcrette J-J, Park B-K (2011) The ERA-Interim reanalysis: configuration and performance of the data assimilation system. Q J R. Meteorol Soc 137(656):553–597. http://dx.doi.org/10.1002/qj.828
30. Dirmeyer PA, Dolman AJ, Sato N (1999) The pilot phase of the global soil wetness project. Bull Am Meteorol Soc 80(5):851–878. http://dx.doi.org/10.1175/1520-0477(1999)080¡0851:TPPOTG¿2.0.CO;2
31. Dirmeyer PA, Guo Z, Gao X (2004) Comparison, validation, and transferability of eight multiyear global soil wetness products. J Hydrometeorol 5(6):1011–1033. http://dx.doi.org/10.1175/JHM-388.1
32. Dorigo W, de Jeu R, Chung D, Parinussa R, Liu Y, Wagner W, Fernández-Prieto D (2012) Evaluating global trends (1988–2010) in harmonized multi-satellite surface soil moisture. Geophys Res Lett 39(18). http://dx.doi.org/10.1029/2012GL052988. (l18405)
33. Dutra E, Viterbo P, Miranda PMA (2008) ERA-40 reanalysis hydrological applications in the characterization of regional drought. Geophys Res Lett 35(19):1–5. http://dx.doi.org/10.1029/2008GL035381. (l19402)
34. Dutra E, Magnusson L, Wetterhall F, Cloke HL, Balsamo G, Boussetta S, Pappenberger F (2013) The 2010–2011 drought in the Horn of Africa in ECMWF reanalysis and seasonal forecast products. Int J Climatol 33:1720–1729. http://dx.doi.org/10.1002/joc.3545

35. Dutra E, Wetterhall F, Di Giuseppe F, Naumann G, Barbosa P, Vogt J, Pozzi W, Pappenberger F (2014) Global meteorological drought part 1: probabilistic monitoring. Hydrol Earth Syst Sci 18(7):2657–2667. http://dx.doi.org/10.5194/hess-18-2657-2014
36. EACS (2014) East African community facts and figures-2014. Report. East African Community Secretariat. (Retrieved on 15, 2015)
37. Elagib NA (2013) Meteorological drought and crop yield in Sub-Sahara Sudan. Int J Water Res Arid Environ 2(3):164–171
38. Entin JK, Robock A, Vinnikov KY, Zabelin V, Liu S, Namkhai A, Adyasuren T (1999) Evaluation of global soil wetness project soil moisture simulations. J Meteorol Soc Jpn Ser II 77(1B):183–198
39. FAO (2017) Millions of people face food shortages in the Horn of Africa. http://www.fao.org/news/story/en/item/468941/icode/. [Accessed 25 2017]
40. Fan Y, van den Dool H (2004) Climate prediction center global monthly soil moisture data set at 0.5 resolution for 1948 to present. J Geophys Res 109:D10102. http://dx.doi.org/10.1029/2003JD004345
41. Farahmand A, AghaKouchak A (2015) A generalized framework for deriving nonparametric standardized drought indicators. Adv Water Res 76:140–145. http://dx.doi.org/10.1016/j.advwatres.2014.11.012
42. Forina M, Armanino C, Lanteri S, Leardi R (1988) Methods of varimax rotation in factor analysis with applications in clinical and food chemistry. J Chemom 3:115–125. http://dx.doi.org/10.1002/cem.1180030504
43. Funk C, Hoell A, Shukla S, Blade I, Liebmann B, Roberts JB, Robertson FR, Husak G (2014) Predicting East African spring droughts using Pacific and Indian Ocean sea surface temperature indices. Hydrol Earth Syst Sci 18:4965–4978. http://dx.doi.org/10.5194/hess-18-4965-2014
44. Funk C, Peterson P, Landsfeld M, Pedreros D, Verdin J, Shukla S, Husak G, Rowland J, Harrison L, Hoell A, Michaelsen J (2015) The climate hazards infrared precipitation with stations - a new environmental record for monitoring extremes. Sci Data 2(150066):1–21. http://dx.doi.org/10.1038/sdata. 2015.66
45. Geladi P, Kowalski BR (1986) Partial least-squares regression: a tutorial. Anal Chem Acta 185:1–17
46. Guan K, Wood E, Caylor K (2012) Multi-sensor derivation of regional vegetation fractional cover in Africa. Remote Sens Environ 124:653–665. http://dx.doi.org/10.1016/j.rse.2012.06.005
47. Hane DC, Pumphrey FV (1984) Crop water use curves for irrigation scheduling. Special Report 706. Agricultural Experiment Station Oregon State University, Corvallis. (Retrieved on 15, 2015)
48. Hannachi A, Jolliffe IT, Stephenson DB (2007) Empirical orthogonal functions and related techniques in atmospheric science: a review. Int J Climatol 27:1119–1152. http://dx.doi.org/10.1002/joc.1499
49. Helland I (2004) Partial least squares regression. Wiley. http://doi.org/10.1002/0471667196.ess6004.pub2
50. Hong Y, Hsu K-l, Moradkhani H, Sorooshian S (2006) Uncertainty quantification of satellite precipitation estimation and Monte Carlo assessment of the error propagation into hydrologic response. Water Res Res 42(8). http://dx.doi.org/10.1029/2005WR004398. (n/a–n/a, w08421)
51. IFRC (2011) Drought in the Horn of Africa: preventing the next disaster. In: International federation of red cross and red crescent societies, Geneva. Retrieved on March 15, 2015
52. Jolliffe IT (1995) Rotation of principal components: choice of normalization constraints. J Appl Stat 22(1):29–35. http://dx.doi.org/10.1080/757584395
53. Jolliffe IT (2002) Principal component analysis. 2nd edn. Springer series in statistics. Springer
54. Kaiser HF (1958) The Varimax Criterion for analytic rotation in factor analysis. Psychometrika 23(3):187–200. http://dx.doi.org/10.1007/BF02289233

55. Kasnakoglu H, Mayo R (2004) FAO statistical data quality framework: a multilayered approach to monitoring and assessment. Conference on Data Quality for International Organizations, Wiesbaden, Germany, 27 and 28 2004. (Accessed on 4/10/2016)
56. Katz RW, Glantz MH (1986) Anatomy of a rainfall index. Mon Weather Rev 114:764–771. http://dx.doi.org/10.1175/1520-0493(1986)114¡0764:AOARI¿2.0.CO;2
57. Khaki M, Awange J (2021) The 2019–2020 rise in Lake Victoria monitored from space: exploiting the state-of-the-art grace-fo and the newly released ERA-5 reanalysis products. Sensors 21(13):4304. https://doi.org/10.3390/s21134304
58. Kogan F (1995) Application of vegetation index and brightness temperature for drought detection. Adv Space Res 15:91–100. http://dx.doi.org/10.1016/0273-1177(95)00079-T
59. Kurnik B, Barbosa P, Vogt J (2011) Testing two different precipitation datasets to compute the standardized precipitation index over the Horn of Africa. Int J Remote Sens 32(21):5947–5964. http://dx.doi.org/10.1080/01431161.2010.499380
60. Kusche J (2007) Approximate decorrelation and non-isotropic smoothing of timevariable GRACE-type gravity field models. J Geod 81:733–749. http://dx.doi.org/10.1007/s00190-007-0143-3
61. Kusche J, Schmidt R, Petrovic S, Rietbroek R (2009) Decorrelated GRACE time variable gravity solution by GFZ, and their validation using a hydrological model. J Geod 83:903–913. http://dx.doi.org/10.1007/s00190-009-0308-3
62. Landerer FW, Swenson SC (2012) Accuracy of scaled GRACE terrestrial water storage estimates. Water Res Res 48. http://dx.doi.org/10.1029/2011WR011453. (n/a–n/a, w04531)
63. Loewenberg S (2011) Humanitarian response inadequate in Horn of Africa crisis. The Lancet 378(9791):555–558. http://dx.doi.org/10.1016/S0140-6736(11)61276-2
64. Long D, Scanlon BR, Longuevergne L, Sun AY, Fernando DN, Save H (2013) GRACE satellite monitoring of large depletion in water storage in response to the 2011 drought in Texas. Geophys Res Lett 40:3395–3401. http://dx.doi.org/10.1002/grl.50655
65. Lorenz EN (1956) Empirical orthogonal function and statistical weather prediction. Statistical forecasting project: Scientific Report No. 1. Department of Meteorology, MIT., (Retrieved on March 15, 2015)
66. Lough JM (1997) Regional indices of climate variation: temperature and precipitation in Queensland, Australia. Int J Climatol 17:55–66. http://dx.doi.org/10.1002/(SICI)1097-0088(199701)17:1¡55::AID-JOC109¿3.0.CO;2-Z
67. Lyon B (2014) Seasonal drought in the Greater Horn of Africa and its recent increase during the March-May long rains. J Clim 27:7953–7975. http://dx.doi.org/10.1175/JCLI-D-13-00459.1
68. Lyon B, DeWitt DG (2012) A recent and abrupt decline in the East African long rains. Geophys Res Lett 39:L02702. http://dx.doi.org/10.1029/2011GL050337
69. Masih I, Maskey S, Mussá FEF, Trambauer P (2014) A review of droughts on the African continent: a geospatial and long-term perspective. Hydrol Earth Syst Sci 18(9):3635–3649. http://dx.doi.org/10.5194/hess-18-3635-2014
70. McKee TB, Doesken NJ, Kleist J (1993) The relationship of drought frequency and duration to time scale. In: Conference proceedings. (eighth Conference of Applied Climatology, Anaheim, California)
71. McNally A, Shukla S, Arsenault KR, Wang S, Peters-Lidard CD, Verdin JP (2016) Evaluating ESA CCI soil moisture in East Africa. Int J Appl Earth Obs Geoinf 48:96–109. http://dx.doi.org/10.1016/j.jag.2016.01.001
72. Michael C (2006) Climate change impacts on East Africa. A review of the scientific literature. WFF-World Wide Fund For Nature, Gland, Switzerland. (Retrieved on March 15, 2015)
73. Mo KC, Chen L-C, Shukla S, Bohn TJ, Lettenmaier DP (2012) Uncertainties in North American land data assimilation systems over the contiguous United States. J Hydrometeorol 13(3):996–1009. http://dx.doi.org/10.1175/JHM-D-11-0132.1
74. Morgan B, Awange JL, Saleem A, Hu K (2020) Understanding vegetation variability and their "hotspots" within Lake Victoria Basin (LVB: 2003–2018), 122: https://doi.org/10.1016/j.apgeog.2020.102238

75. Mpelasoka F, Awange JL, Zerihun A (2018) Influence of coupled ocean-atmosphere phenomena on the Greater Horn of Africa droughts and their implications. Sci Total Environ 610–611:691–702. https://doi.org/10.1016/j.scitotenv.2017.08.109
76. Mwangi E, Wetterhall F, Dutra E, Di Giuseppe F, Pappenberger F (2014) Forecasting droughts in East Africa. Hydrol Earth Syst Sci 18(2):611–620. http://dx.doi.org/10.5194/hess-18-611-2014
77. Naresh Kumar M, Murthy CS, Sesha Sai MVR, Roy PS (2009) On the use of Standardized Precipitation Index (SPI) for drought intensity assessment. Meteorol Appl 16(3):381–389. http://dx.doi.org/10.1002/met.136
78. Naumann G, Dutra E, Pappenberger F, Wetterhall F, Vogt JV (2014) Comparison of drought indicators derived from multiple data sets over Africa. Hydrol Earth Syst Sci 18:1625–1640. http://dx.doi.org/10.5194/hess-18-1625-2014
79. Nicholson SE (2014) A detailed look at the recent drought situation in the Greater Horn of Africa. J Arid Environ 103:71–79. http://dx.doi.org/10.1016/j.jaridenv.2013.12.003
80. OEA (2011) Eastern Africa Drought humanitarian report. OCHA Eastern Africa, number 3, 01-31 May 2011. (Retrieved on 17, 2015)
81. OEA (2011) Eastern Africa Drought humanitarian report. . OCHA Eastern Africa, number 4, 01 June–15 July 2011., (Retrieved on 17, 2015)
82. Omute P, Corner R, Awange JL (2012) The use of NDVI and its derivatives for monitoring Lake Victoria's water level and drought conditions. Water Res Manag 26:1591–1613. https://doi.org/10.1007/s11269-011-9974-z
83. Peters AJ, Waletr-Shea EA, Ji L, Vin-a A, Hayes M, Svoboda MV (2002) Drought monitoring with NDVI-based standardized vegetation index. Photogramm Eng Remote Sens 68:71–75
84. Pinzon JE, Tucker CJ (2014) A non-stationary 1981–2012 AVHRR NDVI3g time series. Remote Sens 6(8):6929. http://dx.doi.org/10.3390/rs6086929
85. Preisendorfer RW (1988) Principal component analysis in meteorology and oceanography. Development in Atmospheric Science 17. Elsevier, Amsterdam
86. Pricope NG, Husak G, Lopez-Carr D, Funk C, Michaelsen J (2013) The climate-population nexus in the East African Horn: emerging degradation trends in rangeland and pastoral livelihood zones. Glob Environ Chang 23(6):1525–1541. http://dx.doi.org/10.1016/j.gloenvcha.2013.10.002
87. Quiring SM (2009) Developing objective operational definitions for monitoring drought. J Appl Meteorol Climatol 48(6):1217–1229. http://dx.doi.org/10.1175/2009JAMC2088.1
88. Quiring SM, Ganesh S (2010) Evaluation of utility of vegetation condition index (VCI) for monitoring meteorological drought in Texas. Agric. For Meteorol 150:330–339. http://dx.doi.org/10.1016/j.agrformet.2009.11.015
89. Raziei T, saghafian B, Paulo AA, Pereira LS, Bordi I (2009) Spatial patterns and temporal variability of drought in Western Iran. Water Res Manag 23:439–455. http://dx.doi.org/10.1007/s11269-008-9282-4
90. Rhee J, Im J, Carbone GJ (2010) Monitoring agricultural drought for arid and humid regions using multi-sensor remote sensing data. Remote Sens Environ 114:2875–2885. http://dx.doi.org/10.1016/j.rse.2010.07.005
91. Richman MB (1986) Rotation of principal components. J Climatol 6(2):293–335
92. Rienecker MM, Suarez MJ, Gelaro R, Todling R, Bacmeister J, Liu E, Bosilovich MG, Schubert SD, Takacs L, Kim G-K, Bloom S, Chen J, Collins D, Conaty A, da Silva A, Gu W, Joiner J, Koster RD, Lucchesi R, Molod A, Owens T, Pawson S, Pegion P, Redder CR, Reichle R, Robertson FR, Ruddick AG, Sienkiewicz M, Woollen J (2011) MERRA: NASA's modern-era retrospective analysis for research and applications. J Clim 24(14):3624–3648. http://dx.doi.org/10.1175/JCLI-D-11-00015.1
93. Rodell M, Houser PR, Jambor U, Gottschalck J, Mitchell K, Meng CJ, Arsenault K, Cosgrove B, Radakovich J, Bosilovich M, Entin JK, Walker JP, Lohmann D, Toll D (2004) The global land data assimilation system. Bull. Am. Meteorol. Soc. 85(3):381–394. http://dx.doi.org/10.1175/BAMS-85-3-381

94. Rojas O, Vrieling A, Rembold F (2011) Assessing the drought probability for agricultural areas in Africa with coarse resolution remote sensing imagery. Remote Sens Environ 115:343–352. http://dx.doi.org/10.1016/j.rse.2010.09.006
95. Rouault M, Richard Y (2003) Intensity and spatial extension of drought in South Africa at different time scales. Water SA 29(4):489–500. http://dx.doi.org/10.4314/wsa.v29i4.5057
96. Rousel JW, Haas RH, Schell JA, Deering DW, Harlan JC (1974) Monitoring the vernal advancement and retrogradation (green wave effect) of natural vegetation. NASA/GSFC Type III Final Report. (Greenbelt, MD)
97. Rui H, McNally A (2016) FEWS NET land data assimilation system Version 1 (FLDAS-1) Products README. NASA/GSFC/HSL 1–18. (Retreaved from ftp://hydro1.sci.gsfc.nasa.gov/data/s4pa/FLDAS/FLDAS_NOAH01_A_EA_M.001/doc/, on September 23 2016)
98. Santos JF, Pulido-Calvo I, Portela MM (2010) Spatial and temporal variability of drought in Portugal. Water Res Res 46(3). http://dx.doi.org/10.1029/2009WR008071. (n/a–n/a, w03503)
99. Schneider U, Becker A, Finger P, Meyer-Christoffer A, Ziese M, Rudolf B (2014) GPCC's new land surface precipitation climatology based on quality-controlled in situ data and its role in quantifying the global water cycle. Theor Appl Climatol 115(1):15–40. http://dx.doi.org/10.1007/s00704-013-0860-x
100. Schönemann PH (1958) Varisim: a new machine method for orthogonal rotation. Psychometrika 31(2):235–248. http://dx.doi.org/10.1007/BF02289510
101. Shapiro SS, Wilk MB, Chen HJ (1968) A comparative study of various tests for normality. J Am Stat Assoc 63(324):1343–1372
102. Sheffield J, Wood EF (2008) Global trends and variability in soil moisture and drought characteristics, 1950–2000, from observation-driven simulations of the terrestrial hydrologic cycle. J Clim 21(3):432–458
103. Shukla S, McNally A, Husak G, Funk C (2014) A seasonal agricultural drought forecast system for food-insecure regions of East Africa. Hydrol Earth Syst Sci 18(10):3907–3921. http://dx.doi.org/10.5194/hess-18-3907-2014
104. Sigdel M, Ikeda M (2010) Spatial and temporal analysis of drought in Nepal using standardised precipitation index and its relationship with climate indices. J Hydrol Meteorol 7(1):59–74. http://dx.doi.org/10.3126/jhm.v7i1.5617
105. Spinage C (2012) African Ecology: Benchmarks and historical perspectives. Springer geography. Springer, Berlin Heidelberg
106. Stampoulis D, Andreadis KM, Granger SL, Fisher JB, Turk FJ, Behrangi A, Ines AV, Das NN (2016) Assessing hydro-ecological vulnerability using microwave radiometric measurements from WindSat Remote Sens Environ 184:58–72. http://dx.doi.org/10.1016/j.rse.2016.06.007
107. Steduto P, Hsiao TC, Fereres E, Raes D (2012) Crop yield response to water. FAO Irrigation and Drainage paper No. 66. FAO, Rome. (Retreaved on March 15, 2015)
108. Svoboda M, Hayes M, Wood D (2012) Standardized precipitation index user guide. World meteorological organization, WMO - No. 1090. (Retreaved on March 15, 2015)
109. Swenson S, Wahr J (2009) Monitoring the water balance of Lake Victoria, East Africa, from space. J Hydrol 370:163–176. http://dx.doi.org/10.1016/j.jhydrol.2009.03.008
110. Tapley B, Belabour S, Watkins M, Reigber C (2004) The gravity recovery and climate experiment: mission overview and early results. Geophys Res Lett 31:1–4. http://dx.doi.org/10.1029/2004GL019920
111. Tierney JE, Smerdon JE, Anchukaitis KV, Seager R (2013) Multidecadal variability in East African hydroclimate controlled by the Indian Ocean. Nature 493:389–392. http://dx.doi.org/10.1038/nature11785
112. Tucker C, Pinzon J, Brown M, Slayback D, Pak E, Mahoney R, Vermote E, El Saleous N (2005) An extended AVHRR 8-km NDVI dataset compatible with MODIS and SPOT vegetation NDVI data. Int J Remote Sens 26(20):4485–4498. http://dx.doi.org/10.1080/01431160500168686
113. Tucker CJ (1979) Red and photographic infra-red linear combinations for monitoring vegetation. Remote Sens Environ 8:127–150. http://dx.doi.org/10.1016/0034-4257(79)90013-0

114. van den Dool H, Huang J, Fan Y (2003) Performance and analysis of the constructed analogue method applied to U.S. soil moisture over 1981–2001. J Clim 108(D16): 8617. http://dx.doi.org/10.1029/2002JD003114
115. van Huffel S (1997) Recent advances in total least squares techniques and errors-in–variables modeling. In: SIAM Proceedings in applied mathematics series. Society for Industrial and Applied Mathematics
116. Verdin J, Funk C, Senay G, Choularton R (2005) Climate science and famine early warning. Philos Trans R Soc Lond B: Biol Sci 360(1463):2155–2168. http://dx.doi.org/10.1098/rstb.2005.1754
117. Viste E, Korecha D, Sorteberg A (2013) Recent drought and precipitation tendencies in Ethiopia. Theor Appl Climatol 112:535–551. http://dx.doi.org/10.1007/s00704-012-0746-3
118. von Storch H, Zwiers FW (1999) Statistical analysis in climate research. Cambridge University Press, Cambridge
119. Wahr J, Molenaar M, Bryan F (1998) Time variability of the Earth's gravity field: hydrological and oceanic effects and their possible detection using GRACE. J Geophys Res Solid Earth 103(B12):30205–30229. http://dx.doi.org/10.1029/98JB02844
120. Wilks DS (2006) Statistical methods in atmospheric sciences, 2nd edn. Academic, Amsterdam
121. Williams AP, Funk C (2011) A westward extension of the warm pool leads to a westward extension of the Walker circulation, drying eastern Africa. Clim Dyn 37(11):2417–2435. http://dx.doi.org/10.1007/s00382-010-0984-y
122. Williams AP, Funk C, Michaelsen J, Rauscher SA, Robertson I, Wils THG, Koprowski M, Eshetu Z, Loader NJ (2012) Recent summer precipitation trends in the Greater Horn of Africa and the emerging role of Indian Ocean Sea surface temperature. J Clim Dyn 39:2307–2328. http://dx.doi.org/10.1007/s00382-011-1222-y
123. Wold S, Sjstrm M, Eriksson L (2001) PLS-regression: A basic tool of chemometrics. Chemom Intell Lab Syst 58(2):109–130. http://dx.doi.org/10.1016/S0169-7439(01)00155-1
124. Wouters B, Bonin JA, Chambers DP, Riva REM, Sasgen I, Wahr J (2014) GRACE, time-varying gravity, earth system dynamics and climate change. Rep Prog Phys 77. http://dx.doi.org/10.1088/0034-4885/77/11/116801. (41pp)
125. Wu H, Hayes MJ, Weiss A, Hu Q (2001) An evaluation of the standardized precipitation index, the China-Z index and the statistical Z-Score. Int J Climatol 21:745–758. http://dx.doi.org/10.1002/joc.658
126. Yang W, Seager R, Cane MA (2014) The East African long rains in observations and models. J Clim 27:7185–7202. http://dx.doi.org/10.1175/JCLI-D-13-00447.1
127. Yang Y, Long D, Guan H, Scanlon BR, Simmons CT, Jiang L, Xu X (2014) GRACE satellite observed hydrological controls on interannual and seasonal variability in surface greenness over mainland Australia. J Geophys Res Biogeo 119:2245–2260. http://dx.doi.org/10.1002/2014JG002670
128. Yilmaz MT, Anderson MC, Zaitchik B, Hain CR, Crow WT, Ozdogan M, Chun JA, Evans J (2014) Comparison of prognostic and diagnostic surface flux modeling approaches over the Nile River basin. Water Res Res 50(1):386–408. http://dx.doi.org/10.1002/2013WR014194
129. Yuan P, Hunegnaw A, Alshawaf F, Awange J, Klos A, Teferle FN, Kutterer H (2021) Feasibility of ERA5 integrated water vapor trends for climate change analysis in continental Europe: An evaluation with GPS (1994–2019) by considering statistical significance. Remote Sens Environ 260. https://doi.org/10.1016/j.rse.2021.112416
130. Ziese M, Schneider U, Meyer-Christoffer A, Schamm K, Vido J, Finger P, Bissolli P, Pietzsch S, Becker A (2014) The GPCC drought index a new, combined and gridded global drought index. Earth Syst Sci Data 6(2):285–295. http://dx.doi.org/10.5194/essd-6-285-2014

Chapter 15
Drought Monitoring: Topography and Gauge Influence

> The failure of drought studies in the region to analyze and/or incorporate the impacts of topography and/or gauge density on the analysis results could lead to confusion and/or reduced confidence on drought analysis results due to inconsistencies between various indicators. These inconsistencies could arise from propagation of varying impacts of topography and gauge density on different products during drought characterization.—[6].

15.1 Summary

The negative impact of Upper Greater Horn of Africa's (UGHA) complex topography on drought characterization exacerbated by gauge density and model forcing parameters has not been investigated. In order to fill this gap, this chapter presents the work of [6] who employ a combination of remotely sensed, in-situ, and model products (1982–2013); precipitation (CHIRPS, GPCC, and CHIRP), soil moisture (ERA-Interim, MERRA-2, CPC, GLDAS, and FLDAS), vegetation condition index (VCI), and total water storage products (GRACE and MERRA-2) to (i) characterize drought, (ii) explore the inconsistencies in areas under drought due to *topographical variations*, *gauge density*, and *model forcing parameters*, and (iii), assess the effectiveness of various drought indicators over Ethiopia (a selected UGHA region with unique topographical variation). A 3-month time scale that sufficiently captures agricultural drought is employed to provide an indirect link to food security situation in this rain-dependent region. The spatio-temporal drought patterns across all the products are found to be dependent on topography of the region, at the same time, the inconsistencies in characterizing drought is found to be mainly driven by topographical variability (directly) and gauge density (inversely) for precipitation products while for soil moisture products, precipitation forcing parameters plays a major role. In addition, the inconsistencies are found to be higher under extreme and moderate droughts than severe droughts. The mean differences in the percentage of

J. Awange, *Food Insecurity & Hydroclimate in Greater Horn of Africa*,
https://doi.org/10.1007/978-3-030-91002-0_15

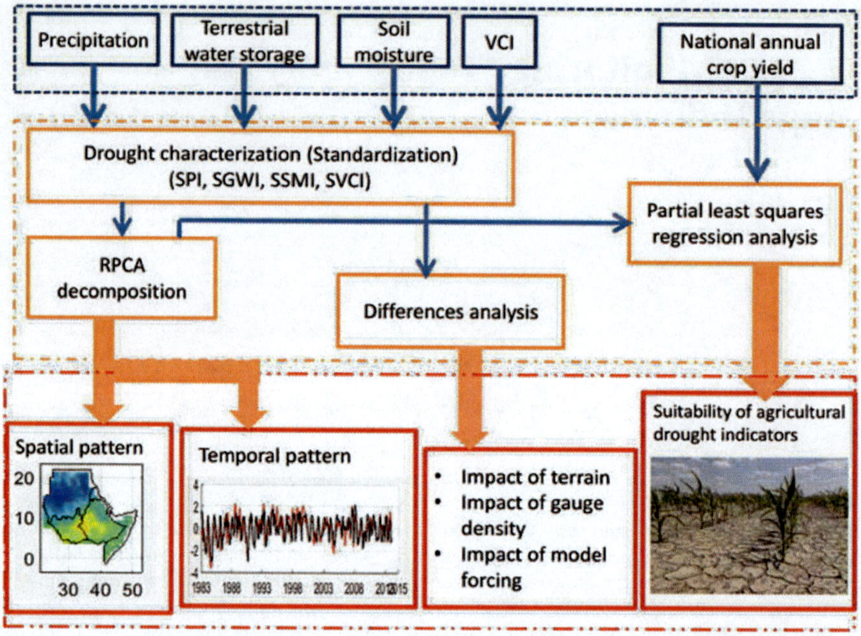

Fig. 15.1 Graphic abstract summarizing the chapter's work. *Source* [6]

areas under drought and different drought intensities over the region are on average 15.87% and 6.16% (from precipitation products) and 12.65% and 5.20% (from soil moisture products), respectively. On the effectiveness of various indicators, for the duration studied, the following are found to be most suitable over Ethiopia; VCI, GPCC, ERA, CPC, and FLDAS. These results are critical in putting into perspective drought analysis outcomes from various products. Figure 15.1 summarizes the chapter's findings.

15.2 Topographical and Rain Gauge Distribution

A majority of the population of the Upper Greater Horn of Africa (UGHA; Djibouti, Somalia, Ethiopia, Sudan, Eritrea, and South Sudan) relies heavily on subsistence agriculture, an activity that has been frequently impacted by droughts (see Chap. 1), leaving the population vulnerable to famine [43, 52, 78]. Further consequences of drought in UGHA include; population displacement, rise in food prices, malnutrition and health related complications [32, 52, 100, 106]. In addition, due to the fact that agriculture is largely rain-fed in Ethiopia and combination of rain-fed and irrigated in Sudan and Somalia, e.g., [15, 34, 65], drought occurrence frequently leads to large-scale crop failures and losses of livestock. The magnitude and severity of the drought impacts in the region emphasize the need for drought indicators that provide a clear, accurate, and consistent picture of the drought extent.

Several studies have characterized and analysed drought impacts in the region from meteorological, agricultural to hydrological, see e.g., [8, 9, 30, 32, 43, 44, 60, 68, 77, 106]. These studies have analyzed droughts based on soil moisture (model and re-analysis), rainfall (satellite-derived, in-situ, and a combination of both), Gravity Recovery and Climate Experiment (GRACE) terrestrial water storage (TWS), and normalised difference vegetation index (NDVI). Due to the limitations associated with in-situ rainfall, see e.g., [76, 77, 89], most of the studies have preferred satellite or satellite—in-situ merged products which provide homogeneous, consistent, and wide coverage, e.g., [22]. However, these satellite precipitation and the other aforementioned products do have inherent errors and uncertainties, e.g., errors arising from retrieval algorithms, data acquisition and post-processing, estimation from cloud top reflectance, model limitations, and infrequent satellite overpasses, see e.g., [2, 22, 50, 76, 89]. Even though the above uncertainties can be reduced through standardization (e.g., Standardized Precipitation Index) or percentiles over time [71], the impacts of UGHA topography on these products and subsequent propagation of these impacts on agricultural drought characterization is largely unknown.

The complex terrain-precipitation relationship in the UGHA region, especially over Ethiopia, has been reported in several research, e.g., [25, 26, 90]. Terrain has been found to be a major factor influencing rainfall distribution, and as such, has been a problem to both satellite-derived and gauge gridded (dependent on gauge density and topography) rainfall products. Satellite derived rainfall products employ either infrared or passive microwave techniques to determine the amount of rainfall [25, 26], both of which are impacted by topography especially for convective rainfall as is found in the UGHA. Soil moisture being an integration of rainfall anomalies over time [95] is expected to be influenced by this complex terrain-precipitation relationship.

Furthermore, the quality of model/reanalysis soil moisture is majorly influenced by the precipitation product it is forced by, and the individual model characteristics such as different operating soil wetness thresholds, i.e., differing variances and mean values; and critical hydrological thresholds, e.g., beginning of surface run-off or levels of evaporation at the potential rates [28]. Despite the significant impacts of topography and gauge density on rainfall and soil moisture products, majority of drought studies in the UGHA region (including the aforementioned) using these products have not addressed their impacts on agricultural drought analysis results.

The failure of drought studies in the region to analyze and/or incorporate the impacts of topography and/or gauge density on the analysis results could lead to confusion and/or reduced confidence on drought analysis results due to inconsistencies between various indicators. These inconsistencies could arise from propagation of varying impacts of topography and gauge density on different products during drought characterization. To this end, this chapter considers the impacts of *terrain*, *gauge density*, and *model forcings* on agricultural drought analysis using soil moisture, vegetation condition index, rainfall, and terrestrial water storage products over the period 1982–2013. Specifically, the chapter explores (i) spatio-temporal agricultural drought patterns, (ii) the inconsistencies in areas under agricultural drought due to topographical variations, gauge density, and model forcing parameters, and (iii), the effectiveness of various drought indicators in capturing agricultural drought over Ethiopia as evidenced from national annual crops such as maize and wheat. The first two objectives are carried out over the entire UGHA region while the third objective is carried out only over Ethiopia. This is because Sudan, South Sudan, and Somali carry out both rain-fed and irrigated agriculture [34, 63] and therefore their annual crop production does not tally with natural water changes in the environment as represented by the indicators.

The remainder of the chapter is organized as follows. In Sect. 15.3, a brief description of UGHA and the drought indicator products used are presented, followed by the drought characterization statistical approaches in Sect. 15.4. In Sect. 15.5, the results and discussion are presented. Finally, the chapter is concluded in Sect. 15.8.

15.3 Climatology of Upper GHA and Drought Indicators

15.3.1 The Upper GHA

The UGHA region comprises of Djibouti, Somalia, Ethiopia, Sudan, Eritrea, and South Sudan (Fig. 15.2). Ethiopia has a complex topography (consisting mostly of high plateaus and chain of mountains) highlighted by the Great Rift Valley, which divides the country roughly along the centre in the northeast to southwest direction [90] (see also Chap. 6). Ethiopian highlands occur on both sides of the Great Rift Valley with the highest elevation on these highlands being in excess of 4500 m (Fig. 15.2a). The topography on the southeastern side of the highlands descend and levels off towards the border with Somalia (Fig. 15.2a; [90]). Topography plays a major role on Ethiopian climate resulting in a diverse microclimate ranging from cold highlands to hot desert over southeastern lowlands [26]. Ethiopia has annual rainfall ranging from over 2000 mm over the highlands to about 300 mm over the low semi-arid lowlands (Fig. 15.2b), and has seen a rise in frequency of drought from one in every 10–15 years a century ago to one in every 5 years or less in 2010 [32, 73]. Sudan and South Sudan consist mostly of flat and/or undulating plains with a south-north gradient (Fig. 15.2a). It has annual rainfall varying from over 1200 mm in the extreme south west, reducing gradually to approximately 25 mm in the North [33], with climate that varies from equatorial in the southern most parts, savannah in the center, and continental in the north [34]. Somalia has a weak bimodal rainfall with the main season running from April to July and a second and/or minor season running from September to November [65]. Agriculture is both rain-fed and irrigated, with crops grown during both the main and minor rainfall seasons. The major rain producing feature in the UGHA region is the Inter Tropical Convergence Zone (ITCZ; [26]). While the climate in the UGHA region is dependent on altitude, the vegetation types

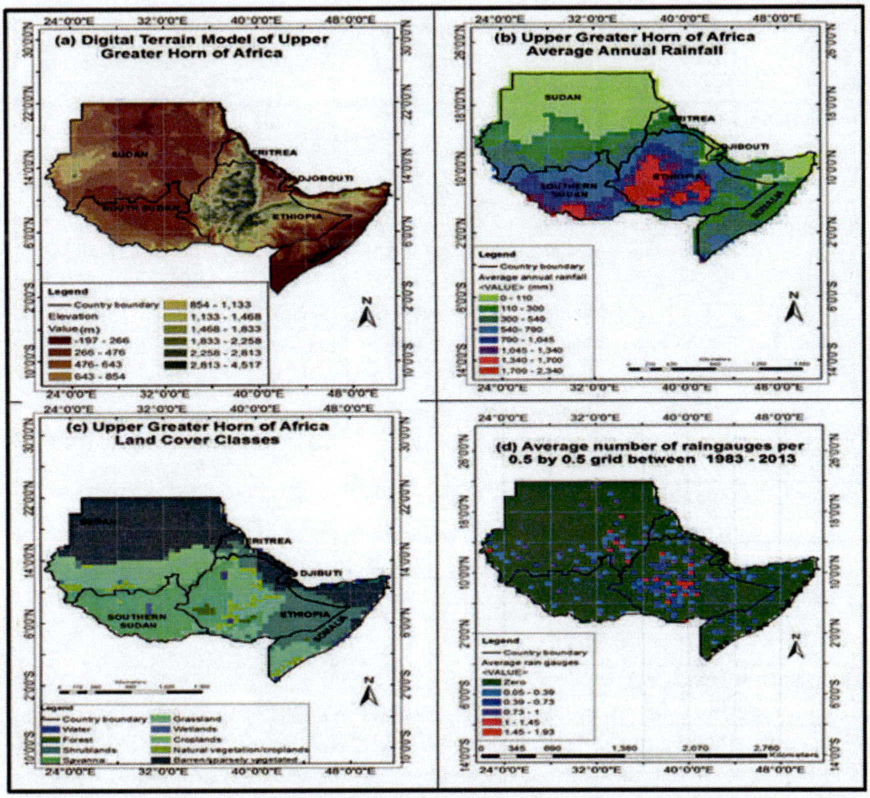

Fig. 15.2 The Upper Greater Horn of Africa (UGHA); **a** Elevation (derived from Shuttle Radar Topographical Mission; SRTM, *source* http://www.cgiar-csi.org/data/srtm-90m-digital-elevation-database), **b** Average annual rainfall, **c** Land cover types (modified from: http://e4ftl01.cr.usgs.gov/MOTA/MCD12Q1.051/), and **d**, Average number of rain gauges per 0.5° by 0.5° grids between 1983–2013, used in deriving Global Precipitation Climatology Centre (GPCC) rainfall. *Source* [6]

mirrors the rainfall zones, see e.g., Figs. 15.2b and c; [60]. The region has relatively high concentration of rain gauges over Ethiopia highlands (Fig. 15.2d).

15.3.2 Drought Indicators: Description of the Products

Drought indicators of the Upper Greater Horn of Africa (UGHA) used in this chapter are precipitation, soil moisture, total water storage changes and vegetation products. These, together with the national annual crop production used are discussed below and summarized in Table 15.1.

Table 15.1 Overview of drought products employed in this chapter. *Source* [6]

	Data	Temporal resolution	Spatial resolution	Period used	References/Studies used
Precipitation	GPCC	Monthly	0.5° × 0.5°	1982–2013	[41, 60, 94]; [31, 116]
	CHIRPS	Monthly	0.05° × 0.05°	1982–2013	[42, 82, 97]; [71]
	CHIRP	Monthly	0.05° × 0.05°	1982–2013	[42]
Soil moisture	MERRA-2	Monthly	0.625° × 0.5°	1982–2013	[17, 18]
	ERA-Interim	Monthly	0.25° × 0.25°	1982–2013	[7, 14, 24]; [23, 30, 106]
	GLDAS	Monthly	1° × 1°	1982–2013	[8, 88, 115]; [71]
	FLDAS Noah	Monthly	0.1° × 0.1°	1982–2013	[72, 92, 115]; [5]
	FLDAS VIC	Monthly	0.25° × 0.25°	1982–2013	[72, 92, 115]; [5]
	CPC	Monthly	0.5° × 0.5°	1982–2013	[37, 104]; [28]
TWS	GRACE	Monthly	1° × 1°	2003–2013	[19, 101, 112]; [9, 20, 64]
	MERRA-2	Monthly	0.625° × 0.5°	1982–2013	[17, 18]
VCI	NDVI	15 d	5-arc-minute	1982–2013	[80, 103]; [29, 59, 89, 105]; [21, 47]

Precipitation Products

CHIRPS monthly 0.05° version 2 precipitation (1982–2013)[1] produced by merging in-situ gauge observation and satellite-based cold cloud duration (CCD) observations (see [42] for more details) is used. It was designed for seasonal drought monitoring, and has been evaluated and used by several studies over the region, e.g., [11, 42, 71, 82, 97].

In addition, Climate Hazard Group InfraRed Precipitation(CHIRP), a satellite only product based on CCD observations is used. The monthly 0.05°, precipitation (1982–2013)[2] is used to characterize drought (see, [42] for more details).

Finally, Global Precipitation Climatology Centre (GPCC) [94] monthly 0.5° version 7 precipitation for the duration 1982–2013[3] is also used. GPCC has been used in the region, e.g., in [10, 11].

Soil Moisture Products

Due to the complex terrain and high rainfall variability in the region, major soil moisture products (model and re-analysis) are used for agricultural drought analysis/characterization in order to decipher and compare the impacts of terrain on agricultural drought characterization consistency using these products. For MERRA-2 (second Modern-Era Retrospective analysis for Research and Applications), root zone soil moisture product is used while for ERA-Interim (European Centre for

[1] Obtained from ftp://ftp.chg.ucsb//.edu/pub/org/chg/products/CHIRPS-2.0/.

[2] Obtained from ftp://ftp.chg.ucsb.edu/pub/org/chg/products/CHIRP/.

[3] From ftp://ftp.dwd.de/pub/data/gpcc/html/fulldata_v7_doi_download.html.

Medium-Range Weather Forecasts Interim Re-Analysis), FLDAS (Famine Early Warning System Network (FEWS NET) Land Data Assimilation System), and GLDAS (Global Land Data Assimilation System), individual soil moisture layers from 0 to 1 m (approximately equal to root zone depth) are aggregated, and CPC (Climate Prediction Center) is used the way it is since it is a single depth bucket product.

ERA-Interim monthly soil moisture for the duration 1982–2013, at 0.25° spatial (grid) resolution[4] is used (note that the current state-of-the-art product is ERA5).

MERRA-2, an upgraded version of the MERRA reanalysis [23, 87] employing Goddard Earth Observing System Model, version 5.12.4 (GEOS 5.12.4) data assimilation system [17], is used. The monthly root zone soil moisture at spatial resolution of 0.625° by 0.5°, for the duration 1982–2013 is downloaded from NASA's website.[5]

GLDAS [88] NOAH-model variant, version 2, is used. The 1° by 1° monthly soil moisture for the period 1982–2010 is obtained from http://disc.sci.gsfc.nasa.gov/services/grads-gds/gldas. Like ERA-Interim's case, GLDAS moisture layers (0–1 m) are aggregated before agricultural drought analysis. The Noah model variant has been used widely by the atmospheric and modeling communities thus model parameters are adequately tested [71]. Besides [71], several studies (e.g., [8, 115]) have used the model over the region.

FLDAS [72, 92], Noah and VIC model variants at 0.1° by 0.1°, and 0.25° by 0.25° spatial resolutions, respectively, are used. The monthly soil moisture for the duration 1982–2013 is obtained from https://ldas.gsfc.nasa.gov/.FLDAS/FLDASdownload.php. The Noah and VIC moisture products are from FLDAS simulation run forced by CHIRPS precipitation, and soil moisture and state fields from MERRA-2. It has been used in the region by several studies, e.g., [5, 8, 71, 115], among others.

Finally, Climate Prediction Center (CPC, [37, 104]) global,version 2 soil moisture is used. The monthly 0.5° by 0.5° spatial resolution mean soil moisture, from 1982 to 2013 is downloaded.[6]

Total Water Storage (TWS)

The Centre for Space Research's (CSR) release five (RL05) monthly GRACE, see e.g., [101, 112] spherical harmonic coefficients for the duration 2003–2013[7] are used. These spherical harmonic coefficients are processed following [108] and detailed description of the processing can be found in Sect. 3.3.3 and also in [5].

The resulting GRACE-derived TWS over upper GHA comprises changes from biomass/canopy water content (assumed to be negligible), surface water, accumulated soil moisture, and groundwater. The surface water variations corresponding to Lake Tana (dominant water body in the region) is removed through integration of satellite altimetry lake level changes and lake Kernel function following [74].

---

[4] obtained from http://apps.ecmwf.int/datasets/data/interim-full-moda/levtype=sfc/.

[5] From https://gmao.gsfc.nasa.gov/reanalysis/MERRA-2/data_access/.

[6] From NOAA's Earth System Research Laboratory (http://www.esrl.noaa.gov/psd/data/gridded/data.cpcsoil.html).

[7] obtained from http://icgem.gfz-potsdam.de/ICGEM/shms/monthly/csr-rl05/.

Assuming the contribution of other surface water bodies in the region, e.g., from the Sudd wetland to be negligible, the removal of the dominant Lake Tana's surface contribution enable the remaining TWS changes to be associated with groundwater and soil moisture. Further, due to shorter duration under consideration and limited exploitation of groundwater in the region, most of the variation is assumed to occur in the unsaturated zone (soil moisture) as it responds more to climate variability than the saturated zone, e.g., [114]. The remaining components is henceforth referred to as GTWS. Its utility in drought and hydrology related analyses has been demonstrated in a number of research, see e.g., [9, 20, 64].

In addition, MERRA-2 Monthly 0.5° by 0.625° (spatial resolution) total land water storage (TWLAND) for the duration 1982–2013[8] is also employed. It is hereafter referred to as MTWS. Unlike GTWS, MTWS does not contain groundwater and canopy water content, see e.g., [86].

Vegetation Condition Index (VCI)
National Oceanic and Atmospheric Administration's Advanced Very High-Resolution Radiometer (AVHRR) long-term series NDVI data set, see e.g., [80, 103] from 1982–2013, 15 d composite at $\simeq$ 0.0833° (8 km) spatial resolution[9] is utilized in computing VCI [59]. VCI is preferred for agricultural drought analysis due to its ability to isolate and/or emphasize weather related vegetation stress, see e.g., [59, 84, 89] that correspond to water availability within the study area. It was computed following [5, 59]. AVHRR NDVI has been used extensively globally and over Africa for drought and other related studies, see e.g., [21, 29, 47, 89, 105].

National Annual Crop Production
The national annual crop (maize and wheat) production data for Ethiopia obtained from Food and Agriculture Organization (FAO) data portal (http://www.fao.org/faostat/en/#data/QC) is employed in evaluating how well the various drought indices capture agricultural droughts.

15.4 Drought Characterization: Statistical Analysis

A Standardised Index, e.g., Standardised Precipitation Index (SPI) [70] is employed to characterize agricultural drought using Rainfall (CHIRPS, CHIRP, and GPCC), soil moisture (MERRA, CPC, ERA-Interim, GLDAS, and FLDAS), and TWS (MERRA-derived). In addition, standardized anomalies (SA) [113] is used to characterize agricultural drought using GRACE-derived TWS since its (GRACE-derived TWS) length is too short to fit SPI. Both standardizations are carried out on a 3-month time frame to which SPI has been associated with agricultural drought in several research, see e.g., [33, 60, 91, 99]. The SI/SA are then decomposed through rotated principal component analysis for spatio-temporal drought patterns. All the

[8] Obtained from https://gmao.gsfc.nasa.gov/reanalysis/MERRA-2/data_access/.

[9] Obtained from http://ecocast.arc.nasa.gov/data/pub/gimms/3g.v0/.

products other than GLDAS and GRACE TWS are resampled to 1° by 1° spatial resolution before standardization.

15.4.1 Agricultural Drought Characterization

Standardized Index (SI)

Standardised precipitation index (SPI) [70] is a form of standardisation in which precipitation is expressed in terms of its deviation from the long-term mean. It is computed by fitting a parametric distribution to precipitation data then transforming parametric distribution function to the standard normal distribution with a mean of 0 and standard deviation of 1, see e.g., [38, 75]. It is one of the most preferred drought indices due to its standardised nature, simplicity (only requires one input), and flexibility of use on different time scales e.g., 1, 3, 6, 9, 12-month, e.g., [1, 99, and the references therein]. A non-parametric approach of fitting SPI, e.g., [1] is adopted due to the dependence of the SPI values on the choice of parametric distribution employed, especially at the tail end of the distribution, see e.g., [83]. This approach involves the use of empirical Gringorten plotting position formula to compute empirical probability, $p(x)$, [46] as

$$p(x_i) = \frac{i - 0.44}{n + 0.12}, \tag{15.1}$$

where $p(x_i)$ is the empirical probability, n the sample size, and i denotes the rank of non-zero precipitation data from the smallest. The empirical probability from Eq. 15.1 is then transformed into standardized index as

$$SI = \phi^{-1}(p), \tag{15.2}$$

where SI is SPI, SSI (standardised soil moisture index), SVCI (standardised vegetation condition index), or STWSI (standardised total water storage index) depending on the variable under consideration, ϕ is the standard normal distribution function, and p is the probability derived from Eq. 15.1. This chapter employs Standardized Drought Analysis Toolbox (SDAT, [38]) in computing SPI values and adopts SPI drought limit categories of [4] in Table 15.2. For this chapter, a drought episode begins any time SPI is continuously less than −0.84 for a period of at least three months, and ends when SPI value exceeds −0.84 (see [9, 13, 73] for alternative definition). The various drought intensities (moderate, severe, and extreme) are then said to occur when the values in Table 15.2 are attained.

Standardized Anomalies (SA)

SA [113], also referred to as Z-scores, is a form of standardization in which a variable is expressed in terms of its deviation from its long-term mean. It is computed by dividing the monthly anomalies by the multi-year standard deviation, see e.g., [5, 79] for formulation. It has been applied to study droughts in various studies, e.g., [3, 57, 67, 113]. Due to the short duration of GTWS, which does not permit standardization

Table 15.2 Drought intensities as per SPI values [4]. *Source* [6]

SPI	Drought category
>1.65	Extremely wet
>1.28	Severely wet
>0.84	Moderately wet
>−0.84 and < 0.84	Normal
< −0.84	Moderate drought
< −1.28	Severe drought
< −1.65	Extreme drought

using SPI approach, the 3-month time series of GTWS is standardised using Z-scores. The 3-month time series cumulation is computed in a manner similar to SPI cumulations, e.g., 3-month time series for January would involve summing from November of the previous year. Positive Z-score values represent wet conditions, zero values represent normal conditions, while negatively values represent dry conditions [113].

Principal Component Analysis (PCA)

PCA [12, 55, 66, 81, 109] has been used to study the spatio-temporal drought variability in various regions across the world, see e.g., [85, 93, 98] through the decomposition of spatio-temporal fields, e.g., SPI, SSI, STWSI, SVCI etc., into spatial and corresponding temporal patterns. For formulation, see e.g., [12, 48, 53, 62, 107].

The significant PCA decomposed SI/SA components (based on log-eigenvalue diagrams; [55] are further subjected Varimax rotation [39, 54, 56]) in order to spread the variance explained and improve interpretability. In addition, the rotation has served to localize the results from regional (GHA) to local (country) level. The resulting spatial pattern, called the scores, properly normalized represents the correlation (relationship) between the SI/SA and the rotated principal components (RPC) while the normalized RPC represents the SI/SA, see e.g., [16].

15.4.2 Agricultural Drought and Their Consistencies

The percentage of areas under agricultural drought and different drought intensities are evaluated for Ethiopia, Sudan, South Sudan, and Somalia through determination of the percentage of pixels under various agricultural drought categories following Table 15.2. Differences in percentage areas under agricultural drought and different drought intensities between various products during 1983–1984, 1987, 1999, 2009, and 2010–2011 drought events are then evaluated for each country.

In order to evaluate the impact of gauge density, the differences between CHIRPS and GPCC derived percentages of areas under drought are evaluated (Eq. 15.3). The in-situ stations in CHIRPS (global historical networks and global teleconnection networks) are all in GPCC plus other more stations, see [42, 94], though the number of stations available on monthly basis differs hence the variations in areas under drought. For the mean differences due to topography, the following differences are evaluated:

$$CHIRPS_{(\%\ area\ under\ drought)} - GPCC_{(\%\ area\ under\ drought)} = Influence\ of\ gauge\ density \tag{15.3}$$

$$CHIRP_{(\%\ area\ under\ drought)} - CHIRPS_{(\%\ area\ under\ drought)} - Influence\ of\ gauge\ density = Influence\ of\ topography \tag{15.4}$$

Equation 15.4 arises from the fact that CHIRP, a satellite only product, determines precipitation from cold cloud duration (CCD) [42] and the regional topography has been known to greatly hamper the satellite products [25, 26, 90]. CHIRPS on the other hand has gauge derived rainfall estimates in addition to CCD estimates; the inclusion of gauge derived estimates serve to correct/scale the CCD estimates to their proper values. This implies that the difference between CHIRPS and CHIRP contain the influence of both gauge and topography.

In addition, the following differences are evaluated for soil moisture products; FLDASN (FLDAS NOAH)–FLDASV (FLDAS VIC), FLDASN–MERRA2, MERRA2–ERA, and FLDASN–GLDAS. The differences between the soil moisture products are taken to explore issues related to the role of forcing precipitation vis-a-vis model differences among other factors in contributing to the inconsistencies in areas under agricultural drought.

These differences are then analyzed using ANOVA and Bonferroni multiple comparison methods, for further details see [58, 61] and [49, 102], respectively. The analyses are used to explore how the following factors contribute to the differences in percentages of areas under agricultural drought across the region; (i) gauge density, (ii) physical characteristics (topography of each country), (iii) products (rainfall vis-a-vis moisture; different rainfall and soil moisture products), and (iv), forcing precipitation and model types.

15.4.3 Effectiveness of Drought Indicators over Ethiopia

As the national annual crop production is given at country level, the SI/SA (only) over Ethiopia are decomposed through rotated principal component analyses. Then the temporal SI/SA values are extracted and regressed through partial least squares regression (PLSR; for further details, see [45, 111] on national annual crop production data (maize/wheat). Detailed explanation of this process can be found in [5]. The evaluation is carried out at 1, 3, and 6 month SI/SA time scales.

15.5 Drought Analysis

Long-term dataset (rainfall, moisture, vegetation condition index, and MTWS) are standardised following SPI, while GTWS is standardised using Z-score. The resulting standardised variables are regionalised through rotated principal component analysis in order to provide spatio-temporal agricultural drought patterns. In addition, the percentage of areas affected by agricultural droughts are analysed and the consistency of percentage of areas under drought as influenced by terrain, gauge density, and soil moisture model related parameters explored across the region. Finally, the effectiveness of various drought indicators are analyzed over Ethiopia. Only Ethiopia is considered since it is the only country in UGHA practicing rain-fed agriculture hence there is a link between crop production and drought indicators. The other countries do both rain-fed and irrigated agriculture.

15.5.1 Agricultural Drought Characterization

The regionalization of SI/SA through rotated principal component analysis results in 4 largely consistent and significant spatio-temporal patterns across majority of the agricultural drought indicators (Figs. 15.3, 15.4, 15.5 and 15.6). The geographical coverage of these spatial, from RPCA clustering of SI/SA values, based on Fig. 15.3 is summarised in Table 15.3. When assessed in relation to Fig. 15.2, the spatial patterns of the precipitation products (i.e., CHIRPS, CHIRP, and GPCC) are largely similar over the regions and follows the region's topography, while those of VCI and MTWS largely follow land cover classes, and GTWS patterns follow both rainfall and land cover classes (Fig. 15.3, compared to Fig. 15.2a–c). All the products show consistent patterns over Region 3, a region of relatively lowland flat topography with low-moderate rainfall (see, Fig. 15.2a, b). For the soil moisture products, the spatial patterns roughly resemble those of rainfall and/or land cover patterns except for CPC spatial patterns (Fig. 15.4, compared to Fig. 15.2b, c). Also, Region 4's spatial patterns are different across the soil moisture products (Fig. 15.5), with MERRA2's spatial pattern resembling that of MTWS in Region 4 (Fig. 15.4). Similar to the patterns of the rainfall group (Fig. 15.4), there is consistency among the soil moisture products over Region 3 (i.e., lowland region of low to moderate rainfall) though not as close as in the rainfall group.

The proportion of variances of the SI/SA explained by these components is tabulated in Table 15.4. The majority of the rainfall products had the lowest variability in Region 4 while the majority of the soil moisture products had the lowest variability in Region 3. Region 4, from rainfall products, comprises mostly Ethiopian highlands; a region with lots of rainfall most of the year hence low variability in SPI while Regions 3 comprises relatively dry regions with low soil moisture most of the year hence lowest variability in SSI. On the other hand, maximum variability as seen in Table 15.4 differed across products with no clear majority in any particular region.

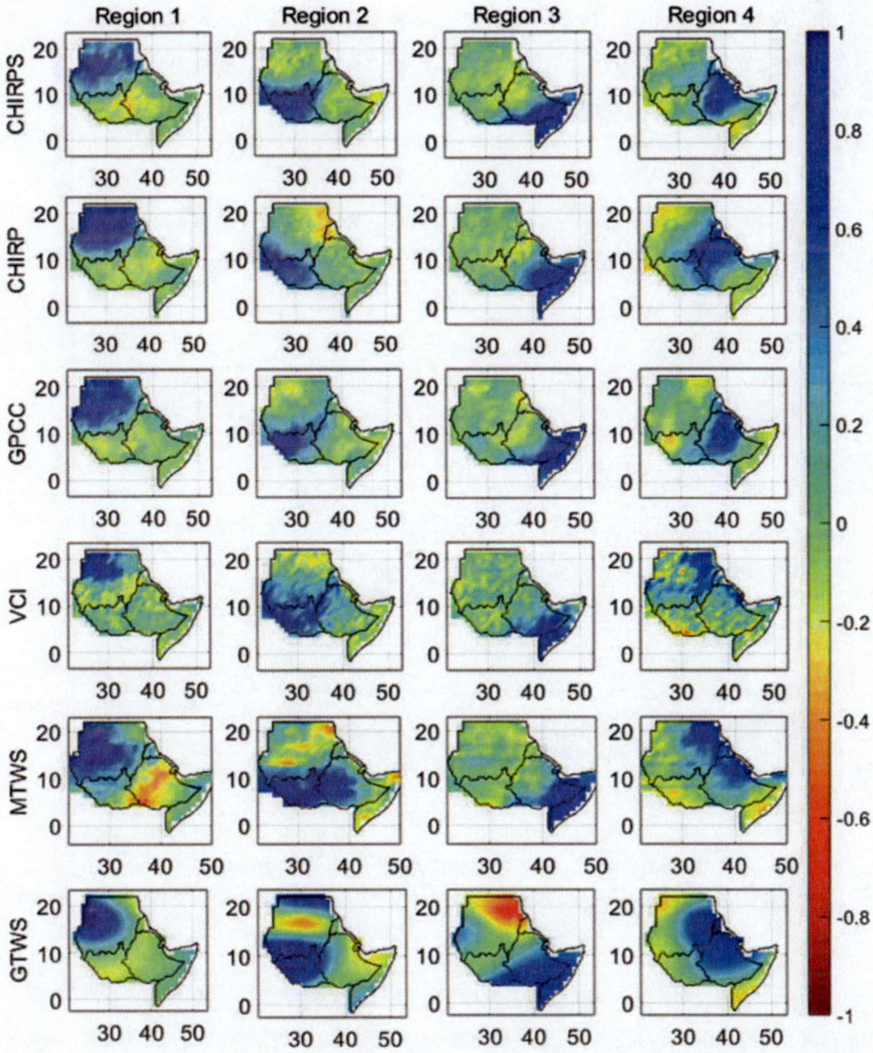

Fig. 15.3 Standardized Indices/Anomalies (SI/SA) spatial patterns. The patterns result from rotated principal component analyses decomposition of SI/SA. They are interpreted jointly with Fig. 15.5, and represent agricultural drought spatial extents whenever Fig. 15.5 falls below −0.84, as in Table 15.2, continuously for at least three months. Rows and columns represent drought indicators and regions, respectively. (Y-axis are latitudes and x-axis longitudes). *Source* [6]

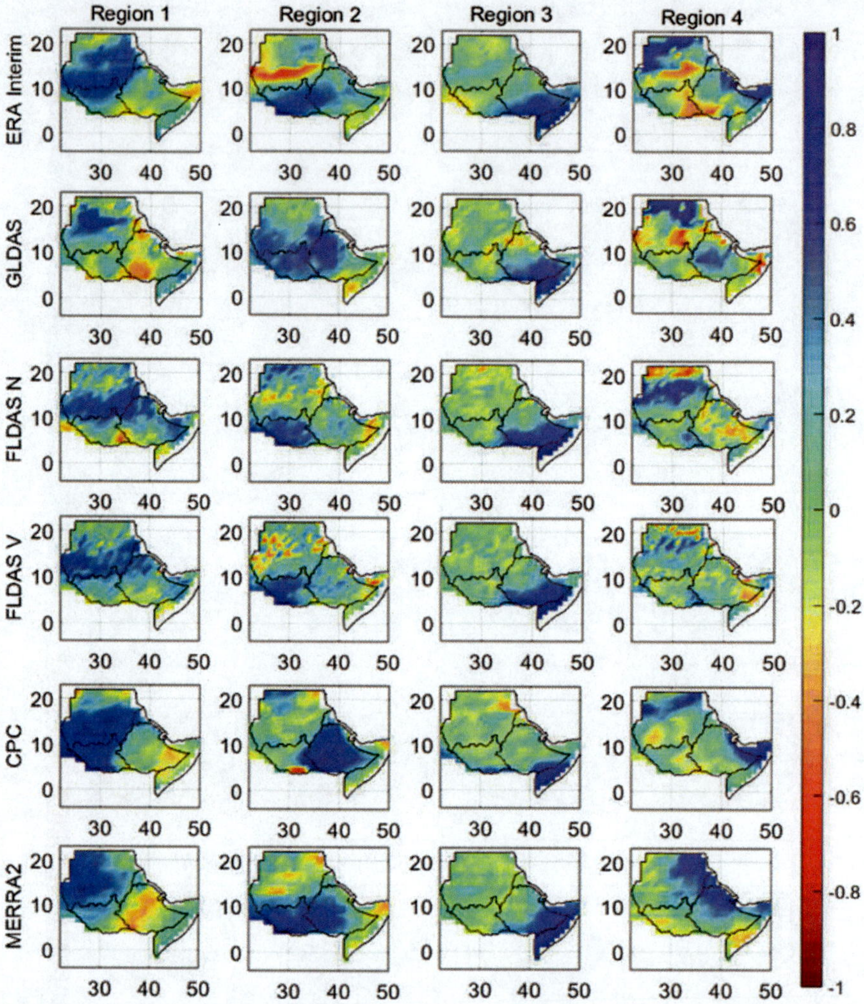

Fig. 15.4 Standardized Indices/Anomalies (SI/SA) spatial patterns. The patterns result from rotated principal component analyses decomposition of SI/SA. They are interpreted jointly with Fig. 15.6 and represent agricultural drought spatial extents whenever Fig. 15.6 falls below −0.84, as in Table 15.2, continuously for at least three months. Rows and columns represent drought indicators and regions, respectively. (Y-axis are latitudes and x-axis longitudes). *Source* [6]

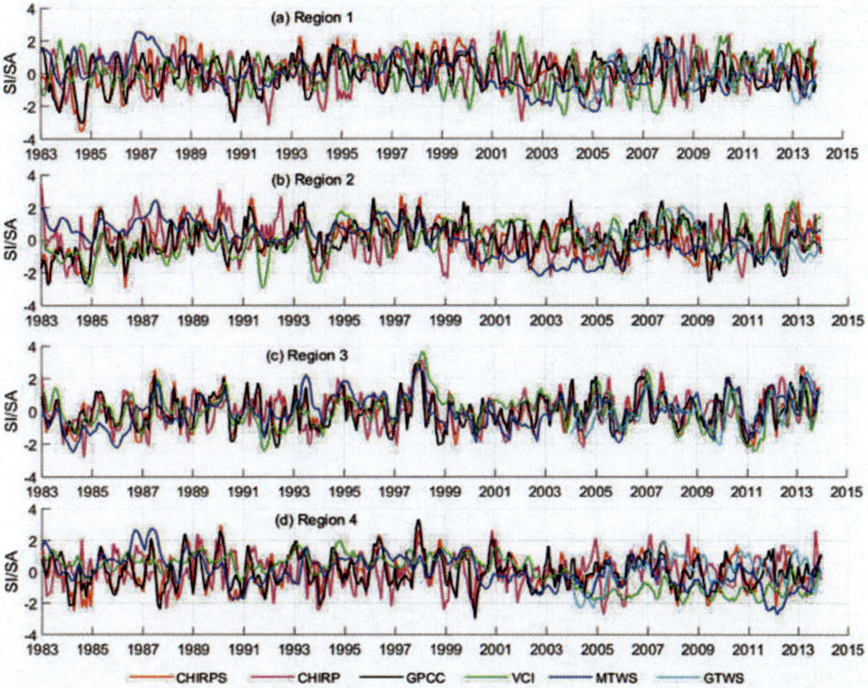

Fig. 15.5 Standardized Indices/Anomalies (SI/SA) temporal evolution. They are the corresponding temporal evolution of spatial patterns in Fig. 15.3, from rotated principal component analyses decomposition of SI/SA. Agricultural drought conditions occur when SI/SA falls below −0.84, as in Table 15.2, continuously for at least three months. (Y-axis anomaly values are unitless). *Source* [6]

Table 15.3 Approximate geographical coverage of SI/SA spatial patterns, based on Fig. 15.3. *Source* [6]

Region	Countries/areas
1	Sudan
2	South Sudan
3	Eastern Ethiopia and Somalia
4	Ethiopian highlands

VCI had Regions 1 and 3 explaining equal magnitudes of variability probably due to similar vegetation types in the vast areas of the two locations (Fig. 15.3 vis-a-vis Fig. 15.2c). On the basis of the total variabilities explained, CHIRPS and GPCC are almost equal but lower than CHIRP while the soil moisture products fall into 3 categories i.e., FLDASV (43%); GLDAS, ERA-Interim, and FLDASN (50–55%); and CPC and MERRA2 (66–67%), see Table 15.4.

Table 15.4 Proportions (%) of variances explained in each Region by each spatial pattern. Regions are as shown in Figs. 15.3 and 15.4. *Source* [6]

Region	CHIRPS	CHIRP	GPCC	VCI	MERRA TWS	GRACE TWS	ERA-Interim	GLDAS	CPC	MERRA2	FLDASN	FLDASV
Region 1	10.03	15.78	13.68	8.08	17.50	14.28	15.84	9.95	30.47	18.01	17.53	13.85
Region 2	10.16	8.86	8.41	12.07	20.45	24.35	12.72	15.97	17.52	19.70	10.66	11.27
Region 3	11.38	11.10	11.00	8.05	12.66	23.34	10.91	10.39	8.56	13.20	12.06	10.81
Region 4	8.73	11.81	8.21	10.74	14.52	21.32	16.23	14.44	10.71	15.22	11.93	7.09
Total	40.03	47.55	41.3	38.94	65.13	83.29	55.70	50.75	67.26	66.13	52.18	43.02

FLDASN—FLDAS NOAH, and FLDASV—FLDAS VIC.

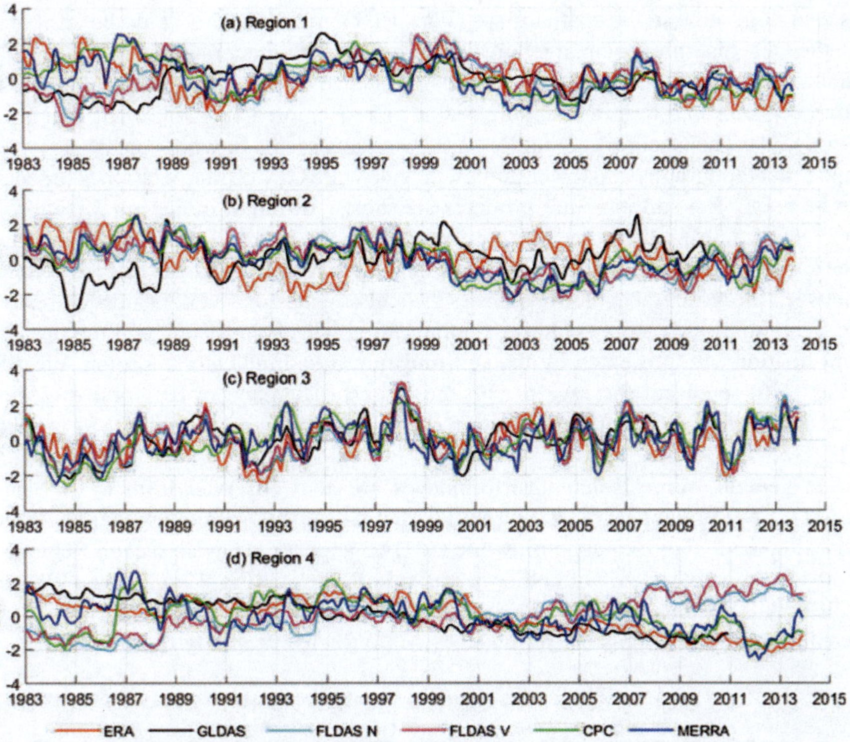

Fig. 15.6 Standardized Indices/Anomalies (SI/SA) temporal evolution. They are the corresponding temporal evolution of spatial patterns in Fig. 15.4, from rotated principal component analyses decomposition of SI/SA. Agricultural drought conditions occur when SSI falls below −0.84, as in Table 15.2, continuously for at least three months. (Y-axis anomaly values are unitless). *Source* [6]

The SI/SA temporal evolutions show region-wide agricultural drought events in the years 1983/1985, 1986/1987, 1990/1991, 2000, 2005/2006, 2009/2010, and 2010/2011 (see Figs. 15.5 and 15.6). These drought episodes and others shown in the figures are consistent with what is reported by previous studies across GHA, e.g., [34, 43, 69, 77, 106, 110].

Of the precipitation products, GPCC and CHIRPS are similar across all the regions with little differences evident in Regions 1 and 2 (Fig. 15.5a and b) though they are significantly different from CHIRP across the region (Fig. 15.5). This could be attributed to both CHIRPS and GPCC containing in-situ rainfall (rain gauge products) while CHIRP is wholly satellite derived hence differences depends on how well these products estimate rainfall across the region, which is a function of topographical changes, see e.g., [25, 26, 90]. Also, FLDASN and FLDASV are largely close across the whole region (Fig. 15.6).

Over the duration considered, all the indicators are found to be consistent in Region 3 more than the rest of the UGHA regions (Figs. 15.5 and 15.6), a pattern that is also noticeable in the spatial maps (Figs. 15.3 and 15.4). This could be attributed to the facts that this region, a relatively flat region with moderate rainfall (Fig. 15.2a and b) does not have much topographical influence on the products as compared to high rainfall/rapid terrain changing regions like Region 4 (Fig. 15.3) and Region 2 (Fig. 15.4). The satellite and model-based products seems to work consistently well under flat topography and lower rainfall ranges while GPCC interpolation also seems to be good. The soil moisture products are mostly different in Region 2 (Fig. 15.4 and Fig. 15.6b), which could be attributed to the failure of the models to capture the complex topography-rainfall relationship in the region. Also, the large difference between the temporal evolutions across the soil moisture-based indicators compared to the rainfall indicators could be a pointer to greater inconsistency in the drought information as represented by the soil moisture based indicators. Region 4 of SSI (Fig. 15.6) represent the region with the largest difference in temporal evolution possibly due to different spatial patterns as observed from the spatial SSI patterns (Fig. 15.4).

In general, some isolated performances are also observed from MTWS and GLDAS. MTWS appears to have a shift with pre-1999 being predominantly wet and dry afterward (Fig. 15.5a and b), while GLDAS appears to have issues in Regions 2 and 4 though consistent with other products in Region 3 (Fig. 15.6). Also, CPC and MERRA-2 are closer over the region. Further, there is a lag in drought events from rainfall to VCI/soil moisture followed by MTWS and eventually GTWS.

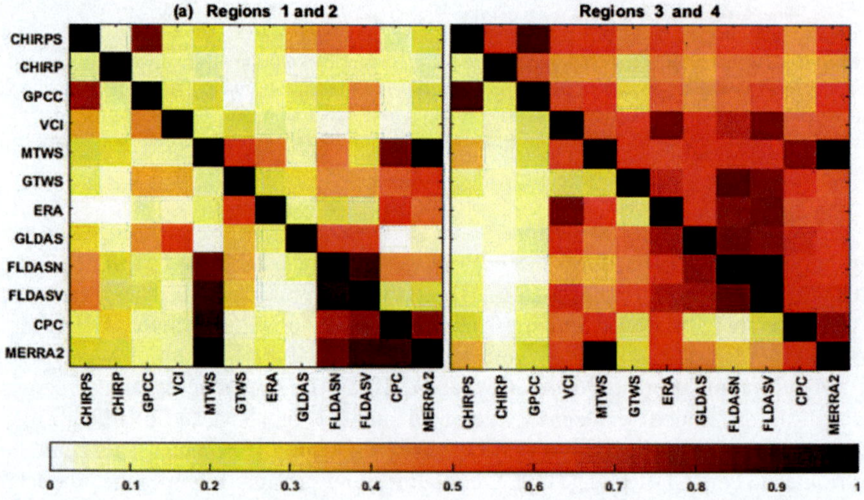

Fig. 15.7 Correlation shaded plot between drought indices by region; **a** upper triangle—region 1, lower triangle—region 2; and **b**, upper triangle—region 3, and lower triangle—region 4. On average, region 3 has higher correlations while the following products registered high correlation across the region; CHIRPS and GPCC, FLDASN and FLDASV, MTWS and MERRA2, and MTWS and CPC. *Source* [6]

Finally, the relationships between the temporal patterns of the SI/SA (Figs. 15.5 and 15.6) analyzed through Pearson correlation are shown in Fig. 15.7. From the correlation coefficients, there is high degree of consistency across the region between the following products; CHIRPS and GPCC, FLDASN and FLDASV, MTWS and MERRA-2, and MTWS and CPC. In addition, the majority of the products have on average highest correlations in Region 3 signifying consistent performance. These are consistent with the results outlined above (see Figs. 15.3, 15.4, 15.5, and 15.6). For the rainfall group, while CHIRPS and GPCC have the lowest correlation in Region 2, CHIRPS and CHIRP, and GPCC and CHIRP have the lowest correlation in Region 1. However, the soil moisture group do not have any particular region in which the majority of products had lowest correlations as values varied with product pairs and region.

15.6 Topographical and Rain-Gauge Density Influence

15.6.1 Consistency of Areas Under Agricultural Drought

In order to analyze the consistency of areas under drought from rainfall and soil moisture products, this section considers the mean differences in areas under drought between various products during 1983–1984, 1987, 1999, 2009, and 2010–2011 drought events. These drought events are considered as they had significant impacts on the region, see e.g., [34, 43, 69, 106]. In addition, spatial correlations of the respective Standardized Indices (SPIs and SSIs) during these drought events are used to support the analysis of the mean differences.

Overall, the rainfall products are found to have higher mean differences in the percentage of areas under drought than soil moisture products, i.e., 15.7% vis-a-vis 12.6% at $F(1,2338) = 28.5176$, with $p < 0.0001$. However, the reverse would be true if only gauge based precipitation products (CHIRPS and GPCC) are considered (overall mean percentage difference would be 9.7% as opposed to 15. 9%). The larger mean difference is attributed to the inclusion of CHIRP, a satellite only product that seems to have problems capturing the regional precipitation well.

From the rainfall products, the mean differences in percentage of areas under drought due to the gauge density (spatial distribution and temporal availability) is lowest in Ethiopia, followed by South Sudan and Sudan, and highest in Somalia (Fig. 15.8). This is inversely related to the gauge density of both GPCC and CHIRPS (see, Fig. 15.2d and [42]) and is further supported by relatively high correlation coefficients over Ethiopia vis-á-vis lower coefficients over Somalia (Fig. 15.8b). Finally, the mean differences in percentage of areas under drought due to topographical variations are lowest over Somalia (an area of low topographical variation), and highest over South Sudan with Ethiopia and Sudan being in between Fig. 15.8a. This is supported by relatively higher correlation coefficients over Somalia compared to the rest of the regions (Fig. 15.8c, d).

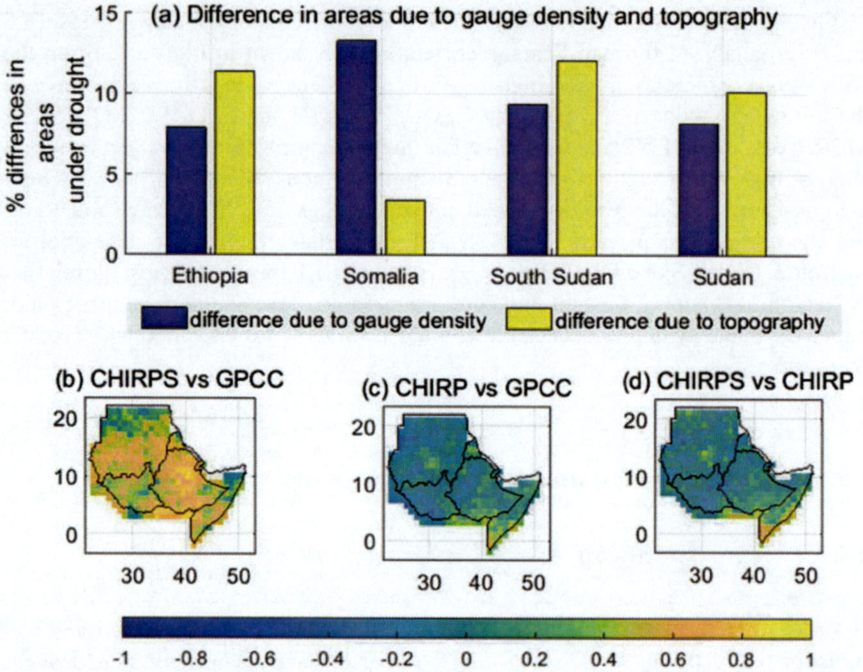

Fig. 15.8 Rainfall products; **a** Mean percentage differences in areas under drought due to gauge density and topography over UGHA, **b** Spatial distribution of SPI correlations for the 1983–1984, 1987, 1999, 2009, and 2010–2011 drought events. The correlations are a function of gauge density (**b**), and topography (**c** and **d**). *Source* [6]

For the soil moisture products, the mean differences in percentage of areas under drought between various products depend on the products under consideration ($F(3, 1292) = 61.92$, $p < 0.0001$) and the specific region (country) being considered ($F(3, 1292)=28$, $p < 0.0001$; Fig. 15.9). Different models with similar forcing precipitation translate into lower mean percentage difference in areas under drought across the entire region than similar models with different forcing precipitations (FLDASN-FLDASV vis-a-vis FLDASN-GLDAS; Fig. 15.9). This is also reflected by the higher correlation between FLDASN and FLDASV SSIs' compared to FLDASN and GLDAS SSIs' during the considered drought events (Fig. 15.9b, e). The mean difference in the percentage of areas under drought between reanalysis products is lower than that between a reanalysis and a normal model product except over Somalia (MERRA-ERA vis-a-vis FLDASN-MERRA) as also shown by higher correlations between MERRA and ERA SSI than between FLDASN and MERRA (Fig. 15.9a, c, d) during the considered drought events. The relatively higher difference in mean areas under drought between MERRA and ERA over Somalia, also reflected by low correlations (Figs. 15.9a, d) could be attributed to the overestimation tendencies of ERA-interim, especially over dry areas as was shown by [7]. The

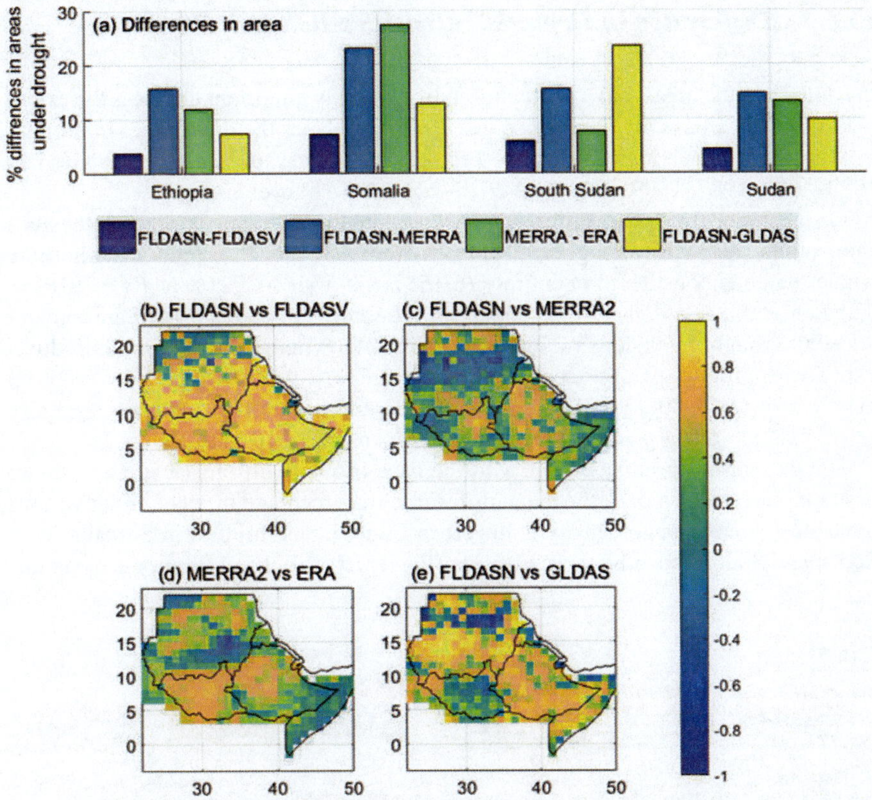

Fig. 15.9 Soil moisture products; **a** Mean percentage differences in areas under drought. Lower mean percentage differences are observed from models forced by same precipitation products (e.g., FLDASN and FLDASV), **b** Spatial distribution of SSI correlations for 1983–1984, 1987, 1999, 2009, and 2010–2011 drought events. *Source* [6]

differences involving different models with different forcing precipitation products (FLDASN-MERRA and MERRA-ERA) have the highest mean difference in the percentage of areas under drought while different models forced with similar precipitation (FLDASN-FLDASV) have the lowest mean difference in the percentage of areas under drought across the region. This is supported by higher correlation between FLDASN and FLDSV SSIs' in comparison to the relatively lower correlations between FLDASN and MERRA-2, and MERRA-2 and ERA SSIs' (Fig. 15.9b–d). Finally, concerning the regional variability of the mean percentage differences in areas under drought, Somalia has the highest mean differences while Ethiopia has the least mean differences on the majority of product differences considered (Fig. 15.9a).

15.6.2 Difference in Drought Intensities Between Products

In addition to the areas under agricultural drought, it is important to assess the extent to which each area is affected hence this section analyses the differences in drought intensity between various products of Sect. 15.6.1 for the same drought episodes i.e., 1983–1984, 1987, 1999, 2009, and 2010–2011 drought events.

On average, the mean difference in areas under different drought intensities from rainfall products is higher than those from soil moisture products when the whole region is considered in entirety (6.164% vis-a-vis 5.2021% at $F(1, 7018) = 33.91$, $p < 0.0001$; Figs. 15.10). The mean differences in percentage of areas under different drought intensities (moderate, severe, and extreme) from rainfall products varies with product pairs under consideration ($F(2, 3088) = 41.13$, $p < 0.0001$), the country ($F(3, 3088) = 4.79$, $p < 0.0001$), and the drought intensity category ($F(2, 3088) = 91.6$, $p < 0.0001$) being considered (Fig. 15.10a–d).

For the rainfall products, similar to the case of mean difference in areas under drought (see, Sect. 15.6.1), the mean difference in percentage of areas under various intensities due to gauge density is lowest in Ethiopia and highest in Somalia, with Sudan and South Sudan being in between (Fig. 15.10a–d). In addition, the mean dif-

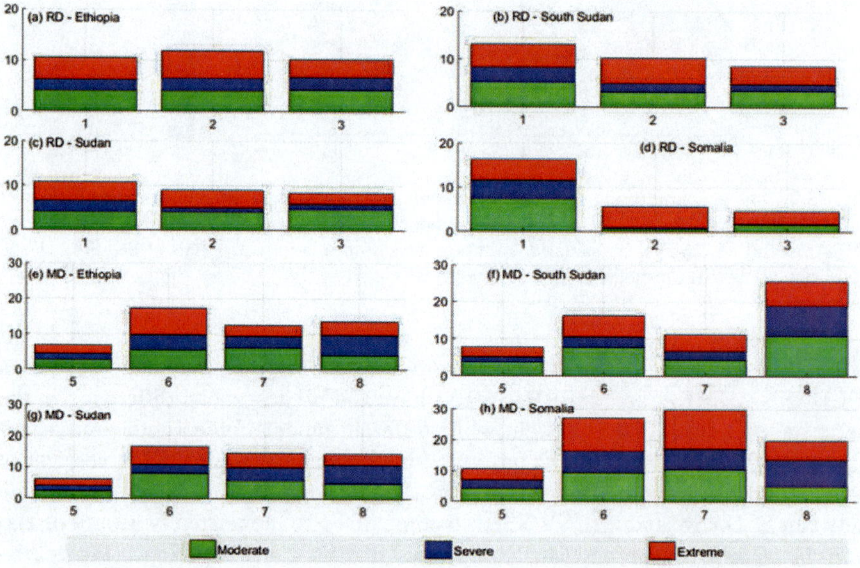

Fig. 15.10 Mean differences in percentage of areas under different drought intensities during 1983–1984, 1987, 1999, 2009, and 2010–2011 drought events; **a–d** Rainfall-derived Differences (RD) due to gauge density (1) and topography (2 and 3, are CHIRPS and GPCC derived, respectively), **e–h** Soil moisture-derived differences (MD; 5 = FLDASN–FLDASV, 6 = FLDASN–MERRA2, 7 = MERRA–ERA, and 8 = FLDASN–GLDAS, FLDASN–FLDAS NOAH, and FLDASV–FLDAS VIC). *Source* [6]

ference in percentage of areas under various drought intensities due to topography is lowest over Somalia and largely the same over the remaining countries (Fig. 15.10a–d).

Similar to the case of the precipitation products, the mean differences in percentage of areas under different drought intensities from soil moisture products are dependent on products pairs ($F(3, 3915) = 77.16$, $p < 0.0001$), the country ($F(3, 3915) = 49.5$, $p < 0.0001$), and the intensity category ($F(2, 3915) = 30.06$, $p < 0.0001$) under consideration (Fig. 15.10e–h). The mean differences in the percentage of areas under different drought intensities are lower for FLDASN-FLDASV compared to the rest of the soil moisture pairs across the regions. The mean differences in the percentage of areas under various drought intensities are lowest in Ethiopia and highest over Somalia (Fig. 15.10 e and h). Like in the case of rainfall products, the mean differences in areas under different drought intensities are lower under severe droughts compared to moderate and extreme, across the countries for all the soil moisture product differences except for FLDASN–GLDAS (8 in Fig. 15.10). As with the case of differences in the percentage of areas under drought, similar precipitation forcing with different models contributes to lower differences in intensity information compared to similar models with different forcing precipitations (FLDASN-FLDASV vis-a-vis FLDASN-GLDAS; Fig. 15.10e–h).

15.6.3 Agricultural Drought: Effectiveness of the Indicators

This section extends the work of [5], who evaluated the effectiveness of drought indicators in capturing agricultural drought over East Africa (Kenya, Uganda, and Tanzania; see Chap. 14), to Ethiopia. Though the rest of the study involved the whole of UGHA, this section is only limited to Ethiopia as Sudan, South Sudan, and Somali carry out both rain-fed and irrigated agriculture [34, 63] and therefore their annual crop production does not tally with natural water changes in the environment as represented by the indicators.

Both SI (computed using approximately 30 years length of data) and SA (computed using approximately 10 years length of data) are regressed with national annual crop (maize and wheat) production, and the model with least mean prediction error has its proportion of variability explained (R^2) reported (Fig. 15.11). As in [5], SA is necessitated by the need to compare how GTWS performs in relation to other indicators.

Based on SI regression with national annual crop production, all the indicators explain over 50% of variability in the national annual crop production (in most instances) except CHIRP and GLDAS (Fig. 15.11a). These results are largely consistent with those of [15, 40] who found a good correlation between crops (teff, wheat, and maize) production and monthly rainfall anomalies over Ethiopia. VCI performs exceptionally well explaining between 67% and 93% of national annual crop production variability. GPCC explains higher percentage of national annual crop production variability than CHIRPS, which explaines higher variability than

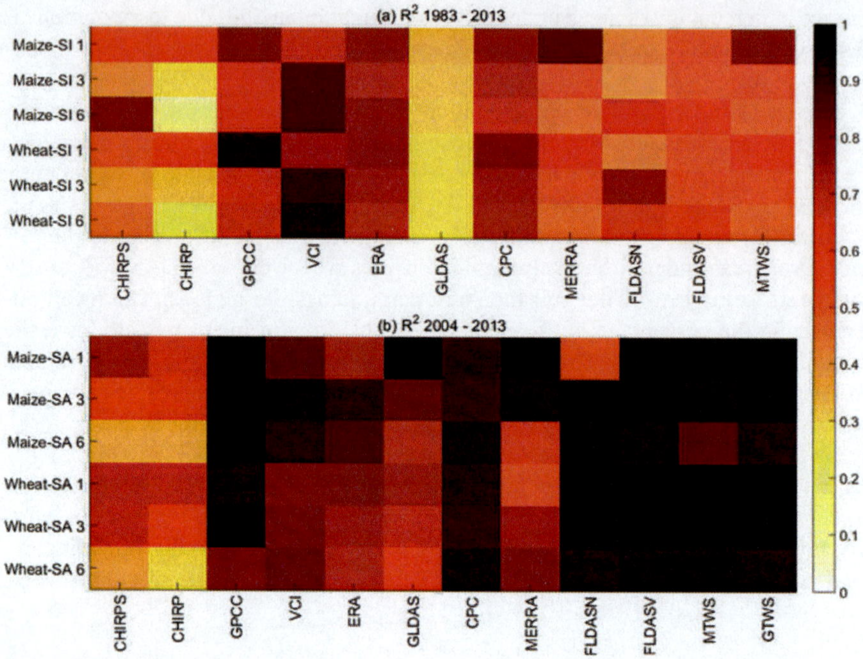

Fig. 15.11 Ethiopian national annual crop (maize and wheat) production variability (R^2) reflected by various drought indices. VCI, GPCC, ERA, CPC, FLDASN, and FLDASV explained relatively higher crop production variabilities, while GLDAS explained lowest variability. The y-axis indicates the crop (maize/wheat), SI (for **a**), and SA (for **b**) while 1, 3 and 6 indicate the standardization time scales for the indicators on the x-axis. *Source* [6]

CHIRP. In addition, ERA and CPC are seen to perform better than the rest, FLDASN and FLDASV explains largely equal variabilities while GLDAS explains the least national annual crop production variability.

SA regression generally explains higher annual crop production variabilities than SI (Fig. 15.11b), possibly due to the shorter duration considered. GTWS performs exceptionally well explaining between 92% and 99% of the variability in national annual crop production. In addition, FLDASN, FLDASV, CPC, VCI, and GPCC also explain high variabilities. Like in SI regression, GPCC outperformed CHIRPS, which performed better than CHIRP while FLDASN and FLDASV explain almost equal variabilities. Also, FLDASN explain more variability than GLDAS. However, the SA regression results should be interpreted with caution due to the shorter duration of the data considered.

The regression results, as pointed out in [5], should not be generalized to other areas and time periods since the association of production (yield) with climatic conditions only holds if production factors, e.g., areas under cultivation, and farm management practices, remain constant.

15.7 Summarized Overview

The spatial patterns of rainfall, VCI, TWS, and majority of the soil moisture products derived drought indicators have been seen to mimic rainfall distribution and/or land cover patterns (Figs. 15.3 and 15.4 vis-a-vis Fig. 15.2b–c) both of which are majorly influenced by topographical changes (terrain variations). Topography is a major factor influencing rainfall distribution, see e.g., [25, 26, 90] and to a large extent the land cover classes over the region [60]. The spatial variability of soil moisture is closely related to the scales under consideration, i.e., at large scales, soil moisture spatial variability is associated with soil types, topography, and vegetation types while at small scales, precipitation and evapotranspiration play major roles [36, 51]. This explains the closeness of the majority of the SSI spatial patterns to the land cover classes and/or spatial rainfall distribution (Fig. 15.4 compared to Fig. 15.2a–c).

Region 3 (southeastern Ethiopia and Somalia) had high consistency in spatial and temporal patterns, largely equal percentage of variabilities explained by rainfall and moisture products, and high average correlation among different products (Figs. 15.3, 15.4, 15.5c, 15.6c and 15.7, and Table 15.4). These could be attributed to relatively flats topography of the region coupled with low rainfall. The combination of flat topography and low rainfall does not seem to give much challenge to the different products in capturing spatio-temporal drought information from the region, hence the high consistency.

The variation in the mean differences in percentage of areas under drought due to gauge density was inversely related to the gauge density of both GPCC and CHIRPS (Figs. 15.8, 15.2d, and [42]). The mean percentage differences were lowest over Ethiopia and highest over Somalia (7.9% vis-a-vis 13.3%). Ethiopia has the highest gauge density from both GPCC and CHIRPS while Somalia has the lowest (Fig. 15.2d, and [42]). Even though GPCC and CHIRPS have in-situ gauge measurements from global historical networks and global teleconnection networks [42, 94], the fluctuation in monthly number of stations available to each (see, e.g., variation in number of available stations for CHIRPS in [42]) is the major source of this difference. In addition, this variation could also be contributed by the additional in-situ stations incorporated in GPCC but absent in CHIRPS, see [42, 94].

Other than gauge density, topographical factors were also found to contribute to the mean difference in percentage areas under drought (Figs. 15.8, 15.2a). Somalia had the lowest mean difference in percentage of areas under drought (3.4%) while Ethiopia (11.3%) and South Sudan (12%) had highest mean differences. This could be due to the fact that Somalia with low topographical variation has low influence on how various products characterize rainfall while Ethiopia and South Sudan with rapid changing topography presents difficulty to various products (Fig. 15.2a). In general, the performance of GPCC is affected by systematic errors (associated with systematic gauge-measurements) and sampling errors (associated with gauge densities), see e.g., [94]. The sampling errors are dependent on orography, season and type of rainfall, of which topography plays a large role hence differing performance across the regions. CHIRPS being a combination of satellite-based precipitation and in-situ observations, has its performance dependent on how well the infrared cold cloud duration (CCD) observed precipitation estimates correspond to the actual rainfall and how effective the merging is done, while CHIRP purely depends on CCD [42]. The degree of correspondence between the CCD estimated rainfall and the actual rainfall over a place is influence by topographical factors. It is the topography related difficulties in rainfall representation that translates to mean differences in percentage areas under drought on drought characterization employing these products. Similarly, the influence of gauge density and topographical variations on the mean differences in percentage of areas under different drought intensities (Fig. 15.10) follows the same logic as above.

The closer performance between FLDASN and FLDASV from temporal variability, higher correlations, lower mean differences in percent of areas under drought and different drought intensities, and explained amount of variabilities in crop production (Figs. 15.6, 15.9a, 15.10e–h, and 15.11) could be attributed to FLDASN and FLDASV being forced by CHIRPS rainfall. However, they are FLDAS versions driven by different models, Noah and VIC, respectively. Further, their lower mean differences compared to FLDASN and GLDAS, and the higher percentage of variabilities in crop production explained by FLDASN over GLDAS implies that forcing precipitation plays a bigger role in the resulting soil moisture than model difference. This is supported by the fact that both FLDASN and GLDASN are ouputs of the same model, Noah, but differing precipitation forcings; FLDASN is forced by CHIRPS while GLDASN is forced by Princeton global meteorological forcing data, see, e.g., [72, 96]. The role of forcing precipitation in influencing the quality of resulting soil moisture has also been recognized in previous studies, e.g., [27, 28, 35]. In addition, different model characteristics such as different operating soil wetness thresholds, i.e., differing variances and mean values; and critical hydrological thresholds, e.g., beginning of surface run-off or levels of evaporation at the potential rate [28], contribute to the differences. Therefore, the largest differences occurred when different models with different forcing precipitations were considered, e.g., FLDAS and MERRA.

Like gauge density based mean differences, the spatial distribution of the moisture mean differences in percentage of areas under drought and different drought intensities were lower over Ethiopia and higher over Somalia (Figs. 15.9a and 15.10e–h). However, for the soil moisture products, this could be linked more to the amount of rainfall and/or moisture levels and the individual model thresholds over the region but not directly on topography since Ethiopia has more topography related issues than Somalia (see Sect. 15.3.1). The influence of topography on the percentage of area differences from soil moisture product is indirectly through the forcing precipitation.

The larger mean differences in areas under moderate and extreme droughts than severe droughts observed in all the differences (both rainfall and soil moisture; Figs. 15.10) across the region could be linked to differences in extreme drought conditions, i.e., lower (moderate) and higher (extreme) as captured by these products. For the soil moisture products, it could be related to the drying thresholds implemented by different models.

Unlike in [5], over East Africa where CHIRPS performed better than GPCC, here GPCC aided by a higher density of gauge distribution (and CHIRPS impeded by low and high varying number of incorporated in-situ stations) explained higher variabilities in national annual crop production. Similar to [5], FLDAS explained higher variabilities than GLDAS while MTWS and MERRA2 registered similar performances. The shorter duration regression results explain higher amount of variability in national annual crop production than the long duration probably due to the fact that under the short duration, the factors of production especially the areas under cultivation are constant. The influence of increase in areas under cultivation could be responsible for relatively lower percentage of variabilities explained, see e.g., [100] who found a link between the increase in crop production and areas under cultivation over Ethiopia. The relationship between crop production (yields) and moisture availability (climate variables) holds subject to the area under cultivation and production factors (e.g., pesticides, fertilizers, and crop cultivars) being constant.

15.8 Concluding Remarks

The chapter characterized agricultural drought over the upper GHA countries (Ethiopia, Sudan, South Sudan, and Somalia) using rainfall (CHIRPS, CHIRP, and GPCC), moisture (ERA-Interim, CPC, GLDAS, FLDAS Noah, FLDAS VIC, and MERRA-2), and TWS (MERRA2 and GRACE) products with bias on the influence of topography and gauge density. Further, it considered the consistency of differences in the percentage of areas under drought and different drought intensities from precipitation and selected moisture products. Finally, it evaluated the effectiveness of various drought indicators in explaining agricultural drought over Ethiopia using national annual crop production. The following were the major results:

(i) The spatial-temporal drought patterns were found to be influenced by topography over the region. All the products were highly consistent over the low land low-medium rainfall regions of Eastern Ethiopian and Somalia while lowest consistency differed across products and regions.
(ii) The mean differences in percentages of areas under drought (15.87%) and different drought intensities (6.16%) from precipitation products were determined by gauge density (distribution and availability) and topography. The mean differences attributed to gauge density were low in areas of high gauge density (Ethiopia) and higher in areas of low gauge density (Somalia) while the mean differences attributed to topography were minimum in low varying topographical areas (Somalia) and higher in rapid varying topographical areas (Ethiopia and South Sudan).
(iii) The mean differences in percentages of areas under drought (12.65%) and different drought intensities (5.20%) from soil moisture products was determined by the differences in the forcing precipitation, models pairs under consideration, and the regions.

(iv) In evaluating the utility of various indicators in explaining agricultural drought over Ethiopia, the following were identified as suited for agricultural drought monitoring during the 1982–2013 period; VCI, GPCC, ERA, CPC, and FLDAS (Noah and VIC).

The information on the consistency of percentage of areas under drought and different drought intensities is critical in understanding and putting into perspective drought analysis results from different products over the region by various stakeholders. This is important for policy and decision makers as it could inform their decision on the number of people affected and the extent to which they are affected without worrying about particular product, e.g., soil moisture used in any particular analysis. These particular decisions are essential for resource mobilization, aiding mitigation of drought impacts, improving drought response plans and early warning systems, and quantifying drought impacts among others.

References

1. AghaKouchak A (2015) A multivariate approach for persistence-based drought prediction: Application to the 2010–2011 East African Drought. J Hydrol 526:127–135. https://doi.org/10.1016/j.jhydrol.2014.09.063
2. AghaKouchak A, Nasrollahi N, Habib E (2009) Accounting for Uncertainties of the TRMM Satellite Estimates. Remote Sens 1:606–619. https://doi.org/10.3390/rs1030606
3. Agnew C, Chappell A (1999) Drought in the Sahel. GeoJ 48(4):299–311. https://doi.org/10.1023/A:1007059403077
4. Agnew CT (2000) Using the SPI to identify drought. Drought Netw News 12:6–12
5. Agutu N, Awange J, Zerihun A, Ndehedehe C, Kuhn M, Fukuda Y (2017) Assessing multi-satellite remote sensing, reanalysis, and land surface models' products in characterizing agricultural drought in East Africa. Remote Sens Environ 194:287–302. https://doi.org/10.1016/j.rse.2017.03.041
6. Agutu NO, Awange JL, Ndehedehe C, Mwaniki MW (2020) Consistency of agricultural drought characterization over Upper Greater Horn of Africa (1982–2013): Topographical, gauge density, and model forcing influence. Science of the Total Environment. 709. https://doi.org/10.1016/j.scitotenv.2019.135149
7. Albergel C, de Rosnay P, Balsamo G, Isaksen L, Muñoz-Sabater J (2012) Soil moisture analyses at ECMWF: Evaluation using global ground-based *in-situ* observations. J Hydrometeorol 13(5):1442–1460. https://doi.org/10.1175/JHM-D-11-0107.1
8. Anderson WB, Zaitchik BF, Hain CR, Anderson MC, Yilmaz MT, Mecikalski J, Schultz L (2012) Towards an integrated soil moisture drought monitor for East Africa. Hydrol Earth Syst Sci 16:2893–2913. https://doi.org/10.5194/hess-16-2893-2012
9. Awange J, Khandu M, Schumacher E, Forootan Heck B (2016) Exploring hydrometeorological drought patterns over the Greater Horn of Africa (1979–2014) using remote sensing and reanalysis products. Adv Water Res 94:45–59. https://doi.org/10.1016/j.advwatres.2016.04.005
10. Awange JL, Ferreira VG, Forootan E, Khandu Andam-Akorful, SA, Agutu NO, He XF, (2016) Uncertainties in remotely sensed precipitation data over Africa. Internatonal Journal of Climatolology 36(1):303–323. https://doi.org/10.1002/joc.4346
11. Awange JL, Hu K, Khaki M (2019) The newly merged satellite remotely sensed, gauge and reanalysis-based Multi-Source Weighted-Ensemble Precipitation: Evaluation over Australia

and Africa (1981–2016). Sci Total Environ 670:448–465. https://doi.org/10.1016/j.scitotenv.2019.03.148
12. Awange JL, Palancz B, Völgyesi L (2020) Hybrid Imaging and Visualization. Employing Machine Learning with Mathematica - Python, Springer Nature International, Berlin. 978-3-030-26152-8, DOI: https://doi.org/10.1007/978-3-030-26153-5
13. Awange JL, Mpelasoka F, Goncalves RM (2016) When every drop counts: analysis of droughts in Brazil for the 1901–2013 period. Sci Total Environ 566–567:1472–1488
14. Balsamo G, Beljaars A, Scipal K, Viterbo P, van den Hurk B, Hirschi M, Betts AK (2009) A revised hydrology for the ECMWF model: verification from field site to terrestrial water storage and impact in the integrated forecast system. J Hydrometeorol 10(3):623–643. https://doi.org/10.1175/2008JHM1068.1
15. Bewket W (2009) Rainfall variability and crop production in Ethiopia: Case study in the Amhara region. In: Ege S, Aspen H, Teferra B, Bekele S (eds) Proceedings of the 16th international conference of Ethiopian studies, vol 3, pp 823–836, Department of Social Anthropology, Norwegian University of Science and Technology, Trondheim
16. Bordi I, Fraedrich K, Petitta M, Sutera A (2006) Large-scale assessment of drought variability based on NCEP/NCAR and ERA-40 Re-Analyses. Water Res Manag 20(6):899–915. https://doi.org/10.1007/s11269-005-9013-z
17. Bosilovich M, Lucchesi G, Suarez M (2016) MERRA-2: File Specification, GMAO Office Note No. 9 (Version 1.1), 73 pp, http://gmao.gsfc.nasa.gov/pubs/office_notes
18. Bosilovich, M.G., S. Akella, L. Coy, R. Cullather, C. Draper, R. Gelaro, R. Kovach, Q. Liu, A. Molod, P. Norris, W. Chao, R. Reichle, L. Takacs, R. Todling, Y. Vikhliaev, S. Bloom, A. Collow, G. Partyka, S. Firth, G. Labow, S. Pawson, O. Reale, S. Schubert, and M. Suarez (2015), Merra-2: Initial evaluation of the climate, Technical Report Series on Global Modeling and Data Assimilation NASA/TM-2015-104606/Vol. 43, NASA:GSFCG, Available online at https://gmao.gsfc.nasa.gov/reanalysis/MERRA-2/docs/
19. Chen JL, Wilson CR, Tapley BD, Ries JC (2004) Low degree gravitational changes from GRACE: validation and interpretation. Geophys Res Lett 31(22). https://doi.org/10.1029/2004GL021670, l22607. n/a–n/a
20. Chen JL, Wilson CR, Tapley BD, Yang ZL, Niu GY (2009) 2005 drought event in the Amazon River basin as measured by GRACE and estimated by climate models. J Geophys Res: Solid Earth 114(B5). https://doi.org/10.1029/2008JB006056. b05404
21. Chen T, de Jeu R, Liu Y, van der Werf G, Dolman A (2014) Using satellite based soil moisture to quantify the water driven variability in NDVI: A case study over mainland Australia. Remote Sens Environ 140:330–338. https://doi.org/10.1016/j.rse.2013.08.022
22. Damberg L, AghaKouchak A (2014) Global trends and patterns of drought from space. Theor Appl Climatol 117:441–448. https://doi.org/10.1007/s00704-013-1019-5
23. Decker M, Brunke MA, Wang Z, Sakaguchi K, Zeng X, Bosilovich MG (2012) Evaluation of the reanalysis products from GSFC, NCEP, and ECMWF using flux tower observations. J Clim 25:1916–1944. https://doi.org/10.1175/JCLI-D-11-00004.1
24. Dee, D.P., Uppala, S.M., Simmons, A.J., Berrisford, P., Poli, P., Kobayashi, S., Andrae, U., Balmaseda, M.A., Balsamo, G., Bauer, P., Bechtold, P., Beljaars, A.C.M., van de Berg, L., Bidlot, J., Bormann, N., Delsol, C., Dragani, R., Fuentes, M., Geer, A.J., Haimberger, L., Healy, S.B., Hersbach, H., Hólm, E.V., Isaksen, L., Kållberg, P., Köhler, M., Matricardi, M., McNally, A.P., Monge-Sanz, B.M., Morcrette, J.-J., Park, B.-K., Peubey, C., de Rosnay, P., Tavolato, C., Thépaut, J.-N., Vitart, F (2011) The ERA-Interim reanalysis: configuration and performance of the data assimilation system. Q J R Meteorolog Soc 137(656):553–597. https://doi.org/10.1002/qj.828
25. Dinku T, Ceccato P, Grover-Kopec E, Lemma M, Connor SJ, Ropelewski CF (2007) Validation of satellite rainfall products over East Africa's complex topography. Int J Remote Sens 28(7):1503–1526. https://doi.org/10.1080/01431160600954688
26. Dinku T, Chidzambwa S, Ceccato P, Connor SJ, Ropelewski CF (2008) Validation of high resolution satellite rainfall products over complex terrain. Int J Remote Sens 29(14):4097–4110. https://doi.org/10.1080/01431160701772526

27. Dirmeyer PA, Dolman AJ, Sato N (1999) The pilot phase of the global soil wetness project. Bull Am Meteorol Soc 80(5):851–878. https://doi.org/10.1175/1520-0477(1999)080<0851:TPPOTG>2.0.CO;2
28. Dirmeyer PA, Guo Z, Gao X (2004) Comparison, validation, and transferability of eight multiyear global soil wetness products. J Hydrometeorol 5(6):1011–1033. https://doi.org/10.1175/JHM-388.1
29. Dorigo W, de Jeu R, Chung D, Parinussa R, Liu Y, Wagner W, Fernández-Prieto D (2012) Evaluating global trends (1988–2010) in harmonized multi-satellite surface soil moisture. Geophys Res Lett 39:(18). https://doi.org/10.1029/2012GL052988.l18405
30. Dutra E, Magnusson L, Wetterhall F, Cloke HL, Balsamo G, Boussetta S, Pappenberger F (2013) The 2010–2011 drought in the Horn of Africa in ECMWF reanalysis and seasonal forecast products. Int J Climatol 33:1720–1729. https://doi.org/10.1002/joc.3545
31. Dutra E, Wetterhall F, Di Giuseppe F, Naumann G, Barbosa P, Vogt J, Pozzi W, Pappenberger F (2014) Global meteorological drought – part 1: Probabilistic monitoring. Hydrol Earth Syst Sci 18(7):2657–2667. https://doi.org/10.5194/hess-18-2657-2014
32. Edossa DC, Babel MS, Gupta AS (2010) Drought analysis in the Awash River Basin, Ethiopia. Water Res Manag 24:1441–1460. https://doi.org/10.1007/s11269-009-9508-0
33. Elagib NA (2013) Meteorological drought and crop yield in Sub-Sahara Sudan. Int J Water Res Arid Environ 2(3):164–171
34. Elagib NA, Elhag MM (2011) Major climate indicators of ongoing drought in Sudan. J Hydrol 409:612–625. https://doi.org/10.1016/j.jhydrol.2011.08.047
35. Entin JK, Robock A, Vinnikov KY, Zabelin V, Liu S, Namkhai A, Adyasuren T (1999) Evaluation of global soil wetness project soil moisture simulations. J Meteorolog Soc Jpn Ser II 77(1B):183–198
36. Entin JK, Robock A, Vinnikov KY, Hollinger SE, Liu S, Namkhai A (2000) Temporal and spatial scales of observed soil moisture variations in the extratropics. J Geophys Res: Atmos 105(D9):11865–11877. https://doi.org/10.1029/2000JD900051
37. Fan Y, van den Dool H (2004) Climate Prediction Center global monthly soil moisture data set at 0.5 resolution for 1948 to present. J Geophys Res 109:D10,102. https://doi.org/10.1029/2003JD004345
38. Farahmand A, AghaKouchak A (2015) A generalized framework for deriving nonparametric standardized drought indicators. Adv Water Res 76:140–145. https://doi.org/10.1016/j.advwatres.2014.11.012
39. Forina M, Armanino C, Lanteri S, Leardi R (1988) Methods of Varimax rotation in factor analysis with applications in clinical and food chemistry. J Chemom 3:115–125. https://doi.org/10.1002/cem.1180030504
40. Funk C, Steffen P, Senay GB, Rowland J, Verdin J (2003) Estimating Meher crop production using rainfall in the ‘long cycle’ region of Ethiopia, Special report, FEWS-NET, USGS/FEWS/USAID. Accessed from: http://reliefweb.int/sites/reliefweb.int/files/resources on March 15, 2015
41. Funk C, Hoell A, Shukla S, Blade I, Liebmann B, Roberts JB, Robertson, F.R., Husak, G., 2014. Predicting East African spring droughts using Pacific and Indian Ocean sea surface temperature indices. Hydrol. Earth Syst. Sci. 18, 4965–4978. https://doi.org/10.5194/hess-18-4965-2014
42. Funk C, Peterson P, Landsfeld M, Pedreros D, Verdin J, Shukla S, Husak G, Rowland J, Harrison L, Hoell A, Michaelsen J (2015) The climate hazards infrared precipitation with stations – a new environmental record for monitoring extremes. Sci Data 2(150066):1–21. https://doi.org/10.1038/sdata.2015.66
43. Gebrehiwot T, van der Veen A, Maathuis B (2011) Spatial and temporal assessment of drought in the Northern highlands of Ethiopia. Int J Appl Earth Obs Geoinf 13(3):309–321. https://doi.org/10.1016/j.jag.2010.12.002
44. Gedif B, Hadish L, Addisu S, Suryabhagavan KV (2014) Drought risk assessment using remote sensing and gis: the case of southern zone, Tigray Region, Ethiopia. J Nat Sci Res 4(23):87–94. ISSN 2225-0921

45. Geladi P, Kowalski BR (1986) Partial least-squares regression: a tutorial. Anal Chem Acta 185:1–17
46. Gringorten II (1963) A plotting rule for extreme probability paper. J Geophys Res 68(3):813–814. https://doi.org/10.1029/JZ068i003p00813
47. Guan K, Wood E, Caylor K (2012) Multi-sensor derivation of regional vegetation fractional cover in Africa. Remote Sens Environ 124:653–665. https://doi.org/10.1016/j.rse.2012.06.005
48. Hannachi A, Jolliffe IT, Stephenson DB, Trendafilov N (2006) In search of simple structures in climate: simplifying EOFS. Int J Climatol 26:7–28. https://doi.org/10.1002/joc.1243
49. Hochberg Y, Tamhane A (1987) Multiple comparison procedures, Wiley series in probability and mathematical statistics: applied probability and statistics. Wiley
50. Hong Y, Hsu K-L, Moradkhani H, Sorooshian S (2006) Uncertainty quantification of satellite precipitation estimation and monte carlo assessment of the error propagation into hydrologic response. Water Res Res 42(8). https://doi.org/10.1029/2005WR004398. w08421
51. Huang Y, Chen L, Fu B, Huang Z, Gong J, Lu X (2012) Effect of land use and topography on spatial variability of soil moisture in a gully catchment of the Loess Plateau. China. Ecohydrology 5(6):826–833. https://doi.org/10.1002/eco.273
52. Ibrahim F (1988) Causes of the famine among the rural population of the Sahelian zone of the Sudan. GeoJournal 17(1):133–141. https://doi.org/10.1007/BF00209083
53. Jennrich RI (1970) Orthogonal rotation algorithms. Psychometrika 35(2): 299–235. https://doi.org/10.1007/BF02291264
54. Jolliffe IT (1995) Rotation of principal components: choice of normalization constraints. J Appl Stat 22(1):29–35. https://doi.org/10.1080/757584395
55. Jolliffe IT (2002) Principal component analysis. Springer series in statistics, 2nd edn. Springer
56. Kaiser HF (1958) The Varimax Criterion for analytic rotation in Factor analysis. Psychometrika 23(3):187–200. https://doi.org/10.1007/BF02289233
57. Katz RW, Glantz MH (1986) Anatomy of a rainfall index. Mon Weather Rev 114:764–771. https://doi.org/10.1175/1520-0493(1986)114<0764:AOARI>2.0.CO;2
58. King B (2010) Analysis of variance, in International Encyclopedia of Education, edited by P. Peterson, E. Baker, and B. McGaw, third edition ed., pp. 32 - 36, Elsevier, Oxford, https://doi.org/10.1016/B978-0-08-044894-7.01306-3
59. Kogan F (1995) Application of vegetation index and brightness temperature for drought detection. Adv Space Res 15:91–100. https://doi.org/10.1016/0273-1177(95)00079-T
60. Kurnik B, Barbosa P, Vogt J (2011) Testing two different precipitation datasets to compute the standardized precipitation index over the Horn of Africa. Int J Remote Sens 32(21):5947–5964. https://doi.org/10.1080/01431161.2010.499380
61. Kutner M (2005) Applied linear statistical models. McGraw-Hill Irwin, McGrwa-Hill international edition
62. Kutzbach JE (1967) Empirical eigenvector of sea-level pressure, surface temperature and precipitation complexes over North America. J Appl Meteorol 6:791–802. https://doi.org/10.1175/1520-0450(1967)006
63. Larsson H (1996) Relationship between rainfall and sorghum, millet and sesame in the Kassala Province, Eastern Sudan. J Arid Environ 32(2):211–223. https://doi.org/10.1006/jare.1996.0018
64. Long D, Scanlon BR, Longuevergne L, Sun AY, Fernando DN, Save H (2013) GRACE satellite monitoring of large depletion in water storage in response to the 2011 drought in Texas. Geophys Res Lett 40:3395–3401. https://doi.org/10.1002/grl.50655
65. Longley C, Jones R, Ahmed MH, Audi P (2001) Supporting local seed systems in southern Somalia: a developmental approach to agricultural rehabilitation in emergency situations, Network paper 115, Agricultural Research and Extension Network, 20 pp
66. Lorenz EN (1956) Empirical Orthogonal Function and Statistical Weather Prediction, Statistical forecasting project: Scientific report no. 1, Department of Meteorology, MIT, Retrieved from: http://eaps4.mit.edu/research/Lorenz, on 15, 2015

67. Lough JM (1997) Regional indices of climate variation: temperature and precipitation in Queensland, Australia. Int J Climatol 17, 55–66. https://doi.org/10.1002/(SICI)1097-0088
68. Lyon B (2014) Seasonal drought in the greater horn of africa and its recent increase during the March–May long rains. J Clim 27:7953–7975. https://doi.org/10.1175/JCLI-D-13-00459.1
69. Masih I, Maskey S, Mussá FEF, Trambauer P (2014) A review of droughts on the African continent: a geospatial and long-term perspective. Hydrol Earth Syst Sci 18(9):3635–3649. https://doi.org/10.5194/hess-18-3635-2014
70. McKee TB, Doesken NJ, Kleist J (1993) The relationship of drought frequency and duration to time scale. In: Conference Proceedings, eighth conference of applied climatology, Anaheim, California
71. McNally A, Shukla S, Arsenault KR, Wang S, Peters-Lidard CD, Verdin JP (2016) Evaluating ESA CCI soil moisture in East Africa. Int J Appl Earth Obs Geoinf 48:96–109. https://doi.org/10.1016/j.jag.2016.01.001
72. McNally A, Arsenault K, Kumar S, Shukla S, Peterson P, Wang S, Funk C, Peters-Lidard CD, Verdin JP (2017) A land data assimilation system for sub-Saharan Africa food and water security applications. Sci Data 4(170012):1–21. https://doi.org/10.1038/sdata.2017.12
73. Mpelasoka F, Awange JL, Zerihun A (2018) Influence of coupled ocean-atmosphere phenomena on the Greater Horn of Africa droughts and their implications. Sci Total Environ 610–611:691–702. https://doi.org/10.1016/j.scitotenv.2017.08.109
74. Moore P, Williams SDP (2014) Integration of altimetric lake levels and GRACE gravimetry over Africa: Inferences for terrestrial water storage change 2003–2011. Water Res Res 50:9696–9720. https://doi.org/10.1002/2014WR015506
75. Naresh Kumar M, Murthy CS, Sesha Sai MVR, Roy PS (2009) On the use of Standardized Precipitation Index (SPI) for drought intensity assessment. Meteorol Appl 16(3):381–389. https://doi.org/10.1002/met.136
76. Naumann G, Dutra E, Pappenberger F, Wetterhall F, Vogt JV (2014) Comparison of drought indicators derived from multiple data sets over Africa. Hydrol Earth Syst Sci 18:1625–1640. https://doi.org/10.5194/hess-18-1625-2014
77. Nicholson SE (2014) A detailed look at the recent drought situation in the Greater Horn of Africa. J Arid Environ 103:71–79. https://doi.org/10.1016/j.jaridenv.2013.12.003
78. Olsson L (1993) On the causes of famine: drought, desertification and market failure in the Sudan. Ambio 22(6):395–403. http://www.jstor.org/stable/4314110
79. Peters AJ, Waletr-Shea EA, Ji L, Via A, Hayes M, Svoboda MV (2002) Drought monitoring with NDVI-based standardized vegetation index. Photogram Eng Remote Sens 68:71–75
80. Pinzon JE, Tucker CJ (2014) A non-stationary 1981–2012 AVHRR NDVI3g time series. Remote Sens 6(8):6929. https://doi.org/10.3390/rs6086929
81. Preisendorfer RW (1988) Principal component analysis in meteorology and oceanography, Development in atmospheric science 17. Elsevier, Amsterdam
82. Pricope NG, Husak G, Lopez-Carr D, Funk C, Michaelsen J (2013) The climate population nexus in the East African Horn: Emerging degradation trends in rangeland and pastoral livelihood zones. Global Environ Change 23(6):1525–1541. https://doi.org/10.1016/j.gloenvcha.2013.10.002
83. Quiring SM (2009) Developing objective operational definitions for monitoring drought. JD Appl Meteorol Climatol 48(6):1217–1229. https://doi.org/10.1175/2009JAMC2088.1
84. Quiring SM, Ganesh S (2010) Evaluation of utility of Vegetation Condition Index (VCI) for monitoring meteorological drought in Texas. Agric. For Meteorol 150:330–339. https://doi.org/10.1016/j.agrformet.2009.11.015
85. Raziei T, Saghafian B, Paulo AA, Pereira LS, Bordi I (2009) Spatial patterns and temporal variability of drought in Western Iran. Water Res Manag 23:439–455. https://doi.org/10.1007/s11269-008-9282-4
86. Reichle R (2012) The MERRA-land data product, global modelling and assimilation Office, version 1.0 ed., http://gmao.gsfc.nasa.gov
87. Rienecker MM, Suarez MJ, Gelaro R, Todling R, Bacmeister J, Liu E, Bosilovich MG, Schubert, S.D., Takacs, L., Kim, G.-K., Bloom, S., Chen, J., Collins, D., Conaty, A., da Silva,